D0860377

ENCYCLOPEDIA OF
GLOBAL WARMING

ENCYCLOPEDIA OF GLOBAL WARMING

Volume 1

Abrupt climate change-Energy Policy Act of 1992

Editor

Steven I. Dutch

University of Wisconsin—Green Bay

Salem Press

Pasadena, California   Hackensack, New Jersey

Cover photo: (©Jan Martin/Dreamstime.com)

∞ The paper used in these volumes conforms to the American National Standard for Permanence of Paper for Printed Library Materials, Z39.48-1992 (R1997).

Library of Congress Cataloging-in-Publication Data

Dutch, Steven I.
Encyclopedia of global warming / Steven I. Dutch.
p. cm.
Includes bibliographical references and index.
ISBN 978-1-58765-563-0 (set : alk. paper) — ISBN 978-1-58765-564-7 (vol. 1 : alk. paper) — ISBN 978-1-58765-565-4 (vol. 2 : alk. paper) — ISBN 978-1-58765-566-1 (vol. 3 : alk. paper)
1. Global warming—Encyclopedias. 2. Climatic changes—Encyclopedias. I. Title.
QC981.8.G56D88 2010
551.603—dc22

2009031763

Mixed Sources
Product group from well-managed forests, controlled sources and recycled wood or fiber
www.fsc.org Cert no. SGS-COC-006193
© 1996 Forest Stewardship Council

PRINTED IN THE UNITED STATES OF AMERICA

Table of Contents

Publisher's Note

The *Encyclopedia of Global Warming* (3 volumes) provides comprehensive coverage of the questions of global warming and climate change, including scientific descriptions and explanations of all factors, from carbon dioxide to sunspots, that might contribute to climate change. The set includes 540 essays, all of which were written specifically for it.

• Scope of Coverage

The *Encylcopedia of Global Warming* is designed to provide students at the high school and undergraduate levels with a convenient source of information on the fundamental science and sociopolitical issues, including the debates and controversies, surrounding climate change. The study of climate change involves not only scientists but also politicians, policy makers, businesses, government and nongovernment agencies, and the general public. A student attempting to understand both the environmental science and social issues and controversies will encounter not just scientific terms and concepts but political organizations, geographic areas, social concepts, persons, countries, organizations, and laws as well. This encyclopedia comprises entries from all those areas, including topics on both sides of the debate and entries from scholars who believe in human-driven global climate change as well as those who are less convinced. Our objective is to provide factual and objective information—not advocacy.

The essays in the set fall into one or more of the following broad categories: animals (16 essays); Arctic and Antarctic (7); astronomy (10); chemistry and geochemistry (41); climatic events and epochs (18); conferences and meetings (6); cryology and glaciology (19); diseases and health effects (7); economics, industries, and products (44); energy (29); environmentalism, conservation, and ecosystems (51); ethics, human rights, and social justice (8); fossil fuels (9); geology and geography (25); laws, treaties, and protocols (37); meteorology and atmospheric sciences (98); nations and peoples (39); oceanography (33); organizations and agencies (70); physics and geophysics (10); plants and vegetation (30); pollution and waste (24); popular culture and society (22); science and technology (23); transportation (4); and water resources (14).

• Essay Length and Format

Essays in the encyclopedia range in length from 400 to 2,000 words. They appear in one of six major formats:

- *Term* essays begin by defining a term and then explain its significance for climate change.
- *Overview* essays provide broad overviews of a scientific, policy, or social phenomenon or debate. They include a list of key concepts related to the topic, followed by background, several topical sections, and a concluding section explaining the climatological context of the topic.
- *Organization* essays provide the date of establishment of the organization, the URL of its official Web site, and discussions of its mission and significance for climate change.
- *Biographical* essays provide the profession, birth date, birthplace, death date, and death place of their subjects, followed by descriptions of their life and climate work.
- *Top-Twenty Emitter* essays provide information on each of the twenty nations with the highest annual emissions of greenhouse gases. These essays begin with a list of key facts, including the population, area, gross domestic product, and annual greenhouse gas emissions of the country, as well as its Kyoto Protocol status. They then provide discussions of the historical and political context of the nation's climatic impact, continue with the nation's relevant contributions to global warming and to international action, and conclude with a summary and foresight of the nation's future commitments and likely actions.
- *Law and Treaty* essays provide the date of passage or ratification of their subjects, as well as lists of all participating nations. They continue with discussions of background, summary of provisions, and significance for climate change.

Each essay longer than 400 words concludes

with an annotated bibliography of suggestions for further reading, and all essays include an author byline and a list of cross-references to other related essays in the set.

• Special Features

Several features distinguish this series as a whole from other biographical reference works. The front matter includes the following aids:

- *Abbreviations and Acronyms:* Each volume includes a list explaining the abbreviations and acronyms used in essays throughout the set.
- *Common Units of Measure:* Measurements in the body of the set are in metric units only. Each volume's front matter includes a table converting metric/SI units into imperial units for the user's convenience.
- *List of Tables, Maps, and Sidebars:* The set includes more than 150 textual sidebars, tables, graphs, charts, maps, and other elements that illustrate or expand upon the essays with key supplementary information. A list of all such elements appears in the front matter of each volume.
- *Complete Table of Contents:* This list of the contents of the entire set appears in all three volumes.
- *Categorized List of Contents:* Each volume also includes a complete list of contents by category, to aid the reader in finding all essays relevant to a particular broad topic.

The back matter to Volume 3 includes several appendixes and indexes:

- *Biographical Dictionary of Key Figures in Global Warming:* A compendium of the people most influential in shaping discoveries, debates, and actions involving climate change.
- *Popular Culture About Global Warming:* A list of major books, films, television programs, and other mass media portraying global warming for a popular audience.
- *Time Line:* A chronology of all major events relating to human understanding of and response to climate change.
- *Glossary:* A complete glossary of technical and other specialized terms used throughout the set.
- *General Bibliography:* A comprehensive list of works on climate change for students seeking more information on the subject.
- *Web Sites:* A list of Web resources, including the official Web sites of key organizations, as well as online databases and other sources of information.
- *Subject Index:* A comprehensive index to all concepts, terms, events, persons, places, phenomena, and other topics of discussion.

• Acknowledgments

Salem Press would like to extend its appreciation to all involved in the development and production of this work. The essays have been written and signed by scholars of history, the sciences, and other disciplines related to the essays' topics.

Special thanks go to Professor Steven I. Dutch, of the Geosciences Program, University of Wisconsin—Green Bay, who developed the contents list and coverage notes for this set. A geologist, Professor Dutch is familiar with the methods, findings, and uncertainty of paleoclimatology. Because he is uninvolved in policy debates over climate change, and at the same time is familiar enough with the science to evaluate claims about climate, he was able to take on the task of assisting Salem Press in compiling a reference encyclopedia that addresses a broad variety of climate-change issues. Moreover, his research focus on the use and abuse of scientific evidence in public controversies has sensitized him not only to the controversies in this field but also to the need to create a contents list that addresses a broad variety of topics and issues. Hence, readers will find not only atmospheric science concepts but also topics like conspiracy theories, junk science, and pseudoscience covered in these pages, along with basic political approaches to climate policy (in essays such as "Conservatism," "Liberalism," and "Libertarianism"), as well as the inclusion of organizations active on both sides of the debate.

Without all the expertise of both Professor Dutch and the many contributing writers, a project of this nature would not be possible. A full list of the contributors' names and affiliations appears in the front matter of this volume.

Contributors

Tomi Akanle
University of Dundee

Emily Alward
Henderson, Nevada, District Libraries

Anita Baker-Blocker
Ann Arbor, Michigan

Melissa A. Barton
Westminster, Colorado

Raymond D. Benge, Jr.
Tarrant County College—Northeast Campus

Alvin K. Benson
Utah Valley University

Cynthia A. Bily
Adrian, Minnesota

Henry Bokuniewicz
SUNY at Stony Brook

Howard Bromberg
University of Michigan

Kenneth H. Brown
Northwestern Oklahoma State University

Jeffrey C. Brunskill
Bloomsburg University of Pennsylvania

Ralph Buehler
Virginia Tech

Michael A. Buratovich
Spring Arbor University

Ewa M. Burchard
California State University, Long Beach

Michael H. Burchett
Limestone College

Douglas Bushey
University of California, Berkeley

Richard K. Caputo
Yeshiva University

Roger V. Carlson
Jet Propulsion Laboratory

Rebecca S. Carrasco
University of New Orleans

Jack Carter
University of New Orleans

Dennis W. Cheek
Ewing Marion Kauffman Foundation

Jongnam Choi
Western Illinois University

Thomas Coffield
Webster, Minnesota

Raymond D. Cooper
Eckerd College

Anna M. Cruse
Oklahoma State University

Robert L. Cullers
Kansas State University

Loralee Davenport
Butte Central Catholic High School

Joseph Dewey
University of Pittsburgh—Johnstown

Steven I. Dutch
University of Wisconsin—Green Bay

Victoria Erhart
Strayer University

Justin Ervin
Northern Arizona University

Johanna C. Estanislao
University of Colorado at Boulder

Thomas R. Feller
Nashville, Tennessee

Danilo D. Fernando
SUNY, College of Environmental Science & Forestry

Femi Ferreira
Hutchinson Community College

K. Thomas Finley
SUNY, College at Brockport

Alan S. Frazier
University of North Dakota

C R de Freitas
The University of Auckland

Dan Gainor
Business & Media Institute

Yongli Gao
East Tennessee State University

Michael Getzner
Klagenfurt University

James S. Godde
Monmouth College

Nancy M. Gordon
Amherst, Massachusetts

Kenneth P. Green
Western Washington University

Phillip Greenberg
San Francisco, California

Joyce M. Hardin
Hendrix College

C. Alton Hassell
Baylor University

Paul A. Heckert
Western Carolina University

Jennifer Freya Helgeson
National Institute of Standards and Technology

Jane F. Hill
Bethesda, Maryland

John L. Howland
Bowdoin College

Patrick Norman Hunt
Stanford University

Raymond Pierre Hylton
Virginia Union University

Solomon A. Isiorho
Indiana University—Purdue University

Jerome A. Jackson
Florida Gulf Coast University

Sikina Jinnah
American University

Bruce E. Johansen
University of Nebraska at Omaha

Edward Johnson
University of New Orleans

Karen N. Kähler
Pasadena, California

Susan J. Karcher
Purdue University

David Kasserman
Rowan University

George B. Kauffman
California State University, Fresno

Amber C. Kerr
University of California, Berkeley

David T. King, Jr.
Auburn University

Samuel V. A. Kisseadoo
Hampton University

Grove Koger
Boise State University

Narayanan M. Komerath
Georgia Institute of Technology

Padma Komerath
SCV, Inc.

Steven A. Kuhl
V & R Consulting

Nicholas Lancaster
Desert Research Institute

Raymond P. LeBeau, Jr.
University of Kentucky

Denyse Lemaire
Rowan University

Josué Njock Libii
Purdue University Fort Wayne

Victor Lindsey
East Central University

Donald W. Lovejoy
Palm Beach Atlantic University

Chungu Lu
NOAA Earth System Research Laboratory

Yiqi Luo
University of Oklahoma

R. C. Lutz
CII Group

Gregory J. McCabe
U.S. Geological Survey

Howard Maccabee
Stanford University Medical School

Elizabeth A. Machunis-Masuoka
Midwestern State University

Marianne M. Madsen
University of Utah

Sergei Arlenovich Markov
Austin Peay State University

W. J. Maunder
Tauranga, New Zealand

Randall L. Milstein
Oregon State University

Robin Kamienny Montvilo
Rhode Island College

Otto H. Muller
Alfred University

Robert P. Murphy
Institute for Energy Research

M. Marian Mustoe
Eastern Oregon University

Alice Myers
Bard College at Simon's Rock

Arpita Nandi
East Tennessee State University

Terrence R. Nathan
University of California, Davis

To N. Nguyen
University of Guelph

Anthony J. Nicastro
West Chester University of Pennsylvania

Eugene E. Niemi, Jr.
University of Massachusetts, Lowell

Oladele A. Ogunseitan
University of California, Irvine

Kevin L. O'Hara
University of California, Berkeley

Zaitao Pan
Saint Louis University

Robert J. Paradowski
Rochester Institute of Technology

Suzanne E. Paulson
University of California, Los Angeles

John R. Phillips
Purdue University Calumet

Nancy A. Piotrowski
Capella University

George R. Plitnik
Frostburg State University

Victoria Price
Lamar University

Maureen Puffer-Rothenberg
Valdosta State University

Cynthia F. Racer
New York Academy of Sciences

Gail Rampke
Purdue University Calumet

P. S. Ramsey
Brighton, Minnesota

Mariana L. Rhoades
St. John Fisher College

Edward A. Riedinger
Ohio State University

Gina M. Robertiello
Felician College

Charles W. Rogers
Southwestern Oklahoma State University

Kathryn Rowberg
Purdue University Calumet

Somnath Baidya Roy
University of Illinois at Urbana-Champaign

Joseph R. Rudolph, Jr.
Towson University

Wayne Allen Sallee
Burbank, Illinois

Virginia L. Salmon
Northeast State Technical Community College

Elizabeth D. Schafer
Loachapoka, Alabama

Jürgen Scheffran
University of Illinois

John Richard Schrock
Emporia State University

Jason J. Schwartz
Los Angeles, California

Miriam E. Schwartz
University of California, Los Angeles

Manoj Sharma
University of Cincinnati

Marlyn L. Shelton
University of California, Davis

Martha A. Sherwood
Kent Anderson Law Associates

R. Baird Shuman
University of Illinois at Urbana-Champaign

Carlos Nunes Silva
University of Lisbon

Paul P. Sipiera
William Rainey Harper College

Adam B. Smith
University of California, Berkeley

Billy R. Smith, Jr.
Anne Arundel Community College

Dwight G. Smith
Southern Connecticut State University

Encyclopedia of Global Warming

Roger Smith
Portland, Oregon

Carolyn P. Snyder
Stanford University

Richard S. Spira
American Veterinary Medical Association

Glenn Ellen Starr Stilling
Appalachian State University

Alexander R. Stine
University of California, Berkeley

Marshall D. Sundberg
Emporia State University

Paul C. Sutton
University of Denver

Rena Christina Tabata
University of British Columbia

Katrina Darlene Taylor
Northern Arizona University

John M. Theilmann
Converse College

Roger Dale Trexler
Southern Illinois University

Robert C. Tyler
Fort Collins, Colorado

Oluseyi Adewale Vanderpuye
Albany State University

Travis Wagner
University of Southern Maine

C. J. Walsh
Mote Marine Laboratory

Donald A. Watt
Dakota Wesleyan University

Shawncey Webb
Taylor University

Thomas A. Wikle
Oklahoma State University

Sam Wong
University of Leeds

Jeffrey V. Yule
Louisiana Tech University

Ming Y. Zheng
Gordon College

Sandie Zlotorzynski
Purdue University Calumet

Abbreviations and Acronyms

AMO: Atlantic multidecadal oscillation
AQI: Air Quality Index
C^{14}: carbon 14
CCS: carbon capture and storage
CDM: clean development mechanism
CER: certified emissions reduction
CFCs: chlorofluorocarbons
CH_4: methane
CITES: Convention on International Trade in Endangered Species
CMP: Conference of the Parties to the United Nations Framework Convention on Climate Change, functioning as the meeting of the Parties to the Kyoto Protocol
CO: carbon monoxide
CO_2: carbon dioxide
CO_2e: carbon dioxide equivalent
COP: Conference of the Parties [to a treaty, such as the Framework Convention on Climate Change or the Convention on Biological Diversity]
COP/MOP: Conference of the Parties to the United Nations Framework Convention on Climate Change, functioning as the meeting of the Parties to the Kyoto Protocol
COP-1: First Conference of the Parties
CSD: Commission on Sustainable Development
DNA: deoxyribonucleic acid
EEZ: exclusive economic zone
ENSO: El Niño-Southern Oscillation
EPA: Environmental Protection Agency
ERU: emission reduction unit
FAO: Food and Agriculture Organization
GCM: general circulation model
GDP: gross domestic product
GHG: greenhouse gas
GWP: global warming potential
H_2: hydrogen (molecular)
HCFCs: hydroclorofluorocarbons
HFCs: hydrofluorocarbons
IAEA: International Atomic Energy Agency
IGY: International Geophysical Year
IMF: International Monetary Fund
INQUA: International Union for Quaternary Research
IPCC: Intergovernmental Panel on Climate Change
ITCZ: Inter-Tropical Convergence Zone
IUCN: International Union for Conservation of Nature
LOICZ: Land-Ocean Interactions in the Coastal Zone
MOC: meridional overturning circulation
MOP: meeting of the Parties [to a treaty, such as the Kyoto Protocol]
MOP-1: first meeting of the Parties
N_2O: nitrous oxide
NAAQS: National Ambient Air Quality Standards
NAM: Northern annular mode
NAO: North Atlantic Oscillation
NASA: National Aeronautics and Space Administration
NATO: North Atlantic Treaty Organization
NGO: nongovernmental organization
NOAA: National Oceanic and Atmospheric Administration
NO_x: nitrogen oxides
NRC: National Research Council
O^{16}: oxygen 16
O^{18}: oxygen 18
O_2: oxygen (molecular)
O_3: ozone
OECD: Organization for Economic Cooperation and Development
OPEC: Organization of Petroleum Exporting Countries
PFCs: perfluorocarbons
QELRCs: quantified emission limitation and reduction commitments
RuBisCO: Ribulose-1,5-bisphosphate carboxylase/oxygenase
SAM: Southern annular mode
SCOPE: Scientific Committee on Problems of the Environment
SF_6: sulfur hexafluoride
SO_2: sulfur dioxide

SSTs: sea surface temperatures
SUV: sports utility vehicle
THC: Thermohaline circulation
UNCED: United Nations Conference on Environment and Development
UNDP: United Nations Development Programme
UNEP: United Nations Economic Programme
UNESCO: United Nations Educational, Scientific, and Cultural Organization
UNFCCC: United Nations Framework Convention on Climate Change
UV: ultraviolet
VOCs: volatile organic compounds
WHO: World Health Organization
WMO: World Meteorological Organization

Common Units of Measure

Common prefixes for metric units—which may apply in more cases than shown below—include *giga-* (1 billion times the unit), *mega-* (one million times), *kilo-* (1,000 times), *hecto-* (100 times), *deka-* (10 times), *deci-* (0.1 times, or one tenth), *centi-* (0.01, or one hundredth), *milli-* (0.001, or one thousandth), and *micro-* (0.0001, or one millionth).

Unit	*Quantity*	*Symbol*	*Equivalents*
Acre	Area	ac	43,560 square feet 4,840 square yards 0.405 hectare
Ampere	Electric current	A *or* amp	1.00016502722949 international ampere 0.1 biot *or* abampere
Angstrom	Length	Å	0.1 nanometer 0.0000001 millimeter 0.000000004 inch
Astronomical unit	Length	AU	92,955,807 miles 149,597,871 kilometers (mean Earth-Sun distance)
Barn	Area	b	10^{-28} meters squared (approx. cross-sectional area of 1 uranium nucleus)
Barrel (dry, for most produce)	Volume/capacity	bbl	7,056 cubic inches; 105 dry quarts; 3.281 bushels, struck measure
Barrel (liquid)	Volume/capacity	bbl	31 to 42 gallons
British thermal unit	Energy	Btu	1055.05585262 joule
Bushel (U.S., heaped)	Volume/capacity	bsh *or* bu	2,747.715 cubic inches 1.278 bushels, struck measure
Bushel (U.S., struck measure)	Volume/capacity	bsh *or* bu	2,150.42 cubic inches 35.238 liters
Candela	Luminous intensity	cd	1.09 hefner candle
Celsius	Temperature	C	1° centigrade
Centigram	Mass/weight	cg	0.15 grain
Centimeter	Length	cm	0.3937 inch
Centimeter, cubic	Volume/capacity	cm^3	0.061 cubic inch
Centimeter, square	Area	cm^2	0.155 square inch

Unit	*Quantity*	*Symbol*	*Equivalents*
Coulomb	Electric charge	C	1 ampere second
Cup	Volume/capacity	C	250 milliliters 8 fluid ounces 0.5 liquid pint
Deciliter	Volume/capacity	dl	0.21 pint
Decimeter	Length	dm	3.937 inches
Decimeter, cubic	Volume/capacity	dm^3	61.024 cubic inches
Decimeter, square	Area	dm^2	15.5 square inches
Dekaliter	Volume/capacity	dal	2.642 gallons 1.135 pecks
Dekameter	Length	dam	32.808 feet
Dram	Mass/weight	dr *or* dr avdp	0.0625 ounce 27.344 grains 1.772 grams
Electron volt	Energy	eV	$1.5185847232839 \times 10^{-22}$ Btus $1.6021917 \times 10^{-19}$ joules
Fermi	Length	fm	1 femtometer 1.0×10^{-15} meters
Foot	Length	ft *or* ′	12 inches 0.3048 meter 30.48 centimeters
Foot, cubic	Volume/capacity	ft^3	0.028 cubic meter 0.0370 cubic yard 1,728 cubic inches
Foot, square	Area	ft^2	929.030 square centimeters
Gallon (British Imperial)	Volume/capacity	gal	277.42 cubic inches 1.201 U.S. gallons 4.546 liters 160 British fluid ounces
Gallon (U.S.)	Volume/capacity	gal	231 cubic inches 3.785 liters 0.833 British gallon 128 U.S. fluid ounces
Giga-electron volt	Energy	GeV	$1.6021917 \times 10^{-10}$ joule
Gigahertz	Frequency	GHz	—
Gill	Volume/capacity	gi	7.219 cubic inches 4 fluid ounces 0.118 liter

Common Units of Measure

Unit	*Quantity*	*Symbol*	*Equivalents*
Grain	Mass/weight	gr	0.037 dram 0.002083 ounce 0.0648 gram
Gram	Mass/weight	g	15.432 grains 0.035 avoirdupois ounce
Hectare	Area	ha	2.471 acres
Hectoliter	Volume/capacity	hl	26.418 gallons 2.838 bushels
Hertz	Frequency	Hz	$1.08782775707767 \times 10^{-10}$ cesium atom frequency
Hour	Time	h	60 minutes 3,600 seconds
Inch	Length	in *or* ″	2.54 centimeters
Inch, cubic	Volume/capacity	in^3	0.554 fluid ounce 4.433 fluid drams 16.387 cubic centimeters
Inch, square	Area	in^2	6.4516 square centimeters
Joule	Energy	J	$6.2414503832469 \times 10^{18}$ electron volt
Joule per kelvin	Heat capacity	J/K	$7.24311216248908 \times 10^{22}$ Boltzmann constant
Joule per second	Power	J/s	1 watt
Kelvin	Temperature	K	–272.15 Celsius
Kilo-electron volt	Energy	keV	$1.5185847232839 \times 10^{-19}$ joule
Kilogram	Mass/weight	kg	2.205 pounds
Kilogram per cubic meter	Mass/weight density	kg/m^3	$5.78036672001339 \times 10^{-4}$ ounces per cubic inch
Kilohertz	Frequency	kHz	—
Kiloliter	Volume/capacity	kl	—
Kilometer	Length	km	0.621 mile
Kilometer, square	Area	km^2	0.386 square mile 247.105 acres
Light-year (distance traveled by light in one Earth year)	Length/distance	lt-yr	5,878,499,814,275.88 miles 9.46×10^{12} kilometers

Unit	*Quantity*	*Symbol*	*Equivalents*
Liter	Volume/capacity	L	1.057 liquid quarts 0.908 dry quart 61.024 cubic inches
Mega-electron volt	Energy	MeV	—
Megahertz	Frequency	MHz	—
Meter	Length	m	39.37 inches
Meter, cubic	Volume/capacity	m^3	1.308 cubic yards
Meter per second	Velocity	m/s	2.24 miles per hour 3.60 kilometers per hour
Meter per second per second	Acceleration	m/s^2	12,960.00 kilometers per hour per hour 8,052.97 miles per hour per hour
Meter, square	Area	m^2	1.196 square yards 10.764 square feet
Metric. *See* unit name			
Microgram	Mass/weight	mcg *or* µg	0.000001 gram
Microliter	Volume/capacity	µl	0.00027 fluid ounce
Micrometer	Length	µm	0.001 millimeter 0.00003937 inch
Mile (nautical international)	Length	mi	1.852 kilometers 1.151 statute miles 0.999 U.S. nautical miles
Mile (statute or land)	Length	mi	5,280 feet 1.609 kilometers
Mile, square	Area	mi^2	258.999 hectares
Milligram	Mass/weight	mg	0.015 grain
Milliliter	Volume/capacity	ml	0.271 fluid dram 16.231 minims 0.061 cubic inch
Millimeter	Length	mm	0.03937 inch
Millimeter, square	Area	mm^2	0.002 square inch
Minute	Time	m	60 seconds
Mole	Amount of substance	mol	6.02×10^{23} atoms or molecules of a given substance

Unit	*Quantity*	*Symbol*	*Equivalents*
Nanometer	Length	nm	1,000,000 fermis 10 angstroms 0.001 micrometer 0.00000003937 inch
Newton	Force	N	x 0.224808943099711 pound force 0.101971621297793 kilogram force 100,000 dynes
Newton meter	Torque	N·m	0.7375621 foot-pound
Ounce (avoirdupois)	Mass/weight	oz	28.350 grams 437.5 grains 0.911 troy or apothecaries' ounce
Ounce (troy)	Mass/weight	oz	31.103 grams 480 grains 1.097 avoirdupois ounces
Ounce (U.S., fluid or liquid)	Mass/weight	oz	1.805 cubic inch 29.574 milliliters 1.041 British fluid ounces
Parsec	Length	pc	30,856,775,876,793 kilometers 19,173,511,615,163 miles
Peck	Volume/capacity	pk	8.810 liters
Pint (dry)	Volume/capacity	pt	33.600 cubic inches 0.551 liter
Pint (liquid)	Volume/capacity	pt	28.875 cubic inches 0.473 liter
Pound (avoirdupois)	Mass/weight	lb	7,000 grains 1.215 troy or apothecaries' pounds 453.59237 grams
Pound (troy)	Mass/weight	lb	5,760 grains 0.823 avoirdupois pound 373.242 grams
Quart (British)	Volume/capacity	qt	69.354 cubic inches 1.032 U.S. dry quarts 1.201 U.S. liquid quarts
Quart (U.S., dry)	Volume/capacity	qt	67.201 cubic inches 1.101 liters 0.969 British quart
Quart (U.S., liquid)	Volume/capacity	qt	57.75 cubic inches 0.946 liter 0.833 British quart
Rod	Length	rd	5.029 meters 5.50 yards

Unit	*Quantity*	*Symbol*	*Equivalents*
Rod, square	Area	rd^2	25.293 square meters 30.25 square yards 0.00625 acre
Second	Time	s *or* sec	1/60 minute 1/3600 hour
Tablespoon	Volume/capacity	T *or* tb	3 teaspoons 4 fluid drams
Teaspoon	Volume/capacity	t *or* tsp	0.33 tablespoon 1.33 fluid drams
Ton (gross or long)	Mass/weight	t	2,240 pounds 1.12 net tons 1.016 metric tons
Ton (metric)	Mass/weight	t	1,000 kilograms 2,204.62 pounds 0.984 gross ton 1.102 net tons
Ton (net or short)	Mass/weight	t	2,000 pounds 0.893 gross ton 0.907 metric ton
Volt	Electric potential	V	1 joule per coulomb
Watt	Power	W	1 joule per second 0.001 kilowatt $2.84345136093995 \times 10^{-4}$ ton of refrigeration
Yard	Length	yd	0.9144 meter
Yard, cubic	Volume/capacity	yd^3	0.765 cubic meter
Yard, square	Area	yd^2	0.836 square meter

List of Tables, Maps, and Sidebars

Volume 1

Volume 2

Volume 3

Complete Table of Contents

Volume 1

Volume 2

Volume 3

Categorized List of Contents

Animals

Arctic and Antarctic

Astronomy

Chemistry and geochemistry

Climatic events and epochs

Conferences and meetings

Cryology and glaciology

Diseases and health effects

Economics, industries, and products

Energy

Environmentalism, conservation, and ecosystems

Ethics, human rights, and social justice

Fossil fuels

Geology and geography

Laws, treaties, and protocols

Meteorology and atmospheric sciences

Nations and peoples

Oceanography

Organizations and agencies

Physics and geophysics

Plants and vegetation

Pollution and waste

Popular culture and society

Science and technology

Transportation

Water resources

ENCYCLOPEDIA OF
GLOBAL WARMING

Editor's Introduction

The debate over whether global warming is occurring has lessened over the past decade, and the focus of the debate has shifted. More and more participants on both sides agree that climate data show a warming trend. The debate now revolves around how long-term and significant that trend is, whether it is due mainly to human (anthropogenic) causes, and what course of action society should take to deal with climate change. Any reference work on global warming and climate change will inevitably, therefore, be controversial. If it is dominated too much by believers in climate change, legitimate climate change skeptics might dismiss it as simply another expression of the "party line." If it incorporates too many skeptical viewpoints, climate change activists will see it as taking a weak stance or possibly even being a covert attempt to advance the skeptics' cause.

Salem Press's *Encyclopedia of Global Warming* strives to incorporate the views of academically qualified scientists, historians, economists, and political scientists, presenting objective information without bias or a polemic tone. Some of the contributors are scientists who accept the anthropogenic basis for recent climate change; some are more reserved in their judgment, cognizant of the complexity of climate change and the limitations of computer models used in predicting climate; and some doubt the role of human action in climate change. Salem Press made a conscious effort to seek out and incorporate the views of academically qualified skeptics, because they would be able to articulate the positions of skeptics most accurately and fairly.

A reader unfamiliar with the climate change debate should understand not only the science itself but also the social and political landscape surrounding the debate. Someone reading about climate change in the newspapers or online will inevitably encounter terms such as "liberal," "conservative," "conspiracy," and so on. For that reason, the encyclopedia includes not just articles on scientific terms and concepts but also overviews of organizations active on both sides of the controversy and essays on basic political terms—such as "liberalism," "libertarianism," and "conservatism"—as they pertain to the debate over climate change. Also, some people on both sides in the climate change debate have accused the other side of promoting poor or even false science and of having political or financial motivations. To provide some help in dealing with such claims, articles are included on topics such as conspiracy theories and junk science and pseudoscience.

Quantitative climate data are available only for the last century or so. To determine earlier climatic conditions, scientists must rely on proxy measures of climate. For example, diaries and news accounts of bodies of water freezing over that normally do not freeze, unusual frosts, rainfall, or drought can furnish clues about climate variation but cannot be directly converted into present numerical values. All the methods used to derive temperatures from pre-instrument times are subject to error and subjective interpretation. In the absence of written records, climatologists must rely on physical indicators such as tree rings, pollen, marine microfossils, and oxygen isotope ratios. Measurements of such proxies can be calibrated against written and instrumental records but nevertheless introduce still another level of uncertainty into the results. Further back in geologic time, there are climatic indicators such as coal beds, evaporites, and glacial deposits, and there are geochemical indicators of temperature and atmospheric oxygen and carbon dioxide concentration. These records require their own methods of interpretation. It is neither scientifically nor socially irresponsible to ask how reliable the climate record is or how one can separate anthropogenic from natural climate change. Indeed, even a cursory browse through climatological journals will turn up many papers that ask precisely those questions. Intellectually honest activists and skeptics alike agree that the surest way to make their case is to make the science as rigorous and reliable as possible.

The late U.S. senator Ed Muskie of Maine once famously lamented that there were not "more one-armed scientists," because scientists were always saying "On the other hand. . . ." One of the things that most separates scientists from nonscientists is that scientists can tolerate ambiguity. Nonscientists find ambiguity in science extremely frustrating. Unfortunately, many people are willing to fill that void with promises of certainty, or to point to ambiguity as proof that science has no answers, or to consider some other interpretation as preferable to the scientific view.

Ambiguity in science might be likened to the strike zone in baseball. To a skilled pitcher or hitter, the strike zone is a big target. Its boundaries are a bit fuzzy. Different umpires might call balls and strikes a bit differently. To someone sitting in the stands, the strike zone is a speck. Nobody familiar with baseball would try to claim that a wild pitch into the stands was a strike merely because the strike zone has fuzzy edges. Despite the ambiguities in science, it is able to categorize some theories as definitely right or definitely wrong. There is a very limited zone of strikes—correct ideas—and a vast realm of wild pitches—incorrect ideas. Regardless of whether or not anthropogenic climate change is occurring, pointing to some local anomaly such as an unusually early frost or a lake where ice breaks up unusually late is simply wrong. It is fallacious reasoning, because it tries to draw a general conclusion about a complex subject from an isolated piece of data. Pointing to unresolved questions as evidence that science in general is unreliable, or using unresolved questions to justify belief in some alternative theory, is also simply wrong.

However, at some point practical decisions have to be made about matters on which existing scientific evidence remains ambiguous. When a food additive or environmental chemical poses an unknown level of risk, someone must decide whether the risk seems low enough to be tolerable or high enough to warrant controlling the substance and incurring the costs of doing so. Someone must decide whether a given penalty deters crime, even if the social science data are unclear. Someone must decide whether an emerging flu strain warrants a response, even before there are many cases. Someone must decide whether the threat posed by anthropogenic climate change warrants a certain level of response. Although there are many uncertainties about global warming, there is an overall consensus among scientists that it is occurring. The real debate is about prudence. What is the prudent response? Should society accept change as inevitable, make limited attempts to reduce carbon dioxide emissions to maximize effect while minimizing economic costs, or assume the worst and mount a strenuous global campaign to reduce carbon emissions, even at huge cost?

The average citizen has a role to play in the decision-making process, principally in the voting booth. Even with the best information on climate change, voters have to weigh a candidate's stance on climate against his or her stance on other important issues. Nevertheless, it is important that voters be as accurately informed about climate change as possible and also that they be skilled at detecting misleading arguments. The intent of this encyclopedia is to help its readers do that.

Steven I. Dutch
University of Wisconsin-Green Bay

Abrupt climate change

• **Category:** Meteorology and atmospheric sciences

Abrupt climate change entails drastic warming or cooling, regionally or globally, that takes place within a few years or decades and that persists for at least a few decades. Such a transformation would have a lasting effect on human institutions and infrastructure.

• Key concepts

albedo: the fraction of radiation reflected by a surface

feedback: a process in which any change accelerates further changes of the same type (positive feedback) or counteracts itself (negative feedback)

greenhouse gases (GHGs): atmospheric gases, such as carbon dioxide, water vapor, and methane, that trap heat radiation from Earth's surface by absorbing it and reemitting it

proxy: remnant physical evidence from which past climatic conditions can be inferred

thermohaline cycle: the "great conveyor belt" of ocean currents powered by density gradients created by heat and relative salt content

tipping point: the point at which the transition from one state in a system to another becomes inevitable

• Background

In 1840, Louis Agassiz published his theory that the Earth had passed through an ice age. As a result of Agassiz's work, the corollary idea that the globe's climate could change dramatically for extended periods entered scientific thinking. Scientists assumed, however, that such change occurred very slowly and smoothly over many millennia. When, in 1922, meteorologist C. E. P. Brooks first proposed that climate can change swiftly, he was largely ignored. During the early 1990's, however, a steady accumulation of data from four main sources strongly supported Brooks's hypothesis. These data suggested that in past epochs the atmosphere went from warm to cool or from cool to warm within decades, perhaps even within a few years.

• Proxy Evidence for Abrupt Change

The theory of abrupt change rests on proxy data from ice cores taken from the ice sheets covering Greenland and Antarctica, as well as from tree rings, sediments in oceans and lakes, and coral. In each of these proxies, layers of material are laid down annually and vary in thickness in accordance with annual atmospheric conditions. In tree rings, for instance, wet years foster greater growth in trees, which is reflected in wider rings than those produced during dry years. In addition to such evidence, gases in bubbles trapped in ice reveal the relative abundance of elements at the time they were trapped, which in turn provides clues to atmospheric temperatures at that time.

Taken together, proxy evidence demonstrates not only sudden climate change in past epochs but also frequent change. The most recent of four ice ages lasted from 120,000 to 14,500 years ago. Even during that frigid period, there were twenty-five periods of abrupt warming, called Dansgaard-Oeschger events, and six extended plunges in temperature, called Heinrich events; in all of them, change took place within decades.

The most studied example of abrupt change is the period known as the Younger Dryas, which began about 12,800 years ago. As the Northern Hemisphere was warming from the ice age, it suddenly relapsed into ice-age temperatures and stayed cold until 11,500 years ago, when temperatures over Greenland rose by 10° Celsius within a decade.

• Mechanisms for Abrupt Change

The United States National Research Council defines abrupt climate change as occurring when the climate system is forced to cross some threshold, triggering a transition to a new state at a rate determined by the climate system itself and faster than the cause.

Mechanisms for such change are poorly understood. It appears that some physical process forces an aspect of the climate system to pass a tipping point—for instance, in the albedo, average cloud cover, or salinity of ocean water. After the tipping point, positive feedback in the system accelerates the warming or cooling trend.

In the case of the Younger Dryas, scientists know that the water of the North Atlantic suddenly be-

An iceberg calved from Greenland's Jacobshavn glacier. Abrupt climate change could drastically increase the melting of this and other glaciers. (Konrad Steffen/MCT/Landov)

came less salty, which slowed or altered the course of the thermohaline cycle. The warm waters of the Gulf Stream no longer flowed north of Iceland and back down along the European coast, causing the continent to relapse to ice-age temperatures. The freshening probably resulted from a sudden outflow of water from a freshwater inland sea, Lake Agassiz, in north-central North America. The physical event that led to this forcing is a point of controversy. Scientists have proposed the breaking of an ice dam after gradual warming or possibly a meteor impact. Forcings for other abrupt changes in past climates include alterations in the salinity of the tropical Atlantic Ocean, evaporation and cloud cover in the South Pacific Ocean, melting of methane clathrates (frozen methane in the ocean beds), and the periodic warming of the South Pacific known as the El Niño-Southern Oscillation (ENSO).

• Anthropogenic Global Warming

Scientists worry that increasing levels of greenhouse gases (GHGs) in the atmosphere, much of them released by the burning of fossil fuels, have trapped radiant energy from the Sun in the atmosphere and increased average global temperatures in both the atmosphere and the oceans. This greenhouse effect could lead to abrupt climate change in several ways.

The vast ice sheets in the Arctic and Antarctica have the highest regional albedo on Earth, but they are shrinking rapidly, especially in the Arctic. There, the ice rests primarily on water, which is darker than ice and absorbs more heat. As the ice disappears, there is more exposed ocean surface to absorb solar energy, and the warmed water in turn helps melt the ice faster, creating a positive feedback loop. This melted ice will not affect ocean levels or salinity, but if ice sheets melt off the land of

Antarctica or Greenland, ocean levels could rise by dozens of meters within a century, lowering ocean salinity enough to stall the thermohaline cycle, which could cool Europe rapidly and drastically even while the rest of the world warmed. Should ocean water heat up too much, clathrates could melt and send billions of metric tons of methane into the atmosphere, accelerating global warming further. The augmented thermal energy in the atmosphere is likely to redistribute wind and rainfall patterns, plunging some regions into drought while making others wetter; catastrophic storms, such as hurricanes and tornados, could become more frequent and severe.

• Context

Some scientists argue than the Earth is entering a new geological age, the Anthropocene, because humanity itself now takes part in shaping Earth's overall surface conditions, climate in particular. Particulate pollution (especially soot), waste heat, release of GHGs, water consumption, and alteration of soil and plant cover affect not only the land, water bodies, and atmosphere but also modern civilization. If human effects on the environment trigger abrupt climate change, the onset of icy conditions in the Northern Hemisphere, droughts, superstorms, or rising sea level—all of which are possible according to computer models of climate change—would require radical, swift, and comprehensive measures to adapt or relocate much of Earth's human population. Not only would that be an expensive undertaking, but it would also mark a shift in the course of human history as profound as the Industrial Revolution.

Roger Smith

• Further Reading

Cox, John D. *Climate Crash: Abrupt Climate Change and What It Means for Our Future.* Washington, D.C.: Joseph Henry Press, 2005. A history of climate change, its effects on human culture, and its scientific study; provides exceptionally clear explanations of fundamental concepts.

Flannery, Tim. *We Are the Weather Makers: The Story of Global Warming.* Rev. ed. London: Penguin, 2007. A well-explained review of environmental science, the evidence for global warming, and what people can do about it; written by a scientist who was once a skeptic.

Lynas, Mark. *Six Degrees: Our Future on a Hotter Planet.* Washington, D.C.: National Geographic, 2008. Based entirely upon scientific research and computer models; describes the specific rapid climate changes that are possible for each degree Celsius's rise in average world temperature.

Pearce, Fred. *With Speed and Violence: Why Scientists Fear Tipping Points in Climate Change.* Boston: Beacon Press, 2007. An earnest, dramatic, and thorough survey of climate change studies and the possible results of abrupt climate change by a noted environmental journalist.

See also: Agassiz, Louis; Albedo feedback; Antarctica: threats and responses; Arctic; Climate change; Climate reconstruction; Thermohaline circulation; Younger Dryas.

Adiabatic processes

• **Category:** Meteorology and atmospheric sciences

• Definition

Adiabatic processes are those in which no heat transfer takes place. In an atmospheric adiabatic process, a parcel of air undergoes changes in its internal temperature but does not lose or gain heat to its ambient (surrounding) environment. According to Boyle's law, air, like any gas, experiences changes in pressure proportionate to changes in volume: When a parcel of air expands to fill more space, it decreases in pressure. When it contracts to fill less space, the air parcel increases in pressure.

As air rises, it expands, and as it descends, it is compressed. This is due to ambient pressures around the parcel of air: A given parcel of air that is near the Earth's surface has a great deal of air above it, weighing it down and increasing its total pressure. By contrast, a parcel of air that is higher up in the atmosphere has less air above it and thus less pressure upon it. As air parcels circulate in the atmosphere, they expand or contract until they reach a state of pressure equilibrium with other air parcels around them. At the same time, their internal temperatures decrease as they expand and increase as they contract. These adiabatic changes in temperature can be observed and mathematically averaged into what are known as "lapse rates."

There are three lapse rates to be considered. First, the environmental lapse rate is the most common baseline rate of change of the overall atmosphere as one ascends into it. This rate is 4° Celsius per 1,000 meters of altitude. Thus, for every 1,000 meters one ascends off the ground, one observes a drop in temperature, on the average, of 4° Celsius. (The atmosphere can also be in a state of inversion, in which temperatures increase rather than decrease as one ascends.)

The next lapse rate is the dry adiabatic lapse rate, which applies to air parcels that are not saturated with moisture. This rate is 10° Celsius per 1,000 meters. As an air parcel rises and cools, its temperature may approach its dew point, meaning that it may become saturated with moisture, causing water vapor to condense into liquid water. In the process of condensation, heat is released, and cooling begins to take place at a slower pace (because of the injection of heat into the ambient environment). This slower rate of cooling, known as the wet adiabatic lapse rate, is about 5° Celsius per 1,000 meters.

• Significance for Climate Change

The lifting condensation level (LCL) of a station's location to some degree can be looked at on a climatological basis. The LCL is the point to which a parcel of air ascends while it cools adiabatically at the dry adiabatic rate. The altitude of the LCL can be calculated by taking the known values of the parcel's air temperature and dew point temperatures in Celsius at the base of its rise, subtracting these values, and then dividing them by 8. That value is multiplied by 1,000 to arrive at the LCL in meters. If conditions are right, condensation will begin to take place and clouds will form at the LCL, and precipitation may be initiated. From that point on, if the parcel were to continue its rise past the LCL, it would expand and cool at the wet adiabatic rate.

Seasonal shifts occur in the average altitude of the LCL. Summertime LCLs are generally higher than wintertime LCLs. Similarly, changes in the climate could impact the average elevations of the LCL. Theoretically, if conditions were to warm as well as dry, some studies suggest the LCL would increase in altitude. The result of this would be to make cloud bases higher, with the possibility of shifting precipitation zones.

It has been suggested that deforestation could also play a role in modifying the altitude of the LCL. In the Monteverde region of Costa Rica, lowland moisture plays an important part in the positioning of the cloud base as winds blow moist air up the coastal range. Deforestation in the region may have lowered humidity, pushing the cloud base to higher altitudes. Consequently, various biological zones have shifted upward in altitude, forcing a shift in flora and fauna.

In contrast, if moister conditions prevail, soils may increase in moisture levels, which can lead to conditions conducive to an increase in evapotranspiration by plants. According to some studies, this increase in humidity would have the effect of lowering the LCL in regional areas. In any case, the adiabatic process does not change, only the positioning of the LCL.

M. Marian Mustoe

• Further Reading

Aguado, Edward, and James E. Burt. *Understanding Weather and Climate.* 4th ed. Upper Saddle River, N.J.: Pearson Prentice Hall, 2007. Survey of the elements of meteorology. Adiabatic processes and the establishment of the lifting condensation level are discussed.

Ahrens, C. Donald. *Meteorology Today.* 9th ed. Pacific Grove, Calif.: Thomson/Brooks/Cole, 2009. Provides explanations of adiabatics and temperature inversions.

Blanchard, D. C. *From Raindrops to Volcanoes: Adventures in Sea Surface Meteorology.* Mineola, N.Y.: Dover, 1995. Explains how precipitation forms and provides examples of how to measure precipitation.

See also: Atmospheric dynamics; Climate feedback; Clouds and cloud feedback; Dew point; Humidity; Rainfall patterns.

Advancement of Sound Science Center

- **Category:** Organizations and agencies
- **Date:** Established 1993

• Mission

The Advancement of Sound Science Center (TASSC), previously the Advancement of Sound Science Coalition, was a lobbying group run by the public relations firm APCO and founded by the tobacco industry. Originally established to provide scientific arguments against the dangers of secondhand smoke, TASSC expanded its mission to also advance the idea that warnings about other environmental dangers, including global warming and chemical exposure, were based on what the group labeled "junk science." TASSC's chief spokesman was libertarian Steven Milloy, junk science commentator for Fox News and creator of the Web site junkscience.com. Other advisers to TASSC included former New Mexico governor Garrey Carruthers and global warming skeptics S. Fred Singer, Michael Fumento, and Patrick J. Michaels. TASSC received financial support from Amoco, Exxon, Dow Chemical, Occidental Petroleum, Philip Morris, and other corporations.

• Significance for Climate Change

It is difficult to separate the activities of Steven Milloy from those of TASSC. TASSC was operated out of Milloy's home, and there was no other paid staff. Initially, TASSC was listed as the sponsor of the Web site junkscience.com, but after the Web site's corporate ties were exposed, the site ran under the sponsorship of Citizens for the Integrity of Science, another group run by Milloy using the same mailing address and phone number that TASSC had used.

TASSC played a role in raising early questions about the scientific consensus about the causes of global warming. In 1997, for example, the group invited citizens to read about global warming on its Web site and then to send emails to President Bill Clinton about proposed reductions in greenhouse gas (GHG) emissions. Articles available on the Web site were unanimously dismissive of the seriousness of global warming. One, titled "Kangaroo Court: The Working Group on Public Health and Fossil Fuel Combustion," argued that "A kangaroo court of junk scientists predicts that 8 million people will die during 2000 to 2020 from particulate matter air pollution associated with fossil fuel, unless the world limits GHG emissions to levels advocated by European nations."

In 2004, Milloy was among the public critics of the Arctic Climate Impact Assessment. In 2008, his Web site offered "The Ultimate Global Warming Challenge," a half-million-dollar prize to the first person to prove, "in a scientific manner, that humans are causing harmful global warming."

Cynthia A. Bily

See also: Cato Institute; Competitive Enterprise Institute; Fraser Institute; Libertarianism; Pseudoscience and junk science; Skeptics.

Aerosols

- **Category:** Chemistry and geochemistry

Aerosols are among the least well understood influences on global climate, but anthropogenic aerosols, especially sulfate aerosols released by fossil fuel combustion, seem to exert a cooling influence on the climate. This cooling effect, however, appears insufficient to counteract the warming caused by GHGs.

• Key concepts

cloud condensation nuclei: atmospheric particles such as dust that can form the centers of water droplets, increasing cloud cover

Dust Veil Index: a numerical index that quantifies the impact of a volcanic eruption's release of dust and aerosols

global dimming: the effect produced when clouds reflect the Sun's rays back to space

stratosphere: part of the atmosphere just above the troposphere that can hold large amounts of aerosols produced by volcanic eruptions for many months

troposphere: location in the lower atmosphere where the majority of aerosols form a thin haze before being washed out of the air by rain

• Background

Effects of aerosol pollutants such as volcanic dust have been debated for a long time. A 1783 eruption of a volcanic fissure in Iceland seemed related to an unusually cool summer in France that year. In 1883, the volcanic dust from the explosion of Krakatoa in the East Indies dimmed the sunlight for months, as had the 1815 eruption of Tambora. Some scientists perceived a pattern of temporary cooling from such events. Others asked if pollutants should be expected to warm, rather than cool, the atmosphere.

Aerosols are minute airborne solid or liquid particles suspended in the atmosphere, typically measuring between 0.01 and 10 microns. They may be of either natural or anthropogenic origin. Natural aerosol sources include salt particles from sea spray; clay particles from the weathering of rocks; volcanically produced sulfur dioxide, which oxidizes to form sulfuric acid molecules; and desert dust. Anthropogenic (human-produced) aerosol sources include industrial pollutants such as sulfates, created by burning oil and coal; smoke from large-scale burning of biomass, such as occurs in slash-and-burn clearing of tropical forests; and pollution from naval vessels' smokestacks.

Normally, most aerosols rise to form a thin haze in the troposphere; rain washes these out within about a week's time. Some aerosols, however, are found in the higher stratosphere, where it does not rain. They can remain in this atmospheric layer for months. Aerosols may influence climate in several ways: directly, through scattering and absorbing radiation, and indirectly, by acting as cloud condensation nuclei or by modifying the optical properties and lifetimes of clouds.

On windy days, bubbles created by breaking waves toss salt into the air when they burst, forming aerosols. Salt aerosols scatter sunlight, lessening the amount of energy that reaches Earth's surface and cooling the climate. The interaction of sea salt with clouds also causes cooling. The resulting whitening of the Earth further reduces the amount of sunlight that can reach the ground. Oceans cover over 70 percent of Earth's surface, and sea salt is a major source of aerosols in areas far distant from land.

Wind also helps form aerosols over land. Particles carried by the wind push and bounce over one another, abrading the surfaces of rock and other landforms. These particles wear rocks down progressively over time, converting their surfaces into dust and other particles. When these particles are incorporated into the air, they too form aerosols.

Following major volcanic eruptions, sulfur dioxide gas vented during the eruptions is converted into sulfuric acid droplets. These droplets form an aerosol layer in the stratosphere. Winds in the stratosphere scatter the aerosols over the entire globe, and they may remain in the atmosphere for about two years. Since they reflect sunlight, these aerosols reduce the amount of energy that reaches the troposphere and the Earth's surface, resulting in cooling.

Another significant natural aerosol is desert dust. "Veils" of dust stream off deserts in Asia and Africa, and they have also been observed on the American continent. These particles fall out of the atmosphere after a short flight, but intense dust storms often blow them to altitudes of 4,500 meters or higher. Since the dust is made up of minerals, the particles both absorb and scatter sunlight. Absorption warms the layer of the atmosphere where they are located, possibly inhibiting the formation of storm clouds and contributing to desertification.

• Early Speculation About Aerosols

Long before there was much interest in aerosols as a factor in climate change or any equipment capable of adequately analyzing aerosol data, a few individuals speculated about a possible aerosol-climate connection. The first man credited with reporting his ideas was Mourgue de Mondtredon, a French naturalist who in 1783 documented the eight-month-long Laki eruption in southern Iceland. The eruption caused the grass to die: Three-quarters of the region's livestock and one-quarter of its people starved to death. For months, a haze hovered over western Europe. When Benjamin Franklin was visiting in France in 1783, he experienced an unseasonably cold summer and speculated that the Laki volcanic "fog" had noticeably dimmed the sunlight.

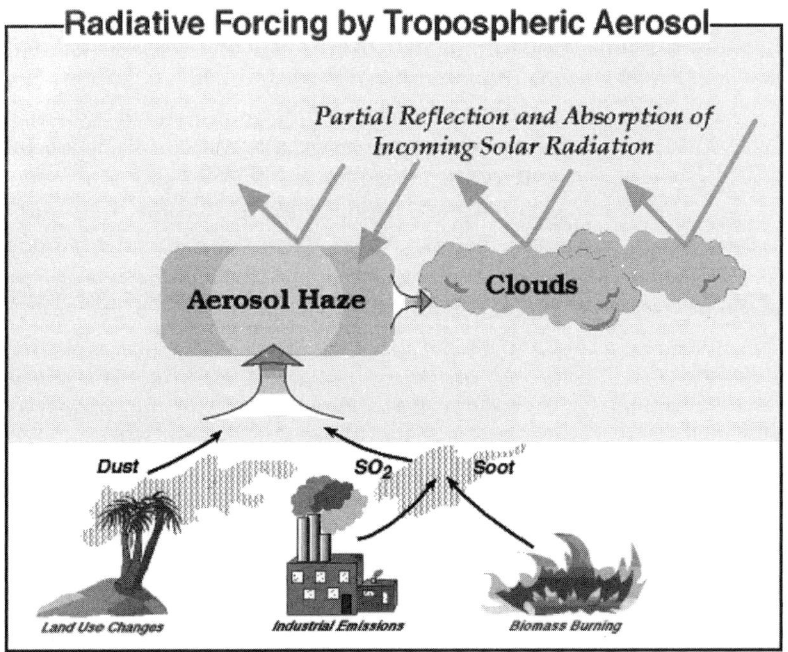

Aerosols can both reflect and absorb solar radiation, making their role in the greenhouse effect particularly complex. (NOAA)

A century later, in 1883, the eruption of the Indonesian volcano Krakatoa (Krakatau) sent up a veil of volcanic dust that reduced sunlight globally for months. Scientists were unable to determine what effect the eruption might have had on the average global temperature, but scientists thereafter acknowledged volcanoes as a possible natural influence on Earth's climate.

A few scientists who examined temperatures after major volcanic eruptions between 1880 and 1910 perceived a pattern of temporary cooling. Only later would older records reveal that the 1815 eruption of Tambora in Indonesia had affected the climate more severely than had the Krakatoa eruption. Speculation led some to ask if volcanic eruptions had precipitated ice ages or had cooled the Earth to the extent that dinosaurs became extinct.

• Early Twentieth Century Aerosol Research

Throughout the first half of the twentieth century, it was known that volcanic aerosols could affect climate. As a result, some scientists suspected that other kinds of dust particles could have similar climatic effects. Physics theory seemed to support the notion that these particles should scatter radiation from the Sun back into space, thereby cooling the Earth. These ideas remained largely speculative, though some researchers began to focus on the possibility that human activity might be a major source of atmospheric particles.

In the 1950's, nuclear bomb tests provided improved data on aerosol behavior in the stratosphere. It was determined that stratospheric dust would remain for some years, but would stay in one hemisphere. Research in the early 1960's indicated that large volcanic eruptions lowered average annual temperatures. Some researchers, however, deemed those results enigmatic, since temperatures had fallen during a period of few eruptions. Meteorologists acknowledged that other small, airborne particles could influence climate, but throughout the first half of the century, speculation fell short of conclusion.

Gradually, scientists shifted their focus to anthropogenic atmospheric particles. Measurements by ships between 1913 and 1929 noted that sea air showed an extended decrease in conductivity, apparently caused by stack smoke and gases from ships and possibly from industry on land. Even in 1953, however, scientists were uncertain about the significance of the pollution.

During the 1950's, some scientists asked whether aerosols might affect climate by helping form clouds. Since cloud condensation nuclei are essential for providing a surface for water droplets to condense around, the notion of seeding clouds with silver iodide smoke to make rain was widespread. By this time, aerosol science was just coming into its own as an independent field of study, having been given impetus by the concern that disease-carrying aerosols and poisonous gas could be lethal. Public concern over urban smog also fueled studies by aerosol experts. By and large, however, scientists avoided the study of cloud formation. Field testing often produced contradictory results and was extremely expensive, and many researchers believed that aerosols' effects on clouds were too complex to comprehend.

• Aerosol Research in the Later Twentieth Century

By the early 1960's, the scientific community was beginning to pay more attention to the possibility that humans influenced clouds. One noted astrophysicist had long had an interest in aerosols after seeing the effects of the Dust Bowl in the 1930's. He noticed changes in the skies over Boulder, Colorado, and pointed out jet airplane contrails, predicting correctly that they would spread, thin, and become indistinguishable from cirrus clouds. The apparent ability of aircraft to create cirrus clouds revealed the possibility that they might be causing climate changes along major air routes. Others questioned the possibility of anthropogenic activity as the source of pollution settling on polar ice caps. At the time, the theory did not receive much credence.

Around 1970, the British meteorologist Hubert Horace Lamb's Dust Veil Index established a connection between dust and lower temperatures. While scientific studies at this time did not yet find strong evidence for an increase in global turbidity, they did document regional hazes that spread in a radius of up to one thousand kilometers or more from industrial centers. The scientific debate shifted from the existence of anthropogenic dust to the effects of that dust. It remained a subject of controversy whether and under what circumstances dust would cool or heat the climate, especially after a spacecraft on Mars in 1971 found that a large dust storm had caused substantial warming of the Martian atmosphere.

Deadly droughts in Africa and South Asia in 1973 caused public concern about climate change, but it was not confirmed that sulfate pollution had contributed to the Sahel drought until the end of the century. Scientific publications in the mid- to late 1970's discussed warming or cooling effects without reaching accord, although a majority felt that greenhouse warming would dominate. At this time, only a few researchers noted that aerosol pollution might cancel out some greenhouse warming and thus temporarily mask its effects. Others denied that industrial pollution could mitigate the enhanced greenhouse effect caused by carbon dioxide (CO_2) emissions.

The 1980's brought the realization that additional factors contributed to climate and climate change. For example, climate scientists generally treated aerosols as a globally uniform background, largely of natural origin, when in fact different aerosol properties obtained in different regions based on relative humidity. Many questions remained.

By 1990, it was acknowledged that from one-fourth to one-half of all tropospheric aerosol particles were anthropogenic. These included industrial soot and sulfates, smoke from forest clearing fires, and dust from overgrazed or semiarid land turned to agriculture. Impressive advances in laboratory instrumentation made possible much more sophisticated satellite observations, greatly increasing the resolution of climate models. The key paper establishing the net effect of aerosols on Earth's heat balance was published in the early 1990's; it concluded that radiation scattering due to anthropogenic sulfate emissions was counterbalancing CO_2-related greenhouse warming in the Northern Hemisphere.

It became apparent that earlier climate projections might be erroneous, because they had not factored in sulfate aerosol increases. Climatologists redoubled their efforts to produce accurate models and projections of Earth's climate. In 1995, for the first time, new results that took into account aerosol influence yielded a consistent and plausible picture of twentieth century climate. According to this picture, industrial pollution had temporarily depressed Northern Hemisphere temperatures around the mid-century. A 2008 study found that black carbon aerosols had exerted a much greater warming effect than had been earlier estimated, because the combined effects of black carbon with sulfate aerosols had not been taken into account. It seemed clear that reducing sooty emissions would both delay global warming and benefit public health.

• **Context**

A number of aerosol specialists have questioned whether they have underestimated the cooling effect of aerosols. If they had, they would have underestimated those aerosols' restraint of greenhouse warming, significantly underestimating the extent of global warming in the absence of anthropogenic aerosol pollution. Much uncertainty remains, and each new study introduces new complexities. It seems clear that reducing sooty emissions would both delay global warming and benefit public health, yet nagging questions remain: Since aerosol and clouds, unlike gases, are not distributed evenly throughout the atmosphere, uniform samples cannot be obtained. Further, the properties of clouds and aerosols are incompletely understood, and scientists are only beginning to understand some of the interactions that take place between aerosols, clouds, and climate. Thus, these interactions have not yet been incorporated into their models.

Victoria Price

• **Further Reading**

Levin, Z., and William R. Cotton, eds. *Aerosol Pollution Impact on Precipitation: A Scientific Review.* New York: Springer, 2009. Discusses the principles of cloud and precipitation formation, the sources and nature of atmospheric aerosols and their distribution, techniques for measuring aerosols, effects of pollution and biomass aerosols on clouds and precipitation, and parallels and contrasts between deliberate cloud seeding and aerosol pollution effects.

Massel, Stanislaw R. *Ocean Waves Breaking and Marine Aerosol Fluxes.* Sapot, Poland: Institute of Oceanology of the Polish Academy of Sciences, 2007. Addresses the basic processes and mechanics of steep and breaking waves, experimental insights into wave-breaking mechanisms, wave-breaking criteria, and various aspects of marine aerosols and marine aerosol fluxes, especially in the Baltic Sea.

Spury, Kvetoslav R., ed. *Aerosol Chemical Processes in the Environment.* New York: Lewis, 2000. Five sections treat general aspects of aerosols, laboratory studies of aerosols, the synthetic chemistry of aerosols, aerosol deposits on buildings, and aerosols in the atmosphere.

See also: Atmospheric chemistry; Atmospheric dynamics; Carbonaceous aerosols; Chemical industry; Clouds and cloud feedback; Ozone; Stratosphere; Sulfate aerosols.

Agassiz, Louis

Swiss naturalist, paleontologist, geologist, and glaciologist

Born: May 28, 1807; Môtier-en-Vuly, Switzerland
Died: December 14, 1873; Cambridge, Massachusetts

Agassiz was the originator of the concept of the "Great Ice Age," an idea of fundamental importance to understanding the phenomenon of cyclical change in Earth's climate.

• **Life**

Louis Agassiz was born and raised in French Switzerland, descended on his father's side from six generations of Protestant clergy. Breaking from the family tradition of a career in the ministry, he studied the sciences, receiving a doctorate in philosophy from the University of Erlangen, Bavaria, in 1829 and a doctorate in medicine from the University of Munich, Bavaria, in 1830. Journeying to Paris, he came to know and be admired by two of the leading scientists of the time, Georges Cuvier and Alexander von Humboldt. The latter obtained an appointment for Agassiz in 1832 as instructor in the natural sciences at the newly founded Academy of Neuchâtel (now the University of Neuchâtel) in Switzerland.

Visiting the United States in 1846, Agassiz turned the focus of his research to North America. The following year, Harvard University appointed him professor of zoology and geology. During his tenure at Harvard, Agassiz became the father of science education in the United States, forming a vanguard generation of professional scientists. He was widely and popularly renowned as a lecturer and writer, and he realized one of his focal ambitions with the founding in 1857 of the Museum of Comparative Zoology at Harvard. In opposition to

Charles Darwin, he disputed the theory of evolution, believing species were complete as they had been created in a divine plan. He married Cecile Braun (sister of the noted botanist, Alexander Braun) in 1833; the couple had three children together. Widowed in 1848, Agassiz married Elizabeth Cabot Cary two years later. Elizabeth later became the first president of Radcliffe College.

• **Climate Work**

To understand the phenomenon of global warming, one must realize that the Earth is capable of long-term, extreme changes of temperature and climate. The first scientific understanding of such change did not emerge until the nineteenth century. Agassiz was among a small group of natural scientists who first postulated that there had been some period in the Earth's history when the environment had been so cold that it sustained a vast sheet of ice over much of the planet.

Agassiz arrived at this perception through an indirect but steady course of research in various fields. As a university student, he had received a singular assignment. The renowned naturalist, Carl Martius, who had recently returned from an expedition in Brazil, gave Agassiz responsibility for classifying a large collection of fish fossils gathered during the trip. Agassiz completed this project and published his findings, as well as another work on fossil fish. He then advanced a new method for zoological classification. In 1836, he received the Wollaston Medal from the Geological Society of London.

From his field research in Switzerland, Agassiz became interested in glaciers and their influence on molding ancient terrains. His observations led him to be among the first to maintain that at one time, amid a series of catastrophes, vast sheets of ice had covered much of the Earth. He thus proposed that there had been an "ice age." In coming to this conclusion, Agassiz was extending and refining emerging speculations about glaciers and rock formation expressed by several contemporary naturalists.

In the United States, Agassiz continued his formidable research and publishing projects, writing a comparative study of the physical characteristics of Lake Superior, followed by his monumental *Contributions to the Natural History of the United States of America* (1857-1862, four volumes). He also wrote a textbook that would be highly influential in the formation of generations of American scientists, *Principles of Zoology, Part 1: Comparative Physiology* (1848). His teaching method emphasized students' direct engagement with objects of scientific observation. The French Academy of Sciences awarded him the Cuvier Prize in 1852. A culminating achievement of his career was an expedition to Brazil and the Amazon about which he wrote, with his second wife, in *A Journey in Brazil* (1868). Erroneously, he maintained that Brazil had once been covered by ice, basing his conclusion on glaciation in other portions of the Americas, including the Andes and Rocky Mountains and the drainage basin of the Missouri, Mississippi, and Ohio rivers. Agassiz was the intimate of a brilliant generation of naturalists. He brought the knowledge and methodology of that group to the United States, forming the foundation of modern American science and science education and crucially shaping the development of environmental awareness.

Edward A. Riedinger

Louis Agassiz. (Library of Congress)

• **Further Reading**

Agassiz, Elizabeth Cary. *Louis Agassiz: His Life and Correspondence in Two Volumes.* 2 vols. 1885. Reprint. Bristol, Avon, England: Thoemmess Press, 2003. Biography and collection of letters published by Agassiz's second wife. The combination of primary and secondary materials is extremely useful for understanding Agassiz's personal and professional life.

Bolles, Edmund Blair. *The Ice Finders: How a Poet, a Professor, and a Politician Discovered the Ice Age.* Washington, D.C.: Counterpoint, 1999. Interweaves the concurrent findings regarding an ice age by Professor Agassiz; Charles Lyell, an Arctic explorer with a poetic imagination; and Elisha Kent Kane, a Scottish geologist who organized a scientific political constituency.

Huxley, Robert. *The Great Naturalists.* London: Thames & Hudson, 2007. Traces the development of natural history as reflected in the lives, observations, and discoveries of some of the world's leading naturalists, including Aristotle, Alexander von Humboldt, and Charles Darwin. Places Agassiz among his fellow naturalists.

See also: Glacial Lake Agassiz; Glaciations; Glaciers; Ground ice; Ice shelves.

Agenda 21

• **Category:** Conferences and meetings
• **Date:** Adopted June 14, 1992; reaffirmed September 4, 2002

Agenda 21 is a comprehensive plan of action to address global carbon emissions. It mandates cooperation at the global, national, and local levels to reduce the emissions of industrialized nations and to slow the rate of increase of emissions in developing nations.

• **Background**

From June 3 to June 14, 1992, the United Nations hosted the Conference on Environment and Development, informally known as the Earth Summit, in Rio de Janeiro, Brazil. Representatives of 172 nations participated, along with representatives of twenty-four hundred nongovernmental organizations (NGOs). The goal of the conference was to find new ways for nations to conserve natural resources and drastically reduce pollution while still developing economically. To achieve these goals, the world would need to study and improve industrial production (paying particular attention to the handling of toxics), find alternative energy sources to reduce reliance on carbon-emitting fossil fuels, encourage mass transit, and protect increasingly scarce sources of freshwater.

Five documents were produced at the 1992 Earth Summit: the Rio Declaration on Environment and Development, the Convention on Biological Diversity, an agreement on forest principles, the U.N. Framework Convention on Climate Change, and Agenda 21. Drafting of Agenda 21 had begun in 1989, and during the subsequent two years of negotiations several specific reduction targets and funding plans were deleted. The final draft was presented at the summit.

The 1992 Agenda 21 plan called for a five-year review of progress, which was conducted at a special session of the United Nations General Assembly in 1997. In 2002, at the World Summit on Sustainable Development held in Johannesburg, South Africa, participants affirmed their commitment fully to implement Agenda 21.

• **Summary of Provisions**

The nine-hundred-page Agenda 21 document is divided into forty chapters in four major sections. Section 1, "Social and Economic Dimensions," covers programs to reduce poverty and help guide developing nations in building their economies sustainably. Section 2, "Conservation and Management of Resources for Development," addresses atmospheric protection, deforestation, desertification, conservation of biological diversity, and other issues. Section 3, "Strengthening the Role of Major Groups," describes programs undertaken by international NGOs, by women and children, by workers and unions, and by business and industry. The fourth section, "Means of Implementation," addresses financial resources, transfer of technology, science, and international cooperation.

A Defining Moment in History

The opening paragraph of the Preamble to Agenda 21 presents an unusually stark statement of the challenges facing humanity at the beginning of the twenty-first century and the need for international cooperation to meet those challenges.

Humanity stands at a defining moment in history. We are confronted with a perpetuation of disparities between and within nations, a worsening of poverty, hunger, ill health and illiteracy, and the continuing deterioration of the ecosystems on which we depend for our well-being. However, integration of environment and development concerns and greater attention to them will lead to the fulfilment of basic needs, improved living standards for all, better protected and managed ecosystems and a safer, more prosperous future. No nation can achieve this on its own; but together we can—in a global partnership for sustainable development.

While some 98 percent of the nations on Earth signed on to Agenda 21, it is not a legally binding document but simply a plan for future action. The 1992 plan included programs in developing nations that were expected to cost billions of dollars annually, and industrialized nations agreed to contribute approximately $125 billion per year toward those costs. The plan also created a new body of the United Nations Economic and Social Council (ECOSOC), the Commission on Sustainable Development, to oversee and coordinate activities that further the goals of Agenda 21.

• Significance for Climate Change

Most observers believe that the goals of Agenda 21 have not been achieved, nor has adequate progress been made. One serious problem has been funding. At the 1992 Earth Summit, nations made nonbinding agreements to contribute funding for specific projects, including phasing out the use of chlorofluorocarbons (CFCs) and supporting the sustainable development efforts of underdeveloped nations. However, few countries have contributed the amounts agreed to, in part because economic recessions in industrialized nations, including the United States, have led these nations to shift their spending commitments to protect their own short-term domestic stability. There has also been resistance to Agenda 21 from those who believe that it undermines state sovereignty. Although the program was intended in part to draw together an international community of concerned citizens, few people, at least in the United States, are aware that Agenda 21 exists.

Agenda 21 has been successful, however, in inspiring national, regional, and local actions. These smaller programs, known as "Local Agenda 21" or "LA-21" programs, have been adopted in Cambridge and Manchester in the United Kingdom; Seattle, Washington; Chicago, Illinois; Whyalla, Australia; and cities in Finland, the Netherlands, Spain, and South Africa. Other national and state governments have created legal requirements or advisory bodies to address relevant parts of Agenda 21.

Cynthia A. Bily

• Further Reading

Dodds, Felix, ed. *The Way Forward: Beyond Agenda 21.* London: Earthscan, 1997. Collection of essays by experts in several disciplines, examining the successes and the remaining challenges Agenda 21 poses internationally, nationally, and at the grassroots level.

Picolotti, Romina, and Jorge Daniel Taillant. *Linking Human Rights and the Environment.* Tucson: University of Arizona Press, 2003. Considers environmental degradation as a human rights issue affecting mostly poor and vulnerable people and looks to Agenda 21 and other environmental policies and treaties as potential remedies for ongoing abuses.

Robinson, Nicholas A. *Strategies Toward Sustainable Development: Implementing Agenda 21.* Dobbs Ferry, N.Y.: Oxford University Press, 2005. Covers all of the action plans generated by Agenda 21; annotations and commentary.

Sitarz, Dan. *Agenda 21: The Earth Summit Strategy to Save Our Planet.* Boulder, Colo.: EarthPress, 1993. Sitarz, an environmental attorney, reworks the language of the nine-hundred-page Agenda 21 agreement to make it readable by the general public.

World Bank. *Advancing Sustainable Development: The World Bank and Agenda 21.* Washington, D.C.: Author, 1997. The official World Bank analysis of the successes and shortfalls of Agenda 21 in its first five years.

See also: Annex B of the Kyoto Protocol; Convention on Biological Diversity; Earth Summits; International agreements and cooperation; Kyoto Protocol; United Nations Climate Change Conference; United Nations Conference on Environment and Development; United Nations Framework Convention on Climate Change.

Agriculture and agricultural land

• **Categories:** Economics, industries, and products; plants and vegetation

Modern, large-scale agriculture has led to increased GHG emissions, primarily resulting from high energy inputs, land clearing, soil degradation, and overgrazing by livestock. The massive conversion of forests into farms has reduced the land's capacity to function as a carbon sink. As a result, more GHGs are emitted into the atmosphere, contributing to global warming and climate change.

• Key concepts

carbon cycle: processes through which carbon atoms circulate among Earth's atmosphere, terrestrial biosphere, oceans, and sediments—including fossil fuels

climate change: a statistically significant variation in either the mean state of the climate or its variability

energy-intensive agriculture: a method of farming that involves working on a large scale, utilizing significant resources, energy, and mechanization; often also referred to as industrialized agriculture

greenhouse effect: phenomenon in which certain gases in a planet's atmosphere trap heat that would otherwise escape into outer space, thereby increasing the planet's surface temperature

• Background

There is now a consensus that the mean temperature of the Earth will increase by an average of 2.0° to 5.8° Celsius in the twenty-first century. Some environmental models, taking into account likely changes in vegetation cover, predict an even higher rise of 8° Celsius during the century. The resulting elevated temperatures (even at the lower end of the estimates) will have significant effects on Earth's biosphere, including human life. Many factors are associated with this predicted temperature rise, but agriculture is among the major contributors.

• Direct Impacts of Agriculture on Climate Change

Agricultural activity is a significant source of greenhouse gases (GHGs). GHG levels are affected by land clearing, high energy inputs, soil degradation, and intensive animal husbandry. Based on current estimates, agriculture contributes to 25 percent of the world's carbon dioxide (CO_2) emissions, 60 percent of methane gas emissions, and 80 percent of nitrous oxide emissions. Agriculture's high energy input results primarily from manufacturing chemical fertilizers, herbicides, and pesticides; operating farm machinery; irrigating farmland using pumps and other machines; and transporting products over long distances. Collectively, these activities account for more than 90 percent of the total energy expenditure in agriculture.

The burning of fossil fuels releases CO_2 into the atmosphere. The CO_2 concentration in the atmosphere has increased from 277 parts per million to 382 parts per million since the beginning of the Industrial Revolution in the mid-eighteenth century. Industrialized agriculture is believed to have contributed to 25 percent of that increase.

Overuse of fertilizers, in addition to energy inputs in fertilizer manufacturing, contributes significantly to climate change. More than half of all synthetic fertilizers applied to the soil either end up in local waterways or emit to the atmosphere. A portion of the excess nitrogen fertilizers in the soil is converted into nitrous oxide, which is 296 times more potent than CO_2 in trapping heat and which has a long atmospheric lifetime of 114 years. Each year, nitrous oxide emissions alone account for the equivalent of 1.9 billion metric tons of CO_2 emissions.

The second greatest GHG emission by agriculture is methane, released in small amounts by rice paddies and in much larger amounts by livestock. As the demand for meat increases, more livestock are raised and are fed higher-protein diets. Both the number of livestock and their protein-rich diets increase the amount of methane they emit. Methane gas is fourteen times more potent than CO_2 in trapping heat. Its concentration has almost tripled since the Industrial Revolution, from 600 parts per billion to 1,728 parts per billion.

• Indirect Impacts of Agriculture on Climate Change

Agriculture also contributes indirectly to climate change. Clearing trees and other natural stands to make land suitable for agricultural uses removes important carbon sinks, so less carbon is returned to the terrestrial biosphere and more CO_2 finds its way into the atmosphere, where it contributes to climate change.

The effect of land clearing on climate change is evident from the consequences of the destruction of tropical rain forests. For instance, large areas in Brazil have been cleared to facilitate soybean production. This clearing disrupts the local water cycle, which in turn alters Brazil's climate. In rain forests, water circulates as a result of evaporation, which greatly increases humidity. Natural tree stands act to buffer extremes of heat, cold, and drought. When the trees are removed, the buffer disappears. Moreover, the amount of water vapor in the air decreases, causing shifts in rainfall patterns, moisture levels, air temperature, and weather patterns generally.

The conversion of forests into agricultural lands has significantly altered Earth's vegetation cover. Such changes in the land surface affect Earth's albedo—that is, the proportion of incident radiation reflected by the planet's surface. Changes in the albedo in turn can affect the surface energy budget, which affects local, regional, and global climates. Changes in vegetation also produce changes in the global atmospheric concentration of CO_2. Agricultural landscape ranks among the lowest in carbon sequestration. Thus, as more land is devoted to agricultural uses, more of Earth's carbon is converted to CO_2 and emitted to the atmosphere, contributing to global warming.

• Effects of Climate Change on Agriculture

Potential climate changes associated with elevated GHGs and an altered surface energy budget include an increased incidence of heat waves, severe storms, and floods, as well as elevated sea levels. Some 30 percent of the agricultural lands worldwide could be affected by these changes. Global warming alone is projected to have considerable effects on agriculture. A warming of 2° Celsius or more could reduce global food supplies and aggravate world hunger. The impact on crop yields will vary considerably across different agricultural regions. Warm regions, such as tropics and subtropics, will be threatened by climate change, while cooler regions, mainly in temperate or higher latitudes, may benefit from warming.

Global climate change may have significant effects upon livestock systems as well. First, the productivity and quality of rangelands may be adversely affected. This in turn will affect the quality and productivity of livestock. Second, higher grain prices resulting from the disruption of crop production will lead to higher costs for livestock products. Third, increased severity and frequency of storms may intensify soil erosion and decrease the productivity of rangelands. Fourth, global warming could result in changes in the distribution and severity of livestock diseases and parasites, which may threaten the health of animals, especially those in intensively managed livestock systems.

• Possible Solutions

Unlike any other industrial GHG emitters, agriculture has the potential to change from being one of the largest GHG sources to being a net carbon sink, reversing its role in climate change. Several practical measures can be taken to mitigate the climate change caused by intensive agriculture. These include the reduced and more efficient use of chemical fertilizers, protection of soil, improvement of paddy rice production, and reduction of demand for meat.

A Chinese farmer works drought-afflicted fields in Shanxi province. Drought is among the greatest threats to world agriculture. (Reuters/Landov)

Precision farming can reduce the need for chemical fertilizers. In precision farming, fertilizers and other agrochemicals are applied based on crops' needs, in precise amounts and on a carefully managed schedule. The reduced application of these chemicals not only cuts GHG emissions but also alleviates other environmental problems such as water pollution and eutrophication of waterways.

As a result of intensive farming, agricultural soils have some of the lowest carbon contents of all land types. If these soils can be modified to absorb more of Earth's carbon, the result will be a net reduction in atmospheric carbon. Low soil carbon content can be reversed through a number of measures, including planting cover crops, fallowing, and engaging in conservation tillage. These practices will increase the amount of organic matter (and thus the carbon content) in the soil. They will also reduce soil erosion and surface runoff, thereby reducing the need for chemical fertilizers. Collectively, these measures can turn agricultural soils into carbon sinks, changing the nature of their impact on climate change.

To reduce methane emissions from rice production, better cultivation techniques will need to be adapted. For example, rather than continuously flooding rice paddies, farmers could supply water to the paddies only when it is needed during the growing season and keep the paddies dry during the nongrowing season. Such measures could reduce methane emission from rice fields significantly.

Livestock raising is the second largest source of GHGs in agriculture. The most efficient way to cut methane emission due to livestock is simply to reduce the number of farm animals. As an ever-increasing demand for meat and dairy products drives increasing animal husbandry, one effective approach to cut methane emission is to reduce the demand for meat, especially in developed countries where consumers have tremendous buying power. Reduced meat and dairy consumption would go a long way toward curbing methane emissions.

• Context

Agriculture and climate change are interlocked processes, in that each exerts effects on the other in a complex fashion. Climate changes, especially shifts in precipitation and temperature, are widely believed to have significant effects on agriculture, because these two factors determine the carrying capacity of any ecosystem. At the same time, modern agriculture is a major contributing factor to global warming, as altered land cover and the emission of CO_2, methane gas, and nitrous oxide from intensive farming increase the GHG content of the atmosphere. However, it remains possible to transform industrialized agriculture, using techniques that could render it more sustainable and mitigate its effects upon global and local climates.

Ming Y. Zheng

• Further Reading

Bazzaz, Fakhri, and Wim Sombroek, eds. *Global Climate Change and Agricultural Production: Direct and Indirect Effects of Changing Hydrological, Pedological, and Plant Physiological Processes.* New York: Wiley, 1996. Assesses the effects of global climate change on the agricultural production of crops and livestock. Covers issues such as the CO_2 fertilization effect, the adverse effects of elevated levels of ultraviolet-B (UVB) radiation and ozone on plant growth and productivity, and the environmental effects of livestock.

Marland, Gregg, et al. "The Climatic Impacts of Land Surface Change and Carbon Management, and the Implications for Climate-Change Mitigation Policy." *Climate Policy* 3 (2003): 149-157. Reviews evidence on how changes in land use and altered vegetation cover affect surface energy budget in addition to GHG emission, which in turn affects local, regional, and global climate.

Palo, Matti, and Heidi Vanhanen, eds. *World Forests from Deforestation to Transition?* Boston: Kluwer Academic, 2000. Addresses global and subnational issues concerning the world's forests, societies, and environment from an independent and nonpolitical point of view. Emphasizes the importance of developing a scientific understanding of the interconnectedness between forests, human activity, and the environment, and of the consequences of environmental change for societies' development and growth.

Paustian, Keith, et al. *Agriculture's Role in Greenhouse Gas Mitigation.* Arlington, Va.: Pew Center on Global Climate Change, 2006. Well-researched and clearly written paper. Offers reasonable measures and reachable goals for mitigating GHG emission by agriculture.

See also: Aerosols; Air travel; Albedo feedback; Amazon deforestation; Animal husbandry practices; Anthropogenic climate change; Carbon cycle; Carbon dioxide; Carbon dioxide equivalent; Carbon dioxide fertilization; Civilization and the environment; Climate change; Composting; Deforestation.

Agroforestry

• **Categories:** Plants and vegetation; economics, industries, and products

• Definition

Agroforestry encompasses a broad range of land-use practices involving the integration of trees with annual crops. Though as old as agriculture itself, agroforestry has only recently attracted scientific attention. Many agroforestry practices have the potential to contribute to climate change mitigation, adaptation, or both.

A land-use system is defined as agroforestry if it includes at least one perennial species (generally a tree) and one crop species that interact on the same land. Usually, the role of the trees is to provide the crop with nutrients, shade, improved soil quality, physical support, or protection from wind, water, or pests. The goal is to achieve a net benefit as compared to growing the species in monocultures. The complementarity can be economic (for example, income diversification) as well as biophysical. Agroforestry systems may also provide fuelwood, timber, fruit, and other products. However, when trees are grown solely for these purposes and are not intended to improve cropland, the practice is usually not considered agroforestry.

• Significance for Climate Change

Agroforestry trees can contribute to climate mitigation by sequestering carbon in their biomass, in the soil, and in wood products. Although agroforestry systems usually store less carbon per unit area than do forests, they have the advantage of allowing the land to remain in use for the production of food or other crops. Additionally, a large total land area is suitable for agroforestry practices, implying a potential for large-scale carbon sequestration—similar in magnitude to that of reforestation, according to a 2000 estimate by the Intergovernmental Panel on Climate Change (IPCC). However, the distributed and diverse nature of agroforestry poses challenges for carbon accounting. Carbon sequestration alone is unlikely to drive the adoption of agroforestry, but it can provide a coincentive.

Agroforestry can also play a role in climate adaptation. Trees can moderate local microclimate, protecting crops by lowering temperatures and reducing soil evaporation. They can ameliorate drought by improving soil water-holding capacity. They can reduce damage from extreme climate events, such as floods and wind storms. Able to access deep supplies of water and nutrients, mature trees are generally less susceptible to climatic fluctuations than are annual crops and can provide alternative sources of food and income under adverse climates.

Achieving these benefits requires a careful choice of species and of management practices, since biophysical interactions in an agroforestry system can be complex. Agroforestry projects often must also consider socioeconomic issues such as land tenure, gender roles, and risk aversion. The potential of agroforestry for climate mitigation and adaptation is likely significant, but still largely unexplored.

Amber C. Kerr

See also: Agriculture and agricultural land; Carbon cycle; Carbon 4 plants; Carbon 3 plants; Deforestation; Forestry and forest management; Forests; Sequestration; Tree-planting programs.

Air conditioning and air-cooling technology

• **Categories:** Economics, industries, and products; science and technology

Air conditioning and air cooling account for roughly half of residential energy use and some 15 percent of industrial energy use. Some refrigerant chemicals have GWPs more than one thousand times that of CO_2. Changes in air conditioning and cooling technologies can reduce global warming significantly.

• Key concepts

coefficient of performance: a standard measurement of energy efficiency

energy efficiency rating: ratio of heat energy produced or removed to power expended

global warming potential: the climatic impact of a given mass of greenhouse gas, measured as a function of the impact of the same mass of carbon dioxide

total equivalent warming impact: a measurement, in mass of CO_2 equivalent, of the global warming potential of an entire system

• Background

Air conditioning systems use energy to remove heat from a given location, transferring it to the exterior in the form of exhaust. Some of the energy is invariably wasted as heat, and this additional heat must also be exhausted. Thus, air conditioning results in net heat release to the environment. The warming effect is aggravated, moreover, because higher forms of energy such as electricity are used to power air conditioners. Generating and transmitting this electricty entails further production of waste heat, as well as greenhouse gas (GHG) emission. Demand for air conditioning peaks along with other demand for electricity during summer days. Auxiliary generators used to meet peak demand are less efficient and often burn fossil fuels.

In 1998, for example, the total equivalent warming impact (TEWI) of European air conditioning was 156 million metric tons of carbon dioxide (CO_2) equivalent. Of that TEWI, 25.6 million metric tons were attributable to direct hydrofluorocarbon emissions and around 130 million metric tons were attributable to indirect emissions. In the same year, the TEWI of 303 million automotive air conditioning systems worldwide represented 0.14 percent of total anthropogenic TEWI.

• Approaches and Choices

There are two basic approaches to cooling the air in an enclosure: refrigeration and evaporation. In the refrigeration cycle, a fluid is compressed so that its temperature rises and is circulated through pipes over which air or water is forced, thus removing heat. The compressed, cooled refrigerant is then expanded through a nozzle, so that its temperature drops sharply before it absorbs heat in an exchanger from the air in the enclosure. The refrigerant with the heat absorbed is then compressed and its heat removed in the exhaust heat exchanger. This process need not include phase change. When a substance evaporates, it absorbs a great deal of heat from the environment, called the latent heat. In large industrial systems, the hot coil heat is removed using flowing water, some of which evaporates into the air flowing through a cooling tower.

• Choice of Refrigerants

Refrigerants enable the exchange of a large amount of heat with the least expenditure of work. Desirable properties for these substances include low boiling point, high latent heat of vaporization,

On Dyess Air Force Base, in Abilene, Texas, a chilled glycol slurry is used to cool buildings in an effort to decrease the base's environmental footprint. (Ralph Lauer/MCT/Landov)

high specific heat, and high critical temperature. Ammonia (R-717) is used in industrial systems. Sulfur dioxide, being toxic, has been abandoned in favor of Freon, a fluorocarbon. Early chlorofluorocarbon (CFC) refrigerants such as R-12 for cars and R-22 for homes were phased out in the early 1990's, because they deplete the ozone in the upper atmosphere. A popular replacement, tetrafluorohydrocarbon R-134, has a global warming potential (GWP) of 1,410 (that is, it contributes 1,410 times as much to the greenhouse effect as does an equivalent mass of CO_2). All gases with a GWP above 150 are slated to be banned in Europe by 2011.

Some modern refrigerants are R-290a (a mixture of isobutane and propane), R-600a (isobutene), and R-744 (CO_2). European automobile manufacturers have announced a switch to CO_2-based systems, but others argue that improving existing R-134a-based systems will prove more effective if the complete system effects are included.

• Alternative Cooling Systems

Ancient Roman mansions were cooled by water flowing through channels in the walls. In areas with significant day-night temperature differences and low humidity (such as deserts), large, modern, industrial systems use ice blocks that freeze overnight as evaporative air coolers. Ground source heat pumps (GSHPs) use the constant temperature 1-2 meters below the ground as a "free" reservoir and heat exchanger to increase efficiency. Subsurface temperature can remain tens of degrees below or above surface air temperature in summer or winter, respectively.

Using solar heat directly to power air conditioning is an ideal solution, because the demand for air conditioning generally corresponds with the presence or availability of solar heat. Approaches for solar air conditioning include using photovoltaic panels to generate the necessary electricity and using evaporation cycles. These cycles require the incoming air to be dry, as in deserts, and are less effective in humid areas.

• Context

Because of the triple pressures of ozone depletion, global warming due to energy use, and global warming due to GHG emissions, air conditioning is poised for a revolution in the first quarter of the twenty-first century. Researchers are developing heat pumps and solar power solutions for residential and industrial air conditioning, as well as CO_2, closed-system cycles for automobiles. Such technologies would reduce energy demand and global warming effects. With improving efficiency and affordability of solar conversion, air conditioning is likely to shift substantially to solar energy, enabling a major reduction of peak power demands and the attendant fossil combustion.

Narayanan M. Komerath

• Further Reading

Bellstedt, Michael, Frank Elefsen, and Stefan S. Jensen. "Application of CO_2 (R744) Refrigerant in Industrial Cold Storage Refrigeration Plant." *EcoLibrium: The Official Journal of AIRAH*, June, 2002, 25-30. Contains data on the global warming potential of various refrigerants. Describes a large facility where CO_2 is used as the secondary-stage refrigerant with a glycol and water mixture as the primary coolant fluid.

Bhatti, M. S. *Global Warming Impact of Automotive Air Conditioning Systems.* Warrendale, Pa.: Society of Automotive Engineers, 1998. Argues the relative merits of R-134 systems versus CO_2 systems.

Chadderon, David. *Air Conditioning: A Practical Introduction.* 2d ed. London: Taylor and Francis, 1997. Systematic introduction to the theory, concepts, and practical applications of air conditioning. Gives system examples and deals with how to design air conditioning systems. Each chapter has learning objectives laid out.

Fischer, S. K. *Comparison of Global Warming Impacts of Automobile Air-Conditioning Concepts.* Washington, D.C.: Office of Scientific and Technical Information, 1995. Compares the global warming potential of vapor-compression air conditioners used in automobiles with four alternatives. Argues that designs using CO_2 and water can reduce the total equivalent warming impact by up to 18 percent.

United Nations. *1998 Report of the Refrigeration, Air Conditioning, and Heat Pumps Technical Options Committee.* New York: United Nations Environment Programme, 1998. Formal U.N. assessment for the Montreal Protocol, the predecessor of the Kyoto Protocol. Gives data on consumption of CFCs and HCFCs and all applications of air conditioning and air cooling, including cold storage and food processing.

See also: Chemical industry; Chlorofluorocarbons and related compounds; Energy efficiency; Hydrofluorocarbons; Industrial greenhouse emissions; Solar energy.

Air pollution and pollutants: anthropogenic

• **Category:** Pollution and waste

Although gaseous air pollutants can pose serious health hazards, only a few gases, such as carbon dioxide, warm the atmosphere. Particulate matter suspended in the air may have the opposite effect, blocking solar radiation and cooling the atmosphere.

• Key concepts

aerosols: minute particles or droplets of liquid suspended in Earth's atmosphere

anthropogenic: deriving from human sources or activities

chlorofluorocarbons (CFCs): chemical compounds with a carbon backbone and one or more chlorine and fluorine atoms

greenhouse effect: global warming caused by gases such as carbon dioxide that trap infrared radiation from Earth's surface, raising atmospheric temperatures

ozone: a highly reactive molecule consisting of three oxygen atoms

parts per million: number of molecules of a chemical found in one million molecules of the atmosphere

• Background

Air pollution has been a problem since humans began burning carbon-based fuels while living in large cities. The first known air-pollution ordinance was passed in London, England, in 1273 in an attempt to alleviate the soot-blackened skies caused by excessive combustion of wood in the heavily populated city. From the mid-eighteenth through the mid-twentieth centuries, the increasingly heavy use of coal for heat, electricity, and transportation resulted in filthy cities and an escalating crisis of respiratory diseases. It was not until the latter half of the twentieth century that governments began attacking the problem by enacting legislation to control noxious emissions at their source.

Before discussing anthropogenic air pollution, one must first define "clean air." Earth's atmosphere is approximately 78 percent nitrogen (N_2), 21 percent oxygen (O_2), and 1 percent argon. These concentrations may be reduced slightly by water vapor, which can make up between 1 percent and 3 percent of the atmosphere. In addition, there are many trace elements present in the atmosphere in concentrations so small that they are measured in parts per million. Among the trace elements near Earth's surface are 0.52 part per million of oxides of nitrogen and 0.02 part per million of ozone, both of which occur both naturally and anthropogenically. This combination of N_2, O_2, argon, water, ozone, and oxides of nitrogen constitutes clean air. Any change in these concentrations or introduction of other compounds into the atmosphere constitutes air pollution, which occurs in one of two forms: gases and particulate matter.

• Gaseous Air Pollutants

The primary gaseous pollutants are oxides of carbon, oxides of sulfur, oxides of nitrogen, and ozone. Carbon oxides occur whenever a carbon-containing fuel is burned; in general, a carbon fuel unites with oxygen to yield carbon dioxide (CO_2) and water vapor. If the combustion is incomplete as a result of insufficient oxygen, carbon monoxide will also be produced. Although CO_2 is a relatively benign compound, the vast amount of fossil fuels (coal, oil, and natural gas) burned since the Industrial Revolution began have raised its atmospheric concentration from about 280 parts per million to about 380 parts per million. CO_2 molecules, while transparent to visible light coming from the Sun, reflect infrared radiation emitted by the Earth when the visible light is absorbed and radiated as heat, thus raising Earth's temperature in proportion to the amount of CO_2 in the atmosphere. As CO_2 concentrations increase, this greenhouse effect will increase Earth's temperature, causing droughts, more severe storms of greater intensity, the shifting of climate zones, and rising sea levels.

Carbon monoxide (CO) is a toxic compound that can cause death by suffocation even when present in relatively small amounts. CO is two hundred times more reactive with hemoglobin than is oxygen; thus CO replaces oxygen in the bloodstream, depriving cells of their necessary oxygen. Deprived of sufficient blood oxygen, an organism will die in about ten minutes.

Since almost all coal contains sulfur, burning coal causes sulfur to react with oxygen to create sulfur dioxide (SO_2), which reacts with water vapor in the atmosphere to produce H_2SO_4, sulfuric acid. This pollutant reaches Earth's surface as a component of rain (acid rain), and it pollutes rivers, lakes, and other bodies of water.

Nitrogen oxides are synthesized whenever air is rapidly heated under pressure and then cooled quickly, as occurs in automobile cylinders and thermoelectric power plants. The two main compounds of this pollution are nitric oxide (NO) and nitrogen dioxide (NO_2); both are toxic, but NO_2 is worse (in equivalent concentrations, it is more harmful than CO). Nitrogen dioxide affects the respiratory system and can lead to emphysema, while nitric oxide often combines with oxygen to form nitric acid (NO_3), another component of acid rain.

NO_2 can also combine with oxygen to form NO and ozone (O_3), a very reactive and dangerous form of oxygen. Combustion-caused ozone is undesirable near Earth's surface, but the compound occurs naturally in the upper atmosphere (about 19 kilometers above the surface) when energetic ultraviolet (UV) light from the Sun interacts with oxygen. Although the ozone composing it constitutes less than 1 part per million of Earth's atmosphere, the ozone layer plays an extremely important role. It prevents most of the Sun's UV light from reach-

A junkyard fire pumps pollutants into the atmosphere. (©iStockphoto.com/Dimitrije Tanaskovic)

ing Earth's surface, a highly desirable effect since it is UV radiation that causes sunburn and skin cancer.

• Chlorofluorocarbons

When first synthesized in the 1930's, chlorofluorocarbon (CFC) was hailed as an ideal refrigerant (Freon), because it was nontoxic, noncorrosive, nonflammable, and inexpensive to produce. Later, pressurized CFCs were used as the propellant in aerosol cans and as the working fluid for air conditioners. In 1974, the chemists Mario Molina and F. Sherwood Rowland proposed that the huge quantities of CFCs released into the atmosphere from aerosol sprays (500,000 metric tons in 1974 alone) and discarded refrigerant units were slowly migrating to the stratosphere. There, the CFCs were decomposed by the highly energetic UV radiation from the Sun, releasing large quantities of ozone-destroying chlorine.

Any decrease of the ozone layer could increase the incidence of skin cancer, damage crops, and decimate the base of the marine food chain. The reduction of ozone was most pronounced over Antarctica, where an "ozone hole," first detected in the early 1970's, was increasing in size annually. Pressured by environmentalists and consumer boycotts, the U.S. government imposed a 1978 ban on aerosol cans and refrigeration units utilizing CFC propellant, forcing the chemical industry to support the ban and to develop alternatives; several other nations soon followed suit. By 1987, the depletion of the ozone layer became so problematic that most CFC-using nations met in Montreal, Canada, to produce an international treaty calling for immediate reductions in all CFC use, with a complete phase-out by 2000. This Montreal Protocol, by 2001, had limited the damage to the ozone layer to about 10 percent of what it would have been had the agreement not been ratified.

• Smog

The word "smog" is a melding of "smoke" and "fog." When a local atmosphere becomes stagnant—for example, during a temperature inversion—pollution levels in the smog can become severe enough to call these smogs "killer fogs." At least three times during the twentieth century, these killer fogs have caused a statistically significant increase in the death rate, particularly among the old and those with respiratory problems. The first documented killer fog occurred in 1948 at Donora, Pennsylvania, when a four-day temperature inversion stagnated a fog that became progressively more contaminated with the smoky effluents of local steel mills. The second documented case occurred in 1952 in London, England, when fog, trapped by another four-day temperature inversion, mixed with the smoke pouring from thousands of chimneys where coal was being burned. Many elderly people and people with respiratory ailments succumbed to these deadly events. Finally, during Thanksgiving, 1966, New York City experienced an increased death rate due to a choking smog.

A second, completely different type of smog is photochemical smog, a noxious soup of reactive chemicals created when sunlight catalyzes reactions of hydrocarbons and nitrogen oxides. This catalysis first occurred in Los Angeles in the late 1940's, when automotive traffic increased drastically, emitting thousands of metric tons of exhaust daily. As mentioned above, car engines, in addition to emitting carbon oxides, emit nitrogen oxides, ozone, and some residual unburned hydrocarbons from the fuel. When light acts on these chemicals, it produces photochemical reactions that create aldehydes (compounds, such as formaldehyde, that are well known for their obnoxious odors) and other dangerous compounds that can induce respiratory ailments, irritate eyes, damage leafy plants, reduce visibility, and crack rubber. Although photochemical smog was first observed in Los Angeles because of the abundant sunlight and heavy automotive traffic, it has since become prevalent in many other large cities.

• Particulates

Particulate matter consists of soot, fly ash, or any other small particles or aerosols suspended in the air that can be breathed into the lungs or ingested with food. It is generated by combustion, dry grinding processes, spraying, and wind erosion. Particulate concentrations in the body can, over time, lead to cancer of the stomach, bladder, esophagus, or prostate.

The human respiratory system has evolved a

Sample Air Pollutants Regulated by the U.S. Environmental Protection Agency

- Acetaldehyde
- Acrylic acid
- Antimony compounds
- Arsenic compounds (inorganic, including arsine)
- Asbestos
- Benzene
- Beryllium compounds
- Cadmium compounds
- Calcium cyanamide
- Carbon disulfide
- Carbon tetrachloride
- Chlorine
- Chloroform
- Chloroprene
- Chromium compounds
- Cobalt compounds
- Coke oven emissions
- Cyanide compounds
- Ethylene glycol
- Fine mineral fibers
- Formaldehyde
- Glycol ethers
- Hydrochloric acid
- Hydrogen fluoride (Hydrofluoric acid)
- Lead compounds
- Manganese compounds
- Mercury compounds
- Methanol
- Nickel compounds
- Parathion
- Phosphorus
- Radionuclides (including radon)
- Selenium compounds
- Styrene

Source: Environmental Protection Agency Technology Transfer Network.

mechanism to filter out and prevent certain sizes of particulates from reaching the lungs. The first line of defense is the nose and nasal passageway, whose mucus membranes and hairs will catch and remove particles larger than 10 microns (one one-hundredth of a millimeter). After passing through the nasal passages, air travels through the trachea, which branches into the right and left bronchi. Each bronchus is divided and subdivided about twenty times, terminating in the small bronchioles located inside the lungs. These end in 300 million tiny air sacs called alveoli, where oxygen is passed to the bloodstream and CO_2 removed for exhalation.

Particles ranging in size from 2 to 10 microns usually settle on the walls of the trachea, bronchi, and bronchioles, before reaching the alveoli. They are eventually expelled by ciliary action, a cough, or a sneeze. Particles smaller than 0.3 micron are likely to remain suspended in inhaled air and then removed from the lungs with exhaled air, similarly failing to enter the bloodstream. Humans thus have evolved a protective mechanism that shields them from particles of all sizes smaller than 0.3 micron and larger than 2 microns. No defense mechanism evolved for this intermediate size range, because during the long course of human evolution there were very few particles of this size in the environment. In recent centuries, however, many particles in this range—including coal dust, cigarette smoke, and pesticide dusts—have been added to the environment. Since no natural defense exists to eliminate these hazards from the human body, they coat the alveoli, causing such illnesses as black lung, lung cancer, and emphysema.

• Context

The issue of whether or not global warming is caused by humans is still being debated, but strong measures were taken in the latter half of the twentieth century to control the noxious gases and particulate emissions known as air pollutants. When it was discovered that the ozone layer was being depleted by CFCs, the Montreal Protocol was ratified by most industrial nations. Both of these historic precedents indicate that strong, effective action and international cooperation is possible when a perceived threat to humanity and the environment is grave enough. Since the preponderance of scientific evidence seems to suggest that global warming is due to humanity's excessive use of fossil fuels, perhaps it would be prudent to err on the side of caution and begin to curtail the disproportionate dependence on nonrenewable resources.

George R. Plitnik

• Further Reading

Elsom, Derek. *Smog Alert: Managing Urban Air Quality.* Covelo, Calif.: Island Press, 1996. Emphasizes the causes of, dangers of, and preventive measures for photochemical smog.

Krupa, S. V. *Air Pollution, People, and Plants: An Introduction.* St. Paul, Minn.: American Phytopathological Society, 1997. Inclusive coverage of air pollution's causes and effects.

Roan, S. *Ozone Crisis: The Fifteen-Year Evolution of a Sudden Global Emergency.* New York: John Wiley, 1989. Traces the science and history of the ozone hole leading up to the Montreal Protocol.

Seinfeld, J., and S. N. Pandis. *Atmospheric Chemistry and Physics: From Air Pollution to Climate Change.* Hoboken, N.J.: Wiley, 2006. Comprehensive coverage of all aspects of air pollution, from chemical reactions and thermodynamics of the atmosphere to global cycles and statistical models.

Vallero, Daniel. *Fundamentals of Air Pollution.* 4th ed. Burlington, Mass.: Academic Press, 2008. Huge, sophisticated tome (more than eleven hundred pages) covering every aspect of air pollution. Not for the technologically timid.

Wilson, Richard, and John Spengler. *Particles in Our Air: Concentrations and Health Effects.* Covelo, Calif.: Island Press, 1996. Provides a practical overview of aerosols and their health effects.

See also: Air pollution and pollutants: natural; Air pollution history; Air quality standards and measurement; Carbon dioxide; Carbon monoxide; Chlorofluorocarbons and related compounds; Coal; Fossil fuel emissions; Greenhouse effect; Ozone.

Air pollution and pollutants: natural

• Category: Pollution and waste

Nature generates pollutants in sufficient volumes to impact Earth's climate. Particulates and sulfur compounds released in volcanic eruptions can lower mean global temperatures for a few years. However, natural air pollutants are generally less persistent in the atmosphere and have a more transient impact on climate than do human-generated ones.

• Key concepts

aerosol: a suspension of solid particles or droplets of liquid in a gas

anthropogenic: generated by humans or human activities

greenhouse effect: global warming caused when atmospheric gases such as water vapor, carbon dioxide, and methane absorb and retain solar energy; Earth's natural greenhouse effect keeps the planet warm enough to support life, and this term is often used to refer to an enhanced, climate-altering greenhouse effect resulting from greenhouse gases contributed by human activity

greenhouse gases (GHGs): a gas in the atmosphere that absorbs infrared radiation, thereby raising temperatures at the planet's surface

ozone: a highly reactive form of oxygen whose molecules are made up of three oxygen atoms (O_3)

stratosphere: the part of Earth's atmosphere that extends from the top of the troposphere to an altitude of 50 kilometers above the planet's surface

troposphere: the densest part of Earth's atmosphere, extending from the planet's surface to altitudes of 8 to 14.5 kilometers; most of Earth's weather is confined to this layer

• Background

The term "air pollution" evokes images of belching industrial smokestacks, rush-hour freeways, and smog-shrouded city skylines: human-caused (anthropogenic) air pollution. Natural processes, however, generate a number of gaseous compounds and particulates that would be regarded as pollutants if human activity had produced them. Among these are oxides of carbon, nitrogen, and sulfur; hydrocarbons; ozone; volatile organic compounds (VOCs); and ash, soot, and other particulates. Nature emits many of these in quantities great enough to affect air quality and global climate.

• Common Natural Air Pollutants

Carbon dioxide (CO_2) occurs naturally in Earth's atmosphere. Only in the late twentieth and early twenty-first centuries did it come to be regarded as a pollutant, when anthropogenic (human-generated) CO_2 was suspected to have a role in global climate change. Along with water vapor, CO_2 is one of the atmosphere's chief absorbers of infrared radiation and is considered a greenhouse gas (GHG)—that is, a gas that keeps solar energy from reradiating into space. The presence of greenhouse gases in Earth's atmosphere allowed the planet to develop a climate conducive to life. However, in the later decades of the twentieth century concerns began to arise that a buildup of greenhouse gases from fossil-fuel use and other human activity could significantly and irreversibly raise temperatures around the globe.

Ozone (O_3) is a highly reactive form of oxygen well known for its role in protecting Earth's surface from damaging ultraviolet (UV) radiation. In the troposphere (lower portion of the atmosphere), however, O_3 acts as a GHG, contributing to increased surface temperatures. At ground level, O_3 is the main component of smog. Sources of O_3 in the troposphere include O_3 that migrated down from the overlying stratosphere and O_3 produced photochemically from nitrogen oxides (NO_x). Higher up, within the stratosphere, is where O_3 performs its UV-absorbing function. Stratospheric O_3 also absorbs visible solar radiation that would otherwise warm the Earth's surface. A decrease in stratospheric O_3 or an increase in tropospheric O_3 results in a rise in surface temperatures.

The hydrocarbon compound methane, CH_4, is classed as a hazardous substance, primarily due to its combustibility. This GHG occurs in the atmosphere at lower concentrations than CO_2, but according to the Intergovernmental Panel on Climate Change (IPCC) its global warming potential over a one-hundred-year period is twenty-five times higher than that of CO_2. Because of CH_4's strong global warming potential, combined with its comparatively short lifetime in the atmosphere (roughly twelve years), curbing CH_4 emissions has the potential to mitigate global warming over the next few decades.

Nitrous oxide, N_2O, by contrast, has an atmospheric lifetime of about 120 years and a 100-year global warming potential 298 times that of CO_2. Other nitrogen oxide compounds (NO_x), while unlikely to contribute directly to climate change, react with volatile organic compounds (VOCs) in the presence of heat and sunlight to form tropospheric O_3. Atmospheric NO_x, which can travel long distances from its source, also causes acid precipitation and is a major component of smog.

Sulfur dioxide, SO_2, another cause of smog and acid precipitation, absorbs infrared radiation. However, its chief climate-altering ability is not as a GHG but as a stratospheric aerosol. Clouds of SO_2 aerosol absorb the Sun's energy and cause a resulting drop in tropospheric temperatures.

VOCs, also found in smog, are carbon-containing compounds that readily become gas or vapor. While they do not directly influence climate, they are important O_3 precursors, especially at the ground level. By enhancing tropospheric O_3 concentrations, they promote global warming.

Carbon monoxide, CO, is a toxic air pollutant. A weak absorber of infrared radiation, it has little direct impact on global climate. However, it contributes to climate change through chemical reactions that boost concentrations of CH_4 and O_3 in the troposphere. CO ultimately oxidizes to CO_2.

Particulate matter includes tiny solid and liquid particles such as dust, salt, smoke, soot, ash, and droplets of sulfates and nitrates. Injected into the atmosphere by anthropogenic or natural processes, these form aerosols, suspensions of particles in air. The length of time these particulates remain in the atmosphere is related to the altitude at which they were introduced and their particle size. Aerosols influence climate directly by reflecting and absorbing atmospheric solar and infrared radiation. While some aerosols cause surface-temperature increases and others cause decreases, the overall effect of aerosols is to lower temperatures. Aerosols influence climate indirectly by serving as condensation nuclei for cloud formation or altering optical properties and lifetimes of clouds.

Major Natural Air Pollutants

- Dust
- Mold spores
- Pollen
- Radon
- Volcanic ash

Smoke billows from Mount Nyiragongo. (©iStockphoto.com/Guenter Guni)

• Volcanic Activity

Volcanic eruptions, the chief source of natural air pollutants, have a demonstrated and complex impact on climate. Gaseous and particulate emissions cause O_3 depletion as well as global atmospheric warming and cooling.

Volcanic eruptions damage O_3 by injecting SO_2 into the stratosphere, where the gas is converted to a sulfate aerosol. The aerosol particles interact with chlorine and bromine in anthropogenic chlorofluorocarbons (once widely used as aerosol can propellants and refrigerants) to produce compounds that break down O_3 molecules. While volcanoes also produce the O_3-degrading compound hydrochloric acid (HCl), it remains largely confined to the troposphere, where it can be washed out by rains.

Volcanism is a significant source of CO_2. According to the United States Geological Survey, subaerial and submarine volcanoes emit an annual total of 130 to 230 million metric tons of CO_2. By comparison, the Carbon Dioxide Information Analysis Center estimates the total global CO_2 emissions from fossil fuel burning in 2007 to have been 8.47 billion metric tons.

Despite their GHG emissions, volcanic eruptions produce a net cooling effect on surface temperatures. This is due in part to dust and ash that remain suspended in the atmosphere after an eruption. These particulates lower mean global temperatures by blocking sunlight, thereby reducing the amount of solar radiation that can reach the planet's surface. The dominant cooling effect, however, results from SO_2 blasted into the stratosphere. The gas combines with water vapor to form droplets of sulfuric acid. This aerosol can remain suspended in the atmosphere, where the droplets absorb solar radiation and cause a decrease in tropospheric temperatures. Global cooling associated with volcanic activity diminishes after a few years as the particles settle out of the atmosphere.

Major volcanic activity in past centuries—notably the 1783 Laki eruption in Iceland, the 1815 eruption of Mount Tambora in Indonesia, and the

1883 eruption of Tambora's neighbor Krakatoa (Krakatau)—have been followed by several months of abnormally cold weather around the globe. More recently, the eruptions of Mount Pinatubo in the Philippines and Mount Hudson in Chile, both during 1991, caused a decrease in mean world temperatures of approximately 1° Celsius over the next two years.

• Wildfires

As wildfires burn, they release carbon stored in vegetation to the atmosphere in the form of CO_2, CO, and CH_4. Researchers Christine Wiedinmyer and Jason Neff estimate that average annual CO_2 emissions from wildfires during the years 2002 through 2006 were 213 (±50 standard deviation) million metric tons for the contiguous United States and 80 (±89 standard deviation) million metric tons for Alaska. This contribution of greenhouse gases—the equivalent of 4 to 6 percent of North America's anthropogenic emissions during that period—has the potential to exacerbate global warming, which can in turn create a hotter, drier environment conducive to larger, more devastating wildfires.

Wildfire combustion products include the GHGs CO_2, CH_4, and N_2O, as well as CO, nitric oxide, and VOCs, all of which promote global warming by enhancing tropospheric O_3. Methyl bromide produced during wildfires can also contribute to global warming by destroying stratospheric O_3.

Wildfires generate large volumes of particulates in the form of smoke, ash, and soot. Clouds of fire-related particulates both absorb and block sunlight, so that their tropospheric effects are both warming and cooling. Particle color influences whether energy is absorbed (dark particles) to produce warming or reflected (light particles) to cause cooling. If soot settles out of the atmosphere onto snow or ice, its dark particles reduce the reflectivity of the frozen surface while enhancing sunlight absorption. Heating, and accelerated melting of the snow or ice, result.

• Other Sources

Oceans and oceanic processes generate a number of natural air pollutants. According to the United Nations Environment Programme (UNEP), oceans produce about 90 billion metric tons of CO_2 annually, emissions that are offset by the estimated 92 billion metric tons that oceans absorb. Oceanic phytoplankton releases dimethyl sulfide, which forms SO_2 as an oxidation product. Oceans also contribute CO, CH_4, N_2O, and NO_x to the atmosphere. Ocean spray sends sea-salt particles aloft, where they decompose in the presence of sunlight to release chlorine molecules that can interact with anthropogenic air pollutants to produce tropospheric O_3.

On land, according to UNEP, vegetation emits 540 billion metric tons of CO_2 but takes in 610 billion metric tons. Another GHG, CH_4, is emitted from a host of terrestrial sources: digestive processes in wild animals and termites; decomposition of wild animal wastes; lakes and wetlands; tundra; and natural oil and gas seeps. Warmer global temperatures could lead to an increase in CH_4 emissions from regions where there is now permafrost.

Other naturally occurring air pollutants include CO from vegetation and the oxidation of naturally occurring hydrocarbons; N_2O and NO_x produced during bacterial denitrification of soil; NO_x created within thunderstorms by lightning; hydrogen sulfide (H_2S) generated during decay underground or under water; SO_2 produced through the oxidation of H_2S; VOCs emitted from coniferous and eucalyptus forests and other vegetation; ammonia released from wild animal wastes; radon gas produced as radium in rock and soil undergoes radioactive decay; and particulate matter in the form of ultrafine soil, dust, pollen, and spores.

• Context

Understanding how naturally occurring substances and processes influence climate is a vital part of comprehending the roles that human activity and anthropogenic materials play. The boundaries between the natural and the anthropogenic are often indistinct. Naturally occurring gases and particulates interact with human-made ones to produce O_3 in the troposphere or damage it in the stratosphere. Deforested areas are left vulnerable to wind erosion that carries soil particulates into the atmosphere. Lightning causes massive wildfires, but so do arsonists and human carelessness.

What is clear is that natural pollutants are not the major concern. Although anthropogenic CO_2 is dwarfed by the natural carbon cycle and there have been episodes where natural emissions have been huge and have had long-lasting effects (such as the possible massive volcanic sulfur emission at the end of the Permian epoch and the possible methane release in the Eocene), human effects can be more important because (1) humans can change the atmosphere on a very rapid time scale compared to natural effects, and (2) humans introduce substances that have a potent effect on the atmosphere and are not normally found in nature, such as chlorofluorocarbons (CFCs). Hence, anthropogenic substances can surpass naturally occurring ones in their current and long-term impacts on the balance between incoming solar radiation and outgoing infrared radiation within Earth's atmosphere.

Karen N. Kähler

• **Further Reading**

Intergovernmental Panel on Climate Change. *Climate Change, 2007—The Physical Science Basis: Contribution of Working Group I to the Fourth Assessment Report of the Intergovernmental Panel on Climate Change.* Edited by Susan Solomon et al. New York: Cambridge University Press, 2007. The "Summary for Policymakers," "Technical Summary," and "Frequently Asked Questions" sections are informative condensations of this nearly thousand-page volume. Table TS.2 of the "Technical Summary" lists GHGs' global warming potentials.

Knipping, E. M., and D. Dabdub. "Impact of Chlorine Emissions from Sea-Salt Aerosol on Coastal Urban Ozone." *Environmental Science and Technology* 37, no. 2 (2003): 275-284. A highly technical paper on sea salt's role in coastal ozone production. Figures, tables, equations, references.

Robock, Alan. "Volcanic Eruptions and Climate." *Reviews of Geophysics* 38, no. 2 (May, 2000): 191-219. This technical paper on the effects of volcanic eruptions on climate is a challenging but not impenetrable read for the nonscientist. Figures, tables, plates, glossary, references.

Wiedinmyer, Christine, and Jason C. Neff. "Estimates of CO_2 from Fires in the United States: Implications for Carbon Management." *Carbon Balance and Management* 2, no. 10 (November 1, 2007). A technical look at the variability of CO_2 emissions from fires across the United States and how they compare to anthropogenic emissions. Figures, tables, references.

Wuebbles, Donald J., and Jae Edmonds. *Primer on Greenhouse Gases.* Chelsea, Mich.: Lewis, 1991. A technical but accessible introduction to climate and climate change. An extensive table (in chapter 6) identifies natural sources of GHGs, where applicable, and summarizes their climate effects.

See also: Aerosols; Air pollution and pollutants: anthropogenic; Air pollution history; Air quality standards and measurement; Atmosphere; Atmospheric boundary layer; Atmospheric chemistry; Atmospheric dynamics; Atmospheric structure and evolution; Carbon dioxide; Carbon dioxide equivalent; Carbon 4 plants; Carbon monoxide; Carbon 3 plants; Carbonaceous aerosols; Chlorofluorocarbons and related compounds; Clouds and cloud feedback; Dust storms; Enhanced greenhouse effect; Fire; Global dimming; Greenhouse effect; Greenhouse gases; Methane; Mount Pinatubo; Nitrous oxide; Ozone; Sulfate aerosols; Volatile organic compounds; Volcanoes.

Air pollution history

• **Category:** Pollution and waste

Air pollution resulting from human action dates primarily from the Industrial Revolution. Humans, however, have had some impact on the environment for at least five thousand years. As industrialized agriculture and manufacturing have increased, humans have added increasing amounts of pollutants to the atmosphere.

• **Key concepts**

air pollution: degradation of air quality through human or natural means

anthropogenic: derived from human actions or sources

forest clearing: destruction of forests in order to create land suitable for human habitation or use

greenhouse gases (GHGs): anthropogenic and natural gases that trap heat within the atmosphere, increasing Earth's surface temperature

health hazard: pollution that produces potential harm to human health

industrialization: development and dissemination of mechanical, mass-producuction technologies

• Background

The history of anthropogenic air pollution is generally traced from the Industrial Revolution of the eighteenth and nineteenth centuries. As by-products of manufacturing processes and through burning the coal and oil necessary to power them, industries added substantial amounts of heavy metals, acidic compounds (such as sulfur dioxide), and carbon dioxide (CO_2) to the atmosphere. The first significant effect on the air occurred, however, when humans cleared land for agriculture. This effect increased over time, as more land was cleared and small-scale industry developed. Thus, before the Industrial Revolution began, human activity was already generating air pollution.

• The Development of Agriculture

At least five thousand years ago, humans began to clear forests for agriculture. Burning of the cleared wood, livestock emissions, and human waste produced methane gas, which is a greenhouse gas (GHG). The gradual increase in Earth's human population contributed to an increasing amount of methane being added to the atmosphere. As more complex agriculture developed, forest clearance increased, leading to a gradual increase in the amount of CO_2 in the atmosphere. The impact of land clearing varied, but the activity continued to add methane and CO_2 to the atmosphere. By the thirteenth century, much of England, for example, had been deforested. Agriculture created pollutants that were not direct health hazards but that added to the levels of GHGs in the atmosphere.

• Ancient Societies

The increase of human population that started several thousand years ago was initially quite slow. As ancient societies became more complex, it became necessary to create larger numbers of tools, to mine and refine the materials used for production, and to develop more complex energy sources. Each of these processes contributed to air pollution, although it is very difficult to quantify their contributions precisely. Aside from smoke and odors, most people were unaware of air pollution.

Smelting metals and creating tools added small amounts of heavy metals and sulfur and nitrogen oxides to the atmosphere. At times, these pollutants were carried by wind currents for considerable distances. For example, it appears that traces of the pollutants derived from Roman metal smelting may be found in Greenland's ice cap. Smelting also required energy; this energy was produced by burning wood. Thus, early proto-industrial concerns added to air pollution, both through cutting down trees and through burning their wood. Ancient Romans and early medieval Europeans were unaware of this pollution except for the occasional downdrafts of smoke from a nearby smelter or the increase of smoke from heating in towns.

• The Middle Ages

By the fourteenth century, an awareness of some negative effects of air pollution began to develop. Some of the awareness focused on unpleasant and noxious odors that resulted from butchering and from human waste in cities such as London. There was also concern over the increasing amount of smoke in the air in urban areas such as London or Paris. The smoke at the time was primarily wood smoke, although some coal was also being used to heat homes and provide power. Official concern tended to focus on the dirt that resulted from burning fossil fuels and the noxious odors that came from burning coal in particular. More harmful pollutants, such as sulfur and nitrogen oxides, were ignored, because people were unaware of their existence. They were equally unaware of the GHGs that were a by-product of burning fossil fuels.

As forests were consumed in the late Middle Ages in Western Europe, producers of goods turned increasingly to coal as an energy source. Burning coal led to an increase in various air pollutants in the atmosphere, although the amounts were still quite small throughout the sixteenth and seventeenth centuries. Increased population led to increased demand for goods, as well as for home heating, food, and building construction. All of these demands contributed to a further clearing of land for agriculture and lumber production, although now the lands cleared were often in North and South America and Asia, rather than in Europe.

• Industrialization

The Industrial Revolution of the eighteenth and nineteenth centuries led to a dramatic increase in atmospheric pollutants. The growth of industrial society was accompanied by an increase in population, which further drove increased industrial and agricultural production. The magnitude of industrialization dating from the early nineteenth century led to a drastic increase in the use of fossil fuels, first coal and then oil. Burning these fossil fuels added CO_2 to the atmosphere, as well as pollutants such as sulfur and nitrogen oxides. Burning fossil fuels and metal fabrication also led to the addition of heavy metals to the atmosphere. Later, other pollutants, such as chlorofluorocarbons, were generated as by-products of industrial society.

• Context

Human activity has affected Earth's atmosphere for thousands of years, and the atmospheric concentration of anthropogenic substances has increased with humanity's population and technological sophistication. The increase in anthropogenic influences on the atmosphere has led to an increased awareness and understanding of air quality and pollution. Such an understanding has led in turn to changing definitions of those terms: Until the late twentieth century, only foreign substances with direct toxic or pathogenic effects were considered pollutants. For example, CO_2—which occurs naturally in the atmosphere and is not directly toxic—was not categorized as a pollutant. The discovery of

Smoke from an iron and steel plant billows into the sky. (Jianan Yu/Reuters/Landov)

the greenhouse effect, and of the role of CO_2 and other gases in that effect, has led many scientists and governments to reconsider their earlier definitions, and many now classify all GHGs as air pollutants. This discovery has also led historians to reconsider the history of air pollution itself and to see in a new light the role that early agriculture played in shaping atmospheric chemistry. By considering the evolution of human technology over several millennia in relation to the global atmosphere and climate, scientists can gain a fuller understanding of the extent to which anthropogenic inputs to the atmosphere may have contributed to changes in the climate.

John M. Theilmann

• Further Reading

Goudie, Andrew. *The Human Impact on the Natural Environment: Past, Present, and Future.* 6th ed. Malden, Mass.: Blackwell, 2006. Provides concise analysis of the nature of air pollution as affected by human action.

Ruddiman, William F. *Plows, Plagues, and Petroleum.* Princeton, N.J.: Princeton University Press, 2005. Ruddiman, a leading climate researcher, argues that human pollution has had a long-term, increasing, and negative impact on the climate.

Somerville, Richard C. J. *The Forgiving Air.* Berkeley: University of California Press, 1996. Good overview of the interaction of technology and environmental change in the atmosphere.

Theilmann, J. M. "The Regulation of Public Health in Late Medieval England." In *The Age of Richard II,* edited by James L. Gillespie. New York: St. Martin's Press, 1997. Good analysis of various forms of pollution, including air pollution, that uses late medieval England as a case study.

Williams, Michael. *Deforesting the Earth.* Chicago: University of Chicago Press, 2003. Comprehensive account of the various effects, including air pollution, of the human deforestation of the Earth.

See also: Agriculture and agricultural land; Air pollution and pollutants: anthropogenic; Air pollution and pollutants: natural; Air quality standards and measurement; Asthma; Carbon dioxide; Clean Air Acts, U.S.; Coal; Emissions standards; Fossil fuels; Industrial emission controls; Industrial greenhouse emissions; Industrial Revolution; Ozone; Population growth.

Air quality standards and measurement

• **Categories:** Pollution and waste; laws, treaties, and protocols

Warmer temperatures and air pollution are interrelated: Rising temperatures result in increases in ozone production and energy utilization. Increased energy demands result in more power plant utilization, which leads to greater emissions.

• Key concepts

Air Quality Index (AQI): a numerical index for reporting air pollution levels to the public

Clean Air Acts: a set of federal laws that form the basis for the United States' air pollution control effort

criteria air pollutant: air pollutants for which acceptable levels of exposure can be determined and ambient air quality standards have been set

National Ambient Air Quality Standards (NAAQS): standards established by the EPA that limit acceptable outdoor air pollution levels throughout the United States

particulate matter: any nongaseous material in the atmosphere

• Background

Climate change affects atmospheric composition and dynamics. In addition to changing global weather patterns, climate change may have a negative impact on air quality. In particular, formation of ozone and particulate matter are influenced by weather conditions such as temperature and precipitation. Air pollution concentrations are also influenced by management strategies that control emissions. The need to reduce anthropogenic influences on the climate, through mitigation of greenhouse gas (GHG) emissions, and to adapt to future climate change is a significant environmental problem. Adapting to global climate change is necessary to protect air quality.

• Climate Change and Air Quality

Air quality is affected by human activities such as driving automobiles or other vehicles, burning coal or other fossil fuels, and manufacturing chemicals. In most countries, transportation is one of the major contributors to air pollution, and it generates many pollutants. Transportation causes wear and tear on cars and on roads, which produces road dust. Heating and cooling residential and commercial buildings requires a great deal of energy, most of which is supplied by burning fossil fuels, constituting another major source of air pollutants and GHGs. Residential heating, particularly from burning wood, also contributes particulate matter and many toxic compounds to the atmosphere.

There are a number of connections and synergistic effects between climate change and air quality. Climate change, including changes to temperature, precipitation, cloud cover, and relative humidity, may affect the atmospheric concentrations of many important chemical species and change the rate at which ozone and particulate matter are formed in the atmosphere, with warmer temperatures increasing ozone and particulate-matter formation. High temperatures also cause the evaporation of toxic substances such as mercury, polycyclic aromatic hydrocarbons (PAHs), and polychlorinated biphenyls (PCBs), from sediments.

Warmer temperatures lengthen pollen and mold seasons and encourage spore production. Climate change may increase the frequency of stagnant air masses, which may cause pollutant buildup in certain regions. Changes in temperature can affect methane emissions. Widespread climate changes may also alter human activity patterns such as agriculture, biomass burning, and energy consumption, including the demand for heating or cooling. This, in turn, would affect the emissions of pollutant gases and particles resulting from these activities. Changes in land use and in fire and drought patterns may affect smoke and mineral dust aerosols in the atmosphere.

• Air Quality Monitoring Techniques

Air quality is measured through direct monitoring or through emissions inventories. Air quality monitoring is the main method for determining concentrations of particular pollutants in the atmosphere at specific points in time. Air quality monitoring is used to determine the Air Quality Index (AQI), a composite indicator of outdoor air quality that is communicated to the public through the media and generally includes recommendations for protection against pollutant-associated health effects.

Emissions inventories estimate the amount of pollutants emitted into the atmosphere from major mobile, stationary, and natural sources over a specific period of time—for example, a day or a year—and form the basis for efforts such as trend analysis, air quality monitoring, regulatory impact assessments, and human exposure modeling. The Environmental Protection Agency (EPA) maintains the National Emission Inventory database, which contains information about sources that emit criteria air pollutants and hazardous air pollutants.

In many countries, air quality monitoring networks provide measurements of pollutant species. Many air quality monitoring programs include both global and regional networks that measure background atmospheric composition at selected remote sites. Other programs include regulatory monitoring networks that analyze day-to-day variations in air quality at numerous sites, primarily in urban areas. Most air quality monitoring sites are focused on heavily populated areas and are designed to determine whether a specific area is in compliance with air quality standards. Urban and regional air quality monitoring networks rely on ground-based sites that sample within the boundary layer.

Remote-sensing or satellite instruments provide global-scale observations of specific pollutants in the atmosphere. A variety of instruments attached to balloons or aircraft are used for atmospheric measurements over a wide range of altitudes. Particulate matter measurements can be conducted in real time, but routine aerosol mass measurements generally rely on particle accumulation over extended sampling times followed by laboratory analyses. Urban areas in the United States are monitored through the EPA's air quality programs, but large data gaps exist in a number of rural areas, and

National Ambient Air Quality Standards for Criteria Pollutants

Pollutant	Averaging Time	Pollutant Level	Effects on Health
Carbon monoxide: colorless, odorless, tasteless gas; it is primarily the result of incomplete combustion; in urban areas the major sources are motor vehicle emissions and wood burning.	1-hour 8-hour	35 ppm 9 ppm	The body is deprived of oxygen; central nervous system affected; decreased exercise capacity; headaches; individuals suffering from angina, other cardiovascular disease; those with pulmonary disease, anemic persons, pregnant women and their unborn children are especially susceptible.
Ozone: highly reactive gas, the main component of smog.	1-hour 8-hour	0.120 ppm 0.080 ppm	Impaired mechanical function of the lungs; may induce respiratory symptoms in individuals with asthma, emphysema, or reduced lung function; decreased athletic performance; headache; potentially reduced immune system capacity; irritant to mucous membranes of eyes and throat.
Particulate matter < 10 microns (PM10): tiny particles of solid or semisolid material found in the atmosphere.	24-hour Annual arithmetic mean	150 µg/m^3 50 µg/m^3	Reduced lung function; aggravation of respiratory ailments; long-term risk of increased cancer rates or development of respiratory problems.
Particulate matter < 2.5 microns (PM2.5): fine particles of solid or semisolid material found in the atmosphere.	24-hour Annual arithmetic mean	65 µg/m^3 15 µg/m^3	Same as PM10 above.
Lead: attached to inhalable particulate matter; primary source is motor vehicles that burn unleaded gasoline and re-entrainment of contaminated soil.	Calendar quarter	1.5 µg/m^3	Impaired production of hemoglobin; intestinal cramps; peripheral nerve paralysis; anemia; severe fatigue.
Sulfur dioxide: colorless gas with a pungent odor.	3-hour 24-hour Annual arithmetic mean	0.5 ppm 0.14 ppm 0.03 ppm	Aggravation of respiratory tract and impairment of pulmonary functions; increased risk of asthma attacks.
Nitrogen dioxide: gas contributing to photochemical smog production and emitted from combustion sources.	Annual arithmetic mean	0.053 ppm	Increased respiratory problems; mild symptomatic effects in asthmatics; increased susceptibility to respiratory infections.

Notes: ppm equals parts per million and µg/m^3 equals micrograms per cubic meter.
Source: United States Environmental Protection Agency (EPA); URL http://www.epa.gov.

insufficient data are available for large regions of the Earth.

• Air Quality Management Methods

Contemporary air quality management plans focus on controlling state and local emissions, although the EPA has also initiated regional air quality management strategies. In general, there are two types of air quality standards. The first type, the National Ambient Air Quality Standards (NAAQS's), are set by the EPA and include target levels for specific pollutants that apply to outdoor air throughout the country. The EPA has set NAAQS's for six principal pollutants, called "criteria pollutants." They are carbon monoxide, lead, nitrogen dioxide, particulate matter (PM_{10} and $PM_{2.5}$), ozone, and sulfur dioxide. The second class of standards constitutes the Air Quality Index (AQI), which uses a scale to communicate the relative risk of outdoor activity to the public.

The Clean Air Acts (1963-1990) require the EPA to set NAAQS's for pollutants from a variety of sources considered harmful to public health and the environment and have significantly strengthened air pollution regulation. The Clean Air Acts set target levels for air pollutants and provide reporting and enforcement mechanisms. Two types of national air quality standards were established through these acts: Primary standards set limits to protect public health, including the health of sensitive populations such as asthmatics, children, and the elderly. Secondary standards set limits to protect public welfare, including protecting against decreased visibility and damage to animals, crops, vegetation, and buildings.

The Clean Air Rules, established in 2004, are a suite of rules focused on improving U.S. air quality. Three of the rules, the Clean Air Interstate Rule (CAIR), the Clean Air Mercury Rule, and the Clean Air Nonroad Diesel Rule, address the transportation of pollution across state borders. CAIR is a management strategy that covers twenty-eight eastern states and the District of Columbia and focuses on the problem of power plant pollution drifting from one state to another. This rule uses a cap-and-trade system to reduce sulfur dioxide and nitrogen oxides.

The Clean Air Mercury Rule (CAMR) represents the first federally mandated requirement that coal-fired electric utilities reduce their emissions of mercury. Coal-fired power plants are the largest remaining domestic source of anthropogenic mercury emissions, and with the CAMR, the United States is the first nation in the world to control emissions from this major source of mercury pollution. Together, the CAMR and the CAIR create a multipollutant strategy to reduce U.S. emissions. The effects of these laws have been very positive, with substantial reductions in emissions of carbon monoxide, nitrogen oxide, sulfur dioxide, particulate matter, and lead. The Clean Air Nonroad Diesel Rule changes the way diesel engines function to remove emissions and the way diesel fuel is refined to remove sulfur. These rules provide national tools to achieve significant improvement in air quality.

• Global Air Quality Management

Air pollution emissions can affect air quality beyond national borders. Westerly winds transport ozone from the eastern United States into Canada and the North Atlantic. Air masses reaching the United States carry pollution originating from many other parts of the world, including Asian industrial pollutants and African dust aerosols. Saharan dust has affected atmospheric particulate-matter concentrations in several inland areas of the American southeast. Long-range transport of gases and of aerosols from burning biomass has also been detected.

Many areas of the world are moving toward mitigating the GHG emissions that contribute to global climate change. As urbanization and industrialization have increased, urban air quality has become a significant public health concern throughout many regions, particularly in developing countries. Particulate air pollution is a chronic problem in much of Asia as a result of coal combustion in factories and power plants and the use of coal and wood for cooking and heating homes.

Automobiles continue to be an increasingly important contributor to air pollution in much of the world, with more than 600 million vehicles in use, a number that is growing exponentially. Motor vehicles are the predominant source of air pollution in many Latin American cities, where automobile use has been restricted in order to manage severe air pollution occurrences. In the United States, air quality has shown steady improvement, partly because of air quality regulatory programs and new, cleaner technologies that have improved both motor vehicles and stationary pollution sources.

The Kyoto Protocol includes provisions that limit GHG emissions from industrialized countries throughout the world. Carbon dioxide (CO_2) is the predominant GHG emitted by most countries, and controlling CO_2 emissions is an essential component of air quality monitoring strategies. Non-CO_2 GHG emissions also have considerable global warming potentials and are important targets for emission reductions. Although ozone and particulate matter also contribute to global warming, control of these species is not currently included in the Kyoto Protocol. Ozone and particulate matter could be effective targets for emission control efforts, however, since many countries already have regulations focused on controlling these species and since reducing ozone and particulate-matter emissions can improve local air quality, health, and agricultural productivity.

• Context

Climate change will alter the extent and nature of air pollution and the general composition of the atmosphere and will be influenced by both natural and anthropogenic factors. While climate change may exacerbate the frequency of smog episodes and related health effects, air pollution is already a serious health concern around the world. Changes in air pollution emissions occurring over the next several decades could affect global health, as emissions may reach areas far beyond their local sources. Efforts under way and contemplated to address climate change may have the related benefit of reducing air pollution in general. However, climate change may also reduce the effectiveness of existing programs: Many of the national control programs currently in place may prove be less effective than was originally expected. As climate change progresses, more stringent air quality standards and management will likely be necessary. Multipollutant approaches that protect human health and climate will require a global perspective to meet air quality objectives.

C. J. Walsh

• Further Reading

Alley, E. Roberts, Lem B. Stevens, and William L. Cleland. *Air Quality Control Handbook.* New York: McGraw-Hill, 1998. Provides comprehensive coverage of the provisions and implications of the 1990 Clean Air Act Amendments and examines the origins and effects of air pollution and the history of air quality control management.

Godish, Thad. *Air Quality.* 4th ed. Boca Raton, Fla.: CRC Press, 2003. For students, this text provides a comprehensive overview of air quality issues, including descriptions of atmospheric chemistry, effects of pollution on public health and the environment, and regulatory practices used to achieve air quality goals.

Griffin, Roger D. *Principles of Air Quality Management.* Boca Raton, Fla.: CRC Press, 1994. Includes a chapter on air quality standards and monitoring and a chapter on global concerns about air quality.

Harrop, Owen. *Air Quality Assessment and Management.* London: Taylor & Francis, 2002. Describes available techniques for air quality assessment, with details on the concepts and methodologies involved and principles of air quality management.

National Research Council. Committee on Air Quality Management in the United States. *Air Quality Management in the United States.* Washington, D.C.: National Academies Press, 2004. Provides an overview of current air quality management in the United States, with recommendations for mitigation efforts and regulations.

See also: Air pollution and pollutants: anthropogenic; Air pollution and pollutants: natural; Air pollution history; Catalytic converters; Clean Air Acts, U.S.; Emissions standards; Industrial emission controls; Motor vehicles.

Air travel

• **Categories:** Economics, industries, and products; transportation

Because they are released so high in the atmosphere, emissions from airplanes have greater relative effects upon air quality than do emissions from ground vehicles. The extent of those emissions' effects upon Earth's climate is a much studied and hotly debated issue.

• Key concepts

aircraft emissions: gases and particulate matter expelled from aircraft engines

greenhouse effect: an atmospheric warming phenomenon by which certain gases act like glass in a greenhouse, allowing the transmission of ultraviolet solar radiation but trapping infrared terrestrial radiation

greenhouse gases (GHGs): gases that tend to hold heat within the atmosphere and contribute to the greenhouse effect

ozone: a greenhouse gas that absorbs ultraviolet radiation

radiative forcing: a change in the balance between incoming and outgoing radiation that increases or decreases overall energy balance

stratosphere: the atmospheric region just above the tropopause that extends up to about 50 kilometers

troposphere: the lowest layer of the atmosphere, in which storms and almost all clouds occur, extending from the ground up to between 8 and 15 kilometers in height

• Background

Powered flight was introduced in 1903, when Wilbur Wright and Orville Wright conducted the first short flight at Kill Devil Hills, North Carolina. The first thirty-five years of powered flight had minimal impacts upon Earth's climate. The advent of widespread aerial cargo and passenger service in the late 1930's and early 1940's arguably marks the first significant influence upon Earth's climate by aircraft. However, beginning in the 1960's, air travel's impact became much more pronounced, as jet aircraft were introduced. The effects of air travel were increased in 1978, when the United States deregulated the airline industry, resulting in a significant increase in the annual number of jet aircraft flights.

• Air Travel Growth Rate

Since 1960, air travel has grown by approximately 9 percent per year. A mode of transportation once regarded as available only to the wealthy quickly became the preferred, affordable method of long-distance travel. With this rapid growth has come a proportionate increase in the impact that air travel has on Earth's environment and climate. It is expected that air travel will continue to grow at an annual rate between 6 and 11 percent during the twenty-first century. This significant anticipated growth suggests that substantial progress must be made in limiting the negative effects that air travel has on the environment.

• Impact of Air Travel on Earth's Climate

Aircraft engines emit gases and particulates into the upper troposphere and lower stratosphere. These gases modify atmospheric composition, resulting in either increases or decreases in radiative forcing. Water vapor and carbon dioxide (CO_2) emissions result in positive radiative forcing. In addition, sulfur and soot emitted by aircraft engines combine with water to form condensation trails, commonly called contrails, resulting in further positive radiative forcing. Both processes warm the climate.

• Impact of Air Travel on Earth's Ozone Layer

Ozone, a greenhouse gas, absorbs ultraviolet radiation. The majority of this absorption occurs in the stratosphere. A wide variety of human interactions with the environment deplete ozone in the atmosphere. Subsonic aircraft engines emit nitrogen oxides in the troposphere and lower stratosphere that increase ozone, reducing the amount of ultraviolet radiation reaching Earth's surface. Supersonic aircraft, flying higher in the stratosphere, have an opposite effect, depleting ozone at those higher altitudes.

Best estimates are that net air-travel-related radiative forcing comprises approximately 3.5 percent of total human atmospheric radiative forcing. Currently, most air travel is conducted by subsonic aircraft operating in the upper troposphere and lower stratosphere. However, significant research and development is under way to introduce supersonic and hypersonic aircraft into the air transportation fleet. If this occurs, much of the ozone-depletion offset generated by subsonic aircraft will be negated by ozone-depleting supersonic aircraft operating primarily in the stratosphere.

An experimental Boeing 747-300 aircraft equipped with a biofuel engine prepares to take off from Tokyo's Haneda airport. (Issei Kato/Reuters/Landov)

With the retirement of the Concorde supersonic transport, there were no operational civil supersonic transport aircraft. However, beginning around 2015, a new civil supersonic aircraft fleet is expected to develop and grow at a brisk rate, resulting in more than one thousand civil supersonic aircraft in service by 2040. If this occurs, air-travel-related radiative forcing will increase by almost 40 percent over 2008 levels.

Interestingly, a disproportionate amount of positive radiative forcing occurs in the delicate upper portion of the Northern Hemisphere as a result of the high volume of jet aircraft traffic in that region. Increases in atmospheric temperatures in the polar region have a potentially significant effect on rising ocean levels resulting from polar-cap and glacial-ice melting.

• Reducing Air Travel Impacts

A number of options exist to reduce air travel's impact upon the environment. Rises in petroleum-based fuel prices reduce the impact of air travel on the environment by making air travel less cost effective and reducing the demand for flights. In addition, airlines are striving to work worldwide with air traffic control organizations to reduce ground holds and increase flight efficiency through more direct routings and highly planned, efficient descents from altitude.

Technological advances in jet-engine design and efficiency also offer great promise for reducing environmental impacts. State-of-the-art turbine engines, utilized on most new transport aircraft, are much more efficient than older versions. Finally, a basic solution to reduce air travel's impact upon the environment is to reduce the need to travel at all. Modern computing technologies, including high-quality video conferencing, make possible complex online meetings. These meetings can be held with participants located throughout the world, reducing the need for business travel.

• **Context**

Air travel is growing at a brisk rate; it has an impact upon global warming, but there are mitigation measures available. It remains unclear, however, to what extent governments, aircraft manufacturers, and airline companies will institute such mitigation measures. Research and development of greener aircraft technologies is an expensive and time-consuming venture. As with all ventures, governments and companies must attempt to strike a compromise between the environmental impacts of aviation and economic factors related to technological innovation. The long-term availability of aircraft fuel and the will of airlines and aircraft manufacturers to adopt green technologies remain unclear.

Alan S. Frazier

• **Further Reading**

Dokken, David J., et al. *Special Report on Aviation and the Global Atmosphere.* Geneva, Switzerland: World Meteorological Organization, Intergovernmental Panel on International Climate Change, 1999. Comprehensive report on the impacts of air travel on Earth's climate.

General Accounting Office, U.S. *Aviation and the Environment: Aviation's Effects on the Global Atmosphere Are Potentially Significant and Expected to Grow.* Washington, D.C.: Author, 2000. A report commissioned by a member of the House of Representatives presenting a summary of air travel's effects upon climate change.

See also: Carbon footprint; Emission scenario; Emissions standards; Fossil fuel emissions; Ozone; Transportation.

Albedo feedback

• **Category:** Physics and geophysics

• **Definition**

Albedo (Latin for "whiteness") is a measure of the amount of incident radation, such as sunlight, that a surface reflects. Arctic ice, for example, has a high albedo, meaning that most of the light hitting it is reflected back into space, so little of the solar energy striking ice is absorbed into the Earth's surface in the form of heat. When the ice melts into liquid water, however, surfaces darken and absorb more light, warming the planetary surface. As the ocean surface warms, more of the remaining ice melts, further increasing the amount of solar heat being absorbed. This cycle in which lower albedo creates conditions that cause a further decrease in albedo is an example of a positive feedback loop.

Albedo, measured as the fraction of solar radiation reflected by a surface or object, is often expressed as a percentage. Snow-covered surfaces have a high albedo, the surface albedo of soils ranges from high to low, and vegetation-covered surfaces and oceans have a low albedo. The Earth's planetary albedo varies mainly through varying cloudiness, snow, ice, leaf area, and land cover. The warming of the Arctic influences weather in the United States. Outbreaks of Arctic cold have become weaker as ice coverage erodes.

Earth's albedo usually changes in the cryosphere (ice-covered regions), which has an albedo much greater (at around 80 percent) than the average planetary albedo (around 30 percent). In a warming climate, the cryosphere shrinks, reducing the Earth's overall albedo, as more solar radiation is absorbed to warm the Earth still further. Cloud cover patterns also could change, resulting in further albedo feedback. Changes in albedo have an important role in changing temperatures in any given location and, thus, the speed with which ice or permafrost melts.

• **Significance for Climate Change**

The rate of Arctic warming around the beginning of the twenty-first century has been eight times the average rate during the twentieth century. Changes in albedo are among the factors contributing to this increase. The number and extent of boreal forest fires have also grown, increasing the amount of soot in the atmosphere and decreasing Earth's albedo. As high latitudes warm and the coverage of sea ice declines, thawing Arctic soils also may release significant amounts of carbon dioxide (CO_2) and methane now trapped in permafrost.

Arctic warming has shortened the region's snow-covered season by roughly 2.5 days per decade, increasing the amount of time during which sunlight is absorbed. Gradual darkening of Arctic surfaces thus produces significant changes in the total amount of solar energy that the area absorbs. Scientists have estimated this increase in surface energy absorption at 3 watts per square meter per decade. This means that, in areas such as the Arctic where albedo has changed markedly, the effect of this change on climate has been roughly equal to the effect of doubling atmospheric CO_2 levels. Moreover, the continuation of contemporary trends in shrub and tree expansion would amplify atmospheric heating by two to seven times.

Changes in albedo over a broad area, such as the Arctic, can produce a significant effect, allowing Earth to be "whipsawed" between climate states. This feedback has been called the "albedo flip" by James E. Hansen, director of the Goddard Institute for Space Studies (GISS) of the National Aeronautics and Space Administration (NASA). The flip provides a powerful trigger mechanism that can accelerate rapid melting of ice. According to Hansen, greenhouse gas (GHG) emissions place the Earth perilously close to dramatic climate change that could run out of control, with great dangers for humans and other creatures. Changes in albedo have been greatest in Earth's polar regions, especially in the Arctic, where snow and ice are being replaced during summer (a season of long sunlight) by darker ocean or bare ground.

Changes in albedo also play a role in increasing Arctic emissions of methane, tropospheric ozone (O_3), and nitrous oxide (N_2O). All of these are GHGs. Tropospheric ozone is the third most influential anthropogenic GHG, after CO_2 and methane.

Black carbon (soot) also has a high global warming potential and deserves greater attention, according to Hansen. Soot's albedo causes massive absorption of sunlight and heat and compounds warming, especially in the Arctic. Increases in soot due in part to combustion of GHGs can play a role

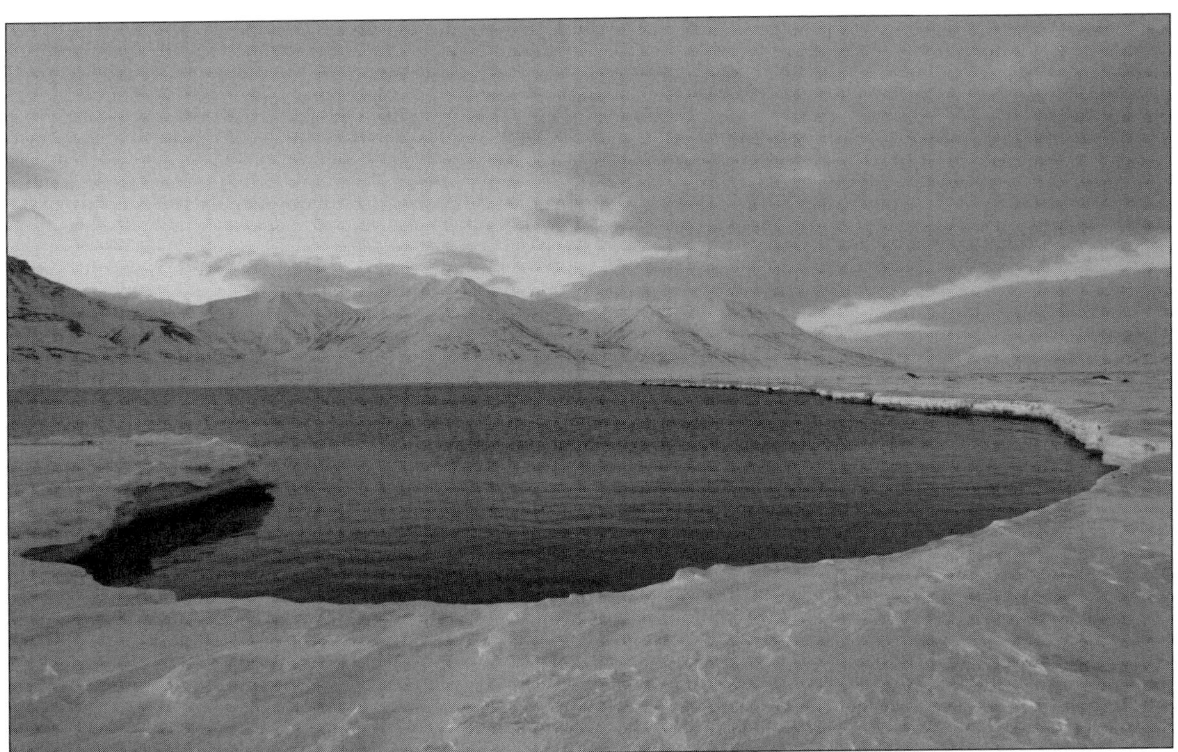

Sun strikes the water of a Norwegian Arctic fjord in April, 2007. The fjord water, normally frozen in April, may be part of a warming-induced albedo feedback loop. (Francois Lenoir/Reuters/Landov)

in accelerating climate change to tipping points, at which feedbacks take control and propel increasing levels of GHGs past a point where human control ("mitigation") is possible.

Bruce E. Johansen

• Further Reading

Alley, Richard B. *The Two-Mile Time Machine: Ice Cores, Abrupt Climate Change, and Our Future.* Princeton, N.J.: Princeton University Press, 2000. Popularization of Alley's scientific work on climate change in the Arctic.

Appenzeller, Tim. "The Big Thaw." *National Geographic,* June, 2007, 56-71. Surveys ice melt in the Arctic, Antarctic, and mountain glaciers. Explains albedo feedback, describing why the process has been accelerating.

Chapin, F. S., III, et al. "Role of Land-Surface Changes in Arctic Summer Warming." *Science* 310 (October 28, 2005): 657-660. A scientific treatment of global warming and albedo's effects.

Foley, Jonathan A. "Tipping Points in the Tundra." *Science* 310 (October 28, 2005): 627-628. Explains the role of Earth's albedo in provoking climate to a tipping point in the Arctic.

See also: Arctic; Climate feedback; Clouds and cloud feedback; Greenhouse effect; Greenhouse gases; Greenland ice cap; Ground ice; Permafrost; Sea ice; Sea surface temperatures.

Alkalinity

• Category: Chemistry and geochemistry

• Definition

Alkalinity is a measurement of the capacity of a solution to neutralize acid by taking up hydrogen atoms. On the pH scale, which rates the alkalinity or acidity of a given substance, alkalines, or bases, are denoted by numbers greater than 7, whereas acids are denoted by numbers less than 7. Some substances that contribute to a solution's alkalinity are dissolved ammonia, borate, hydroxide, nitrate, phosphate, silicate, and sulfide. The alkalinity of a substance can help protect the pH balance of the substance, as naturally basic substances can effectively neutralize or recover from the addition of acids.

• Significance for Climate Change

Rainwater is normally a weak carbonic acid solution. In the atmosphere, the water molecules in rain mix with carbon dioxide (CO_2) molecules, which, because of their weak bonds, can then form hydrogen and bicarbonate ions. Acid precipitation results when pollutants, such as sulfur dioxide or nitrogen oxides, remove low-pH acids from the atmosphere in the form of rain, snow, sleet, or hail. If water or soil where this acidic precipitation falls lacks natural alkalinity, as is the case with soils based on granite or other hard rocks with low carbonate content, the water or soil will be unable to neutralize the acid and the pH balance of the water or soil may be affected, altering the dynamics of the ecosystem.

The alkalinity of a body of water or soil can act as a buffer that can prevent drastic changes in the pH balance and, thus, can more easily recover from the addition of any type of acid. Thus, mildly basic bodies of water protect aquatic life and are less vulnerable to acid rain. The carbon in carbonate rocks, such as limestone, acts as a hydrogen absorber. Often, to increase the alkalinity of a body of water and thus to protect against fluctuations in the pH balance leading to algal bloom, calcium carbonate (also called limestone) is added.

Marianne M. Madsen

• Further Reading

Brimblecombe, P., et al., eds. *Acid Rain: Deposition to Recovery.* New York: Springer, 2007. Compilation of conference proceedings discussing acid rain and its effects on nutrient cycling.

Jenkins, J. C., K. Roy, C. Driscoll, and C. Buerkett. *Acid Rain in the Adirondacks: An Environmental History.* Ithaca, N.Y.: Cornell University Press, 2007. Discusses scientific investigation into the effects of acid rain on a particular ecosystem. Graphs, diagrams, maps.

Schindler, D. W. "Effects of Acid Rain on Freshwater Ecosystems." *Science* 239, no. 4836 (January 8, 1988): 149-157. Discusses the processes contributing to acid increase in bodies of water.

See also: Ocean acidification; pH.

Alleroed oscillation

• **Categories:** Climatic events and epochs; cryology and glaciology

• **Definition**

The Alleroed oscillation was a temperature fluctuation that occurred near the end of the last glaciation period, about thirteen thousand years ago. For several centuries following the oscillation, Europe, the British Isles, and the northern Atlantic warmed to present-day levels. Other areas remained untouched. The Alleroed oscillation is but one of several climate swings affecting the North Atlantic region over a period from 17,700 years ago to 11,500 years ago. These oscillations, revealed in European terrestrial sediments, are known as the Oldest Dryas, Boelling, Older Dryas, Alleroed, and Younger Dryas oscillations. Some oscillations provided warming conditions to circumscribed areas, as during the Alleroed oscillation, while others—such as the Younger Dryas oscillation, the last major cold event—were probably global in scope.

• **Significance for Climate Change**

The climate of the North Atlantic region underwent a series of abrupt cold/warm oscillations when the ice sheets of the Northern Hemisphere retreated. During the Alleroed oscillation, the warm temperatures allowed for a mix of flora and fauna that would look familiar today: Deer, horses, bear, and beaver found a congenial environment in Europe's evergreen and deciduous forests.

Questions arise about the feedback effects of such warming swings. For example, a phenomenon called the thermohaline circulation, or the great ocean conveyor, involves large ocean currents that flow like rivers around the globe, affecting local climates in the process. These currents depend on the ocean's salinity and temperature, so the Alleroed oscillation probably altered their course or strength. However, modeling for the effect of climate warming on the thermohaline circulation is difficult, and definitive conclusions are therefore elusive.

A more likely climate impact from the Alleroed oscillation would have resulted from warmer Arctic temperatures stimulating plant growth. Plant cover darkens the landscape and causes more sunlight to be absorbed, rather than reflected back into the atmosphere. Thus, as plant growth spreads, a feedback effect promotes higher temperatures, which encourages further plant growth over an increasing terrestrial range.

Some climate simulations further support the idea that temperature swings in the North Atlantic Ocean may have wide-ranging climatic effects. These simulations indicate that North Atlantic Ocean cooling causes North Pacific Ocean cooling, which in turn results in a drier climate in western North America.

Finally, current observable increases in global temperature are threatening glacier systems in the Antarctic. This may lead to sea-level increases worldwide, with serious consequences as the sea encroaches on vulnerable landmasses. Whether this occurred during the Alleroed oscillation is subject to speculation, but research indicates that it probably did.

Richard S. Spira

See also: Abrupt climate change; Climate change; Climate feedback; Climate zones; Ground ice; Thermohaline circulation; Younger Dryas.

Alliance of Small Island States

• **Categories:** Organizations and agencies; nations and peoples

• **Date:** Established 1991

- **Web address:** http://www.sidsnet.org/aosis/index.html

- **Mission**

A coalition of small island and low-lying coastal nations, the Alliance of Small Island States (AOSIS) speaks collectively about environmental concerns and about issues affecting economic growth. AOSIS includes Antigua and Barbuda, the Bahamas, Barbados, Belize, Cape Verde, Comoros, the Cook Islands, Cuba, Cyprus, Dominica, the Dominican Republic, the Federated States of Micronesia, Fiji, Grenada, Guinea-Bissau, Guyana, Haiti, Jamaica, Kiribati, the Marshall Islands, Mauritius, Nauru, Niue, Palau, Papua New Guinea, Samoa, São Tomé and Principe, the Seychelles, Singapore, the Solomon Islands, St. Kitts and Nevis, St. Lucia, St. Vincent and the Grenadines, Suriname, Tonga, Trinidad and Tobago, Tuvalu, and Vanuatu. American Samoa, Guam, the Netherlands Antilles, and the U.S. Virgin Islands are officially recognized as AOSIS observers. The alliance's member states, thirty-seven of which are also members of the United Nations, represent about 5 percent of the world's population.

Samoan ambassador Tuiloma Neroni Slade, the president of the Alliance of Small Island States, speaks at the alliance's meeting on climate change in Nicosia, Cyprus in January, 2001. (AP/Wide World Photos)

Because of their small size and their proximity to water, AOSIS nations face several risks in common as global warming progresses. The warming of seawater causes sea levels to rise, increasing the dangers of flooding and of salt water flowing into freshwater supplies. Warmer water also increases the frequency and the intensity of tropical storms and disrupts corals and fish that are important to these nations' economies. In addition to climate change, island and coastal nations are threatened by spilling and dumping from the large freighters operated by larger industrialized nations.

AOSIS has no formal charter or budget and works through collaboration and consensus within the structures of the United Nations. It works to present a unified voice to amplify the influence of its small member states and to educate and persuade larger nations. In 1999, AOSIS hosted the Workshop on the Clean Development Mechanism of the Kyoto Protocol, with fifty participants including guests from the Philippines, Mauritania, the United States, the United Kingdom, Australia, Norway, New Zealand, and Switzerland. A second workshop in 2000 produced a joint statement of cooperation between AOSIS and Italy and was followed by a third workshop in 2001—sponsored by the governments of New Zealand, Norway, and Switzerland—and the 2005 Conference of the Parties to the Climate Change Convention (COP-11).

AOSIS participates in international negotiations on climate change, particularly through the United Nations Framework Convention on Climate Change (UNFCCC). In 2007, AOSIS submitted a proposal for long-term cooperative action to address climate change, underscoring several basic principles: Nations must take care that activities within their control do not harm other nations; precautionary measures must be taken to protect future generations; the most vulnerable parties to the UNFCCC must be protected; and those who create the most environmental damage must assume the greatest amount of responsibility for reversing it. The goal of the

AOSIS proposal was to keep long-term global temperature increases below 2° Celsius.

- **Significance for Climate Change**

AOSIS presented an active and influential voice in the drafting of the UNFCCC at the Earth Summit in Rio de Janeiro, Brazil, in 1992. At the First Conference of the Parties to the UNFCCC (COP-1), held in Berlin in 1995, AOSIS submitted a draft protocol calling for a 20 percent reduction, based on 1990 levels, of greenhouse gas emissions by 2005. Although the specifics of the so-called AOSIS Protocol were not adopted, the language and the vision of the protocol informed subsequent negotiations leading to the Berlin Mandate and the Kyoto Protocol.

With its dozens of members, AOSIS made up one of the largest unified coalitions at COP-1 and succeeded in persuading larger nations that its cause was just. International climate negotiations now recognize the principle that small nations should be represented based on the amount of risk they face, rather than based solely on population or economic power. Although AOSIS continues to participate actively in international education and negotiation, it remains the group of nations most seriously threatened by global warming. While the member nations have benefited internally from projects leading to enhanced energy technologies, the group's repeated calls for industrialized nations to reduce their own emissions in order to slow sea-level rise have largely gone unheeded.

Cynthia A. Bily

- **Further Reading**

Acharya, Anjali. "Small Islands: Awash in a Sea of Troubles." *World Watch* 8, no. 6 (November/December, 1995): 24-33. Describes the ecological challenges faced by small island nations and argues that the formation of AOSIS is a good first step for these nations to take control of their own interests.

Bowness, Nicholas, and Alina Zyszkowski. *Small Islands, Big Issues: Sustainable Development of Islands.* Washington, D.C.: Counterpart International, 1997. This report covers the action plan for the "Sustainable Development of Small Island States," drafted in Barbados in 1994, and profiles the AOSIS member states.

"Not Enough Money to Keep Small Islands Afloat." *Africa News Wire,* December 14, 2007. A news story reporting the AOSIS response to the new Adaptation Fund, meant to help vulnerable nations cope with global warming, negotiated at the United Nations climate conference in Bali in 2007.

Roberts, J. Timmons, and Bradley C. Parks. *A Climate of Injustice: Global Inequality, North-South Politics, and Climate Policy.* Cambridge, Mass.: MIT Press, 2006. Analyzes why underdeveloped southern nations, including small island states, are reluctant to cooperate with wealthier, more polluting northern countries to solve the global climate crisis.

Wilford, Michael. "Law: Sea-Level Rise and Insurance." *Environment* 35, no. 4 (May, 1993): 2-5. Explains how the AOSIS, concerned about rising sea levels, has formed an international insurance pool to share the risks of drastic climate change.

See also: Islands; Kyoto Protocol; Sea-level change; Sea surface temperatures; United Nations Framework Convention on Climate Change.

Amazon deforestation

- **Category:** Plants and vegetation

The Amazon rain forest, sometimes called "the lungs of the world," plays a key role in global climate, and it supports a diverse population of species, many of which exist nowhere else on Earth. Loss of Amazonian forest lands through both human clearing and drought has significant effects upon climate regulation and Earth's biodiversity.

- **Key concepts**

carbon sink: an entity that absorbs and stores carbon, thereby removing CO_2 from the atmosphere

rain forest: a tropical area dominated by evergreen trees whose leaves form a continuous canopy and that receives at least 254 centimeters of rain per year

savanna: grassland with scattered trees, characteristic of tropical areas with seasonal rainfall on the order of 50 centimeters per year

• Background

The Amazon rain forest occupies more than 10,000,000 square kilometers of land in South America, the bulk of which is located in Brazil (60 percent) and Peru (13 percent). Nearly untouched as late as 1970, the region underwent rapid development in the last quarter of the twentieth century. Annual rates of clearing in Brazil peaked at 29,059 square kilometers in 1995 and 27,429 square kilometers in 2004. Between 1970 and 2006, the total area of rain forest in Brazil shrank from 41,000,000 square kilometers to 34,400,254 square kilometers, a decrease of 16 percent, leading to predictions of total annihilation of the Amazon within a century. As a result of international pressure and domestic conservation efforts, the annual clearing rate declined to 13,100 square kilometers in 2006, but it rose again in 2008 as increased world demand for soybeans and ethanol encouraged expansion of Brazilian agriculture.

A forest of this magnitude affects world climate in numerous ways. On a regional level, dense vegetation supports higher temperatures, higher rainfall, less runoff, and lower daily and seasonal temperature fluctuations. In the long term, high global temperatures favor forests. On a geologic time scale, the warmest periods have coincided with the greatest extent of rain forest, whereas much of the area later occupied by the Amazon rain forest was savanna during the height of the last Pleistocene glaciation.

Plants extract carbon dioxide (CO_2) from the atmosphere via photosynthesis. An expanding forest acts as a carbon sink, removing CO_2 from the air and sequestering carbon in its woody parts. A mature forest is in equilibrium, emitting as much carbon through animal consumption and decomposition as it fixes through photosynthesis. Clearing or burning forests releases CO_2 into the atmosphere; however, if tree trunks are converted to lumber and the land is subsequently used to grow crops, the net carbon release may be relatively small.

Workers in Rondonia, Brazil, burn a portion of the Amazon rain forest to clear land for agriculture. (Getty Images)

• Contribution of Global Warming to Amazon Deforestation

Global climate change can affect a forest profoundly. Although warm temperatures in general favor forests, shifts in patterns of prevailing winds brought about by small changes in oceanic temperatures may bring drought to regions accustomed to high rainfall and flooding to formerly arid regions. Although such perturbations are common in the geologic record and the Earth's biota has repeatedly shown a rapid response, the rate of recovery is slow on a human time scale.

Increasing atmospheric CO_2 may actually stimulate forest growth in the tropics. High CO_2 levels favor rapid growth of trees, which tend to crowd out understory species, leaving fewer niches for animal species, particularly insects dependent on specific food plants. In the short term, such highly productive forests may be commercially desirable for lumber production, but ecological diversity and sustainability suffer.

Cycles of the El Niño-Southern Oscillation (ENSO) cause large natural fluctuations in rainfall in the Amazon basin. During the unusually severe drought of 2005-2006, some scientists predicted that tree species would die off and natural fires would destroy significant areas of forest, creating a climate feedback loop that would turn much of the Amazon into savanna. The forest appears to be unexpectedly resilient, however. During a drought year, deep-rooted trees remain green, and they even grow faster than normal, owing to the absence of cloud cover.

Some efforts to address environmental problems elsewhere in the world contribute to Amazon deforestation. Strenuous conservation efforts in the developed world, unaccompanied by reduction in wood-product consumption, increase logging pressure in places like Brazil. The United States' drive to produce and deploy corn-based ethanol opened the way for rapid growth in Brazil's soybean production. Brazil is also a leading producer and exporter of ethanol derived from sugarcane. These crops are rarely planted directly on cleared jungle land, but cattle ranchers displaced by soybeans and sugarcane migrate to the Amazon.

Finally, decreasing levels of sulfur dioxide resulting from more stringent pollution controls in Europe and North America have been implicated in Brazil's devastating 2005 drought. This effect, the subject of a May, 2008, article in *Nature*, was the first firm scientific evidence of the importance of sulfur dioxide emissions in canceling the greenhouse effect of CO_2.

• Contribution of Deforestation to Global Warming

The effects of Amazonian deforestation on world species diversity eclipse its large-scale climatic effects, as loss of the forest may lead to the extinction of thousands of species. Nonetheless, even if only CO_2 emissions are considered, the deforestation's climatic effect is not negligible. CO_2 from slash burning following logging may account for as much as half of Brazil's carbon contribution to the atmosphere, estimated at 90 million tons in 2004. Although Brazil ranks sixteenth in terms of its contribution to world CO_2 pollution, it accounts for only a little over 2 percent of the world total and has a very low per capita level of fossil fuel consumption because Brazilians rely on hydroelectric power and ethanol.

Some of the released carbon is recaptured when land is used for crops or pastureland. However, indiscriminate logging combined with a drier and more uncertain climate due to global warming may ultimately convert large areas of the Amazon to semiarid grassland of minimal value as a carbon sink. This scenario, which appeared to many to be imminent during the 2005-2006 drought, is now thought to be avoidable through feasible management schemes, some of which are already being implemented.

• Context

It is tempting to view environmental threats en masse and to assume that a policy that ameliorates one ecological disaster will have a correspondingly benign effect on others. The interactions between global warming and the deforestation of the Amazon rain forest demonstrate that this is not always the case. The forest has shown itself to be more resilient to drought than scientists anticipated. The principal, immediate, global-warming-related threat to the Amazon rain forest appears to be the rapid expansion of Brazilian agriculture in response to rising world demand for biofuels. Models for controlling this expansion so as to encourage efficient land use and sustainability favor large agricultural businesses over individual farmers and pay inadequate attention to preserving biodiversity.

No discussion of a global-warming issue is complete without mention of population issues. The populations of Brazil and other countries bordering on the Amazon are growing at a very rapid rate. Despite the overall low level of energy consumption in the area, this population growth increases human impact on the environment exponentially. The Amazon ecosystem is apparently robust enough to withstand present levels of global CO_2 emissions, but unless exponential trends are reversed, the grim scenario of degradation to savanna looms in the future.

Martha A. Sherwood

- **Further Reading**

Alverson, Keith D., Raymond S. Bradley, and Thomas Pedersen, eds. *Paleoclimate, Global Change, and the Future.* Berlin, Germany: Springer Verlag, 2003. Collection of scholarly papers comparing past climate changes with present anthropogenic trends.

Cox, Peter M., et al. "Increasing Risk of Amazonian Drought Due to Decreasing Aerosol Pollution." *Nature* 453 (May 8, 2008): 212-215. Groundbreaking letter announcing the finding that aerosol pollution is capable of counteracting the greenhouse effect.

Gash, John H. C. *Amazonian Deforestation and Climate.* New York: John Wiley, 1996. Based on a 1994 symposium; reports on a series of intensive studies of pasture and forest upon which predictions of future climatic trends are based.

Intergovernmental Panel on Climate Change. *Climate Change, 2007—Impacts, Adaptation, and Vulnerability: Contribution of Working Group II to the Fourth Assessment Report of the Intergovernmental Panel on Climate Change.* Edited by Martin Parry et al. New York: Cambridge University Press, 2007. Chapter 13, "Latin America," contains a wealth of statistical information on climate change in the Amazon.

London, Mark. *The Last Forest: The Amazon in the Age of Globalization.* New York: Random House, 2007. Documents continuing destruction and local climatic change in the Amazon; optimistic about Brazil's efforts to encourage more sustainable growth.

See also: Agriculture and agricultural land; Brazil; Carbon cycle; Carbon dioxide; Ethanol; Extinctions and mass extinctions; Forestry and forest management; Forests.

American Association for the Advancement of Science

- **Category:** Organizations and agencies
- **Date:** Established 1848
- **Web address:** http://www.aaas.org

- **Mission**

The American Association for the Advancement of Science (AAAS) was founded in Philadelphia in 1848. With nearly 150,000 members, it is the largest and among the most influential general science organizations in the world. From its inception, the association has admitted both professional and amateur scientists as members. It is organized by subdivisions for all areas of the physical sciences and some social sciences.

The AAAS's multidisciplinary character has fostered one of the most important dynamics of the association, its interdisciplinary and cross-disciplinary characteristics, creating a synergy of knowledge. A further dynamic of the AAAS is its regular engagement and communication with the lay public through publishing and media outlets that divulge the latest scientific findings. Among its most distinguished publications has been *Science* magazine, noted for its tradition of popular diffusion of advanced, cutting-edge scientific research.

- **Significance for Climate Change**

The leadership of the AAAS has often been at the forefront of worldwide scientific discoveries. One of its earliest presidents, Louis Agassiz, was among a group of geologists in the mid-nineteenth century who first discovered that the Earth, over long geological ages, experienced periods of cooling and warming. Investigating the causes of these alterations has engaged many of the association's members. Of paramount importance has become identifying the human role in global warming—the extent to which increased carbon emissions due to human exploitation of coal, gas, and petroleum prompts atmospheric changes that have intensified solar heat and thereby altered fundamental biological processes. This project poses challenges along the entire spectrum of the sciences and demands concerted interaction.

Investigating global warming and its anthropogenic factors has involved not only the research of AAAS members but also communication of their findings through the popular media. *Science* has included numerous articles on global warming, and the association has prepared guidebooks and curricular guidelines regarding it. Moreover, presentations and debates at the association's annual conferences have concentrated on the multifaceted scientific aspects of global warming. This debate culminated in 2007, when the AAAS board of directors issued a statement declaring that the events of global warming comprise "early warning signs of even more devastating damage to come, some of which will be irreversible."

Edward A. Riedinger

See also: American Astronomical Society; American Geophysical Union; American Institute of Physics; American Meteorological Society; Scientific credentials; Scientific proof.

American Association of Petroleum Geologists

- **Categories:** Organizations and agencies; fossil fuels
- **Date:** Established 1917
- **Web address:** http://www.aapg.org

- **Mission**

The American Association of Petroleum Geologists (AAPG) is an international professional organization that supports research in the science of geology and in technologies used to locate underground reservoirs of oil, manage them, and extract their contents. Its approximately thirty thousand members from 116 countries include geologists, geophysicists, oil company executives, university professors, consultants, and students. Of the members, slightly more than half work to find oil reservoirs or develop existing reservoirs. Full membership requires a degree in geological sciences and three years' experience, but there are associate memberships for those lacking practical work experience and student memberships for those pursuing degrees in geology or a related field.

Geographically, the AAPG is divided into six regional sections within the United States and six international regions. It comprises four divisions with distinct missions. The Division of Environmental Geosciences seeks to keep members up to date on the relation of the petroleum industry to environmental problems through education, support of research on the effects of oil exploration, and the sharing of research with governmental agencies. The Division of Professional Affairs sets ethical standards, provides certifications, and helps in career planning in order to promote professionalism

American Association of Petroleum Geologists Policy Statement

The following policy statement expresses the AAPG's official attitude toward climate change.

Although the AAPG membership is divided on the degree of influence that anthropogenic CO_2 has on recent and potential global temperature increases, the AAPG believes that expansion of scientific climate research into the basic controls on climate is important. This research should be undertaken by appropriate federal agencies involved in climate research and their associated grant and contract programs.

among its members. The Division of Energy Minerals fosters research and disseminates information on energy minerals, unconventional hydrocarbons (such as gas hydrates and oil sands), geothermal energy, oil shale, tar sands, gas hydrates, energy economics, and remote sensing. The Division of Student Programs offers publications, lectures, and grants-in-aid for undergraduate and graduate students.

• Significance for Climate Change

Adopted in 2007, the AAPG policy statement posted on the association's Web site expresses doubt about the accuracy of climate simulation models produced on computers and calls for further basic research. Moreover, the AAPG argues that although reducing emissions from fossil fuel use is a worthy goal, the environmental gains must be weighed against the potential economic costs. Accordingly, it advocates using carbon sequestration technologies and energy conservation.

Roger Smith

See also: Fossil fuel emissions; Fossil fuel reserves; Fossil fuels; Geological Society of America; Oil industry.

American Astronomical Society

- **Category:** Organizations and agencies
- **Date:** Established 1899
- **Web address:** http://www.aas.org

• Mission

The American Astronomical Society (AAS) fosters research into climate change both on Earth and on other planets, as well as research into the role of the Sun in global warming. The AAS is the largest professional organization in North America for scientists who conduct research in astronomy and related sciences. Among its approximately sixty-five hundred members are astronomers, mathematicians, geologists, and engineers. Its mission is to promote the advancement of knowledge by awarding grants and prizes for research, publishing journals, arranging conferences, fostering debate and discussion on its Web site, posting news items, and issuing public policy statements on educational and political issues related to astronomy and planetary science.

The Council of the AAS governs the organization and considers recommendations from its thirty-eight committees to support research, award prizes, and set policy. The society comprises the Division for Planetary Science, Division on Dynamical Astronomy, High Energy Astrophysics Division, Historical Astronomy Division, and Solar Physics Division. It sponsors *The Astronomical Journal, The Astrophysical Journal, Bulletin of the AAS,* and *Icarus,* as well as departmental newsletters.

• Significance for Climate Change

Among AAS members are planetary scientists studying climate change on Earth and other planets in the solar system, and AAS publications present their findings. Of particular pertinence are investigations into how variations in the Sun's energy output affect planetary warming on Earth and other bodies, such as Venus and Saturn's moon Triton.

On June 2, 2004, the AAS Web site posted its endorsement of "Human Impacts on Climate," the policy statement of a sister organization, the American Geophysical Union (AGU). Acknowledging the closer involvement of the AGU in scientific subdisciplines directly addressing terrestrial climate change, the AAS endorsement notes that

> . . . the human impacts on the climate system include increasing concentrations of greenhouse gases in the atmosphere, which is significantly contributing to the warming of the global climate. The climate system is complex, however, making it difficult to predict detailed outcomes of human-induced change: there is as yet no definitive theory for translating greenhouse gas emissions into forecasts of regional weather, hydrology, or response of the biosphere.

Accordingly, the AAS calls for further peer-reviewed research. It especially cites the need for improved observations and computer modeling in order to provide governments with a solid basis for making decisions about how best to mitigate the harmful effects of climate change and to help communities adapt to such change.

Roger Smith

See also: American Geophysical Union; Planetary atmospheres and evolution; Sun.

American Chemical Society

- **Category:** Organizations and agencies
- **Date:** Established 1876
- **Web address:** http://www.acs.org

• Mission

The American Chemical Society (ACS) is the world's largest scientific society, with more than 160,000 members worldwide. Its members are professional chemists and engineers. It is a nonprofit organization, headquartered in Washington, D.C., with 33 technical divisions and 189 local sections. It publishes the weekly *Chemical and Engineering News*, the *Journal of the American Chemical Society*, and a number of other scientific journals. The ACS's *Strategic Plan: 2008 and Beyond* provides its mission statement: "to advance the broader chemistry enterprise and its practitioners for the benefit of Earth and its people." The society promotes interest in chemistry, as well as scientific research and education, and maintains a practical focus on solving problems and meeting challenges through chemistry. In addition to providing services for its members, the ACS sponsors scientific meetings and provides resources for educators.

• Significance for Climate Change

The ACS officially endorses the position that climate change is anthropogenic and attributable in part to increases in greenhouse gases (GHGs) and aerosol particles in Earth's atmosphere. It asserts that urgent action is required to respond to these conditions. Recommendations in the ACS position statement on global climate change include increased funding for research to better predict climate changes and their impact; GHG emission reductions; and increased investments in technologies that conserve energy and use nonfossil fuel.

The ACS has formed the Green Chemistry Institute to promote the concept and techniques of green, or environmentally friendly, chemistry. According to the Green Chemistry Institute,

> Green chemistry differs from previous approaches to many environmental issues. Rather than using regulatory restrictions, it unleashes the creativity and innovation of our scientists and engineers in designing and discovering the next generation of chemicals and materials so that they provide increased performance and increased value while meeting all goals to protect and enhance human health and the environment.

The institute hosts annual green chemistry and engineering conferences. As part of its chemistry education activities, ACS plans annual activities in celebration of Earth Day and has created a series of podcasts presenting the problems facing humanity in the twenty-first century and the potential role of chemistry in solving those problems.

Susan J. Karcher

See also: Chemical industry.

American Enterprise Institute

- **Category:** Organizations and agencies
- **Date:** Established 1943
- **Web address:** http://www.aei.org

• Mission

In the ongoing public debate over climate policy, regulatory proposals by governmental agencies and nongovernmental organizations (NGOs) are often challenged by policy analysts who value free markets, small government, economic growth, and individual rights. The American Enterprise Institute (AEI) is a public policy think tank associated with the latter values. A subset of AEI scholars regularly evaluate, criticize, and promote climate-related policies through in-house publication of books and policy articles, as well as external publications and appearances in mainstream media. The organization also holds conferences and roundtable discussions at which policy experts discuss diverse elements of climate-policy-related science. These conferences are advertised to decision makers, corporate leaders, AEI supporters, and subscribers, as well as national media outlets in Washington, D.C. Transcripts of AEI conferences are also posted to the organization's Web site, sometimes accompanied by streaming video excerpts.

Most broadly, AEI, like most think tanks, attempts to influence public opinion and government decision makers in order to favor the development of government policies that the institute supports. AEI's mission includes protecting and supporting democratic capitalism, limited government, private enterprise, individual liberty and responsibility. It also advocates for vigilant and effective defense and foreign policies, political accountability, and open debate. The institute's motto is "Competition of ideas is fundamental to a free society." AEI describes its target audience as "government officials and legislators, teachers and students, business executives, professionals, journalists, and all citizens interested in a serious understanding of government policy, the economy, and important social and political developments."

• Significance for Climate Change

The analysis of AEI researchers is widely cited in the mainstream media, where it influences opinion among the institute's core audience, the broader public, and policy makers seeking to attract the AEI customer base. While AEI does not adopt an institutional point of view on any given policy issue, AEI's analysts have expressed a diverse spectrum of opinion regarding the relationship of greenhouse gases (GHGs) to climate change and an equally broad spectrum regarding what ought to be done about the threat of anthropogenic climate change. Some AEI analysts have sought to disclaim a human influence on climate and resisted GHG mitigation efforts, while others have accepted an anthropogenic influence on climate, and still others have suggested policy options including emission trading, carbon taxation, expanded adaptation efforts, and geoengineering.

Kenneth P. Green

See also: Nongovernmental organizations; United States; U.S. energy policy.

American Geophysical Union

- **Category:** Organizations and agencies
- **Date:** Established 1919
- **Web address:** http://www.agu.org

• Mission

The American Geophysical Union (AGU) is a professional organization whose members conduct research in a wide variety of disciplines devoted to understanding Earth's composition, dynamics, and environment in space. The AGU has a membership of more than fifty thousand researchers, teachers, and students who study the Earth, its atmosphere and oceans, space, and other planets. It is divided into eleven sections, each for a scientific discipline: Atmospheric Sciences; Biogeosciences; Geodesy; Geomagnetism and Paleomagnetism; Hydrology; Ocean Sciences; Planetary Sciences; Seismology; Space Physics and Aeronomy; Tectonophysics; and Volcanology, Geochemistry, and Petrology. An elected council composed of six general officers and the president and president-elect of each section governs the AGU, authorizing programs, controlling finances, and approving policies.

The AGU's mission is to promote the scientific study of Earth and its surrounding space and to pass on discoveries to the public through research projects, professional meetings, the journal *Eos*, books (often available through its Web site), a weekly newsletter, and educational programs for nonscientists. Moreover, the AGU fosters cooperation among scientific organizations supporting geophysics and related fields. Three principles

"Human Impacts on Climate Change"

The AGU's important evaluation of anthropogenic climate change reads, in part, as follows:

Warming greater than 2° Celsius above nineteenth century levels is projected to be disruptive, reducing global agricultural productivity, causing widespread loss of biodiversity, and—if sustained over centuries—melting much of the Greenland ice sheet with ensuing rise in sea level of several meters. If this 2° Celsius warming is to be avoided, then our net annual emissions of CO2 must be reduced by more than 50 percent within this century.

guide its mission: adherence to the scientific method, the free exchange of ideas, and accountability to the public.

- **Significance for Climate Change**

The AGU's policy statement, "Human Impacts on Climate Change," revised and reaffirmed in late 2007, unambiguously declares that Earth's climate is out of balance because of unnatural global warming caused by anthropogenic greenhouse gases and aerosols. Moreover, it warns of civilization-altering consequences that require concerted, dramatic action. The AGU emphasizes the uncertainty of the projections upon which its judgments are based. However, it also stresses that all projections, regardless of specifics, indicate that climate change will have serious consequences for life in the twenty-first century. The uncertainty, according to the union, rests only in the precise nature of those consequences.

The AGU statement goes on to emphasize that, unlike ozone depletion, climate change is the result of fundamental aspects of modern human society, so it cannot be mitigated easily. At base, climate change results from human energy use, which lies at the heart of virtually all aspects of modern civilization. To mitigate anthropogenic influences upon global warming, then, the AGU asserts that government, scientists, industry, and consumers will all need to cooperate in finding and adopting solutions.

Roger Smith

See also: American Astronomical Society; American Chemical Society; American Institute of Physics; Anthropogenic climate change; Climate change; Human behavior change.

American Institute of Physics

- **Category:** Organizations and agencies
- **Date:** Established 1931
- **Web address:** http://www.aip.org

- **Mission**

The membership of the American Institute of Physics (AIP) investigates basic properties of matter and energy, including those that underlie processes causing climate change. A nonprofit corporation, the AIP promotes research in physics and application of the knowledge acquired in the interests of human welfare. It provides professional services to its ten member societies: the Acoustical Society of America, the American Association of Physicists in Medicine, the American Association of Physics Teachers; the American Astronomical Society, the American Crystallographic Association, the American Geophysical Union, the American Physical Society, AVS (Science and Technology of Materials, Interfaces, and Processing; originally the American Vacuum Society), the Optical Society of America, and the Society of Rheology. A forty-two-member governing board, headed by an executive director, oversees the institute's services and financing. Among the membership are more than 125,000 scientists, engineers, and students.

The AIP helps its member societies and individual members by publishing more than twenty-five journals of scientific and engineering societies, magazines, and conference proceedings. It reports on employment and educational trends, encourages interaction between science and industry, and compiles scientific archives. The institute also mentors undergraduate physics programs and advocates for science policy to the public and the U.S. Congress.

- **Significance for Climate Change**

In 2004, the governing board of the AIP endorsed "Human Impacts on Climate Change," the policy statement issued by its member society, the American Geophysical Union, in 2003. Additionally, its Web site offers "A Hyperlinked History of Climate Change Science." In the introduction, Spencer R. Weart, director of the Center for History of Physics, concludes that twenty-first century computer modeling combined with data collected from many sources support the conclusion that climate change is likely to result from human emissions:

> Depending on what steps people took to restrict emissions, by the end of the century we could expect the planet's average temperature to rise anywhere between about 1.4° and 6° Celsius.

The AIP saw some hope, however, in the fact that the media had begun to trust scientists who predicted further climate change based on the apparent validity of their past predictions.

Roger Smith

See also: American Chemical Society; American Geophysical Union; Anthropogenic climate change; Human behavior change.

American Meteorological Society

- **Category:** Organizations and agencies
- **Date:** Established 1919
- **Web address:** http://www.ametsoc.org

- **Mission**

The eleven thousand members of the American Meteorological Society (AMS) are scientists, students, and lay enthusiasts involved in the atmospheric, oceanic, and hydrologic sciences. The AMS supports them by producing and sharing information on these fields and through educational programs. Its elective council ensures that the society's strategic goals are pursued through the activities of its six major divisions: the Commission of the Weather and Climate Enterprise, the Commission on Professional Affairs, the Education and Human Resources Commission, the Publications Commission, the Scientific and Technological Commission, and the Planning Commission.

The AMS publishes nine journals, hosts twelve annual conferences, and provides a Web site for news, policy statements, and professional development. It offers certifications for broadcast and consulting meteorologists, a career center, scholarships and grants for students, and awards for service and research in the pursuit of seven goals: advancing knowledge through publications and meetings, accelerating development of the application of the knowledge, promoting science-based decision making, educating the public, attracting new talent into its ranks, developing cooperation, and supporting national and international programs of benefit to society.

- **Significance for Climate Change**

The extensive policy review on climate change, "An Information Statement of the American Meteorological Society," adopted in 2007, concludes that Earth is undergoing global warming to which human activity is a significant contributing factor. This warming, the statement continues, will continue beyond the next century and will affect animal life and ecosystems, as well as human civilization. The statement emphasizes that government policy must adopt the twin goals of reducing climate change and confronting the reality that some climate change will continue no matter what humans do, so plans will need to be made to enable society to adapt to that change: "Prudence dictates extreme care in managing our relationship with the only planet known to be capable of sustaining human life."

Roger Smith

See also: Anthropogenic climate change; Climate models and modeling; Human behavior change.

American Physical Society

- **Category:** Organizations and agencies
- **Date:** Established 1899
- **Web address:** http://www.aps.org

• Mission

The world's second largest organization of physicists, the American Physical Society (APS) takes a firm stand relating to the belief that global warming is occurring and that human activities play a significant role in changing the atmosphere in ways that affect climate change. The APS was initially organized, in 1899, to advance and diffuse the knowledge of physics, and in its early years the main activity of the society was to hold scientific meetings. In 1913, the APS turned to publishing scientific journals as a major activity. Later, the group became active in public and governmental affairs, as well as in the affairs of the international physics community. The society also provides a number of educational programs and functions as a lobbying and advocacy agency, making official statements made on issues of critical importance to the nation.

Typical of advocacy statements adopted by the APS Council is a strong statement in 2007 that made clear the position of the APS on global warming and climate change:

> Emissions of greenhouse gases from human activities are changing the atmosphere in ways that affect the Earth's climate. . . . The evidence is incontrovertible: Global warming is occurring. If no mitigating actions are taken, significant disruptions in the Earth's physical and ecological systems, social systems, security and human health are likely to occur. We must reduce emissions of greenhouse gases beginning now.

The APS urged scientists to redouble their efforts to understand the relationship between human activity and climate change and to develop new technologies to mitigate anthropogenic warming.

• Significance for Climate Change

The APS engages in significant public advocacy. It prepares personalized letters and e-mails for citizens to send to legislators, provides contact information for various science coalitions, and holds grassroots meetings to educate the public as to government activities and procedures. APS presentations provide an overview of lobbying efforts and governmental actions. In addition to published statements and lobbying efforts made by individuals in the APS, over forty-six thousand members of the society work to educate the public about the gravity of the global warming phenomenon in the hope of changing behaviors that contribute to emission of greenhouse gases.

Victoria Price

See also: American Geophysical Union; American Institute of Physics.

Amphibians

- **Category:** Animals

• Definition

Amphibian species are distributed worldwide and include frogs, toads, salamanders, and newts. They are cold-blooded animals, or ectotherms, and their physiology is affected by their external environment. Amphibian populations face an extinction crisis, as they have experienced dramatic population decreases worldwide since the 1980's. An estimated one-third of amphibian species are currently in decline, with many amphibian species now either threatened or extinct.

In the late 1980's, scientists began to report mass mortalities of amphibians at an alarming rate, with causes not well understood. Although amphibian extinctions have occurred globally, declines have been particularly significant in the western United States, Central and South America, and Australia. Among the amphibian species that experienced dramatic population declines are the golden toad, *Bufo periglenes*, of Costa Rica and many harlequin frog species (*Atelopus*) that were once common in South America. These frog species are listed as critically endangered. These species lived in pristine areas, so their extinctions raised particular con-

A tiny Corroboree frog sits on a zookeeper's thumb. The Corroboree is among Australia's most endangered amphibians. (Mick Tsikas/Reuters/Landov)

cern, because they could not be linked to human activities.

A number of potential explanations for amphibian declines have been proposed, with many of the causes also affecting other organisms. The causes of amphibian declines are likely to be complex; many probably act synergistically. Some causes may include destruction of both terrestrial and aquatic habitats, introduced species, overexploitation, pollution, and pesticides. Many amphibian declines, however, have occurred in pristine habitats where such effects are unlikely. Therefore, although habitat loss is known to have affected amphibians for decades, recent research has focused on the effects of environmental contaminants, increased ultraviolet radiation, emerging diseases, and climate change.

Since the health of amphibian populations is thought to be an indicator of overall environmental health, reports of global amphibian declines have led to considerable public concern. The causes of amphibian declines might also threaten other species of animals and plants. One reason amphibians are thought to be indicator species for environmental health is their sensitivity to the environment. Amphibians' skin is extraordinarily thin, which makes them very sensitive to even small changes in temperature, humidity, and air or water quality. Their skin is also very permeable, which makes them very sensitive to toxins in both land and water environments.

• Significance for Climate Change

Climate change has probably contributed to the observed decline of many amphibian species worldwide. Although there is no simple answer to what is causing amphibian declines, many factors related to global warming are believed to play a role. Global warming is thought to initiate amphibian declines by triggering epidemics such as fungal diseases. Ozone layer depletion as a result of increased pollution leads to greater ultraviolet (UV) radiation exposure, which damages the delicate skin of amphibians, as well as their eyes, eggs, and immune systems.

Amphibian reproduction is affected by climate change: Changes in breeding behavior have been linked to increased temperatures due to global warming. UV radiation may also reduce hatching success and the rate of survival to metamorphosis. Global warming increases the metabolic rate of toads during hibernation and thus affects their body condition. Environmental temperature has a dramatic effect on amphibians' immune systems, so climate change may affect their defenses against invading pathogens. Tropical amphibian species, especially those in higher altitudes, are particularly susceptible to adverse effects related to global warming.

Although other pathogens play a role, chytrid fungi are believed to be responsible for many amphibian declines. About two-thirds of the 110 known harlequin frog species in Central and South America disappeared in the 1980's and 1990's; the cause was thought to be an infectious disease triggered by changes in environmental temperatures. These abundant frog deaths led to the discovery of a pathogenic fungus, *Batrachochytrium dendrobatidis,* which belongs to a family known as chytrids. By moderating temperature extremes that used to keep the growth of this fungus in check, global warming has arguably created ideal conditions to support its growth and reproduction. However, although many scientists support a connection between outbreaks of chytrid fungus, amphibian population declines, and global warming, this link is now controversial, with some questioning the role of temperature changes in fungal outbreaks.

The disease caused by chytrid fungi, chytridiomycosis, is fatal for otherwise healthy animals. Frogs with chytridiomycosis generally exhibit skin lesions that affect respiration across the skin, resulting in mortality. Since its discovery, the fungus has been linked to many amphibian extinctions. Amphibian declines due to chytrid fungus are most common at higher elevations in the tropics, and chytrid fungus is believed to be the cause of decline in three-quarters of frog species in Costa Rica and Panama and in species in the United States as well.

C. J. Walsh

• **Further Reading**

Cherry, Lynne, and Gary Braasch. *How We Know What We Know About Our Changing Climate: Scientists and Kids Explore Global Warming.* Nevada City, Nev.: Dawn, 2008. Written for teenagers and young adults; includes a chapter about the effects of changing climate on tropical rain forests.

Hofricher, Robert. *Amphibians: The World of Frogs, Toads, Salamanders, and Newts.* Toronto: Key Porter Books, 2000. Introductory textbook on the biology of amphibians, including many photographs. The last sections of the book discuss the effects of environmental degradation on amphibians.

Lannoo, Michael, ed. *Amphibian Declines: The Conservation Status of United States Species.* Berkeley: University of California Press, 2005. Includes a comprehensive description of amphibian status, including the life history of every known amphibian species in the United States. Also includes a chapter on the effects of global climate change on amphibian populations.

Linder, Greg, Sherry K. Krest, and D. W. Sparling. *Amphibian Decline: An Integrated Analysis of Multiple Stressor Effects.* Pensacola, Fla.: Society of Environmental Toxicology and Chemistry, 2003. Experts in the field address the role of various stressors, including global warming, in amphibian population declines.

Pounds, J. A., et al. "Widespread Amphibian Extinctions from Epidemic Disease Driven by Global Warming." *Nature* 439 (2006): 161-167. A much-cited scientific article that describes a link between chytrid fungus, global warming, and amphibian declines in Costa Rica.

See also: Endangered and threatened species; Extinctions and mass extinctions; Lichens.

Animal husbandry practices

• **Categories:** Animals; economics, industries, and products

• **Definition**

Humans raise livestock to meet individual and community nutritional and commercial demands for animal products. Livestock farmers range from individuals, who may raise a few animals to supply their families with protein sources or earn income in local markets, to commercial agriculturists who invest in industrial methods to manage numerous livestock simultaneously. Animal husbandry practices reflect diverse human cultures and types of livestock, from nomads in developing countries who tend herds of indigenous livestock that migrate to water and forage resources to farmers in developed countries who often rely on technology to mass-produce genetically standardized animals at centralized locations.

• Significance for Climate Change

Climate issues associated with animals worldwide caused the Food and Agriculture Organization (FAO) of the United Nations to assess livestock's environmental impact. In the early twenty-first century, several billion livestock, including bovines, goats, sheep, swine, and poultry, inhabited the Earth, and 30 percent of the planet's land was used for grazing and to grow livestock feed crops. An estimated 1.3 billion people participated in animal husbandry-related work. Annual meat production totaled 229 million metric tons and milk production was 580 million metric tons in 2001, with experts projecting that those amounts would double in fifty years to meet increasing food demands.

By spring, 2008, U.N. representatives determined that livestock produced 18 percent of the greenhouse gases (GHGs) associated with global warming, noting that those emissions exceeded the amounts produced collectively by land and air transportation vehicles. Media referred to the "carbon hoofprint" when discussing livestock's contribution to global warming. Animal husbandry was found to be responsible for 82 million metric tons of carbon dioxide (CO_2) emissions annually. In addition, each individual beef cow emitted approximately 80 kilograms of methane yearly and each dairy cow emitted 150 kilograms. The European Union (EU) suggested that farmers cull their herds to limit emissions. By 2008, New Zealand became the world's first nation to consider taxing farmers whose livestock emitted more GHGs than legislated allowances. The Pastoral Greenhouse Gas Research Consortium investigated scientific ways to manage livestock emissions.

Animal husbandry practices include clearing land for grazing livestock and planting grain crops to sustain them. Deforestation to establish ranches results in greater amounts of CO_2 reaching the atmosphere. Approximately 70 percent of Amazon rain forests have been cleared to maintain livestock, which often harm fields by stripping them of vegetation, causing exposed soil to become eroded or compacted. Subterranean storage of precipitation is disrupted, sometimes initiating desertification. Clearing land for livestock occasionally involves draining wetlands.

The loss of trees and vegetation due to farmers transforming forests into fields reduces albedo (the proportion of incident radiation reflected by Earth's surface): Bared ground absorbs rather than reflects sunlight and heat, intensifying global warming. Although Earth's Arctic ice is distant from most animal husbandry procedures (except for reindeer farming in northern latitudes), the conversion of ice to seawater due to decreasing albedo affects livestock globally. In some areas, extreme heat associated with lower albedo frequently dries up water resources used for animal husbandry. Elsewhere, rising water levels due to increased ice melting displaces livestock.

Excess CO_2 detrimentally affects indigenous plant growth, particularly that of native forages stifled by GHGs. Woody shrubs exotic to prairies thrive when exposed to CO_2. Weeds, some toxic to livestock, spread and often deprive soil of moisture. Climate change can motivate livestock to seek new food and water sources, and these migrating livestock transport seeds from plants and parasites to new locations when they travel, carrying them in fur, hooves, and intestinal tracts, further altering ecosystems.

Veterinary professionals report cases of livestock diseases previously unknown in specific geographical areas, attributing their spread to changed climates. In 2006, bluetongue, an insect-transmitted viral disease usually confined to southern France, was detected in several thousand northern European livestock, necessitating quarantines. Warm temperatures enable insect populations, including tsetse flies and mosquitoes, to extend into locations where usually cooler climates would have inhibited them. When frosts are infrequent, lungworm larvae survive in grasses, causing respiratory disease in livestock.

An FAO report has estimated that one livestock breed becomes extinct each month during the early twenty-first century. Industrial animal husbandry practices limit production to selected western breeds, which replace many native livestock in developing countries. By 2002, only one-fourth of sows in Vietnam were indigenous breeds. The FAO promoted preserving genetic material from indigenous livestock breeds that possessed such resilient traits as heat tolerance and drought hardiness, which might be crucial to future animal husbandry as global warming alters climatic conditions.

In November, 2006, a U.N. climate conference at Nairobi, Kenya, discussed how global warming threatened nomadic livestock herders in Africa. Early twenty-first century droughts caused an estimated 500,000 nomads to cease raising livestock, as deserts began overtaking fields, disrupting the normal four seasons that herders need to raise livestock. Violence resulted, as herders battled for water and pasture resources limited by climate changes.

Elizabeth D. Schafer

• Further Reading

Morgan, Jack A., et al. "Carbon Dioxide Enrichment Alters Plant Community Structure and Accelerates Shrub Growth in the Shortgrass Steppe." *Proceedings of the National Academy of Sciences* 104, no. 37 (September 11, 2007): 14,724-14,729. Experimental results indicate GHGs cause displacement of indigenous livestock forage.

Seo, Sungno Niggol, and Robert Mendelsohn. "Climate Change Impacts on Animal Husbandry in Africa: A Ricardian Analysis." World Bank Policy Research Working Paper 4261. Washington, D.C.: World Bank, 2007. Economic study examining the influences of temperature, precipitation, and animals' heat tolerance upon farmers' decisions.

Steinfeld, Henning, et al. *Livestock's Long Shadow: Environmental Issues and Options.* Rome: Food and Agriculture Organization of the United Nations, 2006. Comprehensive report discussing animal husbandry and global warming; suggests ways to limit damage and restore resources. Extensive bibliography.

Watson, Paul. "New Zealand Aims for Greener Pastures: Officials Ruminate on How to Curb Methane from the Nation's Livestock, a Culprit in Global Warming." *Los Angeles Times,* June 8, 2008, p. A-6. Describes governmental efforts to reduce animal husbandry emissions, farmers' reactions, and possible scientific solutions.

See also: Agriculture and agricultural land; Amazon deforestation; Carbon dioxide; Carbon footprint; Deforestation; Greenhouse gases; Methane.

Annex B of the Kyoto Protocol

- **Category:** Laws, treaties, and protocols
- **Date:** Adopted December 11, 1997; entered into force February 16, 2005; amended November, 2006

The Kyoto Protocol's second annex allowed countries to collaborate on emissions-reducing projects and to trade and purchase emissions credits. This system gave wealthier nations incentives to reduce emissions and encouraged them to work with developing nations to reduce theirs.

- **Participating nations:** *1997:* Australia, Austria, Belgium, Bulgaria, Canada, Croatia, Czech Republic, Denmark, Estonia, Finland, France, Germany, Greece, Hungary, Iceland, Ireland, Italy, Japan, Latvia, Liechtenstein, Lithuania, Luxembourg, Monaco, Netherlands, New Zealand, Norway, Poland, Portugal, Romania, Russian Federation, Slovakia, Slovenia, Spain, Sweden, Switzerland, Ukraine, United Kingdom, United States; *2006:* Belarus

• Background

The Kyoto Protocol to the Framework Convention on Climate Change, whose goal is to reduce the emissions of greenhouse gases (GHGs) that lead to global warming, was adopted in December, 1997, after nearly thirty months of international negotiations. The text of the protocol included twenty-eight articles; Annex A, listing the six GHGs covered by the treaty and sources of gas emissions; and Annex B, listing countries participating in the emissions credit trading agreement.

Several of the specific requirements of Annex B were crafted by and adopted under pressure from the United States. The emissions credit trading system closely follows the system adopted in the United States under the 1970 Clean Air Act Extension, and emissions from American military marine and aviation operations are not subject to regulation. The United States, however, has not ratified the Kyoto Protocol.

• Summary of Provisions

The Kyoto Protocol regulated the emissions of six specific GHGs. Parties to Annex B agreed to reduce their emissions of these gases between 2008 and 2012. Countries agreed to different reduction amounts, based on factors including the types and quantities of energy they produced and their relative levels of pollution. European Union countries, for example, were required to reduce their emissions to 8 percent less than 1990 levels, and the United States had a reduction target of 7 percent below those levels. Iceland agreed to increase its emissions by no more than 10 percent over 1990 levels. While each country had a different target, the target for the Annex B nations as a whole was 5 percent below 1990 emission levels.

Annex B created a system under which countries that could not immediately meet their emissions goals could purchase permits or credits from other Annex B nations that did not need them. For example, if a country exceeded its target methane emissions, it could purchase methane emission permits from a country that had exceeded its reduction goal; the total amount of methane emitted globally would remain the same. Further, a country could earn credits by financing or otherwise supporting a project in another country. Through another provision called the clean development mechanism, an Annex B party could earn credits by helping develop clean energy projects in developing nations not party to Annex B. These provisions created flexibility, as each country could determine whether it was more cost effective to meet its targets through reducing emissions or through purchasing credits.

Countries participating in Annex B must monitor and report their GHG emissions or be banned from project-based credit trading. If it is learned after the fact that a country has failed to comply with these terms, that country will be stripped of any earned credits.

• Significance for Climate Change

Scientists and economists began to question the potential effectiveness of Annex B immediately after its adoption. They wondered, for example, whether the United States would ever be able to meet its target without tremendous economic cost and whether there would be enough credits available for purchase to make the emissions credit trading system work. In fact, the United States did not ratify the Kyoto Protocol and is not participating in international emissions credit trading, thereby reducing any effectiveness the agreement might have had. No country has passed national legislation that requires it to comply with the terms of Annex B.

Scientists including Tom M. L. Wigley, a senior scientist with the National Center for Atmospheric Research, and Gyeong Lyeob Cho of the Korea Energy Economic Institute have generated computer models to predict how effective the provisions of Annex B will be in reducing global warming. Exam-

Annex B Nations

The following table lists the parties to Annex B of the Kyoto Protocol and the emissions target of each. The emissions target refers to the maximum annual average of greenhouse gas (GHG) emissions each nation has committed to producing during the period between 2008 and 2012.

Parties	*Annex B Commitment*[a]
European Union, Bulgaria, Czech Republic, Estonia, Latvia, Liechtenstein, Lithuania, Monaco, Romania, Slovakia, Slovenia, Switzerland	–8
United States[b]	–7
Canada, Hungary, Japan, Poland	–6
Croatia	–5
New Zealand, Russian Federation, Ukraine	0
Norway	+1
Australia	+8
Iceland	+10

a. Percentage by which 2008-2012 average emissions must be below, or may be above, the Party's 1990 baseline emissions.

b. The United States has not ratified the Kyoto Protocol.

ining scenarios under which the Annex B countries continue to reduce their emissions after 2012, when Annex B expires, they have concluded that small increases in reductions after 2012 will have minimal impact on overall warming. Much greater reductions than those called for under Annex B will be necessary to affect climate change, according to these models.

Cynthia A. Bily

- **Further Reading**

Douma, Wybe, and L. Massai. *The Kyoto Protocol and Beyond: Legal and Policy Challenges of Climate Change.* West Nyack, N.Y.: Cambridge University Press, 2007. Collects presentations from two international conferences; looks at the "flexible mechanisms" provisions of the Kyoto Protocol and how these provisions should be implemented and adapted after 2012.

McKibbin, Warwick J., and Peter J. Wilcoxen. *Climate Change Policy After Kyoto: A Blueprint for a Realistic Approach.* Washington, D.C.: Brookings Institution, 2002. The authors, both economists, argue that the international negotiated approach of the Kyoto Protocol is unrealistic and will not reduce carbon emissions.

Wigley, Tom M. L. "The Climate Change Commitment." *Science* 307 (2005): 1766-1769. Argues that anthropogenic global warming is a reality and that stopping further warming will require reductions of GHGs well beyond the levels required under existing agreements.

See also: Clean Air Acts, U.S.; Emissions standards; International agreements and cooperation; Kyoto Protocol; United Nations Framework Convention on Climate Change; U.S. energy policy.

Antarctic Treaty

- **Categories:** Laws, treaties, and protocols; Arctic and Antarctic
- **Date:** Opened for signature December 1, 1959; entered into force June 23, 1961

The Antarctic Treaty depoliticized Antarctica and created free access for research scientists, protecting the continent and leading to research that improved understanding of climate change.

- **Participating nations:** *1959:* Argentina, Australia, Belgium, Chile, the French Republic, Japan, New Zealand, Norway, the Union of South Africa, Soviet Union (now Russia), United Kingdom, United States; *2000:* Brazil, Bulgaria, China, Ecuador, Finland, Germany, India, Italy, Netherlands, Peru, Poland, South Korea, Spain, Sweden, Ukraine, Uruguay; *Observing nations:* Austria, Belarus, Canada, Colombia, Cuba, Czech Republic, Denmark, Estonia, Greece, Guatemala, Hungary, North Korea, Papua New Guinea, Romania, Slovak Republic, Switzerland, Turkey, Venezuela

- **Background**

Antarctica, the land and ice that surround the South Pole, is the only continent with no indigenous human population. Exceeding 12 million square kilometers, it is nearly 1.5 times larger than the continental United States. An ice layer averaging more than 1.6 kilometers in thickness covers approximately 95 percent of the continent's land area. Although few terrestrial species are found on the continent, the surrounding waters are rich in marine life, supporting large populations of marine mammals, birds, fish, and smaller creatures, some found nowhere else on Earth. Antarctica and its surrounding waters play a key but not yet fully understood role in the planet's weather and climate cycles.

In the early twentieth century, seven nations asserted territorial claims on Antarctica; these claims persisted unresolved for decades. International scientific cooperation among twelve countries during the 1957-1958 International Geophysical Year (IGY) led to the establishment of sixty research stations on the continent. As the IGY drew to a close, the scientific community argued that Antarctica should remain open for continuing scientific investigation, unfettered by territorial rivalries. This led to the negotiation of the Antarctic Treaty, which entered into force in 1961.

• Summary of Provisions

The signatory nations have agreed that "it is in the interest of all mankind that Antarctica shall continue forever to be used exclusively for peaceful purposes and shall not become the scene or object of international discord." Initially, twelve nations signed and became consultative parties; sixteen additional nations were later granted consultative party status, and eighteen more acceded to the treaty as participant-observers only.

The treaty provides that the continent of Antarctica will include all land and ice shelves south of 60° 00′ south. It prohibits military activity on the continent and promotes scientific cooperation among the parties. Antarctica is to be open to scientific investigation and cooperation, as well as free exchange of information and personnel. Signatory nations agreed to freeze existing territorial claims and make no new ones. Nuclear explosions and disposal of radioactive wastes are prohibited. Treaty-state observers are to have free access, including aerial observation, to any area and may inspect all stations, installations, and equipment. Members are to discourage activities by any country in Antarctica that are contrary to the treaty, and disputes will be settled peacefully among the parties or by the Inter-

The Antarctic Treaty

Articles I to III and V of the Antarctic Treaty lay out the major goals and objectives of the agreement and the positive obligations of its signatories.

Article I

1. Antarctica shall be used for peaceful purposes only. There shall be prohibited, inter alia, any measure of a military nature, such as the establishment of military bases and fortifications, the carrying out of military manoeuvres, as well as the testing of any type of weapon.

2. The present Treaty shall not prevent the use of military personnel or equipment for scientific research or for any other peaceful purpose.

Article II

Freedom of scientific investigation in Antarctica and cooperation toward that end, as applied during the International Geophysical Year, shall continue, subject to the provisions of the present Treaty.

Article III

1. In order to promote international cooperation in scientific investigation in Antarctica, as provided for in Article II of the present Treaty, the Contracting Parties agree that, to the greatest extent feasible and practicable:

a. information regarding plans for scientific programs in Antarctica shall be exchanged to permit maximum economy of and efficiency of operations;

b. scientific personnel shall be exchanged in Antarctica between expeditions and stations;

c. scientific observations and results from Antarctica shall be exchanged and made freely available.

2. In implementing this Article, every encouragement shall be given to the establishment of cooperative working relations with those Specialized Agencies of the United Nations and other technical organizations having a scientific or technical interest in Antarctica. . . .

Article V

1. Any nuclear explosions in Antarctica and the disposal there of radioactive waste material shall be prohibited.

2. In the event of the conclusion of international agreements concerning the use of nuclear energy, including nuclear explosions and the disposal of radioactive waste material, to which all of the Contracting Parties whose representatives are entitled to participate in the meetings provided for under Article IX are parties, the rules established under such agreements shall apply in Antarctica.

national Court of Justice. Finally, provisions were made to amend the treaty, leading to subsequent agreements and treaties.

- **Significance for Climate Change**

The impact of this agreement on issues of global warming and climate change has been twofold: First, Article 7 of the treaty made possible free access to the continent by climatologists, atmospheric scientists, biologists, geologists, oceanographers, and other researchers from around the world, and their investigations have revealed much about the impact of rising global temperatures on Antarctic ice shelves, sea-level change, endangered or threatened species, and other changes relating to climate.

Second, consultative meetings, held every two years until 1993 and annually thereafer, have resulted in additional agreements addressing environmental issues, such as species loss and tourism, as well as economic issues. For example, the 1982 Convention on the Conservation of Antarctic Marine Living Resources (CCAMLR), which addresses fishery management, used ecosystem criteria, rather than political boundaries, to define the applicable territory. The Antarctic Treaty Consultative Meeting (ATCM) is now held annually. During each ATCM, there is also a meeting of the Committee of Environmental Protection (CEP). The Scientific Committee on Antarctic Research (SCAR) is an observer at ATCMs and CEPs and provides independent scientific advice as requested in a variety of fields, particularly on environmental and conservation matters.

In 1991, the historic Antarctic Environmental Protocol was adopted. It banned mineral and oil exploration for a minimum of fifty years. Annexes to the protocol contain legally binding provisions regarding environmental assessments, protection of indigenous plants and animals, waste disposal, marine pollution, and designation of protected areas. The protocol entered into force in January, 1998, after ratification by all consultative parties.

Phillip Greenberg, updated by Christina J. Moose

- **Further Reading**

Bastmeijer, Kees. *The Antarctic Environmental Protocol and Its Domestic Legal Implementation.* New York: Kluwer Law International, 2003. Collection of essays providing an overview of the Antarctic Treaty System, including implementation and enforcement. Illustrated, bibliographic references, index.

Joyner, Christopher C. *Governing the Frozen Commons: The Antarctic Regime and Environmental Protection.* Columbia: University of South Carolina Press, 1998. Addresses governance structures, regime dynamics, resource conservation and management, environmental protection and preservation, and science and tourism. Illustrations, maps, references, index.

Stokke, Olav Schram, and Davor Vidas, eds. *Governing the Antarctic: The Effectiveness and Legitimacy of the Antarctic Treaty System.* New York: Cambridge University Press, 1997. Essays contributed by political scientists and international lawyers consider the Antarctic Treaty System's impact on the environment, tourism, fisheries, and mineral activities.

Triggs, Gillian, and Anna Riddell, eds. *Antarctica: Legal and Environmental Challenges for the Future.* London: British Institute of International and Comparative Law, 2007. Papers presented at a conference organized by the British Institute of International and Comparative Law and the Foreign and Commonwealth Office in June, 2006. Bibliographical references.

Watts, Arthur. *International Law and the Antarctic Treaty System.* New York: Cambridge University Press, 1993. Surveys the legal framework for Antarctic activities by examining the 1959 treaty and subsequent additions.

See also: Antarctica: threats and responses; Ice shelves; International agreements and cooperation; International waters; Ocean life; Penguins.

Antarctica: threats and responses

- **Category:** Arctic and Antarctic

Antarctica contains 90 percent of the world's ice and 75 percent of its freshwater. If a rise in the planet's temperature were to cause this store of ice to melt completely, it would result in the world's sea levels rising by approximately 60 meters.

• Key concepts

Antarctic peninsula: a peninsula stretching northward toward South America that contains about 10 percent of the ice of Antarctica

East Antarctic ice sheet: ice sheet located east of the Transantarctic Mountains that stores over 60 percent of the world's total freshwater

glacier: a mass of ice that flows downhill, usually within the confines of a former stream valley

ice sheet: a mass of ice covering a large area of land

ice shelf: a platform of freshwater ice floating over the ocean

mass balance: the difference between the accumulation of snow and the ablation of ice on a given glacial formation

sea ice: frozen ocean water

West Antarctic ice sheet: the smallest ice sheet in Antarctica, located west of the Transantarctic Mountains

• Background

Antarctica, located around the South Pole, is the world's fifth largest continent, with a surface area of 12.4 million square kilometers. Approximately 5,500 kilometers wide at its broadest point, it is surrounded by the southern portions of the Atlantic, Pacific, and Indian Oceans. This immense landmass is covered with an ice sheet larger than the continent itself. At its maximum winter extent, during the month of July, the ice sheet measures about 14 million square kilometers and contains 30 million cubic kilometers of ice.

• The Antarctic Ice Sheet: Unevenly Divided

More than 98 percent of Antarctica is covered with ice of an average thickness of about 2,100 meters. Most of the 2 percent of the continent not covered by ice is in the Antarctic Peninsula. The Antarctic ice sheet reaches a thickness of almost 5,000 meters at its highest point. From a geologic standpoint, Antarctica is made up of two structural provinces: East Antarctica and West Antarctica. East Antarctica is a stable shield separated from the much younger Mesozoic and Cenozoic belt of West Antarctica. The contact zone between these two provinces is the Transantarctic Mountains and the depression separating the Ross Sea and the Weddell Sea.

The Transantarctic Mountains divide the ice-covered continent into two ice sheets, the largest masses of ice known on Earth. The East Antarctic ice sheet, which is mostly situated in the Eastern Hemisphere, comprises 90 percent of the Antarctic ice. It is surrounded by the southern Atlantic Ocean, the Indian Ocean, and the Ross Sea. The South Pole is located in the East Antarctic ice sheet. The East Antarctic continental landmass on which it rests is close to sea level. Besides its ice sheets, Antarctica has many ice streams, glaciers, and ice shelves. While the East Antarctic ice sheet is dome-shaped, the West Antarctic ice sheet is more elongated along the mountains in the center of the peninsula.

As its name implies, the West Antarctic ice sheet is located in the Western Hemisphere. The northernmost part of the West Antarctic ice sheet protrudes in a peninsula that ends beyond the Antarctic Circle, south of South America. The backbone of the peninsula is composed of high mountains, an extension of the Andes Mountain Range, reaching about 2,800 meters. The northernmost latitude of the peninsula is 63°13′ south. The largest part of the West Antarctic ice sheet along the Amundsen Sea flows into the Ross Ice Shelf, a platform of floating ice on which the American research station McMurdo is located.

The peninsula glaciers drain into the Weddell and Bellingshausen seas and the Ronne and Filchner ice shelves. Unlike the East Antarctic ice sheet, this ice sheet sits on a continental platform that in some places is 2,500 meters below sea level. It is therefore more influenced by changes in ocean temperatures than is the East Antarctic ice sheet. The West Antarctic ice sheet experiences warmer temperatures than the East Antarctic ice sheet, both because it extends at lower latitude and because it has a lower average elevation.

• Measuring Temperature Change in Antarctica

Because of its high latitude and high elevation, Antarctica is the coldest continent on Earth. The ice that covers Antarctica results from the transforma-

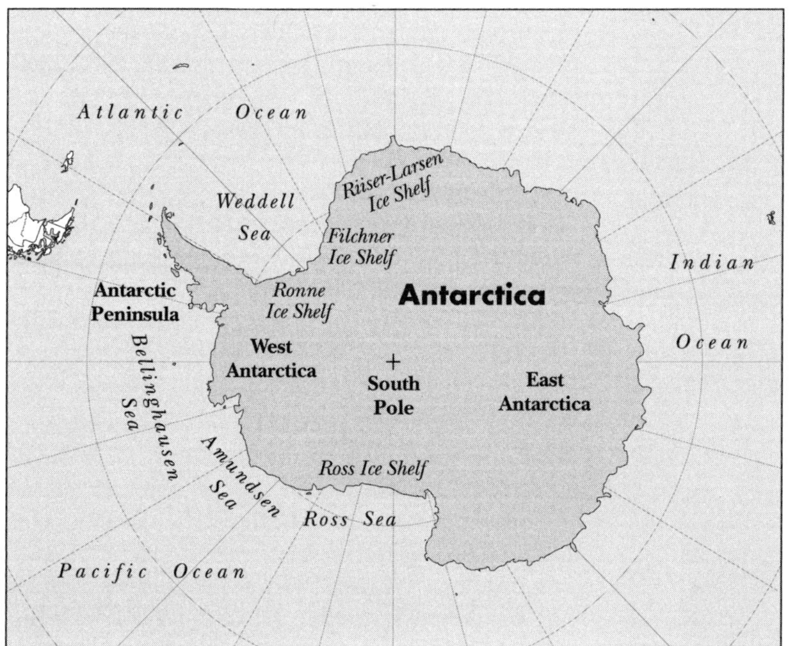

tion of snow into ice. The amount of precipitation that Antarctica experiences is not uniform over the entire continent; the coasts, with lower elevation and higher temperatures, record about six times more annual snow accumulation than does the much higher and colder interior, which receives less than 3 centimeters of water-equivalent precipitation annually. The lowest temperature ever recorded on Earth (–89.6° Celsius) occurred at Vostok, a Russian research station, where only 166 millimeters of precipitation is received on average per year.

As Antarctic snow layers are progressively transformed into ice, they preserve evidence of the temperature at the time the snow fell in the form of isotope ratios within the ice. As a result, ice cores may be drilled from the Antarctic ice and examined to obtain a chronology of Antarctic temperatures. Antarctic weather stations, moreover, have been recording temperature and measuring precipitation for about 150 years, albeit not in a continuous manner. During and after the International Geophysical Year (1957-1958), weather stations were systematically installed at the forty-eight bases created in Antarctica by twelve countries. One of the most challenging tasks that these stations have faced has been the physical maintenance of the devices measuring such a harsh environment. Anemometers, which measure wind speed, are particularly vulnerable to the ferocious katabatic winds that sweep the continent, sublimating (reducing through evaporation) the surface ice as well as damaging the devices that measure them.

• Effects of Global Warming in Antarctica

Antarctica plays an important role in assessing climatic changes. Its ice reveals the variation of temperature of the continent over 800,000 years. It is also the perfect laboratory for studying the effects of human activities on Earth's atmosphere. Considered hostile to humans and unexplored until the beginning of the twentieth century, Antarctica came into the spotlight when the ozone hole over it was discovered and when the world's longest ice core was retrieved at Vostok. The ozone hole threatens the planet by allowing short-wavelength ultraviolet radiation to penetrate the lower atmosphere.

Scientists are concerned over the potentially calamitous effect of the melting of Antarctic ice. In 2006, for example, Eric Rignot, a French glaciologist working at the University of California, computed that Antarctica had lost 178 billion metric tons of ice, mostly from the Antarctic Peninsula. This would result in a rise in sea level of about 0.5 millimeter. Though the amount is relatively small, the trend it indicated was troubling; the loss had increased from the 102 billion metric tons recorded in 1996. From 1996 to 2006, some glaciers in the west began moving more rapidly toward the sea and thus produced more icebergs, as their terminuses have collapsed into the sea.

• Measuring Ice Losses

To understand and compute the mass balance of the ice in Antarctica, one must determine whether the continent loses or gains mass. Antarctica gains

An iceberg is calved from the Antarctic ice sheet. (Reuters/Landov)

mass when snow falls. The entire continent, which is about 1.5 times the size of the United States, contains only about one hundred weather stations, so it is not easy to estimate with a great degree of certainty how much snow accumulates on it. The error margin is likely to be high. Melting is rare in Antarctica, because the temperature tends to remain below the freezing point year-round. In 2008, Rignot estimated that 99 percent of the ice lost in Antarctica forms icebergs. However, the total picture of ice loss is not completely uniform; while Antarctica is losing ice at a greater rate and the area covered by sea ice in the Arctic has steadily decreased, sea ice in Antarctica has slightly increased, particularly along the East Antarctic ice sheet's coast.

• Context

The Intergovernmental Panel on Climate Change, which was awarded the 2007 Nobel Peace Prize for its intensive research on the projected effects of climate change, has stated that the mass balance of the Antarctic ice sheet that could contribute to sea level was 0.21 ±0.35 millimeter per year. From the computation above, there is a possibility that the Antarctic ice sheet could be responsible for a decrease of sea level by 0.14 millimeter per year.

Scientists agree that the northern part of the West Antarctic ice sheet has been melting, but there have been conflicting assessments of the behavior of the East Antarctic ice sheet. Measurements taken by the weather stations in the interior of the East Antarctic ice sheet show a slight decrease in temperature and an increase in precipitation. Other temperature assessments have been performed using satellite information based on the amount of infrared light reflected by the snow covering the continent. These show a slight increase in temperature in the coastal margins of the continent and an increase in temperature on the West Antarctic ice sheet. In the January 22, 2009, issue of *Nature*, Eric Steig, a geochemist and glaciologist at the University of Washington, reported that his study of surface temperatures collected from 1957 until 2007, combined with satellite information, indicates that, even though some portions of the continent have been cooling for a long time, Antarctica as a whole has warmed by 0.5° Celsius.

Knowledge of the processes by which the great ice masses of Antarctica grow and shrink is not yet perfect. Improving that knowledge will be of increasing importance, and understanding the climatological mechanisms at work on the continent will be crucial in assessing the impact of human activity on the health and stability of the Earth.

Denyse Lemaire and David Kasserman

• Further Reading

Alley, Richard B. *The Two-Mile Time Machine: Ice Cores, Abrupt Climate Change, and Our Future.* Princeton, N.J.: Princeton University Press, 2000. Provides ice-core research information and contextualizes the work of geoscientists in relation to the cryosphere. Concise and clear, with superb artwork and photographs.

Copeland, Sebastian. *The Global Warning.* San Rafael, Calif.: Earth Aware Editions, 2007. This abundantly illustrated book about Antarctica tends to focus on the potential calamitous effects of global climate change using clear and simple language.

McGonigal, David. *Antarctica: Secrets of the Southern Continent.* Richmond Hill, Ont.: Firefly Books, 2008. Describes the discoveries of the International Polar Year, 2007-2008, in various fields, including geology, geography, and climatology. Discusses potential effects of global warming.

Trewby, Mary. *Antarctica: An Encyclopedia from Abbott Ice Shelf to Zooplankton.* Toronto: Firefly Books, 2002. General encyclopedia of Antarctica that explains all major features of the continent.

Turney, Chris. *Ice, Mud, and Blood: Lessons from Climates Past.* New York: Macmillan, 2008. Examines various discoveries derived from ice cores and how these discoveries led to a better understanding of paleoclimates.

See also: Antarctic Treaty; Arctic; Cryosphere; Glaciers; Greenland ice cap; Ice cores; Ice shelves; Mass balance; Sea ice.

Anthropogenic climate change

• **Categories:** Economics, industries, and products; environmentalism, conservation, and ecosystems; energy

Since the Industrial Revolution, two types of human activity have contributed significantly to changes in Earth's climate: modifications of the planet's surface and the invention and deployment of new energy technologies.

• Key concepts

aerosols: tiny particles suspended in Earth's atmosphere

albedo: the fraction of incident light reflected from a body such as Earth

anthropogenic: deriving from human sources or activities

fossil fuels: energy sources such as coal, oil, and natural gas that were formed by the chemical alteration of plant and animal matter under geologic pressure over long periods of time

global dimming: a reduction in the amount of sunlight reaching the surface of the Earth

greenhouse gases (GHGs): atmospheric gases that trap heat within a planetary system rather than allowing it to escape into space

urban heat island: a spot on Earth's surface that is significantly warmer than the surrounding area as a result of human alterations to the landscape

• Background

Large-scale human impact on the Earth extends back many centuries. However, at the beginning of the Industrial Revolution, in the mid-eighteenth century, the pace of human effects on the natural environment increased. The invention of power sources such as the steam engine, the internal combustion engine, and systems for delivering electric power sped anthropogenic environmental alterations in three ways. First, these technologies made it easier to modify the landscape; second, they created a vast demand for energy to fuel them; and finally, emissions from steam and internal combustion engines altered the composition of the atmosphere.

• Human Capacity to Affect Climate Change

By far the largest source of energy available to Earth is the Sun. Solar energy is the ultimate source of most other energy on the Earth's surface, including wind, hydroelectric, and biomass energy. After solar energy, the most important energy source in the Earth is geothermal energy, generated mostly by the decay of radioactive elements within the Earth. Tidal energy is also generated from interactions between the Earth, Moon, and Sun. The amount of solar energy available on Earth is about twenty-seven hundred times the amount of energy provided by geothermal heat and twenty-nine thousand times the amount of tidal energy.

Humans can affect climate by modifying the Earth's ability to absorb or reflect heat from the Sun and by modifying the atmosphere's ability to retain heat. Compared to those effects, the direct production of heat by human activities is insignificant. The annual supply of solar energy is more than eight thousand times the total amount of human energy use. In other words, human energy use in one year is equal to roughly one hour of global sunlight. By comparison, the total energy in the world's nuclear arsenals is about one-ten thousandth of the annual input from the Sun, or less than an hour's worth of sunlight. A large-scale nuclear war would inject smoke and dust into the atmosphere that would have a far greater effect on climate than would the heat given off by the nuclear explosions themselves. During the 1991 Persian Gulf War, the smoke emitted from burning oil wells in Kuwait had a strong cooling effect on areas under the smoke layer, but the heat from the oil fires had negligible effects on local or regional weather.

• How Human Changes to Earth's Surface Affect Climate

When humans modify the landscape, or surface, of Earth, they trigger climate change in a number of ways. First, by replacing forest with cleared land for use as farmland or pasture, or simply by felling timber for wood or fuel, they increase the albedo of Earth's surface: More sunlight is reflected back into space and less is absorbed and retained by the Earth. Old forests, especially conifer forests, are generally dark, whereas younger growth or cleared land is much lighter. Generally, making the surface lighter reflects more solar energy back into space and contributes to climatic cooling, and making the surface darker contributes to warming.

Many human changes to the landscape affect local climate. One of the most familiar examples is the urban-heat-island effect. Cities impede the flow of air, trapping heat, and they also contain large areas of materials such as asphalt and concrete that absorb heat. As a result, cities tend to be significantly warmer than nearby countryside. Other effects associated with urban heat islands include increased rainfall and stalling of intense rainstorms.

Among the most extreme climatic effects associated with human changes to Earth's surface are those due to the drying of the Aral Sea in central Asia. Since 1960, diversion of water for agriculture has reduced the area of the Aral Sea (actually a vast lake) by more than 80 percent. The original size of the Aral Sea was about 68,000 square kilometers, making it the fourth largest lake in the world—large enough to have significant moderating effects on the regional climate. These effects have almost entirely disappeared with the drying of the lake, and the climate has become more continental, with hotter summers, colder winters, and much less rainfall. Similarly, increased drought has been linked to the vast Three Gorges Dam hydroelectric project in China, which, according to Wang Hongqi, a

Global Anthropogenic GHG Emissions, 2004

Gas	*Percent of Global Emissions*
Carbon dioxide (burning fossil fuels)	57
Carbon dioxide (deforestation and biomass decay)	17
Methane	14
Nitrous oxide	8
Carbon dioxide (other sources)	3
F-gases	1
Total	100

Data from Intergovernmental Panel on Climate Change.

Beijing-based atmospheric physicist, "has artificially altered the natural terrain, elongated the vent channel, and disturbed water vapor circulation, resulting in an imbalance of temperature" and leading to drought.

• Indirect Human Modifications of the Atmosphere

Human changes to the landscape also affect the atmosphere. Human activities may create dust and smoke, for example, which can reflect sunlight back into space or block sunlight from reaching Earth's surface. Such activities thus increase Earth's albedo and cool the planet, but they can also prevent radiation from being reflected back into space from the surface and thus trap heat, warming the planet. For example, studies of condensation trails, or contrails, of aircraft have shown that they reflect sunlight back to space but also prevent heat from the surface from escaping, so that their overall effect is to warm the Earth. Tiny particles, or aerosols, can also serve as nuclei for the condensation of water droplets and affect fog, cloud cover, or precipitation.

The degree to which human dust and smoke alter visibility is remarkable. In preindustrial times, it was normal in most places for visibility to exceed 100 kilometers. The Great Smoky Mountains were so named precisely because the persistent haze in the valleys, due to natural emissions by the forests, was unusual. Persistent haze unrelated to local weather was so unusual in preindustrial times that it was recorded by chroniclers and has been used by geologists to pinpoint the dates of large volcanic eruptions in remote areas of the world. Visibility in heavily populated contemporary industrial regions is often only a few kilometers, and even remote national parks in the western United States are threatened with diminished visibility. Some studies have suggested that global dimming, the reduction of sunlight reaching the surface of the Earth by dust, haze, and smoke, may have masked the effects of global warming.

Human changes to the landscape often release greenhouse gases (GHGs). These changes generally result in the destruction of biomass, either by burning or by decay, thereby adding carbon dioxide (CO_2), one of the six major GHGs, to the atmosphere. Drainage of wetlands for agriculture can also result in the decomposition of organic material and also adds CO_2 to the atmosphere.

Human activities release other GHGs, particularly methane. Modification of the land can release methane trapped in the soil. Agriculture increases the amount of methane in the atmosphere in several ways. Livestock produce large amounts of methane in their digestive tracts, so increased numbers of cattle lead to increased methane emissions. Clearing of forest lands for agriculture reduces the ability of soils to absorb and oxidize methane. Certain types of agriculture, notably rice production, create oxygen-poor conditions for the decay of organic materials and thus emit methane. Finally, burial of waste rather than incineration results in methane emission.

According to a controversial theory by William Ruddiman, if it were not for human activities, the Earth would already have passed the peak of the present interglacial period and would be on the way to the start of the next glacial advance. Ruddiman argues that, while clearing forests for agriculture increased the CO_2 content of the atmosphere, increased methane production—especially that due to rice cultivation—is the more important climatic change agent.

• Direct Human Changes to the Atmosphere

Beginning with the Industrial Revolution, human activities began modifying the composition of the atmosphere directly on a large scale. The burning of fuels—first wood and then fossil fuels—released increasing amounts of smoke and gases directly into the atmosphere. Among the most important emissions were CO_2, nitrogen oxides, sulfur dioxide, and ozone-depleting chemicals.

CO_2 is the most important anthropogenic GHG contributing to the enhanced greenhouse effect. The latter effect supplements Earth's already significant natural greenhouse effect, in the context of which water vapor is the single most important GHG in Earth's atmosphere.

Nitrogen oxides are the result of high-temperature combustion during which atmospheric nitrogen and oxygen combine. Ordinary fires are not hot enough to cause reactions between nitrogen and oxygen, but at temperatures above 1,600° Celsius the two gases can react. Lightning is a natural source of nitrogen oxides, but human activities also create large amounts, especially in internal combustion engines. Nitrogen oxides react with hydrocarbons to cause smog as well as ozone in the lower atmosphere. Ozone high in the atmosphere protects Earth's surface from ultraviolet light, but at ground level ozone is a pollutant that contributes to respiratory problems. Finally, nitrogen oxides combine with water vapor to form nitric acid and contribute to acid precipitation.

Sulfur dioxide is emitted naturally by volcanoes but also is produced by smelting of sulfide ores or burning fossil fuels that contain sulfur. Sulfur dioxide is a GHG, but its most important environmental effect is that it combines with water vapor to create sulfuric acid. Tiny droplets or aerosols of sulfuric acid can aggravate respiratory problems, contribute to atmospheric haze, and make rain and snow more acidic, contributing to acid precipitation.

Ozone-depleting chemicals include a large number of synthetic chemicals, mostly organic chemicals containing chlorine or bromine. Both of these elements are highly effective at destroying ozone, and ozone-depleting chemicals are extremely stable, enabling them to survive long enough to reach high altitudes. International controls on ozone-depleting chemicals, such as the Montreal Protocol and the Kyoto Protocol, have slowed the depletion of stratospheric ozone, but the existing chemicals in the atmosphere will continue to have an effect for a long time. Although ozone-depleting chemicals also act as GHGs, in most respects global warming and ozone depletion are separate problems.

• Context

Although the amount of CO_2 and other GHGs in the atmosphere has increased and strong evidence exists that the global climate has become warmer in the recent past, the actual web of cause and effect relating to climate change is extremely complex, and many unanswered questions remain. The consequences of making a wrong decision about climate change could be serious. If human activities are responsible for climate change, then failure to act could lead to catastrophic environmental, social, and economic changes. If human activities contribute only insignificantly to climate change, then attempting to halt climate change through government policies could have catastrophic economic effects without ameliorating climate change. The enormous stakes and the sheer complexity of climate are the reasons that the debate about global warming is so fierce.

Steven I. Dutch

• Further Reading

Broecker, Wallace S., and Robert Kunzig. *Fixing Climate: What Past Climate Changes Reveal About the Current Threat—and How to Counter It.* New York: Hill and Wang, 2008. Broecker, one of the world's most experienced climate researchers, summarizes the evidence for climate change. Among his observations: "If you're living with an angry beast, you shouldn't poke it." Illustrations, references, index.

Fagan, Brian. *The Great Warming: Climate Change and the Rise and Fall of Civilizations.* New York: Bloomsbury, 2008. Writing from an anthropologist's perspective, Fagan examines the impact of the Medieval Warm Period on regional climates and their civilizations. Though weakened by deriving its temperature graphs from the disputed hockey stick graph, this study is useful for its descriptions of the potential impact of climate change—including intense and abrupt global warming—on human civilizations. Illustrations, maps.

Leroux, Marcel. *Global Warming, Myth or Reality? The Erring Ways of Climatology.* New York: Springer, 2005. French climatologist Leroux questions global warming and anthropogenic contributions by noting the limitations of climate models' abilities to predict the impact of the greenhouse effect and postulating alternate causes of climate change, suggesting that there may be some benefits to climate change. Figures, bibliography, index.

Ruddiman, William F. *Plows, Plagues, and Petroleum: How Humans Took Control of Climate.* Princeton, N.J.: Princeton University Press, 2005. Summarizes human impacts on climate and presents the case that agriculture delayed the onset of the next glaciation. Illustrations, figures, tables, maps, bibliography, index.

Sagan, Carl, Owen B. Toon, and James B. Pollack.

"Anthropogenic Albedo Changes and the Earth's Climate." *Science* 206, no. 4425 (December 21, 1979): 1363-1368. Sagan is best known as an astronomer, popularizer of science, and spokesperson for the nuclear winter hypothesis, but this article is a pioneering study of how human activities can affect global climate.

See also: Aerosols; Air pollution and pollutants: anthropogenic; Air pollution and pollutants: natural; Climate and the climate system; Climate change; Deforestation; Enhanced greenhouse effect; Greenhouse effect; Greenhouse gases.

Anthropogeomorphology

• **Categories:** Economics, industries, and products; environmentalism, conservation, and ecosystems

Anthropogeomorphology, the alteration of the Earth's surface by human activity, has profound effects on the environment, including land, air, and water, and has the potential to alter local climates as well.

• Key concepts

additive processes: processes that enhance surface topography

alluviation: soil erosion that results in the filling of watercourses or harbors

dew point: the temperature at which water vapor condenses into liquid water

orographic precipitation: precipitation caused by changes in topography that drive air higher, where it cools and condenses

polders: Dutch land areas reclaimed from sea or marsh

subtractive processes: processes that reduce surface topography

• Background

Human civilizations alter natural landscapes, both land and water. They partially remove or scrape away natural landforms (subtractive processes), and they build up new topographic features (additive processes). Humans have long reshaped existing bodies of water, created new bodies of water, redirected water in canals or aqueducts, changed the paths of existing rivers, scraped away surfaces, and reclaimed land from seas or marshes behind dikes and seawalls. There is some evidence that these modifications of the landscape affect local climates.

Anthropogenic changes in the landscape easily surpass the scope of natural erosional processes globally. Overconsumption of water resources in such places as the Jordan Valley in the Levant causes bodies of water such as the Sea of Galilee and the Dead Sea to shrink. It also increases desertification by changing the watershed and reducing evaporation with increased land temperature. Replanting trees in arid zones can mitigate surface temperatures and increase orographic precipitation at lower elevations, where evaporation is reduced with dew point and condensation is reached more easily. This has occurred, for example, in Israel's Judaean Hills, where rainfall increased dramatically over thirty documented years after reforestation.

• Anthropogeomorphology and Climate Change

One anthropogenic subtractive process has been documented for millennia. Plato noted that severe soil erosion in fourth century B.C.E. Greece was causing much lost topsoil. Ancient observers were sometimes able to determine causes of topsoil loss, such as aggressive deforestation to clear farmland or destruction of native plant roots by animal grazing. Sheep and goat herds were particularly destructive, because they consumed root systems as well as surface plants, effectively killing the plant cover.

Once there were no longer roots to hold the soil, erosion often removed devastating volumes of surface soil. As soil cover moved downward, the resulting alluviation filled river bottoms, lakes, and harbors, which were eventually silted up, changing surface landscapes. Coastlines also changed when alluviated river deltas encroached into bodies of water, as, for example, the Rhone River encroached into the Mediterranean Sea. Erosion is normally a subtle, gradual process, but it can be accelerated by extreme deforestation through human agency. Erosion was exacerbated in the late Roman Empire and afterward in North Africa, where Atlas Mountain deforestation by humans resulted in alluviated coastal watersheds and silted up harbors. This erosion combined with rising land temperatures when rainfall ceased, partly as a result of deforestation, to render the climate much less hospitable. Ultimately, once great North African cities such as Sabratha and Leptis Magna were abandoned as a result of these climate changes.

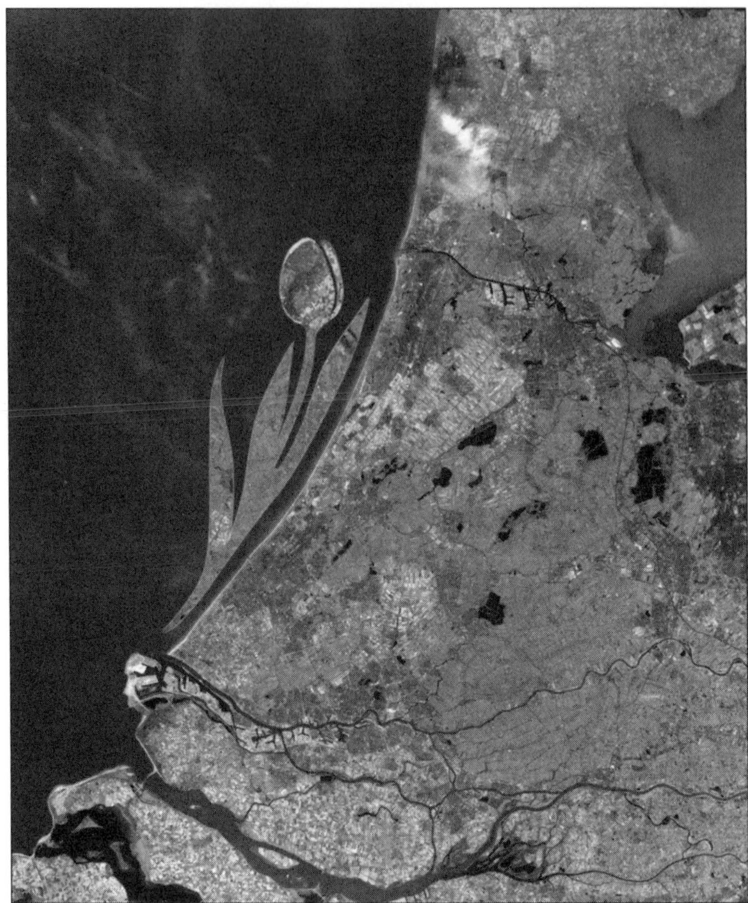

Artist's rendering of a proposed artificial island off the coast of the Netherlands. Tulip Island would increase the nation's living space and also protect the coastline from rising seas. (Reuters/Landov)

• Alteration of Seacoasts

Another example of ancient anthropogenic land change along seacoasts occurred at Tyre (now in Lebanon), beginning in 332 B.C.E., when Alexander the Great built a stone causeway out to the then-island city in a siege. Over millennia, the causeway trapped enough marine-transported sandy alluvium that the land bridge—originally around 7 meters across—widened into today's peninsula, which is so much broader that a casual observer would not guess that there was once an open water channel of almost 400 meters between the mainland and island. Prevailing currents from the north built up far more seaborne sandy alluvium on the curved northern side of the peninsula, whereas the southern side of the artificial peninsula remains more contiguous to the original causeway. The volume of sand and eventual structures added over time now approximate about 200 hectares and millions of metric tons of alluvium. This land extension has changed local water and air circulation patterns along and over the coast of Lebanon.

• Dams and Reservoirs

Humans have also created many artificial, interior bodies of water, such as reservoirs and artificial lakes. Water storage in artificial lakes fills millions of hectares of land surface on every continent, with concomitant climatic impacts ranging from temperature changes to increased evaporation. Additional anthropogenic aquatic change includes the construction of major canals, such as the Suez Canal linking the Mediterranean and Red Seas and the Panama Canal linking the Pacific Ocean and the Caribbean Sea. Construction of such canals includes the creation or exploitation of connecting lakes and locks to accommodate sea-level differences.

One dramatic human engineering project, China's Three Gorges Dam, is already threatening to offset potential hydroelectric economic gains. Water has seeped into steep lands along the dam's perimeter, causing more than 35 kilometers of banks to cave in, resulting in more than 20 million cubic meters of rockslide since 2003. In addition, there is mounting evidence than regional rainfall has been decreased, leading to drought and loss of biodiversity, while fault activity has increased where the dammed lake sits across two major, active fault lines.

• Land Reclamation

Conversely, the Zuider Zee's extensive dikes in the Netherlands reclaimed from the North Sea millions of hectares of land slightly below or just at sea level. The reclaimed land was used extensively for farming. The major dike (Afsluitdijk) created new land polders that gradually reduced the former early twentieth century Zuider Zee by about 38 percent.

A similar phenomenon exists around urban New Orleans, where former swamps and Mississippi River delta wetlands were drained, and dense human settlements were generally protected by extensive levees. Hurricane Katrina's surge in summer, 2005, however, emphasized the fragility of such land reclamation and urbanization. Levee failure caused devastating flooding of 80 percent of New Orleans at great local, regional, and national cost.

Channeling former coastal rivers into stone or concrete storm drains is a major surface change, mostly aimed at reducing flooding. Precedents for such projects can be found in pre-Columbian Inca Peru, along the Urubamba River in the Yucay Valley. Around 1400, Inca engineers not only took out natural oxbows and straightened the river but also created farmland from the natural floodplain, where there had been no prior extensive agriculture, thus humidifying the air in the Yucay Valley.

• Anthropogeomorphology and the Mining Industry

Perhaps the most extensive modern anthropogeomorphologic surface change in North America has been produced by the Canadian oil sands industry, which engages in open pit mining of the Athabaska River Valley in Alberta. Boreal forest and earth are scraped away to depths of about 30 meters in many places over a 390-square-kilometer area. This mining is creating vast, toxic, mine-tailing sludge lakes and growing pollutant containment problems. Naphthenic acid and polycyclic aromatic hydrocarbons, which are not easily degradable for centuries, leak from the mines into water tables, as well as the Athabaska River.

The oil sands industry extracts enormous quantities of bitumen-laced sands, heats them, and cleans them using hot water and other agents. This process has resulted in a huge spike in aerosol CO_2 emissions over northern Canada, at a much higher rate than the emission rate of conventional oil production. Oil produced in Alberta's oil sands is mostly consumed by the United States; this single source supplies 10 percent of total U.S. foreign oil.

It takes 3.6 metric tons of earth to produce one barrel of oil, and 750,000 barrels of synthetic crude oil are produced per diem in Alberta. This nonstop daily process mixes 907,000 metric tons of crushed oil sands with 181,000 metric tons of water to be heated to boiling temperatures for steam. It requires considerable energy, generated from natural gas, and yields a vast landscape of stored toxic waste. Environmental scientists are beginning to link accelerated Arctic ice-cover loss with the northern Canadian oil sand industry. The industry may raise local temperatures through deforestation, as well as adding steam to the air.

• Context

Anthropogeomorphology as a by-product of human activity is a growing concern, as evidenced by a burgeoning global environmental response. Quantitative analyses of climatic changes resulting from anthropogeomorphology have yet to be produced, but growing attention to the subject will render such analyses of great scientific and political interest. Once the phenomenon is better understood, it may be possible not only to prevent further projects from creating negative climatic effects but also to launch projects to mitigate local and global climate trends.

Patrick Norman Hunt

• Further Reading

Atlas of Israel: Cartography, Physical and Human Geography. New York: Macmillan, 1985. Documents and maps all geophysical data available in Israel for over forty years of quantitative measurements from meteorology, oceanography, demography, and other disciplines.

Hvistendahl, Mara. "China's Three Gorges Dam: An Environmental Catastrophe?" *Scientific American*, March 25, 2008. Assessment of China's initial failure to consider the dam's ecological impacts and recent acknowledgment of its possible consequences for the environment.

Kunzig, Robert. "Scraping Bottom: The Canadian Oil Boom." *National Geographic*, March, 2009. Fair but controversial report on the Canadian oil industry's bringing about both an economic boom and environmental degradation.

LaFreniere, Gilbert. *The Decline of Nature: Environmental History and the Western Worldview.* Bethesda, Md.: Academica Press, 2008. Examines human policies that have affected nature through history, detailing both philosophical attitudes underlying those policies and their possible catastrophic results.

Nikiforuk, Andrew. *Tar Sands: Dirty Oil and the Future of a Continent.* Berkeley, Calif.: Greystone Books, 2008. Examines potential long-term environmental effects of Canada's oil sands industry in light of possible climatic changes and the ethics of fossil fuel consumption.

See also: Climate engineering; Flood barriers, movable; Land use and reclamation; Levees.

Aral Sea desiccation

- **Category:** Water resources

Water availability has fluctuated in the Aral Sea as a result of natural and anthropogenic forces. Anthropogenic effects in the second half of the twentieth century brought about an environmental disaster, as the sea shrank drastically, with deadly effects on agriculture, vegetation, and human and animal populations in the region.

• Key concepts

anthropogenic: caused or produced by humans
hazardous: poisonous, corrosive, flammable, explosive, radioactive, or otherwise dangerous to human health
herbicides: substances or preparations for killing plants, such as weeds
pesticides: chemical preparations that kill pests, including unwanted animals, fungi, and plants
Pleistocene epoch: first half of Quaternary period, beginning about two million years ago and ending about ten thousand years ago
Pliocene epoch: Tertiary period, beginning about ten million years ago and ending about two million years ago, known for its cool climate, mountain building, and increased mammal populations

• Background

The dramatic drying of Central Asia's Aral Sea is sometimes called one of the greatest ecological disasters of the twentieth century. Over the past ten thousand years, the area and volume of this internal lake have greatly fluctuated as a result of both natural and anthropogenic forces. Anthropogenic forces in particular greatly reduced the sea at the end of the twentieth century, from 67,500 square kilometers in 1960 to 17,382 square kilometers in 2006.

The Aral Sea is located in an area of cold temperatures and deserts: the Karakumy is to the south, the Kyzylkum Desert is to the southeast. The sea's 1.8-million-square-kilometer drainage basin encompasses six central Asian countries: Iran, Turkmenistan, Kazakhstan, Afghanistan, Tajikistan, and Uzbekistan (including the Karakalpak Autonomous Republic). Kazakhstan and Uzbekistan are physically adjacent to the Aral Sea. Although nine streams flow within the drainage basin, the Syr Dar'ya and the Amu Dar'ya are the major rivers. *Dar'ya* translates from the Turkic languages of central Asia as "river."

In 1918, the Soviet Union decided to develop the area around the Aral Sea to grow cotton. The decision was economic: Cotton, "white gold," provided revenue for the government. Herbicides, pesticides, and fertilizers were heavily used to bolster crop production, and the Amu and Syr Dar'ya were diverted for irrigation.

The Soviet irrigation programs were inefficient, with open waterways and irrigation basins subject to evaporation. Open-air channels were dug through sandy deserts with the thirteen-hundred-kilometer Karakam Canal diverting between 20 and 30 percent of the Amu Dar'ya's flow west to Turkmenistan.

Between 1987 and 1989, the Aral Sea divided into the small Aral Sea in the north, fed by the Syr Dar'ya, and the large Aral Sea in the south, supplied by the Amu Dar'ya. Using the years 1960 to January, 2006, as a baseline, the water level of the little Aral fell by 13 meters, the large Aral by 23 meters.

• Displaced Fishing Industry

Historically, the Aral Sea fishing industry employed several thousand workers and provided, according to commercial fishing reports, one-sixth of the Soviet fish supply. The lowered lake level reduced the industry and increased the distance between the lake and fishing ports. Decreased water flow to the river deltas and wetlands diminished fish spawning and feeding, so that of the thirty-two fish species formerly existing in the lake, only six survived. Those remaining survived by inhabiting small water areas of river deltas—areas that play a large role in regenerating lake fish supplies. Commercial fisheries did not exist after the mid-1980's.

When lake water is reduced by evaporation and freshwater input is negligible, salts within the water are concentrated and approach the salinity of a typical ocean, 35 grams of salt per liter. Since the Aral Sea was becoming more saline in the 1970's, a saltwater fish, the Black Sea flounder (*Platichthys flesus lulscus*) was introduced into the sea. The intent was to enable the lake's fishing industry to survive, but by 2003 the flounder no longer existed in the Aral Sea, whose salinity had reached greater than 70 grams per liter.

• Hazardous Lakebed Deposits

As the Aral Sea shrinks, calcium sulfate, calcium carbonate, sodium chloride, sodium sulfate, and magnesium chloride are deposited on the exposed seafloor. In addition to these salts, pesticide residues of organochlorines, dicholorodiphenyl-tricholoethanes

Ruined ships lie on sand that was once the bed of the Aral Sea. (Shamil Zhumatov/Reuters/Landov)

(DDT), hexachloro-cyclohexane compounds (HCH, Lindane), and toxaphene remain. Other toxic materials present are the result of biological weapons testing and failed industrial sites.

The region immediately surrounding the Aral Sea—Uzbekistan, Kazakhstan, and parts of Turkmenistan—are affected by hazardous dust and salt storms. Most major storms occur in a one-hundred-kilometer margin along the north-northeastern coastal zone. Some 60 percent of these storms trend southwesterly, for 500 kilometers, depositing salts, agrochemical dusts, and aerosols on the delta of the Amu Dar'ya. This southern river delta region is densely populated, so toxic storms affect human and animal health and economic stability. Another 25 percent of the storms trend west, moving over and beyond the Ust-Urt plateau, an area of livestock pastures.

Toxic dust storms harm the human food supply and physical health of domestic and wild animals. The human risk associated with airborne salt and dust is high, and greater than average incidences of respiratory illnesses, eye problems, throat and esophageal cancer, skin lesions and rashes, and liver and kidney damage are reported in the Aral Sea region.

• Aral Sea Geology

The Aral Basin has experienced geologic cycles of diversion and desiccation. During the Pliocene epoch, the ocean withdrew from Eastern Europe and Turkestan, leaving remnant basins such as the Aral, Caspian, and Black seas. Late Pliocene continental crust movements created a more permanent depression in the area of the Aral Sea, which was filled with water, some of which came from the ancestral Syr Dar'ya.

The effects of the Pleistocene epoch are recorded by terrestrial sedimentary deposits. Eolian processes operated in the Aral depression during the early and middle Pleistocene. During the late Pleistocene, fluvial processes filled the depression by inflows from the ancestral Amu Dar'ya; then, the basin was filled for a second time with waters from the Syr Dar'ya.

Both rivers affect lake level changes, but when the 2,525-kilometer course of the Amu Dar'ya migrates away from the basin, lake level drops. Diversions of the Amu Dar'ya are natural, resulting from filling fluvial channels during heavy rains or floods. Some river diversions are the result of human actions, such as improper or failing irrigation systems or intentional destruction of river dams and levees during political upheaval or war.

• Aral Sea Restoration

The Aral Sea cannot be reestablished to its pre-1960 status, because to do so would mean curtailing irrigation, which uses 92 percent of all Aral water withdrawals. Curtailment of irrigation would mean crop failure and economic and social collapse in the Aral Basin. After the fall of the Soviet Union in 1991, Kyrgyzstan, Uzbekistan, Turkmenistan, Kazakhstan, and Tajikistan joined together to address the Aral Sea crisis. Two major agencies were formed by these new regional states. with the International Fund for the Aral Sea (IFAS) taking the lead role in 1997. Also, the United Nations, the European Union, and many other international aid agencies operated to improve the region.

In order to regulate flow in the little Aral, Kazakhstan and the World Bank constructed an eighty-five-million-dollar, thirteen-kilometer earthen ditch connected to a concrete dam with gates and spillways. Completed in November, 2005, this system brought early success: The water level increased to 42 meters from 40 meters, and, by summer of 2006, the lake area increased by 18 percent. Salinity decreased by one-half to almost 10 grams per liter in 2006, but future levels will vary, by area, from 3 to 14 grams per liter. Decreased salinity increases fish population, aiding the fishing industry and the Kazakhstan economy. A former fish-processing plant has reopened in Aralsk, Kazakhstan, to process lake carp, Aral bream, Aral roach, Pike perch, and flounder—the top five species caught in autumn, 2007.

• Context

The Aral Sea Basin has become synonymous with irreversible environmental disaster. The entire region demonstrates the potential for humans to act as geomorphic change agents. The area's inhabitants diverted rivers for irrigation; replaced desert vegetation with such crops as melons, cotton, and rice; and altered natural water chemistry to salinities greater than that of ocean water. Anthropogenic environmental degradation has also affected human health in the Aral Basin.

Populations around the sea are in a state of upheaval, dislocation, and poverty as a result of the collapse of the fishing industry and a lack of government support from the former Soviet Union since the early 1990's. Essential medications and adequate hospital facilities are not available when economic conditions are stagnant. Health problems begin in the youngest populations: High infant mortality, low birth weight, growth retardation, and delayed puberty are present in the basin.

Poor quality and insufficient quantities of drinking water in the basin have increased typhoid, hepatitis A, and diarrhea in all age groups. High levels of mineralized water within the basin may contribute to kidney and liver diseases. Also increasing are acute respiratory diseases—killing almost one-half of all children. Dust storms sourced from former seabeds deliver salt and toxic chemical sediments to humans and areas of human habitation.

Possibly the greatest health risk is from pesticides, which contaminate the water and food supplies and infiltrate during dust storms. Pesticides may be applied to crops, especially cotton, several times during a growing season. Lindane (HCH) has also been used biannually to rid sheep's skin and fleece of vermin.

It is difficult to determine the relative importance of each human health issue, especially when population groups may suffer multiple medical problems and live in impoverished conditions. What is clear is that residents of the Aral Sea are experiencing a pronounced health crisis, not unlike the environmental one that surrounds them.

Mariana L. Rhoades

- **Further Reading**

Micklin, P. "The Aral Sea Disaster." *Annual Review of Earth and Planetary Science* 35 (2007): 47-72. Excellent collection of data on the water balance, salinities, and hydrology of the Aral Basin; human and ecological consequences of the disaster; improvement efforts by global aid agencies; and engineering mitigation.

_______. "Desiccation of the Aral Sea: A Water Management Disaster in the Soviet Union." *Science* 241, no. 4844 (September 2, 1988): 1170-1176. Micklin's early work on causes of sea recession and resulting environmental problems; details local water quality improvements and future schemes for preservation of the Aral.

Micklin, P., and N. V. Aladin. "Reclaiming the Aral Sea." *Scientific American* 298, no. 4 (April, 2008): 64-71. Outlines the collapse of the Aral Sea due to wasteful irrigation of the desert; the geography of residual lakes; the 2005 dam construction and future of the Amu Dar'ya; and application of lessons learned from the Aral disaster to other regions with similar risks.

Nihoul, J. C. J., P. O. Zavialov, and P. Micklin, eds. *Dying and Dead Seas: Climatic Versus Anthropic Causes.* Dordrecht, the Netherlands: Kluwer Academic, 2004. Synthesis of sixteen lectures given at the May, 2003, North Atlantic Treaty Organization Advanced Research Workshop in Belgium on dead or dying internal seas and lakes; assesses the past and present roles of natural and anthropogenic causes.

Whish-Wilson, P. "The Aral Sea Environmental Health Crises." *Journal of Rural and Remote Environmental Health* 1, no. 2 (2002): 29-34. Provides general background on the disaster; in-depth health status assessment of the region; list of causes and outcomes of pollution and its effects on humans and the environment; and details of the health community's response.

See also: Diseases; Drought; Fishing industry, fisheries, and fish farming; Floods and flooding.

Arctic

- **Category:** Arctic and Antarctic

Global warming is being felt most intensely in the Arctic, whose ice and snow have been melting away. Arctic Ocean ice cover shrank more dramatically between 2000 and 2007 than at any time since detailed records have been kept.

• Key concepts

albedo: proportion of incident radiation reflected by Earth's surface

drunken forest: forest that leans at an odd angle as a result of melting permafrost

feedback loops: climatic influences that compound or retard each other, accelerating or decelerating the rate of global warming

permafrost thawing: defrosting of previously permanently frozen ground, usually in or near the Arctic

tipping point: the point at which feedback loops take control and propel climate change

• Background

According to the Arctic Climate Impact Assessment Scientific Report, produced by 250 scientists under the auspices of the Arctic Council, Arctic sea ice was half as thick in 2003 as it had been thirty years earlier. The melting of ice in the Arctic accelerated through 2007, advancing the projected date of an ice-free summer to perhaps 2020. During September, 2007, the Arctic ice cap shrank to its smallest extent since records have been kept, 4.12 million square kilometers, versus the previous record low of 5.28 square kilometers in 2005. The shrinkage from 2005 to 2007 represents a loss of more than 20 percent of the Arctic's ice cover, or an area the size of Texas and California combined.

• Retreat of Arctic Ocean Ice

Scientists were shaken by the sudden retreat of the Arctic ice during the summer of 2007, which was much greater than their models had projected. Some said that a tipping point had been reached and that the Arctic could experience ice-free summers within a decade or two. At the annual American Geophysical Union meeting in San Francisco during December, 2007, scientists reported that temperatures in waters near Alaska and Russia were as much as 5° Celsius above average.

Scientists at the University of Washington said the Sun's heat made the greatest contribution to the record melting of the Arctic ice cap at the end of summer, 2007. Sunlight added twice as much heat to the water as was typical before 2000. Relatively warm water entering the Arctic Ocean from the Atlantic and the Pacific Oceans was also a factor, according to Michael Steele, an oceanographer at the University of Washington. Energy from the warmer water delayed the expansion of ice in the winter as it warmed the air.

In addition to the retreating extent of sea ice in the Arctic, a greater proportion of the ice that does remain in the region is thin and freshly formed, unlike older, thicker ice that is more likely to survive future summer melting. The proportion of older, "durable" ice dropped drastically between 1987 and 2007, according to studies by Ignatius G. Rigor of the University of Washington. In addition to a decrease in the extent of Arctic Ocean ice, by 2007 large areas of the ice that remained were only about one meter thick—half what they had been in 2001, according to measurements taken by an international team of scientists aboard the research ship *Polarstern.*

• Walrus Deaths

With ice receding hundreds of kilometers offshore during the late summer of 2007, walruses gathered by the thousands on the shores of Alaska and Siberia. According to Joel Garlich-Miller, a walrus expert with the U.S. Fish and Wildlife Service, walruses began to gather onshore late in July, a month earlier than usual. A month later, their numbers had reached record levels from Barrow to Cape Lisburne, about 480 kilometers southwest, on the Chukchi Sea. Walruses dive from the ice to feed on clams, snails, and other bottom-dwelling creatures. As a result of increased melting, however, the Arctic ice has receded too far from shore to allow walruses to engage in their usual feeding patterns.

A walrus can dive 180 meters, but water under receding ice shelves is now more than a thousand meters deep by late summer. The walruses have been forced to swim much farther to find food, using energy that could cause increased calf mortality. In addition, more calves are being orphaned. Russian research observers have reported many more walruses than usual on shore, tens of thousands in some areas along the Siberian coast. These creatures would have stayed on the sea ice in earlier times.

Walruses are prone to stampedes when they gather in large groups. The appearance of a polar bear or a human hunter, or even noise from a low-flying small airplane, can send thousands of pan-

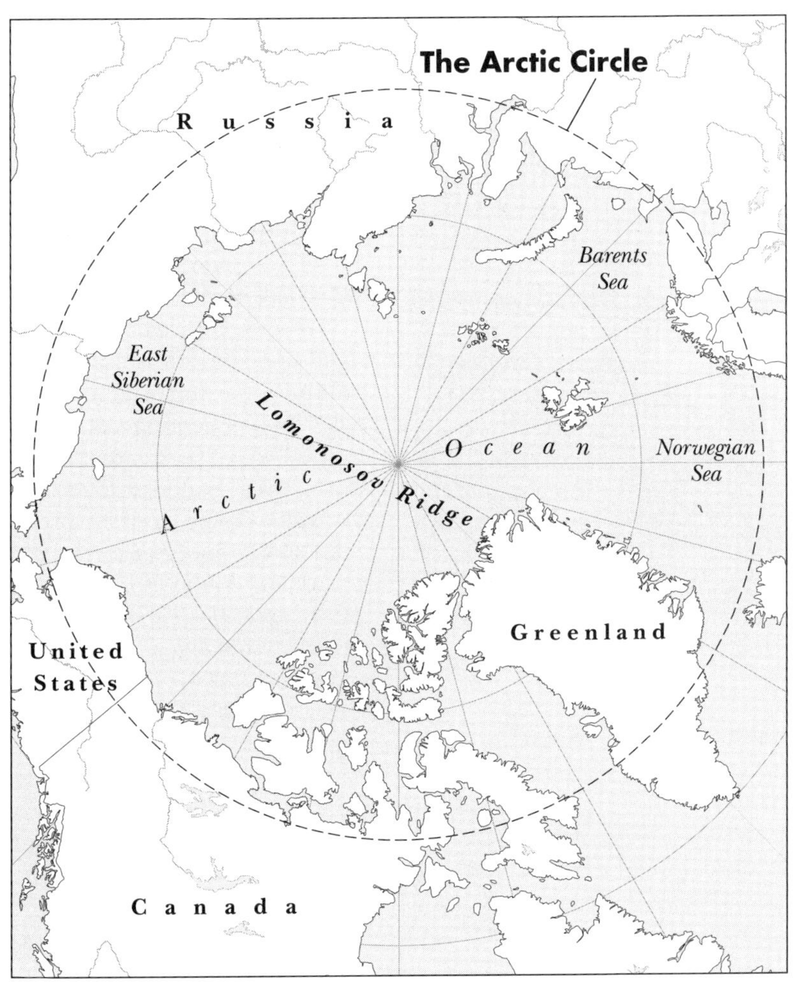

By the end of 2005, scientific projections for the Arctic were becoming more severe. A climate modeling study published in the *Journal of Climate* indicated that if humanity does not address global warming in the near future, irreversible damage may take place. The paper's lead author, Govindasamy Bala of the U.S. Energy Department's Lawrence Livermore National Laboratory, predicted that it might take twenty or thirty years before the scope of anthropogenic changes becomes evident, but that after that the damage would be obvious.

The *Journal of Climate* study projected that the global concentration of atmospheric carbon dioxide (CO_2) would be double that of preindustrial levels in 2070, triple in 2120, and quadruple in 2160—predicting slightly less than present-day rates of increase. It was the first study to assume consumption of all known reserves of fossil fuels. This model anticipates that the Arctic will see the planet's most intense relative warming, with average annual temperatures in many parts of Arctic Russia and northern North America rising by more than 14° Celsius by about 2100.

icked walruses rushing to the water, stampeding one another into bloody pulp. Thousands of Pacific walruses were killed on the Russian side of the Bering Strait during the late summer of 2007, when more than forty thousand hauled out on land at Point Shmidt as ice retreated northward.

• Forecasts

In the past, low-ice years often were followed by recovery the next year, when cold winters or cool summers kept ice from melting further. That kind of balancing cycle stopped after 2002. The year 2004 was the third in a row with extreme ice losses, indicating acceleration of the melting trend. Arctic ice has been declining by about 8 percent per decade as part of this trend.

• Acceleration of Ice-Cap Melting

Julienne Stroeve and her colleagues compared more than a dozen models of Arctic ice-melt and found that nearly all of them underestimated the actual speed of ice-melt, in many cases by large amounts. These findings have two important implications, according to a summary of Stroeve's study in *Science*. First, the effect of increasing greenhouse gases may have been greater than has been believed. Second, future loss of Arctic Ocean ice may be more rapid and extensive than predicted. Within a decade, projections of the first ice-free

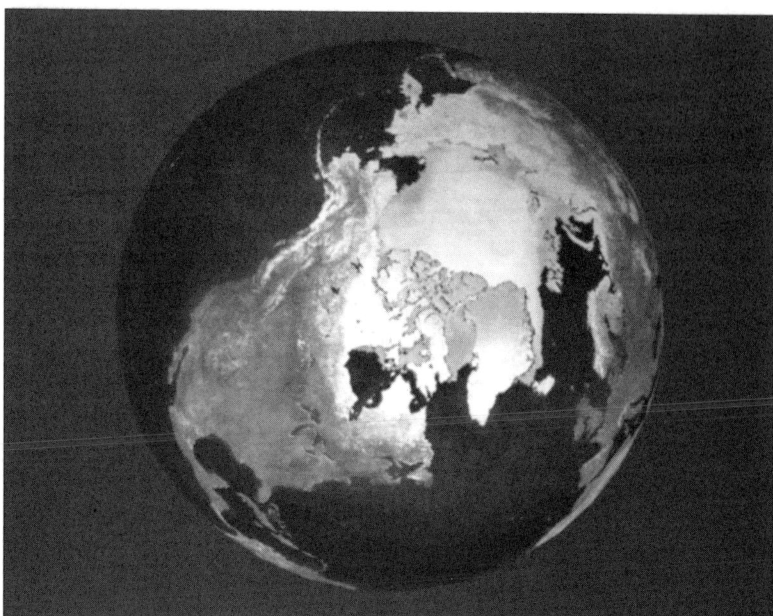

A global view of Arctic sea ice in 2000. (NASA-JPL)

summer in the Arctic have moved from the end of the twenty-first century to roughly the year 2020. Even projections made by the Intergovernmental Panel on Climate Change (IPCC) in its 2007 assessments (forecasting an iceless Arctic in summer between 2050 and 2100) were out of date weeks after they were made public, according to reports by scientists at the National Center for Atmospheric Research and the University of Colorado's National Snow and Ice Data Center. The study, "Arctic Sea Ice Decline: Faster than Forecast?" was published during early May, 2007, in the online edition of *Geophysical Research Letters.*

Beginning in the mid-1990's, scientists observed pulses of relatively warm water from the North Atlantic entering the Arctic Ocean, further speeding ice-melt. Mark Serreze, senior research scientist at the National Snow and Ice Data Center, said that such warm-water pulses represented yet another potential kick to the system that could accelerate rapid sea-ice decline and send the Arctic into a new state. As Arctic ice retreats, ocean water transports more heat to the Arctic, and the open water absorbs more sunlight, further accelerating the rate of warming and leading to the loss of more ice.

The entire Arctic system is thus beset by accelerating feedback loops that intensify climate change. For example, permafrost has been melting, injecting additional CO_2 and methane into the atmosphere. In Alaska, trees that have become destabilized in melting permafrost lean at angles, creating so-called drunken forests.

The speed of ice breakup can sometimes be astonishing. For example, the thirty-meter-thick Ayles shelf of floating ice, a shelf roughly sixty-five square kilometers in area, had extended into the Arctic Ocean from the north coast of Ellesmere Island in the Canadian Arctic for roughly three thousand years. The Ayles shelf was detached during the summer of 2005 by wind and waves in warming water. The break-up was observed by Laurie Weir of the Canadian Ice Service in satellite images of Ellesmere Island during August, 2005. The images showed a broad crack opening and the ice shelf flowing out to sea with a speed that could be observed hour by hour.

In 1906, Arctic explorer Robert Edwin Peary surveyed 26,160 square kilometers of ice shelves. Nine-tenths of these have broken up over the last century, according to Luke Copland, director of the Laboratory for Cryospheric Research at the University of Ottawa.

• Context

The extent of Arctic Ocean ice has been declining year-round for more than three decades, according to an analysis of satellite data compiled during 2006 by the U.S. National Ice Center. Data sets of the Thirty-Year Arctic Sea Ice Climatology show shrinkage in the Arctic Ocean summer ice cover of more than 8 percent per decade, according to Pablo Clemente-Colón, the ice center's chief scientist. While the extent of winter sea ice has also been decreasing, summer shrinkage has become more pronounced.

A tipping point toward an ice-free Arctic in summer probably already has been passed, according to James E. Hansen, director of the National Aeronautics and Space Administration's Goddard Institute for Space Studies. The complete loss may occur rapidly, on the time scale of a decade, once ice loss has reached such a degree that the albedo feedback becomes a dominant process, according to Hansen. The albedo feedback refers to the fact that loss of some sea ice increases the amount of solar energy absorbed by the Arctic because the liquid ocean is darker than the ice, absorbing more of the Sun's heat and thus increasing ice melt. Hansen and his team could not determine the exact level of added CO_2 necessary to cause complete ice loss, but they declared that once such a state were reached, it would be difficult to return to a climate with summer sea ice, because of the long lifetime of atmospheric CO_2.

Bruce E. Johansen

• Further Reading

Alley, Richard B. *The Two-Mile Time Machine Ice Cores, Abrupt Climate Change, and Our Future.* Princeton, N.J.: Princeton University Press, 2000. Popular version of Alley's scientific work on climate change in the Arctic.

Amstrup, S. C., et al. "Recent Observations of Intraspecific Predation and Cannibalism Among Polar Bears in the Southern Beaufort Sea." *Polar Biology* 29, no. 11 (October, 2006): 997-1002. Details conditions under which some polar bears, unable to reach ice masses from which they used to hunt ringed seals, have been eating their own young.

Appenzeller, Tim. "The Big Thaw." *National Geographic,* June, 2007, pp. 56-71. Survey of ice melt in the Arctic, the Antarctic, and mountain glaciers, with explanations describing why the process has been speeding up.

Arctic Climate Impact Assessment Scientific Report. New York: Cambridge University Press, 2006. Detailed scientific reports of Native American peoples' experiences and scientific trends regarding global warming in the Arctic, where temperatures have been rising more quickly than anywhere else on Earth.

Schiermeier, Quirin. "On Thin Ice: The Arctic Is the Bellwether of Climate Change." *Nature* 441 (May 10, 2007): 146-147. Examines the Arctic's crucial role in global climate.

Serreze, Mark C., and Roger G. Barry. *The Arctic Climate System.* New York: Cambridge University Press, 2006. A glaciologist looks at the decline of ice in the Arctic, along with other issues related to climate change.

See also: Albedo feedback; Antarctica: threats and responses; Arctic peoples; Arctic seafloor claims; Ground ice; Ice cores; Ice shelves.

Arctic peoples

• Categories: Arctic and Antarctic; nations and peoples

The lives of approximately 150,000 Inuit and Eskimo who ring the Arctic Ocean in Greenland, Canada's Arctic, Alaska, and Russia have been substantially changed by some of the most rapid warming on Earth.

• Key concepts

Inuit: indigenous people of the Canadian Arctic

Inuit Circumpolar Conference: group representing the interests of 150,000 Inuit aand Eskimo in international forums

pack ice: a solid mass of sea ice formed when smaller masses and fragments are forced together under pressure

• Background

Around the Arctic, in Inuit villages connected by the oral history of traveling hunters as well as by e-mail, weather watchers report striking evidence that global warming is an unmistakable reality. Rapid changes have been evident at all seasons. During the summer of 2004, for example, several yellowjacket wasps (*Vespula intermedia*) were sighted in Arctic Bay, a community of seven hundred people located on the northern tip of Baffin Island, at more than 73° north latitude. Inuktitut, the Inuit language, has no word for the insect.

• Inuit Hunters' Experiences

The Arctic's rapid thaw has made hunting, never a safe or an easy way of life, even more difficult and dangerous. Arctic weather has changed significantly since roughly the middle 1990's. Several hunters have fallen through unusually thin ice and sustained severe injuries or died. Changing weather is only one factor. Inuit hunters' dogs have an ability to detect thin ice, which they may refuse to cross. Snowmobiles, which have replaced dogs for many Inuits, have no such sense. While pack ice was visible from the Alaskan north coast year-round until a few years ago, it has now retreated well over the horizon, forcing hunters to journey more than 160 kilometers offshore to find bearded seals and walruses.

Pitseolak Alainga, an Iqaluit-based hunter, has said that climate change compels caution: One must never hunt alone. Before venturing onto ice in fall or spring, Alainga cautions, hunters should test its stability with a harpoon. Alainga knows the value of safety on the water. His father and five other men died during late October, 1994, after an unexpected ice storm swamped their hunting boat. The younger Alainga and one other companion barely escaped death in the same storm. Alainga believes that more hunters are suffering injuries not only because of climate change but also because basic survival skills are not being passed from generation to generation as they were in years past, when most Inuit lived off the land.

Inuit daily life has changed in many other ways because of the rapidly warming climate. Many Inuit once used ice cellars to store meat in frozen ground. The climate has warmed so much that the cellars are no longer safe, even as far north as Barrow, Alaska. Meat that once could be safely stored in permafrost has been spoiling. With increasing inability to hunt, Inuit are being forced to buy imported food at prices several times those paid for the same goods at lower latitudes. Diabetes and other problems associated with imported food have been increasing. Drinking water obtained by melting ice has become dirtier. A warmer climate increases risks for insect-vector diseases, even West Nile virus.

A young Nenets girl stands next to a sled. The Nenets are indigenous peoples of the Russian Arctic. (Vasily Fedosenko/Reuters/Landov)

• Shishmaref, Alaska

Six hundred Alaskan Native people in the village of Shishmaref, on the far western shore of Alaska about 160 kilometers north of Nome, have been watching their village erode into the sea. The permafrost that once reinforced Shishmaref's waterfront is thawing. Shishmaref residents decided during July, 2002, to move their entire village inland, a project that the U.S.

Army Corps of Engineers estimated would cost more than $100 million. By 2001, seawater was lapping near the town's airport runway, its only long-distance connection to the outside world. By that time, three houses had been washed into the sea. Several more were threatened. The town's drinking water supply also had been inundated by the sea. By fall, 2004, Shismaref's beaches had retreated still farther during vicious storms. The same storms flooded businesses along the waterfront in Nome and damaged power lines, fuel tanks, and roads in at least a half dozen other coastal villages.

• Artificial Hockey Ice and Air Conditioning

By the winter of 2002-2003, warming Arctic temperatures were forcing hockey players in Canada's far north to seek rinks with artificial ice. Canada's *Financial Post* reported that global warming had cut in half the hockey season, which ran from September until May in the 1970's. Instead, the season—dependent in some areas upon natural ice—had come to begin around Christmas and end in March. The northern climate has changed so swiftly that by the summer of 2006, the Inuit were installing air-conditioning in some buildings near the Arctic Circle. Traditional Inuit homes are built with southern exposure to soak up every available hint of the sun's warmth. Windows are usually small, and they rarely open easily. Some Inuit were developing severe heat rashes during the summer.

• Context

The Inuit are very conscious of their pivotal role in a natural world undergoing climate change, as they seek to avoid destruction of an environment that has sustained them for many thousands of years. The Inuit Circumpolar Conference has petitioned the Inter-American Commission on Human Rights regarding violations of fundamental human rights among Arctic peoples. The petition defends the rights of Inuits as a people within evolving international human rights law. It seeks a declaration from the commission that emissions of greenhouse gases (GHGs) from the United States—the source of more than 25 percent of the world's greenhouse GHGs during the last century—is violating Inuit human rights as outlined in the 1948 American Declaration on the Rights and Duties of Man.

The petition asserts that the Inuits' rights to culture, life, health, physical integrity and security, property, and subsistence have been imperiled by global warming: With accelerating loss of ice and snow, hunting, travel, and other subsistence activities have become more dangerous and in some cases impossible. The petition does not seek monetary damages. Instead, it requests cessation of U.S. actions that violate Inuit rights to live in a cold environment—not a small task, since such action would involve major restructuring of the U.S. economic base to sharply curtail emissions of GHGs. The petition anticipates the types of actions that will have to take place on a worldwide scale, including alterations to the rapidly expanding economies of China and India, to preserve the Inuit way of life, as well as a sustainable worldwide biosphere.

Bruce E. Johansen

• Further Reading

Johansen, Bruce E. "Arctic Heat Wave." *The Progressive,* October, 2001, 18-20. On-the-scene reporting from Baffin Island describing the effects of accelerating warming on Inuit hunters.

Lynas, Mark. *High Tide: The Truth About Our Climate Crisis.* New York: Picador/St. Martin's, 2004. Elegantly written travelogue describing effects of climate change around the world, with detailed emphasis on the Arctic's people.

Watt-Cloutier, Sheila, and the Inuit Circumpolar Conference. *Petition to the Inter-American Commission on Human Rights Seeking Relief from Violations Resulting from Global Warming Caused by Acts and Omissions of the United States.* Iqualuit, Nunavut: Author, 2005. This legal brief contains copious detail describing how global warming is changing the lives of Inuit people at a very personal level, with many case studies.

Wohlforth, Charles. *The Whale and the Supercomputer: On the Northern Front of Climate Change.* New York: North Point Press, 2004. Detailed compendium of global warming's intensifying effects on Arctic peoples, focusing on Northern Canada.

See also: Antarctica: threats and responses; Arctic; Diseases; Human behavior change; Maldives; Polar climate.

Arctic seafloor claims

- **Category:** Arctic and Antarctic

Global warming has led to significant declines in Arctic Ocean ice coverage, allowing unprecedented access to seafloor resources. Estimates place approximately one-quarter of the world's fossil fuel reserves beneath the Arctic seafloor, making seafloor competition a likely source of future geopolitical conflict.

• Key concepts

albedo: the proportion of incident light reflected from a surface

continental shelf: a submerged extension of a continental coastline, where the ocean is relatively shallow and the seafloor is built from terrestrial sediments

Gakkel Ridge: an Arctic Ocean spreading zone that continues the Mid-Atlantic Ridge

Lomonosov Ridge: An 1,800-kilometer submerged ridge crossing the geographic North Pole from Russia to Greenland

methyl hydrates: icelike substances formed under pressure beneath permafrost and Arctic Ocean sediments that trap gases such as methane within an ice lattice

placer deposits: accumulations of minerals washed onto the continental shelf from river runoff

United Nations Convention on the Law of the Sea: an international treaty that grants an exclusive economic zone to all countries bordering significant bodies of water

• Background

During the late twentieth and early twenty-first centuries, Arctic Ocean ice was in steady retreat. Ice thickness declined by 40 percent, and ice shrinkage freed over 2.6 million square kilometers of water along the ocean's coastlines. Record rates of ice shrinkage in 2007 and 2008 opened the Northwest Passage to navigation for the first time in recorded history.

Loss of polar ice creates open water that absorbs light rather than reflecting it, reducing the Arctic albedo, delaying yearly ice formation, promoting early spring melt, and thinning the permanent ice shelf. While costly to the Arctic ecosystem, ice loss has allowed unprecedented access to researchers, shippers, and resource explorers from the five countries bordering Arctic waters: the United States, Russia, Canada, Norway, and Danish Greenland.

• Arctic Resources

Research conducted in the Arctic indicates the potential existence of 90 billion barrels of oil and 4.5 trillion cubic kilometers of natural gas buried beneath the seafloor. Most fuel deposits lie close to shore, where they are readily extractable using existing technologies. Placer deposits of diamonds and gold also lie close offshore. Farther from the coast can be found large deposits of oil shale, methyl hydrates, and coal-bed methane.

Current technologies do not allow extraction of these deposits, and some, such as methyl hydrate, remain unproven as a fuel resource. However, increasing global fuel demands will make even these deposits attractive to future exploitation. Moreover, the discovery of undersea volcanoes and hydrothermal vents in the deep Arctic in 2001 added precipitated deposits of gold, silver, and copper to the list of recoverable resources. The lure of seafloor wealth has stimulated intense competition among those countries wishing to establish Arctic seafloor claims.

• Claiming the Seafloor

Territorial jurisdiction within Arctic waters has long been disputed. The United Nations Convention on the Law of the Sea (UNCLOS)—which was concluded in 1982 and entered into force in 1994—was designed specifically to address such disputed seafloor claims. Under UNCLOS, every country bordering an ocean or an inland sea is granted sovereign rights over natural resources within an exclusive economic zone (EEZ) extending outward from its coastline for 370 kilometers.

Most known Arctic resources lie within these EEZs, but the Arctic is a relatively shallow ocean, and resource deposits near the North Pole and the Gakkel Ridge, both considered international territory, have triggered competing claims under Article 76 of UNCLOS. Article 76 allows the extension of a country's EEZ to up to 650 kilometers offshore. To claim this additional territory, the claimant must prove that the continental shelf lying beyond its established 370-kilometer zone is an extension of its already-claimed shelf zone. The new territory also must not include any area more than 185 kilometers beyond the point where the water's depth reaches 2,500 meters. Article 76 could allow almost 90 percent of the Arctic seafloor to become part of the EEZs of the five bordering nations.

• The Lomonosov Ridge

The Lomonosov Ridge, lying between Siberia and Greenland, parallels the Gakkel Ridge, passes through the geographic North Pole, and represents the primary legal battlefield for Article 76 claims. The Lomonosov Ridge has a complicated and unique geological history. Core sampling indicates that it split off of the Eurasian margin of the continental shelf along northern Scandinavia and Russia some 50 million years ago. Coring also indicates that the Arctic was at one time a swampy inland sea, bolstering the likelihood that the ridge and the ocean basin it created are rich in buried fossil fuel reserves. The promise of such riches led the Russia Federation in August, 2007, to claim a portion of the Lomonosov Ridge, the Alpha-Mendeleyev ridge system, as well as part of the Amerasian basin: The Russian submersible *Mir-1* planted a flag 4 kilometers beneath the ice of the geographic North Pole, thereby symbolically claiming more than 1.2 million square kilometers of seafloor.

Under Article 76, such a claim is valid only if Russia can prove that the Lomonosov Ridge is currently a geological extension of its continental shelf. Though core samples indicate that the ridge originated from the Eurasian margin, faulting patterns near Siberia suggest that it has since become detached from the Siberian shelf. Russia's first

A Russian deep-sea submarine is lowered into the Arctic Ocean, where it will conduct a mission designed to bolster Russia's claims to the mineral resources of the sea floor. (AP/Wide World Photos)

claim to the Lomonosov Ridge was rejected by UNCLOS in 2002 based on limited geological evidence.

Canada and Denmark maintain that Russia will be unable to prove that the Lomonosov Ridge is an extension of its shelf. They have filed counterclaims of their own. The stakes are huge: Successful claimants stand greatly to increase their fuel reserves and to generate billions of dollars in revenue. So far, the United States, a nonsignatory to UNCLOS, has stayed out of the debate, but with the potential gains so high it is unlikely that it will remain on the sidelines.

• Context

At one time, research in the Arctic Ocean was nearly impossible. Its 14-million-square-kilometer surface was covered by heavy pack ice almost year-round, making access—even with icebreakers—dangerous and of scant utility. Global warming has changed this situation dramatically. Ice thinning and ice-free periods in the Arctic have made both access and research easier, even though they come at the expense of the ocean and its inhabitants. Given the rate of ice melt and the lure of valuable natural resources, competition over seafloor claims is expected to be fierce.

Decisions made under Article 76 will be binding. Thus, once these decisions are made, the consequent redrawing of undersea maps will have profound effects on regional nationalism and access to resource wealth. Successful claimants will determine access, set the environmental rules governing exploration and exploitation, and control royalties and pricing. However, drilling, pipeline transport, and shipping will all remain dangerous even in an ice-free environment. Moreover, full exploitation of the fuel and mineral resources found in the Arctic seafloor could contribute to global warming, enhancing a warming feedback cycle that may further facilitate research and mining, but only at the expense of global sustainability.

Elizabeth A. Machunis-Masuoka

• Further Reading

Cressey, Daniel. "The Next Land Rush." *Nature* 451 (January 3, 2008): 12-15. Science-news feature discussing surveying efforts to determine the international legitimacy of Arctic seafloor claims. References.

Funk, McKenzie. "Arctic Land Grab." *National Geographic*, May, 2009, pp. 104-121. Excellent story describing international claims to Arctic seafloor resources and the relationship of contemporary and future claims to global warming. Multiple maps depict land claims, drilling operations, and ice loss since 1980.

Krajick, Kevin. "Race to Plumb the Frigid Depths." *Science* 315 (March 16, 2007): 1525-1528. Science-news feature discussing Russia's claim to Arctic seafloor resources.

Moore, T. C., et al. "Sedimentation and Subsidence History of the Lomonosov Ridge." *Proceedings of the Integrated Ocean Drilling Program* 302 (2006): 1-6. One of many papers published online by the IODP on the geology of the Arctic Ocean. Describes the geologic history encoded in core samples from the Lomonosov Ridge.

See also: Antarctic Treaty; Arctic; Arctic peoples.

Arrhenius, Gustaf

Swedish oceanographer and geochemist

Born: September 5, 1922; Stockholm, Sweden

Arrhenius determined that ocean sedimentation can be used to reconstruct paleoclimates. This reconstruction provides a basis for projecting future climates.

• Life

Gustaf Arrhenius—the son of Dr. Olof Vilhelm Arrhenius, a chemist, and the grandson of the Nobel laureate Svante August Arrhenius—was born into a family history of scientists, Arrhenius naturally chose science as his career field. Even in his teens, his fascination with oceanography was evident, when, for a science project, he made a study of the physical and chemical properties of the fjord where he lived. After college, he was offered a position as geochemist on two deep-sea expeditions.

Arrhenius's first expedition was in 1946 on the ship *Skagerak*; the crew tested a new piston corer invented by Borje Kullenberg, as well as new seismic and optical techniques in the Mediterranean and eastern Atlantic. The expedition was a success. The second expedition was an eighteen-month (1947-1948) trip around the world. This trip supplied the cores from the east equatorial Pacific region that Arrhenius studied for his Ph.D. While on the expedition, Arrhenius flew back to England to marry Eugenie de Hevesy, the daughter of George de Hevesy, who worked with radioisotopes. The Arrheniuses would have three children, Susanne, Thomas, and Peter, and settle in La Jolla, California.

During the time that Arrhenius was working on the cores, Roger Revelle invited him to join the Scripps Institution of Oceanography in San Diego, California, as a visiting research oceanographer in time for the Capricorn expedition to the Marshall Islands. After the expedition, Arrhenius studied cores collected by the expedition for the western Pacific region. He remained at Scripps thereafter. Arrhenius received a Ph.D. in 1953 from Stockholm University and became an associate professor of biogeochemistry at Scripps in 1956. During these years, he studied crystal structures of natural and synthetic materials, superconductivity, space chemistry, prebiotic chemistry, and the earliest sedimentary records of Earth, Mars, and the moon. Arrhenius has published over 150 articles and books.

• Climate Work

Arrhenius analyzed the deep-sea cores from the east equatorial Pacific region for carbonate, humus carbon, nitrogen, and phosphorus in his father's laboratory. The cores were 19 meters long and represented the last one million years. Using his data and data from other analyses—such as the analysis of Radiolaria (amoebid protozoa with intricate mineral skeletons), diatoms (unicellular algae containing silica), foraminifera (amoeba-like animals that grow shells), and coccoliths (calcite plates from phytoplankton, a single-celled type of algae)—Arrhenius was able to determine a cyclic nature of the sedimentation.

Sediment is an indication of the condition of the ocean above. The nature of the ocean is an indication of the atmosphere above the ocean. Arrhenius was able to determine the dates of the different sections of the cores from carbon 14 studies and learned that the ice ages had a cyclic nature during the period of time, the Pleistocene era, covered by those deep-sea cores. In later years, that study was expanded to 380 million years and to sediments from all around the world.

By determining Earth's climatological history, scientists are better able to project the planet's climatological future. Arrhenius's grandfather, Svante, had predicted that if a large amount of carbon dioxide were released into the atmosphere, it would cause changes in the climate. Arrhenius's work helps set a baseline for scientists seeking to determine whether modern climate change is anthropogenic or natural.

C. Alton Hassell

• Further Reading

Arrhenius, Gustaf. "Carbon Dioxide Warming of the Early Earth." *Ambio* 26, no. 1 (February, 1997): 12-16. Arrhenius returns to the subject matter of his early work, the climate of ancient times, commenting on it from the point of view of later endeavors.

_______. "Oral History of Gustaf Olof Svante Arrhenius." Interview by Laura Harkewicz. San Diego, Calif.: Scripps Institution of Oceanography Archives, 2006. Arrhenius discusses his career and his association with the Scripps Institution of Oceanography.

"G. Arrhenius' Analyses of Ocean-Floor Sediment Cores Taken on the Swedish Deep-Sea Expedition." *Bulletin of the Geological Society of America* 63 (May, 1962): 515-517. Short note describing Arrhenius's Ph.D. research. Bibliography.

Sears, Mary. *Oceanography: Invited Lectures Presented at the International Oceanographic Congress Held in New York, 31 August-12 September 1959.* Washington, D.C.: International Oceanographic Congress, 1959. Includes a summary of the concept that the climate leaves a record in the sedimentation on the ocean floor in the chapter titled "Geological Record on the Ocean Floor." Illustrations, charts, maps.

Wang, Y., G. Arrhenius, and Y. Zhang. "Drought in the Yellow River: An Environmental Threat to the Coastal Zone." *Journal of Coastal Research* 34

(2002): 503-515. Summarizes climatological effects contributing to Yellow River drought and its consequences for the environment. Bibliography.

See also: Arrhenius, Svante August; Ice cores; Paleoclimates and paleoclimate change; Sea sediments.

Arrhenius, Svante August

Swedish physical chemist

Born: February 19, 1859; Castle of Vik, near Uppsala, Sweden

Died: October 2, 1927; Stockholm, Sweden

Sometimes called the "father of climate change," Arrhenius was the first to quantify the relationship between enhanced or diminished atmospheric CO_2 concentrations and the elevation or depression of Earth's surface temperature, which later inspired others to develop more sophisticated mathematical models of global warming.

• Life

A surveyor's son and a child prodigy who was reading and calculating by age three, Svante August Arrhenius was a brilliant student throughout his education in Uppsala, Sweden, at the elementary, high school, and university levels. After receiving his bachelor's degree from Uppsala University, Arrhenius traveled in 1881 to Stockholm, where, at the Physical Institute of the Swedish Academy of Sciences, he began his lifelong interest in the conductivities of such electrolytes as sodium chloride. Besides an experimental section dealing with electrolytic conductivities, his doctoral dissertation contained a theoretical portion in which he proposed the revolutionary idea that, in very dilute solutions, electrolytes are present as charged atoms or groups of atoms (called ions). Incredulous over this radical theory, his examiners passed him but "without any praise," the lowest category.

After receiving his doctorate in 1884, Arrhenius studied and conducted research with some influential European scientists who were establishing the new field of physical chemistry, and he developed equations describing how rates of chemical reactions increased with temperature. His academic career comprised increasingly prestigious positions at Uppsala University (1884-1891), Stockholm University (1891-1905), and the Nobel Institute for Physical Chemistry (1905-1927). In 1903, he became the first Swede to win the Nobel Prize, which conferred official approval on his electrolytic theory of dissociation. Although his most important contributions were in physical chemistry, he also developed interesting ideas in immunochemistry, regarding the equilibrium between toxins and antitoxins, and geology, regarding the cause of ice ages. In exobiology, his theory of panspermia posited that cosmic microorganisms originated life on Earth. After the end of an early two-year marriage, which produced a son, he married Maria Johansson, with whom he had two daughters and a son. He died in Stockholm in 1927 but was buried in Uppsala, the city of his youth and early accomplishments.

• Climate Work

In his famous 1896 paper on climate change, Arrhenius acknowledged the influence of such precursors as Joseph Fourier and John Tyndall. In 1825, Fourier had grasped that Earth's atmosphere acts "like the glass of a hot-house," letting through high-energy (ultraviolet) radiation but retaining low-energy (infrared) radiation. In 1859, Tyndall, using spectrophotometry, discovered that certain gases, such as water vapor and carbon dioxide (CO_2), had the power to absorb ultraviolet rays. Arrhenius became interested in this phenomenon because of his curiosity about the cause of the ice ages. Earth's ice sheets had expanded when the global temperature was low and retreated when it was high, and Arrhenius speculated that changes in atmospheric CO_2 concentrations might explain these variations.

In 1895, Arrhenius presented a paper to the Stockholm Physical Society titled "On the Influence of Carbonic Acid in the Air upon the Temperature of the Ground." He was not a climatologist, but he had the ability to take the data collected by others and forge a mathematical model of how even small amounts of "carbonic acid" (CO_2) could affect global temperatures. For example, he calculated that a modest decline in atmospheric CO_2 would be sufficient to lower global temperatures by 4° to 5° Celsius, sufficient to cause the spread of ice sheets between the 40th and 50th northern parallels. On the other hand, particularly in his later writings, he understood that doubling concentrations of atmospheric CO_2 would lead surface temperatures to rise by about 5° Celsius.

Svante August Arrhenius. (©The Nobel Foundation)

In Arrhenius's view, atmospheric CO_2 was due to "volcanic exhalations," organic decay, the burning of coal, and the decomposition of carbonates. He also understood that greater human populations with increased energy use would result in higher levels of atmospheric CO_2, with concomitant global warming. He did not see such warming as detrimental, however, for he believed that warmer climates would lead to better harvests and fewer famines.

Some scientists found Arrhenius's calculations implausible, because they were based on a highly oversimplified model of the Earth's extremely complex atmosphere. Geologists rejected his theory as an explanation for the ice ages, since he was unable to find a satisfactory mechanism for the removal of so much CO_2 from the atmosphere (later analyses of many ice cores falsified Arrhenius's theory). Other scientists were unconcerned about anthropogenic increases in atmospheric CO_2, because they felt that oceans and plants would mitigate any CO_2 buildup.

It was not until after World War II, when advanced technologies resulted in improved atmospheric data and larger and faster computers supported more rigorous mathematical models, that climatologists were better able to understand past, present, and future concentrations of atmospheric CO_2 and other greenhouse gases. To many of these scientists, Arrhenius became an admired predecessor, especially when, during the 1970's and 1980's, CO_2 was recognized as the key molecule in climate change. Like his theory of ionic dissociation, which was initially ridiculed then widely accepted, Arrhenius's theory of the influence of CO_2 on global temperatures passed through a long period of rejection before it achieved the scientific consensus that has now become the basis of significant political actions.

Robert J. Paradowski

• **Further Reading**

Arrhenius, Svante. "On the Influence of Carbonic Acid in the Air upon the Temperature of the Ground." *Philosophical Magazine and Journal of Science.* 5th ser. 41 (April, 1896): 237-275. An English translation, also available online, of the paper presented to the Stockholm Physical Society that some see as pivotal in later developments of the study of climate change.

Christianson, Gale E. *Greenhouse: The Two-Hundred-Year Story of Global Warming.* Washington, D.C.: Walker, 1999. Presents an illuminating survey of the history of global warming, including Arrhenius's contributions.

Crawford, Elisabeth. *Arrhenius: From Ionic Theory to the Greenhouse Effect.* Canton, Mass.: Science History, 1996. This biography also contains a historical analysis of Arrhenius's scientific work along with a critical discussion of several misinterpretations of his role in the story of climate change.

Fleming, James Rodger. "Global Environmental Change and the History of Science." In *The Modern Physical and Mathematical Sciences.* Vol. 5 in *The Cambridge History of Science*, edited by Mary Jo Nye. New York: Cambridge University Press, 2003. This updated recapitulation of the author's *Historical Perspectives on Climate Change* (1998) analyzes Arrhenius's precursors, his own contributions, and his successors. Many notes to primary and secondary sources.

See also: Carbon dioxide; Greenhouse effect; Greenhouse gases.

Asia-Pacific Partnership

• **Category:** Laws, treaties, and protocols
• **Date:** Announced July 28, 2005; launched January 12, 2006

The Asia-Pacific Partnership is a pact designed to increase the development and dispersal of new technologies in order to reduce the emission of GHGs.

• **Participating nations:** Australia, Canada, China, India, Japan, South Korea, United States

• **Background**

When the Kyoto Protocol to the United Nations Framework Convention on Climate Change was adopted on December 11, 1997, several industrialized nations had reservations about its fairness and potential effectiveness. The United States, the only developed nation that did not ratify the Kyoto Protocol, believed that complying with the treaty would put undue strain on the U.S. economy. President George W. Bush objected in 2001 to provisions that called for the United States to reduce its carbon emissions while exempting China and India, which with the United States were among the world's largest emitters of greenhouse gases (GHGs). India and China, on the other hand, objected to the idea that they should deny technological advancement to their large and poor populations. Bush promised to develop an alternative plan that would address GHG emissions more effectively than the Kyoto Protocol while at the same time eradicating poverty and protecting human health, American jobs, and American investments.

The Asia-Pacific Partnership on Clean Development and Climate (APP) was announced on July 28, 2005, after months of closed-door negotiations. At the APP's first ministerial meeting in Sydney, Australia, in January, 2006, ministers created a formal charter to provide a structure for the partnership, as well as a work plan. According to the White House, the goals of the APP grew out of the work of earlier initiatives, including the Carbon Sequestration Leadership Forum, the International Partnership for a Hydrogen Economy, and Methane to Markets. With Canada, which joined the APP in 2007, member countries are responsible for about one-half of the world's GHG emissions and contain about one-half of the world's population. Australia, China, India, and the United States are the world's four largest coal-consuming nations.

• **Summary of Provisions**

The formal documents do not include timetables, targets, or dedicated funding; these aspects of the partnership are voluntary and are set individually by each country. The stated goals of the APP are to develop new clean technologies, increase the use of existing clean technologies, address growing energy needs, reduce GHG emissions, protect economic development, enhance international collaboration, and find ways to make use of the private sector.

Eight public-private task forces were established at the Sydney meeting to focus on aluminum, buildings and appliances, cement, cleaner fossil energy, coal mining, power generation and transmission, renewable energy and distributed generation, and steel. In its discussions of the work plan, the Bush administration emphasized the potential of clean coal, coal gasification, and nuclear power, as well as the increased opportunity for investment that would drive private industry innovation.

The APP's work plan describes several specific tasks. Partners will identify possible storage sites for carbon sequestration, develop appropriate power solutions for rural areas, find cleaner ways to manufacture cement and steel, and improve the energy efficiency of buildings. Other projects include creating renewable energy hubs in rural areas of India and China, granting scholarships for studying photovoltaics and solar energy engineering, and developing small wind turbines for remote areas.

- **Significance for Climate Change**

Observers are divided about the possible impact of the Asia-Pacific Partnership. Supporters argue that the flexibility inherent in the partnership will lead to greater compliance and success, while complementing the work of the Kyoto Protocol. Governments and businesses have welcomed the APP as an alternative to the Kyoto Protocol, which some have seen as too restrictive and not cost-effective. Opponents, including many environmental groups, contend that since the partnership's targets are voluntary and since each nation is responsible for monitoring its own compliance, little will be achieved. They believe that the APP was created to supplant, not to complement, the Kyoto Protocol. Some opponents, however, have acknowledged that the formation of APP was a sign of progress. Nonetheless,

From left: Japanese minister of the environment Yuriko Koike, U.S. secretary of energy Samuel Bodman, Australian minister for foreign affairs Alexander Downer, and South Korean minister of commerce Lee Hee-beom participate in the inaugural meeting of the Asia-Pacific Patnership. (David Gray/Reuters/Landov)

in the first two years after the formation of the Asia-Pacific Partnership, none of the member countries succeeded in lowering its GHG emissions.

Cynthia A. Bily

• **Further Reading**

Eichengreen, Barry, Yung Chul Park, and Charles Wyplosz, eds. *China, Asia, and the New World Economy.* New York: Oxford University Press, 2008. Collection of ten essays, including two about Asia's increasing demands for energy and its role in global climate change.

Flannery, Tim. "The Ominous New Pact." *The New York Review of Books* 53, no. 3 (February 23, 2006): 24. Examines the political influence of the Asia-Pacific Partnership and the implications of its reluctance to reduce the consumption of coal.

Gerrard, Michael B., ed. *Global Climate Change and U.S. Law.* Chicago: American Bar Association, 2007. Comprehensive reference book covering legal issues involved in climate change, as well as the Asia-Pacific Partnership and other international frameworks in which the United States participates.

Klare, Michael T. *Rising Powers, Shrinking Planet: The New Geopolitics of Energy.* New York: Metropolitan Books, 2008. Argues that it is increasingly important that the United States cooperate with "petro-superpowers," including Russia, China, and India, in developing alternative energy and working against global climate change.

See also: Byrd-Hagel Resolution; International agreements and cooperation; Kyoto Protocol; United Nations Framework Convention on Climate Change; U.S. energy policy.

Asthma

• **Category:** Diseases and health effects

• **Definition**

Asthma is an inflammatory disease brought on by an allergen or other environmental trigger. This trigger starts a complex interaction among immune system inflammatory cells (mast cells and eosinophils), their mediators (cytokines), the cells lining the airways (the airway epithelium), the smooth muscle that controls the diameter of the airways, and the nervous system. The inflammatory process results in airway hyper-reactivity or bronchial constriction, airway swelling (edema), and an overproduction of mucus, which may form an airway-blocking mucus plug.

Clinically, asthma is characterized by periodic airway obstruction resulting from constriction of lung airways or bronchioles. The increased effort required to overcome this obstruction results in the symptoms of breathlessness, wheezing, coughing, and chest tightness. In severe cases, airway constriction limits the amount of oxygen transported to the bloodstream, leading to hypoxia, which may be life threatening. Importantly, this increase in airway resistance can be reversed with inhalation of medication or a bronchiole dilator that relaxes the airway's smooth muscle and rapidly reduces the symptoms. Airway constriction may occur in response to numerous environmental factors, such as allergens (molds, dust, or pollen), but it may also be induced by cold air, exercise, or occupational exposure. Avoiding allergens can be an effective way of controlling asthma in some patients.

• **Significance for Climate Change**

Around 34.1 million Americans have been diagnosed with asthma by a health care professional. Risk factors for developing asthma include a family history of allergic disease, elevated levels of the allergen-specific immunoglobulin E (IgE), viral respiratory illnesses, exposure to allergen triggers, obesity, and lower income levels. Some studies have shown that exposure to allergens from dust mites during a baby's first year may later lead to development of asthma, while others may develop asthma symptoms following a respiratory-tract viral infection. Increased enviromental temperatures can lead to increased allergens in the environment. Global warming may promote growth of molds and fungi that may increase lung inflammation and trigger asthma attacks. Moreover, higher temperatures and higher carbon dioxide levels, along with an earlier spring, stimulate plant growth, causing more pollens to be released into the air. Increased temperatures are also predicted to increase humidity in urban areas, and each 10 percent increase in indoor humidity has been shown to increase the incidence of asthma symptoms by 2.7 percent. With an epidemic of childhood asthma observed in Western countries, the need for new treatments focusing on reducing the inflammatory disease process, along with the recognition of environmental triggers, will result in better disease control for asthma patients in the future.

Robert C. Tyler

See also: Air pollution and pollutants: anthropogenic; Air pollution and pollutants: natural; Air pollution history; Air quality standards and measurement; Diseases; Mold spores.

Atlantic heat conveyor

• **Categories:** Oceanography; meteorology and atmospheric sciences

• Definition

The Atlantic heat conveyor transfers heat from the equator northward to the polar region of the Atlantic Ocean, through the meridional overturning circulation (MOC). Because the Earth is a sphere, incoming solar radiation heats the surface of the Earth unevenly, with the equator receiving a greater amount of heat per surface area than the polar regions. This uneven heating creates temperature gradients that drive atmospheric currents (wind) and surface currents in the ocean.

In the Atlantic Ocean, the Gulf Stream is the surface current that transports heat from the equator to the polar region. The Gulf Stream is a western boundary current that originates in the Gulf of Mexico and travels north along the east coast of the United States and Newfoundland. Near 50° north, the Gulf Stream splits into two branches: the North Atlantic drift (or North Atlantic current) and the Azores current. The North Atlantic drift (NAD) flows northeastward toward northern Europe, while the Azores current flows east toward the Azores and then south as the Canary current. The Gulf Stream is a large, fast-moving current, transporting between 30 million and 150 million cubic meters of water per second. This flow of water transports approximately 1.4 petawatts (1.4×10^{15} watts) of heat per year. As the Gulf Stream moves north and mixes with cooler surface water from the poles, it releases this heat.

The Atlantic heat conveyor may be responsible for maintaining a milder climate in northern Europe, as compared to Newfoundland, which is located at the same latitude. However, this idea has been challenged by scientists, who hypothesize that atmospheric heat transport is more important in maintaining a mild climate in northern Europe than is oceanic heat transfer.

• Significance for Climate Change

The Atlantic heat conveyor delivers a large amount of heat to the Arctic region from the equator, making it an important component of the global climate system and of any changes in that system. One consequence of global warming is the melting of glaciers in the Arctic, which would lead to a large flux of freshwater into the surface waters of the Greenland Sea. This cap of freshwater could serve as a barrier to the northward flow of the Gulf Stream, thereby blocking the Atlantic MOC and the heat conveyor. Thus, according to this conceptual model, projected global warming would actually lead to a localized cooling of the Northern Hemisphere, particularly Europe.

In 2005, scientists from the National Oceanography Center presented data to suggest that the Atlantic MOC had slowed during the late twentieth century. These data were subsequently challenged by other scientists, who presented different data sets that showed no MOC slowdown. Despite the controversy regarding whether such slowdown has occurred, computer models consistently indicate that a shutdown of the Atlantic heat conveyor could lead to a lesser warming or even cooling in the Northern Hemisphere as a result of global warming. Geologic evidence suggests that the Younger Dryas, a time of global cooling that lasted from 12,800 to 11,500 years before present, may have been caused by MOC collapse in the Atlantic because of a large influx of freshwater to the North Atlantic from the emptying of glacial Lake Agassiz.

The hypothesis of the Atlantic heat conveyor has been criticized by scientists from the Lamont-Doherty Earth Observatory, who believe that the atmospheric transport of heat and the slow response time of the ocean are the main reasons for Europe's mild climate. Additionally, they point to long waves in the atmosphere, created as air masses flow around the Rocky Mountains, as a third potential cause of a mild European climate. According to this hypothesis, global warming will not stop the Atlantic heat conveyor, nor will it lead to a cooling of Europe. This view, however, is not widely held among climate scientists or policy makers. In its Fourth Assessment Report, released in 2007, the Intergovernmental Panel on Climate Change (IPCC) found that there was insufficient evidence to determine if there were trends—either weakening or increasing—in the magnitude of the Atlantic heat conveyor.

Anna M. Cruse

• Further Reading

Alley, Richard B. "Abrupt Climate Change." *Scientific American* 291, no. 5 (November, 2004): 62-69. Article written by one of the pioneers of ice-core records; explains the science behind abrupt climate change and thermohaline circulation.

Schellnhuber, Hans, et al., eds. *Avoiding Dangerous Climate Change.* New York: Cambridge University Press, 2006. Collection of papers presented by leading climatologists at the Avoiding Dangerous Climate Change conference, hosted by the British government in 2005. In particular, chapters 5 and 6 discuss THC in the Atlantic Ocean. Figures, tables, references.

Seager, Richard. "The Source of Europe's Mild Climate." *American Scientist* 94, no. 4 (July/August, 2006): 334-341. Dissenting hypothesis on the role of the Atlantic heat conveyor in maintaining a mild climate in Europe, written by a leading climate scientist.

Vellinga, Michael, and Richard A. Wood. "Global Climatic Impacts of a Collapse of the Atlantic Thermohaline Circulation." *Climatic Change* 54, no. 3 (August, 2002): 251-267. Presents data from a computer model to determine the rate of climate response if thermohaline circulation in the Atlantic were to collapse. Among the variables investigated are air temperature, rainfall patterns, the location of the Inter-Tropical Convergence Zone, and variations in oceanic primary productivity.

See also: Europe and the European Union; Gulf Stream; Intergovernmental Panel on Climate Change; Meridional overturning circulation; Ocean dynamics; Younger Dryas.

Atlantic multidecadal oscillation

• Categories: Oceanography; meteorology and atmospheric sciences

• Definition

The Atlantic multidecadal oscillation (AMO) is a global-scale mode of multidecadal climate variability (that is, it is an example of a cyclical or semicyclical pattern of climate change that repeats on a timescale on the order of several decades). The AMO is based on sea surface temperatures in the North Atlantic Ocean between the equator and 70° north latitude. Generally, the AMO is computed as a detrended ten-year running mean of these sea surface temperatures and represents variability across the entire North Atlantic basin.

The AMO exhibits a long-term, quasi-cyclic variation at timescales of fifty to seventy years. Modeling studies reveal that multidecadal variability in the North Atlantic Ocean is dominated by this single mode of sea surface temperature variability. The range in AMO values between warm and cold extremes is only about 0.6° Celsius; however, because the North Atlantic Ocean is so large, even small differences in sea surface temperatures represent extremely large exchanges in energy between the ocean and atmosphere.

The predominant hypothesis is that the AMO is primarily driven by the thermohaline circulation. This hypothesis is supported by both instrumental and climate-model studies. The thermohaline circulation is the large-scale ocean circulation that moves water among all of the world's oceans and is driven by density differences in ocean water caused by heat and freshwater fluxes. When the thermohaline is fast, warm water is moved from tropical areas into the North Atlantic Ocean and the AMO enters a warm phase. When the thermohaline is slow, warm water is not readily moved into the North Atlantic Ocean and the AMO enters a cool phase. During the twentieth century, the AMO was in a warm phase from 1926 through 1963, and it was in cool phases from 1905 through 1925 and 1964 through 1994. In 1995, it entered another warm phase.

Research has suggested that the North Atlantic Ocean may provide information that explains significant amounts of multidecadal climate variability. For example, when the AMO is in a warm phase, the likelihood of drought in North America increases, and when the AMO is in a cool phase, the likelihood of drought in North America decreases. Analysis of major U.S. droughts during the last century indicates that North Atlantic Ocean surface temperatures were warm during the 1930's and 1950's droughts, as well as during the dry period that began in the late 1990's. In contrast, both the early (1905-1920) and late (1965-1995) twentieth century pluvials in the western United States were associated with cool North Atlantic Ocean surface temperatures. The AMO also has been linked to the occurrence of hurricanes in the North Atlantic Ocean. During warm phases of the AMO, hurricanes are more frequent, whereas when the AMO is in a cool phase hurricane frequency decreases.

• Significance for Climate Change

Because the AMO is an important mode of multidecadal climate variability, there are a number of important implications of the AMO for the study of climate variability and change. Understanding what portion of climate variability is due to multidecadal variability, such as that driven by the AMO, allows the discrimination of anthropogenic changes in climate from natural variability. Knowledge of multidecadal climate variability, such as that indicated by the AMO, also has implications for defining and potentially estimating risks in agriculture, water resources, public health, and nature. Variability of the AMO also has an effect on global temperature, and the beginning of a new warm phase of the AMO in 1995 may have contributed to the strong warming of global temperatures in the years immediately following.

Some scientists suggest that if the AMO shifts into a cool phase, the cooling of North Atlantic Ocean surface temperatures may reduce the amount of global warming and may result in a leveling of global temperatures for about a decade. In addition, research has indicated that North Atlantic Ocean surface temperatures may have predictability on the order of a decade or longer, which has important implications for climate forecasting.

The actual physical mechanisms that explain the associations between the North Atlantic Ocean and global climate are still unknown, but several possible mechanisms have been recognized. North Atlantic Ocean surface temperatures may affect Northern Hemisphere atmospheric circulation, such that the frequency of zonal versus meridional atmospheric flow is modulated. Decadal-to-multidecadal variability of North Atlantic Ocean surface temperatures may be aliasing for low-frequency or lagged variations of the tropical oceans. North Atlantic Ocean surface temperatures may be influencing the location and strength of subtropical high pressures. Finally, the North Atlantic Ocean may be modulating the strength and variability of tropical Pacific Ocean surface temperatures.

Gregory J. McCabe

• Further Reading

Enfield, D. B., A. M. Mestas-Nuñez, and P. J. Trimble. "The Atlantic Multidecadal Oscillation and Its Relation to Rainfall and River Flows in the Continental U.S." *Geophysical Research Letters* 28 (2001): 277-280. Provides an explanation of the effects of the AMO on precipitation and streamflow in North America.

Knight, Jeff R., Chris K. Folland, and Adam A. Scaife. "Climate Impacts of the Atlantic Multidecadal Oscillation." *Geophysical Research Letters* 33, no. 17 (September, 2006). Describes some of the climate effects of variations in the AMO.

McCabe, G. J., M. A. Palecki, and J. L. Betancourt. "Pacific and Atlantic Ocean Influences on Multidecadal Drought Frequency in the United

States." *Proceedings of the National Academy of Sciences* 101 (2004): 4136-4141. Identifies a link between the AMO and drought occurrence in the western United States.

Sutton, R. T., and D. L. R. Hodson. "Climate Response to Basin-Scale Warming and Cooling of the North Atlantic Ocean." *Journal of Climate* 20 (2007): 891-907. Discusses the use of a climate model to assess the effects of basin-wide warming and cooling of the North Atlantic Ocean.

Zhang, Rong, Thomas L. Delworth, and Isaac M. Held. "Can the Atlantic Ocean Drive the Observed Multidecadal Variability in Northern Hemisphere Mean Temperature?" *Geophysical Research Letters* 34, no. 2 (January, 2007). Discusses a link between the AMO and Northern Hemisphere temperature variability.

See also: Atlantic heat conveyor; Ekman transport and pumping; El Niño-Southern Oscillation; La Niña; Meridional overturning circulation; Sea surface temperatures; Thermohaline circulation.

Atmosphere

• **Category:** Meteorology and atmospheric sciences

Earth's climate system has five components: the atmosphere, hydrosphere, cryosphere, lithosphere, and biosphere. Among these components, the atmosphere is the most sensitive to changes in Earth's climate. Thus, current data suggesting global warming are mostly atmospheric, observed as an increase of near-surface air temperature over the last hundred years.

• Key concepts

condensation: the transformation of a substance from its gaseous to its liquid state, accompanied by a release of heat

convection: motion in a fluid that results in the transport and mixing of the fluid's physical properties, such as heat

coupled atmosphere-ocean models: computer simulations of alterations in and interactions between Earth's atmosphere and oceans

El Niño and La Niña: periodic warming and cooling of the eastern tropical Pacific Ocean that affects global weather patterns

greenhouse gases (GHGs): atmospheric trace gases that allow sunlight to reach Earth's surface but prevent heat from escaping into space

latent heat: the heat released or absorbed by a change of state, such as condensation

• Background

The atmosphere is a layer of gas surrounding the Earth. It is a mixture of several components, mostly gas-phased molecules. The total weight of Earth's atmosphere is about 5.08 quadrillion metric tons. Its existence makes possible life on Earth, and changes in its composition may destroy life on Earth.

The atmosphere's various physical properties affect many aspects of human existence. Its optical properties make the sky blue and create rainbows and auroras. It carries sound waves, making aural communication possible. It also propagates heat, and seasonal changes in atmospheric thermal properties result in cooler and warmer temperatures at different times of year. More generally, the weather is a function of the atmosphere: Wind, precipitation, humidity, and storm systems are all atmospheric phenomena.

• Atmospheric Structure

Because of the Earth's gravity, most atmospheric molecules are distributed very close to the Earth's surface. That is why the atmosphere is relatively thin. More than 90 percent of Earth's atmosphere lies within 32 kilometers of the planet's surface, and the lower atmosphere begins to shade into the upper atmosphere and outer space at an altitude of about 80 kilometers. This distance is only a fraction of the Earth's radius, which is about 6,400 kilometers.

The atmosphere's density and pressure decrease steadily with altitude. Altitudinal variations in temperature, on the other hand, are more complex. The temperature decreases with altitude from the surface to about 10-16 kilometers high. From that height to about 20 kilometers, the temperature remains relatively constant. From 20 to 45 kilometers,

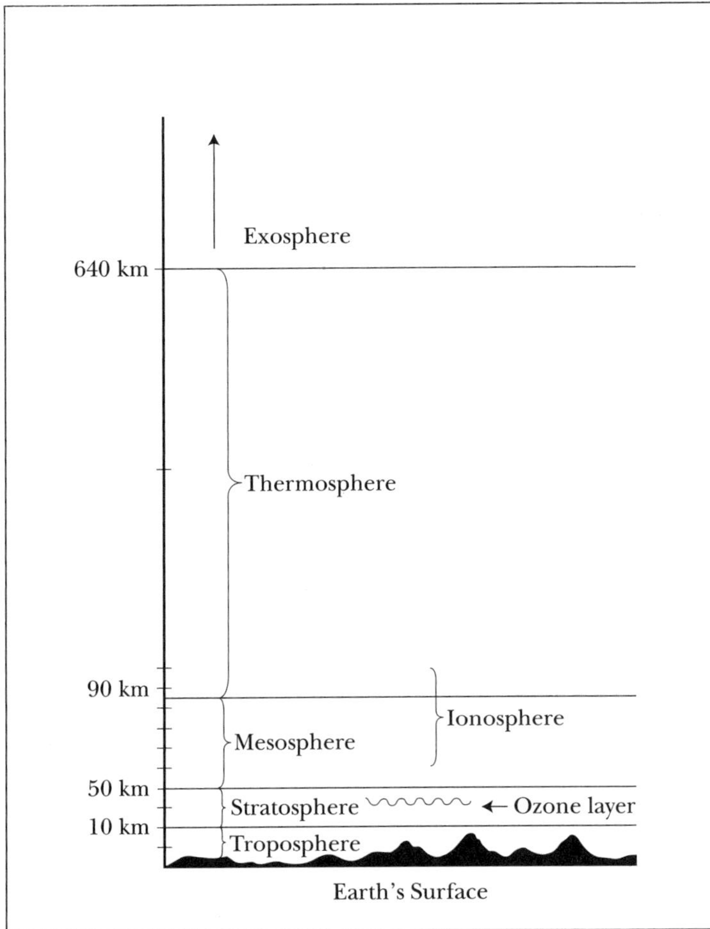

the temperature increases with height, and at altitudes of 45-50 kilometers it again remains constant. Above 50 kilometers, the temperature again begins to decrease with altitude, continuing until reaching a height of about 80 kilometers. From that height to about 90 kilometers, temperature is constant. From 90 kilometers to the end of Earth's atmosphere, temperature increases with altitude. Based on these altitudinal temperature variations, scientists divide the atmosphere into sections: the troposphere, tropopause, stratosphere, stratopause, mesosphere, mesopause, and thermosphere.

The atmosphere can also be classified according to its composition. In the lower atmosphere (below 80 kilometers), weather convection and turbulence mixing keep the composition of air fairly uniform. This area is therefore known as the homosphere. Above the homosphere, collisions among atoms and molecules are infrequent, and the air is unable to keep itself well mixed. This layer is therefore called the heterosphere. Furthermore, above the stratosphere, large concentrations of ions and free electrons exist, in a layer known as the ionosphere.

• The Energy Budget and the Greenhouse Effect

The Sun is the ultimate energy source for the Earth. Solar energy reaches the Earth via solar radiation. During daytime, the Earth is warmed by the Sun, while during nighttime the Earth cools. The sunlight that the Earth receives is a form of electromagnetic radiation with a relatively short wavelength, called "shortwave radiation." When the Earth cools during night, the heat it releases into space is another form of electromagnetic radiation with a relatively long wavelength—"longwave radiation," or infrared radiation (IR).

If Earth did not have an atmosphere, it would simply receive shortwave solar radiation during the day and give out longwave radiation at night. When an equilibrium was reached (meaning that the amount of energy received was equal to the amount of energy given out), the Earth would be balanced at an equilibrium temperature. This temperature would be much lower than is the actual temperature on Earth: Some scientists have estimated it at about –18° Celsius. Earth's surface temperature is thus much warmer than it would be without an atmosphere.

This difference in temperature is a result of the greenhouse effect. Shortwave solar radiation can mostly penetrate Earth's atmosphere and reach the its surface. Infrared radiation, by contrast, is intercepted by the atmosphere, and some of this longwave radiation is reflected back to Earth's surface, increasing the equilibrium temperature of the planet. The strength of the greenhouse effect and the warmth of Earth's equilibrium temperature are dependent on the specific composition and properties of the atmosphere. Changes in atmospheric chemistry can significantly alter Earth's energy balance and equilibrium temperature.

• Atmospheric Composition, Chemistry, and GHGs

The atmosphere is composed of nitrogen (about 78 percent by volume), oxygen (21 percent), and various trace gases (totaling about 1 percent). If all trace gases were removed, these percentages for nitrogen and oxygen would remain fairly constant up to an altitude of about 80 kilometers. At the planet's surface, there is an approximate balance between the destruction (output) and production (input) of these gases.

The two most plentiful components of the atmosphere, nitrogen and oxygen, are of significance to life on Earth. Humans and animals cannot live without oxygen. By contrast, the trace noble gases, such as argon, neon, and helium, are not very active chemically. Water vapor is distributed inconsistently in the lower atmosphere; its concentration varies greatly from place to place and from time to time, and it can constitute from 0 to 4 percent of local air. This variable concentration is one reason that water vapor is so important in influencing Earth's weather and climate.

Water vapor provides the main physical substance of storms and precipitation, and its condensation into liquid water generates the large amount of energy (latent heat) necessary to initiate powerful and violent storms. Water vapor is also a greenhouse gas (GHG): It strongly absorbs longwave radiation and reemits this radiation back to the Earth, causing global warming. Clouds, which are generated from water vapor, also play an extremely important role in climate and climate change.

Another very important GHG is carbon dioxide (CO_2). Observations indicate that the concentration of CO_2 in the atmosphere has been rising steadily for more than a century. The increase of CO_2 concentration indicates that CO_2 is entering the atmosphere at a greater rate than its rate of removal. This rise is largely attributable to the burning of fossil fuels, such as coal and oil. Deforestation also contributes to the increase in atmospheric CO_2 concentration. Estimates project that by sometime in the second half of the twenty-first century, CO_2 levels will be twice as high as they were early in the twentieth century. Other GHGs include methane, nitrous oxide, and chlorofluorocarbons.

Ozone (O_3) is another important gas for Earth's weather and climate. At Earth's surface, O_3 is a major air pollutant, and it is closely monitored for its effects on air quality. However, at upper levels (about 25 kilometers high), O_3 forms a shield for Earth's inhabitants from harmful ultraviolet solar radiation. For this reason, the loss of O_3 high in the atmosphere as a consequence of human activity has become a serious global-scale issue. One of the examples of O_3 depletion is the O_3 hole found over Antarctica. Finally, aerosols, including particulate matter, are also important constituents of the atmosphere, affecting weather formation, air quality, and climate change.

• Weather and Climate

Atmospheric conditions can generally be classified as either weather or climate. Weather is a particular atmospheric state at a given time and place. Climate is an average of weather conditions at a given location over a period of time.

Weather includes many atmospheric phenomena of different scales, including middle latitude cyclones (extratropical cyclones), hurricanes (tropical cyclones), heavy rains and floods, mesoscale convective systems, thunderstorms, and tornadoes. Climate includes atmospheric conditions that are millennial, centennial, decadal, interannual, or seasonal. For example, global warming can occur on a centennial or longer timescale, an El Niño or La Niña episode will generally occur on an interannual timescale, and seasonal changes occur on relatively short timescales of months, weeks, or days.

• Global Warming and Climate Change

Earth's climate has changed constantly over its history. The planet has experienced many cold periods, as well as several warm periods. For example, about ten thousand years ago, the Earth cooled during a period known as the Younger Dryas, when the average global temperature was about 3° Celsius colder than it is today. However, about six thousand years ago, the Earth reached the middle of an interglacial period, known as the Mid-Holocene Maximum. The temperature then was 1° Celsius higher than today's norm. Some of the natural mechanisms causing this climatological variabilty include: drift of plate tectonics, volcanic activities, ocean circulations, variations in Earth's orbit, and solar variability.

The rapid warming that has occurred in the past hundred years seems to coincide with the socioeconomic development and industrialization of humankind. During the past century, human life and societies began to depend heavily on burning fossil fuels. As a result, increasing amounts of CO_2 have been added to the atmosphere. Global temperatures and CO_2 levels evince a consistent upward trend during the same period. Therefore, many scientists believe that human activity may have contributed to global warming.

Various coupled atmosphere-ocean models have projected that Earth will experience an average temperature increase over the course of the twenty-first century of between 1.4° and 5.8° Celsius. Some possible consequences of this global warming include higher maximum and minimum temperatures, more hot days and heat waves, fewer cold days and frost days, more intense precipitation events, more summer drying and drought, increased tropical cyclone intensity, increased Asian summer monsoon precipitation variability, intensified droughts and floods associated with El Niño events, and increased intensity of midlatitude storms. Global warming will also exert some profound effects on many other social and environmental issues. For example, the distribution of water resources and farming may be changed by future warmer climates. Warming-induced sea-level rise can have significant effects on many countries' coasts. Arctic sea-ice melting can also have geopolitical and economic consequences.

• Context

The atmosphere is central to almost all aspects of human existence, constituting not only the source of vital oxygen but also the medium of movement, sound, and weather. It is also the most variable component of Earth's climate system. Because that system is so complex and interconnected, changes in the atmosphere will inevitably result in changes to the rest of the system, some of which are extremely difficult to predict. At the same time, feedback from other components of Earth's environment may exert a significant influence on the atmosphere, including producing positive and negative feedback loops that help alter or maintain Earth's climate.

Chungu Lu

• Further Reading

Ahrens, C. Donald. *Essentials of Meteorology: An Invitation to the Atmosphere.* 5th ed. Belmont, Calif.: Thomson Brooks/Cole, 2008. Widely used introductory textbook on atmospheric science; covers a wide range of topics on weather and climate.

Climate Change Science Program, U.S. *Our Changing Planet.* Washington, D.C.: Author, 2006. Report by a committee established in 2002 to empower the U.S. national and global community with the science-based knowledge to manage risks and capitalize on opportunities created by changes in the climate and related environmental systems.

Intergovernmental Panel on Climate Change. *Climate Change, 2007—Synthesis Report: Contribution of Working Groups I, II, and III to the Fourth Assessment Report of the Intergovernmental Panel on Climate Change.* Edited by the Core Writing Team, Rajendra K. Pachauri, and Andy Reisinger. Geneva, Switzerland: Author, 2008. Comprehensive overview of global climate change published by a network of the world's leading climate change scientists under the auspices of the World Meteorological Organization and the United Nations Environment Programme.

_______. *Climate Change, 2007—The Physical Science Basis: Contribution of Working Group I to the Fourth Assessment Report of the Intergovernmental Panel on Climate Change.* Edited by Susan Solomon, et al. New York: Cambridge University Press, 2007. Companion volume to *The Synthesis Report* that provides more detail on various aspects of climate change and global warming.

Lutgens, Frederick K., and Edward J. Tarbuck. *The Atmosphere.* 10th ed. Upper Saddle River, N.J.: Pearson Prentice Hall, 2007. Introductory textbook that covers a wide range of atmospheric sciences.

See also: Atmospheric boundary layer; Atmospheric chemistry; Atmospheric dynamics; Atmospheric structure and evolution; Climate and the climate system; Climate change; Ekman transport and pumping; El Niño-Southern Oscillation; Greenhouse effect; Greenhouse gases; Hadley circulation; Inter-Tropical Convergence Zone; Mesosphere; Ocean-atmosphere coupling; Oxygen, atmospheric; Stratosphere; Thermosphere; Troposphere.

Atmospheric boundary layer

• **Category:** Meteorology and atmospheric sciences

• **Definition**

Also known as the planetary boundary layer (PBL), the atmospheric boundary layer (ABL) is the lowest 10-20 percent of the troposphere, the lowest layer of the atmosphere. Its contact with a planetary surface directly influences its behavior. The ABL contains a disproportionately large amount of the mass and the kinetic energy of the atmosphere. It is the most dynamically active of the layers of Earth. The phrase "boundary layer" originates in the study of boundary layers in fluid flows: The boundary layer is the layer of fluid that is most influenced by friction with the Earth's surface. A defining characteristic of the ABL is turbulence caused by thermal convection, due to thermal buoyancy, and wind shear, due to frictional forces. The atmospheric boundary layer has three layers: the surface layer, the core, and the entrainment layer, also called the capping inversion layer.

• **Significance for Climate Change**

The atmospheric boundary layer is important for ensuring that Earth's atmospheric composition remains relatively homogeneous throughout, despite external heat and energy inputs. Consequently, it is important for ensuring that life can be sustained on Earth. ABL considerations are particularly important in the area of the urban environment. The development of large cities has changed the ABL in those areas, resulting in surface heating, and artificial boundary layers have developed that trap pollutants. Calculations of the boundary layer can help architects develop urban environments in such a way as to minimize impact on the boundary layer. ABL research can help improve weather forecasts, especially long-term forecasts and forecasts of longtime climate models. Without a properly constituted boundary layer, Earth would lose the unique conditions that make it hospitable to human existence.

The ABL is important meteorologically in the area of assessing convective instability. The entrainment zone (at the top of the ABL) acts as a lid on rising air parcels attributable to temperature inversion. If that entrainment layer is broken, capped air parcels can rise freely, resulting in vigorous convection that produces severe thunderstorms. When sunlight enters the atmosphere, a part of it—known as the albedo—is immediately reflected back to space; the remainder penetrates the atmosphere, and the Earth's surface absorbs it. This energy is then reemitted by the Earth back into the atmosphere in the form of longwave radiation. Carbon dioxide and water molecules absorb this energy, then emit much of it back toward Earth again. This delicate exchange of energy between the Earth's surface and the atmosphere is what keeps the average global temperature from changing drastically from year to year. When this exchange is disrupted, climate problems result.

Victoria Price

See also: Albedo feedback; Atmosphere; Atmospheric dynamics; Atmospheric structure and evolution; Planetary atmospheres and evolution; Stratosphere; Thermosphere; Thunderstorms; Troposphere; Weather forecasting.

Atmospheric chemistry

• **Categories:** Chemistry and geochemistry; meteorology and atmospheric sciences

The Earth's global temperature, as well as the amount of solar radiation reaching its surface, can be significantly influenced by changes in the concentrations of chemicals naturally present in the atmosphere, such as natural GHGs, and by anthropogenic chemicals, such as CFCs.

• Key concepts

anthropogenic: caused by humans

chlorofluorocarbons (CFCs): compounds of chlorine, fluorine, and carbon, popularly known by the trade name Freon

greenhouse effect: result of atmospheric trace gases that allow high-energy sunlight to reach the terrestrial surface but absorb low-energy heat that is radiated back

greenhouse gases (GHGs): tropospheric gases such as carbon dioxide, methane, and water vapor that cause the greenhouse effect

ozone layer: a stratospheric region containing relatively high concentrations of triatomic oxygen (ozone) that prevents much ultraviolet solar radiation from reaching Earth's surface

primary air pollutants: harmful substances that are emitted directly into the atmosphere

secondary air pollutants: harmful substances that result from the reaction of primary air pollutants with principal atmospheric components

stratosphere: an atmospheric region extending from about 17 to 48 kilometers above the Earth's surface

troposphere: an atmospheric region extending from the Earth's surface to about 17 kilometers high over equatorial regions and to about 8 kilometers high over polar regions

• Background

The significance of the Earth's atmosphere is vastly disproportionate to its size. Although its thickness relative to Earth's sphere is comparable to an apple's skin, it is essential for life. It was not until the eighteenth century that scientists began to understand the role of atmospheric gases such as oxygen and carbon dioxide (CO_2) in plant and animal life, and it was not until the end of the nineteenth century that scientists grasped the details of how soil microorganisms utilized atmospheric nitrogen to create compounds necessary for the health of plants and animals. Throughout the twentieth century, climatologists, atmospheric chemists, and others gathered information about how such anthropogenic gases as CO_2, methane, and nitrous oxide were increasing Earth's greenhouse effect and elevating the planet's average global temperature. This enhanced greenhouse effect fosters climate changes that are potentially so devastating that some scholars have called climate change the most important issue of the twenty-first century.

• Chemical Composition of the Earth's Atmosphere

Approximately three-quarters of the Earth's air mass is located in the troposphere, and dry air in this region is 78.1 percent nitrogen, 20.9 percent oxygen, and 0.93 percent argon by volume. The troposphere also contains trace amounts of many other gases, such as methane, various nitrogen oxides, ammonia, sulfur dioxide, and ozone, and these come from both natural and anthropogenic sources. Human activities have not changed the concentrations of the major gases in the atmosphere—nitrogen and oxygen—but scientific evidence accumulated over the past century indicates that human beings, particularly in advanced industrialized societies, are dramatically affecting the concentrations of certain trace gases. Examples of these include CO_2, methane, nitrous oxide, carbon monoxide, chlorofluorocarbons (CFCs), and sulfur dioxide. Some of these atmospheric trace gases, such as CFCs, result from certain industries and their products, such as refrigerants and aerosols. Others, such as CO_2 and sulfur dioxide, are produced by burning fossil fuels. Agricultural practices are also significant sources of such gases as methane and nitrous oxide.

Although the Earth's stratosphere contains much less matter than the troposphere, it contains similar proportions of such gases as nitrogen and oxygen. It differs markedly from the troposphere, however, in its concentrations of water vapor and ozone. Stratospheric water-vapor concentrations are only about one-thousandth of tropospheric conentrations, but ozone concentrations are much higher in the stratosphere. Ozone is localized in a layer ranging from about 15 to 35 kilometers above Earth's surface. This ozone layer, whose molecules are created when oxygen interacts with high-energy solar radiation, prevents about 95 percent of the Sun's ultraviolet radiation from reaching Earth's surface, where it could damage living organisms. The ozone layer also prevents tropospheric oxygen from being converted to ozone, which, in the lower atmosphere, is a dangerous air pollutant.

• Chemical Reactions in the Troposphere

Besides being home to such major gases as nitrogen and oxygen, the troposphere contains hundreds of other distinctive molecules, leading to myriad chemical reactions, some of which have an influence on climate change. Because oxygen is such a reactive species, many of these reactions are oxidations, and some scientists see these reactions as constituting a low-temperature combustion system. Fueling this combustion are chemicals released from both natural and artificial sources. For example, methane enters the troposphere in large amounts from swamp and bog emissions, termites, and ruminant animals. Human activities contribute a large number of organic compounds, and CO_2 and water are the end results of their oxidation. CO_2 and water vapor are powerful greenhouse gases (GHGs).

Atmospheric chemists have also been attempting to work out in detail the influence of chemical radicals on tropospheric gases. Such charged groups of atoms as the hydroxyl radical (composed of hydrogen and oxygen) play an important role in the daytime chemistry of the troposphere, and the nitrate radical (composed of nitrogen and oxygen) is the dominant nighttime oxidant. Fossil-fuel combustion is a significant contributor to tropospheric pollution. Particulates such as soot were a factor in some "killer smogs," and scientists have recently discovered that particulates contribute to global dimming, a lessening of sunlight's ability to penetrate particle-filled hazes and reach the Earth's surface. Sulfur dioxide, which is produced by the combustion of certain kinds of coal and oil, can be a primary air pollutant, since it is toxic to living organisms as well as damaging to buildings. It can also be a secondary air pollutant, because it reacts with water vapor to create sulfuric acid, which is an acid-rain component, causing harm to various lifeforms, including trees and fish.

• Chemical Reactions in the Stratosphere

Just as in the lower atmosphere, chemical reactions in the upper atmosphere exhibit great variety, and some of these reactions have an important influence on climate change. Over the past decades, the chemical species that has received the most attention has been ozone. Scientists paid heightened attention to the chemical reactions in the ozone layer when, in the late 1980's, a hole was discovered in this layer above the Antarctic. During the 1970's scientists had found a threat to the ozone layer when they worked out the reactions between chlorine-containing radicals and ozone. These reactions changed ozone molecules into diatomic oxygen molecules, thus weakening the ability of the ozone layer to protect Earth's surface from high-energy solar radiation.

A primary source of these catalytic, chlorine-containing species turned out to be CFCs. General Motors had introduced CFCs in 1930, and they proved to be successful in such products as refrigerator and air-conditioning coolants, as well as aerosol propellants. Because of the widespread and accelerating use of CFCs, the tropospheric concentrations of these chemicals increased from the 1930's to the 1970's, when Mexican chemist Mario Molina and American chemist F. Sherwood Rowland showed that CFCs, although seemingly inert in the troposphere, became very reactive in the stratosphere. There, ultraviolet radiation split the CFCs into highly reactive radicals that, in a series of reactions, promoted the debilitation of the protective ozone shield.

The exhaust from aircraft and spacecraft also helped deplete stratospheric ozone. Despite attempts, such as the Montreal Protocol (1987), to reduce concentrations of CFCs and other ozone-depleting chemicals in the atmosphere, the Antarctic ozone hole continued to grow in the 1990's and early twenty-first century. This meant that countries near Antarctica began experiencing higher levels of ultraviolet solar radiation.

• Atmospheric Chemistry and Global Climate Change

Humans tend to be most aware of weather—that is, a local area's short-term temperature and precipitation variations. Scientists such as atmospheric chemists tend to concentrate on climate, or a large region's long-term variations in temperature, precipitation, and cloud cover. Because of discoveries revealing the extreme complexity of chemical reactions in the atmosphere, atmospheric chemistry has become a profoundly interdisciplinary field, depending on new facts and ideas found by physicists, meteorologists, climatologists, oceanographers, geologists, ecologists, and other scientists.

Paleoclimatologists have studied changes in Earth's atmosphere over hundreds of millions of years, while other environmental and atmospheric chemists have focused on such pivotal modern problems as global warming. These studies have led to research aimed at understanding the causes of global warming and the development of theories to explain existing data. Particularly useful has been computerized modeling of Earth's atmosphere, through which experiments can be performed to help scientists understand likely future effects of climate change. These theoretical predictions have placed pressure on various governments to make important changes in policy, such as taxing fossil-fuel use to motivate reductions in GHG emissions.

Atmospheric chemists have come to realize that the goal of their research on global climate change is to understand the relevant chemical species in the atmosphere, their reactions, and the role of anthropogenic chemicals, especially GHGs, in bringing about global warming. Many atmospheric chemists believe that the greenhouse effect is a certainty, and they are also highly confident that human activities generating GHGs are a significant element in the recent rise in average global temperatures. Less certain are predictions about the future.

Computer models developed to synthesize and test theories about the complex chemical interactions in the troposphere and stratosphere necessarily involve assumptions and simplifications. For example, the numbers of chemical compounds and their reactions have to be reduced to formulate even a crude working model of the Earth's atmosphere. Despite these problems, many environmental chemists, building on what they are most sure of, have played an important part in several countries in determining governmental policies as they relate to global climate change.

• Context

Atmospheric chemists' discoveries have had a major influence on how environmentalists and other scientists understand the gravity, interrelatedness, and complexity of atmospheric problems. Many atmospheric chemists educate their students and the public about issues relating to global climate change, while others have been carefully monitoring the changes in the Earth's atmosphere. They have also participated in international discussions and agreements about controlling GHG emissions, developing substitutes for CFCs, and passing local and international laws that would lessen the likeli-

Global Sulfur Emissions by Source and Latitude

Atmospheric chemical inputs can vary greatly by region, as illustrated by the table below listing the vastly different sources of atmospheric sulfur in different parts of the globe.

Latitude	*Anthropogenic* %	*Marine* %	*Terrestrial* %	*Volcanic* %	*Biomass Burning* %
90° south	0	0	0	0	0
75° south	0	80	0	19	1
58° south	2	97	0	0	1
45° south	22	72	0	9	1
28° south	67	28	0	1	4
15° south	21	47	1	22	10
0°	21	39	1	33	7
15° north	40	30	1	19	1
28° north	85	6	0	8	1
45° north	88	4	0	7	1
58° north	86	3	0	10	1
75° north	30	40	0	23	7
90° north	0	0	0	0	0

Source: Pacific Marine Environmental Laboratory, National Oceanic and Atmospheric Administration.

hood of some catastrophic scenarios predicted by various computer models. Just as the many components and reactions in the atmosphere make a full understanding of these complexities very difficult, so, too, environmental chemists find themselves in an even more complex milieu in which they have to integrate their understanding with those of other scientists, industrialists, and government officials in both developed and developing countries. Therefore, though global climate change is, at root, a physical and chemical issue, to solve the problem of global climate change will require an integrated, multidisciplinary, and international approach that, though daunting, appears to be increasingly necessary.

Robert J. Paradowski

• Further Reading

Birks, John W., Jack G. Calvert, and Robert E. Sievers, eds. *The Chemistry of the Atmosphere: Its Impact on Global Change—Perspectives and Recommendations.* Washington, D.C.: American Chemical Society, 1993. Intended to be "many things to many different people," including scientists, politicians, and the public, this book grew out of an international conference on atmospheric chemistry. Contains lists of concrete proposals to ameliorate harmful atmospheric changes. Appendixes and index.

Jacob, Daniel J. *Introduction to Atmospheric Chemistry.* Princeton, N.J.: Princeton University Press, 2007. Undergraduate textbook written by a Harvard professor; provides an overview of the new and rapidly growing field of atmospheric chemistry. Illustrations and index.

Makhijani, Arjun, and Kevin R. Gurney. *Mending the Ozone Hole: Science, Technology, and Policy.* Cambridge, Mass.: MIT Press, 1996. Called "the most comprehensive overview of the ozone-depletion problem," this book, accessible to both scientists and general readers, analyzes the problem as well as various solutions. Helpful summaries of chapters, appendixes, notes, twenty-five pages of references, and an index.

Seinfeld, John H., and Spyros N. Pandis. *Atmospheric Chemistry and Physics: From Air Pollution to Climate Change.* New York: John Wiley & Sons, 1998. This massive volume is intended "to provide a rigorous, comprehensive treatment of the chemistry of the atmosphere." Assumes a reader has had introductory courses in chemistry, physics, and calculus, including differential equations. Many figures and tables, appendixes, and index.

Wayne, Richard P. *Chemistry of Atmospheres.* 3d ed. New York: Oxford University Press, 2000. Undergraduate textbook that analyzes the principles of atmospheric chemistry, including both traditional and contemporary developments and both terrestrial and other planetary atmospheres. Illustrated with many graphs and figures; with an extensive bibliography and index.

See also: Atmosphere; Atmospheric dynamics; Atmospheric structure and evolution; Climate models and modeling; Climate prediction and projection; Greenhouse gases; Oxygen, atmospheric; Ozone; Stratosphere; Troposphere.

Atmospheric dynamics

• **Category:** Meteorology and atmospheric sciences

Scientists speculate that, while global warming may not increase the number of storms occurring on Earth, it may increase their average severity.

• Key concepts

air parcel: a theoretical house-sized volume of air that remains intact as it moves from place to place

Coriolis effect: in the Northern Hemisphere, the westward deflection of southward-moving air and the eastward deflection of northward-moving air—caused by Earth's rotation

stratosphere: the atmospheric region just above the tropopause and extending up about 50 kilometers

tropopause: the transition region between the troposphere and the stratosphere

troposphere: the lowest layer of the atmosphere—

in which storms and almost all clouds occur—extending from the ground up to between 8 and 15 kilometers high

• Background

In the eigthteenth century, Edmond Halley, for whom Halley's comet is named, charted the monsoons and the trade winds, making the first known meteorological map. In an effort to understand the trade winds, Halley correctly surmised that the Sun-warmed air over the equator would rise high into the atmosphere and then flow toward the poles. He further supposed that the air would cool off and sink at the poles, and then return to the equator as a surface wind, but he could not explain why the trade winds came from the northeast, or even the east, instead of from the north.

• The Role of the Coriolis Effect

Earth is about 40,000 kilometers around at the equator, and it rotates once in twenty-four hours, so at the equator the land, sea, and air are rushing eastward at nearly 1,700 kilometers per hour. At 45° north latitude, by contrast, Earth is only about 28,000 kilometers in circumference, so a point located at that latitude travels eastward at about 1,200 kilometers per hour.

Consider a parcel of air at rest with respect to the land at the equator. Suppose that it is filled with red smoke so that its location is easily seen. Now let the parcel move northward; because of its eastward momentum, it will also be moving eastward with respect to the land north of the equator. This eastward deflection of northward-moving air parcels is called the Coriolis effect and is named for Gaspard-Gustave de Coriolis, who studied it in 1835. Also as a result of the Coriolis effect, if an air parcel at rest with respect to the land at some point north of the equator begins moving southward toward the equator, it will be deflected westward, because its eastward momentum will be less than that of the ground below it.

Fifty years after Halley's surmises, George Hadley explained the direction of the trade winds by referring to the Coriolis effect. Hadley believed that air parcels heated at the equator rose high and then migrated north to the pole. Cooled during the journey, the parcel would sink to the surface and head back south. Because of the Coriolis effect, the southward moving air would be deflected westward. This would explain the trade winds north of the equator, and this proposed air circulation route was called a Hadley cell.

The American meteorologist William Ferrel pointed out that atmospheric dynamics could not actually be that simple, since the prevailing winds at midlatitudes are westerlies, not easterlies such as the trade winds. Ferrel suggested that the Hadley cell extended only to about 30° latitude north of the equator, where the cooled air sank and returned to the equator. The Ferrel cell lies between about 30° north latitude and 60° north latitude. Air rises at 60°, flows southward, cools and descends at 30°, and flows northward and from the west near the ground—hence the westerlies. The Polar cell extends from 60° to the pole, with air rising at 60° and sinking at the pole. The cells of the Southern Hemisphere mirror those of the Northern Hemisphere.

• Jet Streams

Where the Ferrel cell meets the Polar cell, the temperatures and pressures of the air masses are generally different. These differences give rise to winds blowing north from the Ferrel cell toward the Polar cell, but this wind is soon deflected eastward by the Coriolis effect. Hemmed in between the Polar and Ferrel cells, the wind becomes the polar jet stream—a river of air 160 to 500 kilometers wide, 1 kilometer deep, and generally 1,500 to 5,000 kilometers long. Several discontinuous segments of the jet stream together might come close to circumnavigating the Earth. These segments wax and wane over time and sometimes disappear completely.

The polar jet stream forms at the tropopause, 7 to 12 kilometers above sea level. (The tropopause is the transition region between the troposphere below and the stratosphere above.) The speed of the jet stream averages 80 kilometers per hour in the summer and 160 kilometers per hour in the winter, but it can reach speeds of up to 500 kilometers per hour. While it generally flows eastward, sometimes it also meanders hundreds of kilometers south and then back north.

A second jet stream, the subtropical jet stream, occurs between the Hadley and the Ferrel cells, but since the tropopause is higher there, this jet stream is between 10 and 16 kilometers above ground level. This jet stream tends to form during the winter, when temperature contrasts between air masses are the greatest. Other low-level jet stream segments may form near the equator. The jet streams of the Southern Hemisphere mirror those of the Northern Hemisphere. Studies show that jet streams help carry carbon dioxide (CO_2) from where it is produced to other parts of the world.

There are practical reasons for studying jet streams. Jet streams influence the paths of storms lower in the atmosphere, so meteorologists must take them into account in their forecasts. Pilots flying from Tokyo to Los Angeles can cut their flight times by one-third if they can use the jet stream for a tailwind. One percent of the energy of the world's jet streams could satisfy all of humanity's current energy needs. Someday, it may be possible to use balloons to lift windmills into the jet streams, but they would need to be tethered to the ground to keep them from being blown along with the jet stream.

• Oscillations

India was stricken by a severe famine and drought in 1877 because the monsoons failed. In response, Gilbert Walker headed a team at the Indian Meteorological Department using statistical analysis on weather data from the land and sea looking for a link to the monsoons. They eventually found a link between the timing and severity of the monsoons and the air pressure over the Indian Ocean and over the southern Pacific Ocean. The team found that high pressure over the Pacific meant low pressure over the Indian Ocean, and vice versa. Walker named this alternating of pressures the Southern Oscillation and linked it with the monsoon. It has since been linked to other weather phenomena.

Normally, there is a large region of high pressure in the Pacific just off the coast of South America. The trade winds near the surface blow westward from this high-pressure region to a low-pressure region over Indonesia. The winds pick up moisture as they cross the Pacific and deliver it in the monsoons over Indonesia, India, and so forth. Energy from the condensation of moisture heats the air and causes it to rise higher; then, the air flows back eastward to the South American coast, cools, and sinks to complete the Walker cycle.

Just after Christmas, a warm current flows south by the coasts of Ecuador and Peru. In some years, that current is stronger and warmer, and then it brings beneficial rains to the South American coast—a Christmas gift. The event is called El Niño (little boy) with reference to the Christ Child. It is also called the El Niño-Southern Oscillation, or ENSO. The trade winds weaken, and warm water surges eastward to the South American coast. Air rises as it is heated by the warm coastal waters, and air comes from the west to replace the air that rose. As a result, the trade winds are reversed and blow eastward. The winds that carried moisture from the Indian Ocean toward the equator are weakened, so the monsoons are weakened. Low pressure develops over the Indian Ocean and pulls the subtropical jet stream south. The displaced jet stream brings more rain to East Africa and drought to Brazil. Central Asia, the northwestern United States, and Canada experience heat waves, while Central Europe experiences flooding.

After a few years, the El Niño event weakens and things return to normal—except that two-thirds of the time nature overshoots "normal," and La Niña (little girl) appears. The west-blowing trade winds return but are much stronger than normal. Cool water rises from the deep and forms a cool region off the west coast of South America. Colder-than-normal air blows over the Pacific Northwest and over the northern Great Plains, but the rest of the United States enjoys a milder winter. The Indian monsoons strengthen, and the subtropical jet stream returns to its normal position. Eventually, things quiet down and return to normal.

Since El Niño begins when warm water collects off the western coast of South America, global warming will probably increase the frequency and intensity of El Niño. The accompanying heat waves, flooding rains, and droughts will likely be more severe.

• Fronts and High-Pressure Air Masses

A weather front is the boundary between two air masses of different densities. They normally differ in temperature and humidity. Cold air is denser than warm air, so the air of an advancing cold front

Hurricane Michelle, in November, 2001. (NASA)

wedges beneath and lifts warm air. This upward motion produces low pressure along the front, and as the air lifts and cools, moisture condenses and forms a line of clouds or showers along the front. Light rain may begin 100 kilometers from the front, with heavy rain beginning 50 kilometers from the front.

Cold fronts generally come from the north and head south, while warm fronts generally come from the south and head north. The leading edge of an advancing warm front takes an inverted wedge shape, such that the high cirrus clouds that mark the approach of the front may be hundreds of kilometers ahead of the front's ground-level location. The cloud base continually lowers as the front approaches, and with enough moisture present rain may extend 300 kilometers in front of the ground-level front. Fronts are the principal cause of non-rotating storms.

Sinking air forms a high-pressure area at the surface. This air is usually dry, having delivered its moisture elsewhere, rendering the sky clear. If there were any warm, moist air, it would be prevented from rising by the descending air of the high-pressure area, so it could not form clouds and rain. Many of the world's deserts form where the circulating air of the Hadley cell descends. Any high-pressure area that remains in place for a long time can cause drought. Winds blowing outward from the high-pressure area will be turned by the Coriolis effect in a direction established by a simple rule: If—in the Northern Hemisphere—one stands so that the low-pressure area is on one's left, the wind will be at one's back. In the Southern Hemisphere, the wind would be at one's front.

• Hurricanes and Low-Pressure Air Masses

Rising air forms a low-pressure area near Earth's surface. Higher-pressure air from outside the area will flow toward the low-pressure area (like water running downhill), but it will be turned by the Coriolis effect and slowly spiral inward. This air will eventually be caught up in the rising air currents, will cool off as it rises, will be heated as its moisture condenses, and will then rise higher and contribute to the updraft. At this point, the low-pressure area has become a storm. If its rotation speed is over 120 kilometers per hour, the storm has become a hurricane. Global warming calculations predict that if CO_2 levels increase, the number of hurricanes occurring annually will not increase, but the average intensity of these storms will increase.

• Context

More than a century ago, Sir Gilbert Walker began the process of describing the world's weather in one comprehensive model. The weather in each part of the world is tied in some fashion to the weather elsewhere—through jet streams, circulation cells, or movement of air masses. As Earth's global climate undergoes changes, the interrelationship of the planet's weather systems becomes more important than ever, since the climate change will affect not only individual weather patterns and events but also the way in which those individual events affect one another and Earth's weather generally.

Charles W. Rogers

• **Further Reading**

Lutgens, Frederick K., Edward J. Tarbuck, and Dennis Tasa. *The Atmosphere: An Introduction to Meteorology*. 10th ed. Boston: Prentice Hall, 2006. Textbook covering atmospheric structure, air masses, circulation, storms, oscillations, the changing climate, and more.

Lynch, John. *The Weather.* Buffalo, N.Y.: Firefly Books, 2002. An outstanding book for beginners, lavishly illustrated and well written. Goes one layer deeper than typical introductory books, but it is not difficult to understand. Includes a short chapter on global warming.

Walker, Gabrielle. *An Ocean of Air: Why the Wind Blows and Other Mysteries of the Atmosphere.* New York: Harvest Books, 2007. Well written and easily read. Elucidates how the atmosphere behaves and how people learn about it.

See also: Climate change; Climate feedback; Coriolis effect; Hadley circulation; Jet stream; Polar climate; Stratosphere; Troposphere; Weather vs. climate.

Atmospheric structure and evolution

• **Category:** Meteorology and atmospheric sciences

Earth's atmosphere is a blanket that keeps the planet a beneficial 35° Celsius warmer than it would be otherwise. Human activities, however, are adding GHGs to the atmosphere that may raise the temperature enough to alter the climate.

• **Key concepts**

albedo: percentage of incident light Earth reflects back into space

greenhouse gases (GHGs): gases that allow sunlight to pass through to the ground but trap, at least partially, the infrared radiation that would otherwise escape into space

ice age: a period during which sea ice and glaciers cover a significant fraction of Earth's surface

late heavy bombardment: a period about 3.9 to 4.0 billion years ago, when Earth was pummeled by debris from space at one thousand times the normal rate, heating the atmosphere and melting the crust

Milanković cycles: recurring time periods during which the shape of the Earth's orbit, the tilt of its axis, and the occurrence of its farthest distance from the Sun all change

• **Background**

The atmosphere is an ocean of gases held to the Earth and compressed by gravity. If greenhouse gases (GHGs) were newly introduced into the atmosphere, less energy would leave the Earth than strike it, so the planet would warm until those two rates balanced. (Infrared radiation from a warmer Earth has a shorter wavelength and is therefore more likely to escape into space.) GHGs keep Venus 500° Celsius, Earth 35° Celsius, and Mars 7° Celsius warmer than each planet would be without its atmosphere. Without the greenhouse effect, much of Earth would be permanently covered with snow and ice.

• **Earth's First Atmosphere**

If Earth's atmosphere had formed along with Earth itself, its composition would be expected to reflect the relative abundance of solar elements extant at the time and present in gases heavy enough to be retained by Earth's gravity. Hydrogen and helium are the two most abundant gases, but Earth's gravity is not strong enough to hold them, so they gradually escape into space. Oxygen is the next most abundant gas, but it is so chemically active that, without plant life to replenish it, it would soon disappear from Earth's atmosphere. Neon is next in abundance, is chemically inert, and is heavy enough to be retained by Earth's gravity. The fact that it is not the most abundant gas in Earth's atmosphere provides evidence that Earth's primordial atmosphere escaped into space, probably during a flare-up of the young Sun or during the late, heavy bombardment following the Earth's formation.

Scientists believe that the current atmosphere consists of gases released as rocks and minerals inside the Earth were heated by the energy from radioactive decay. These gases were subsequently emit-

Solar Abundances of the Elements

Element	*Solar Nebula Abundance per Hydrogen Atom*
Hydrogen	1
Helium	0.16
Oxygen	0.00089
Neon	0.0005
Carbon	0.0004
Nitrogen	0.00011
Silicon	0.000032
Magnesium	0.000025
Sulfur	0.000022
Argon	0.0000076

ted by volcanoes. Water vapor is the most abundant volcanic gas. It condensed to form the oceans. The next most abundant volcanic gases are carbon dioxide (CO_2), nitrogen, and argon. CO_2 is removed from the atmosphere when it dissolves in the oceans, where most of it eventually combines with calcium oxide ions, precipitates out, and forms carbonate rocks (limestone). Sulfur dioxide and sulfur trioxide are chemically active and do not stay in the atmosphere very long. Thus, if the Earth's atmosphere came from volcanoes, it should be dominated by nitrogen, followed by CO_2 and then argon. The atmospheric CO_2 would soon be depleted as it dissolved in the oceans. The atmospheres of Venus and Mars are richer in CO_2 than is Earth's atmosphere, because those planets have no oceans to remove their CO_2.

In the absence of oxygen, iron dissolves in water, but when oxygen is dissolved in the same water, iron oxide precipitates out and sinks to the bottom. Ore deposits of this iron compound first appeared about 2.6 billion years ago and peaked about 1.8 billion years ago. This probably reflects the increasing abundance of plants that released oxygen into the atmosphere and the increasing atmospheric concentraton of oxygen that resulted. Earth's atmospheric oxygen concentration reached a maximum of 30 percent 300 million years ago and a minimum of about 12 percent 200 million years ago. It now stands at about 21 percent.

• The Faint Early Sun Paradox

Stars such as the Sun roughly double in brightness over their lifetimes as normal stars. The Sun is already 30 percent brighter than it was when it first became a normal star, 4.56 billion years ago. Based on contemporary values of Earth's albedo and its atmosphere, Earth should not have had liquid water before about 2 billion years ago. There is, however, abundant geological evidence that liquid water has existed on Earth for at least 3.8 billion years. Mars exhibits a similar paradox, since it seems to have had abundant surface water 3.8 billion years ago.

If the Sun had been born 7 percent more massive, the Earth would have been warm enough for liquid water. Because at normal rates the Sun can have lost only 0.05 percent of its mass since it was born, however, this is probably not the case. A sufficiently large amount of CO_2 in the atmosphere could have made the early Earth warm enough for water, but geological evidence for that much atmospheric CO_2 is lacking. However, Philip von Paris and his colleagues at the Aerospace Centre in Berlin have developed a plausible computer model that allows the early Earth to have developed liquid water with only 10 percent of the CO_2 previously thought necessary, an amount not ruled out by geological evidence.

Composition of Air and of Volcanic Gases

Gas	*Percent by Volume*	
	Air	*Volcanic* Gas*
N_2 (nitrogen)	77	5.45
O_2 (oxygen)	21	
H_2O (water vapor)	0.1 to 2.8	70.8
Ar (argon)	0.93	0.18
CO_2 (carbon dioxide)	0.033	14.07
Ne (neon)	0.0018	
CH_4 (methane)	0.00015	
NH_3 (ammonia)	0.000001	
SO_2 (sulfur dioxide)		6.4
SO_3 (sulfur trioxide)		1.92
CO (carbon monoxide)		0.4
H (hydrogen)		0.33

*Kilanea volcano, Hawaii.

• Structure of the Atmosphere

The atmosphere has settled into a series of layers, one on top of the other, like the layers of an onion. The lowest layer of the atmosphere is called the troposphere. It extends from the ground up to about 15 kilometers over the equator and slants down to just 8 kilometers in height over the poles. The word "troposphere" is based on the Greek word *tropos*, which means "turning" or "mixing." Storms are a result of the mixing that takes place in this layer, so storms and almost all clouds are confined to the troposphere. In the troposphere, temperature decreases by about 8° Celsius with each 1 kilometer of altitude. Tourists are often surprised to find that the South Rim of the Grand Canyon, in Arizona, is 12° Celsius cooler than the inner gorge, 1.5 kilometers below.

The atmospheric layer above the troposphere is the stratosphere (from the Latin *stratum*, meaning "horizontal layer"). It extends from the troposphere up to an altitude of about 50 kilometers. Commercial airliners usually cruise in the lower stratosphere, where the reduced air density produces less drag and where they are above clouds and storms. The ozone layer, which protects Earth life from most of the Sun's ultraviolet radiation, is located in the stratosphere, mostly between 20 and 40 kilometers high.

Above the stratosphere lies the mesosphere (from the Greek *mesos*, meaning "middle"). The mesosphere extends from about 50 kilometers up to 80 or 90 kilometers in altitude. Incandescent trails may be observed in the mesosphere. These trails result when meteoroids the size of sand grains or pebbles strike Earth from space: They are heated by the friction of their swift passage through the mesosphere until they disintegrate.

The thermosphere (from the Greek *thermos*, meaning "heat") begins about 90 kilometers above the ground and extends upward to about 600 kilometers high. Gas molecules of the thermosphere absorb solar energy and convert it into kinetic energy (energy of motion). The speed of a molecule is a measure of its temperature, and speeds corresponding to temperatures of up to 15,000° Celsius are expected in thermosphere gas molecules. While that may seem hot, an unprotected person in the thermosphere would soon freeze, because such gas molecules are few and far between, making the average temperature of this layer extremely cold. Auroras form in the lower thermosphere when high-energy particles slam into air atoms.

The International Space Station (ISS) orbits in the thermosphere, about 340 kilometers above Earth's surface. The atmosphere there is thin enough that drag on the ISS is small, but not zero. The advantage to this location is that, although it must periodically be reboosted by a supply ship while in use, when the ISS is finally abandoned it will slowly and naturally deorbit itself.

Beginning in the lower thermosphere, air atoms are far enough apart that when ultraviolet light from the Sun drives electrons away from their parent atoms, they do not recombine for some time. The clouds of free electrons (or ions) that thus form can reflect radio waves. This region is the ionosphere. Depending upon the intensity of incident sunlight, the ionosphere can extend throughout the thermosphere and up into the exosphere. The exosphere begins 500 to 600 kilometers high and extends up to about 10,000 kilometers, where it shades into interplanetary space.

• Earth's Greenhouse Gases

Clouds and water vapor together are responsible for roughly 80 percent of Earth's greenhouse warming. People can affect the amount of water vapor in the atmosphere locally by deforestation, for example, but the global amount is determined by Earth's vast oceans. The global amount of atmospheric water vapor remains quite constant over time, so most atmospheric scientists conclude that it cannot be responsible for recent warming. Water vapor also plays a major role in feedback; for example, if Earth warms, more water will evaporate from the oceans, becoming vapor. More water vapor will produce more clouds, and more clouds blanketing Earth will keep it warmer, but more clouds will also reflect more sunlight back into space, keeping Earth cooler. Experts indicate that the warming effect will be greater than the cooling effect, but it is nevertheless apparent that global warming is quite a complex phenomenon.

CO_2 accounts for most of the rest of Earth's greenhouse effect. The amount of atmospheric CO_2 is increasing by about 0.4 percent per year, with per-

Glacial and Interglacial Periods

Years Before Present (thousands)	*Temperature Deviation**	*Description*
0.2 to 0.4	-1	Little Ice Age (Northern Europe)
0.8 to 1.2	+1	climatic optimum
20 to 500	+2 and -6	4 cold periods and 4 warm periods
Years Before Present (millions)		
3 to 10	+1 to +2	warm period
10 to 25	+3	Antarctica thawing
25 to 35	+1	Antarctica glaciation
35 to 80	up to +6	Eocene optimum
260 to 360	-2 to -5	Karoo glaciation
420 to 450	+2 to -5	Andean-Saharan glaciation
635 to 800	-5	Cryogenian glaciation
2,400 to 2,100	-5	Huronian glaciation

*Rough global averages; deviation is from 17° Celsius, expressed in ° Celsius.

haps the majority of that increase coming from transportation, power generation, and other human activities. It also comes from volcanoes and burning vegetation. CO_2 is removed from the atmosphere by plants and by being dissolved in the ocean. In the ocean, it may combine with calcium to make limestone. It is also removed from the ocean by sea creatures that use limestone to make their shells.

Methane is present in the atmosphere in only trace amounts, but, molecule for molecule, it is about twenty times more effective than CO_2 as a GHG. Atmospheric methane has increased by 11 percent since 1974. Significant amounts of it are released by coal and oil production, cattle and sheep, swamps and rice paddies, and jungle termites.

• Ice Ages

Life on Earth has survived many cooling and warming cycles in the past, but scientists do not fully understand the causes of climate change. Different factors dominate changes at different times. For example, increased atmospheric CO_2 from volcanoes may warm Earth enough to end an ice age, but in other cases atmospheric CO_2 might not increase until centuries after the climate warms. A few of the other factors to consider are Milanković cycles, solar intensity, albedo, and the ability of the oceans to absorb CO_2 and to lock it in limestone.

• Context

At one other time, humanity stood on the verge of having the power to change the climate. When world stockpiles of nuclear weapons were at their highest levels, some people wondered if an all-out nuclear war would result in nuclear winter—months or years of freezing temperatures and little sunlight—followed by ten thousand years of nuclear spring—temperatures several degrees above normal and dangerous levels of ultraviolet light reaching the ground. Although it is doubtful that there were ever enough warheads to cause nuclear winter, stockpiles have been reduced. Humans are again on the verge of being able to change the climates of Earth, and enough is not known to predict all of the results of changing the climate or of taking actions designed to avoid changing it.

Charles W. Rogers

• Further Reading

Archer, David. *Global Warming: Understanding the Forecast.* Malden, Mass.: Blackwell, 2007. Slightly technical but written for nonscientists. Explains various issues relating to global warming, including the interaction of greenhouse gases with other gases, with rocks, and with the ocean.

Barry, Roger G. *Atmosphere, Weather, and Climate.* 8th ed. New York: Routledge, 2003. Comprehensive, nontechnical introduction to these subjects. Treats the composition and structure of the atmosphere, circulation, and storms.

Walker, Gabrielle. *An Ocean of Air: Why the Wind Blows and Other Mysteries of the Atmosphere.* New York: Harvest Books, 2007. Explains how the atmosphere behaves.

See also: Atmosphere; Atmospheric chemistry; Atmospheric dynamics; Carbon dioxide; Earth history; Earth structure and development; Greenhouse effect; Greenhouse gases; Mesosphere; Ocean-atmosphere coupling; Ocean dynamics; Planetary atmospheres and evolution; Stratosphere; Thermosphere; Troposphere.

Australia

- **Category:** Nations and peoples
- **Key facts**

Population: 21,262,641 (July, 2009, estimate)
Area: 7,692,208 square kilometers
Gross domestic product (GDP): $800.5 billion (purchasing power parity, 2008 estimate)
Greenhouse gas (GHG) emissions in millions of metric tons of carbon dioxide equivalent (CO_2e): 408 in 1990; 491 in 2000; 534 in 2004
Kyoto Protocol status: Ratified 2007

• Historical and Political Context

Australia is a federal parliamentary state. The British monarch, Queen Elizabeth II, is the chief of state, although constitutional links with Britain were ended in 1968. Prime Minister Kevin Rudd became the head of state in 2007. Australia achieved federation status in 1901. Despite the nature of its geography, bound by water in all directions, Australia has eschewed isolationism and fought alongside the British in World Wars I and II and with the United States in the Korean and Vietnam Wars. During the 1960's, the country sought to deal more fairly with its indigenous population of aborigines. In 2001, Australia joined the United States in its response to the September 11 terrorist attacks.

• Impact of Australian Policies on Climate Change

When Prime Minister Kevin Rudd was elected, ending the eleven-year ministry of John Howard, he made ratifying the Kyoto Protocol a priority of his ministry. Shortly after his election, Rudd was invited to the United Nations Climate Change Conference in Bali. Australia's pledge to ratify the Kyoto Protocol signaled a policy shift from that of the previous government; Australia had previously signed the protocol but had not ratified it.

Industrialized, Annex I parties to the Kyoto Protocol such as Australia are committed to cut their greenhouse gas (GHG) emissions by an average of 5 percent from their 1990 levels between 2008 and 2012. A heavier burden is placed on industrialized countries than on developing nations, because the

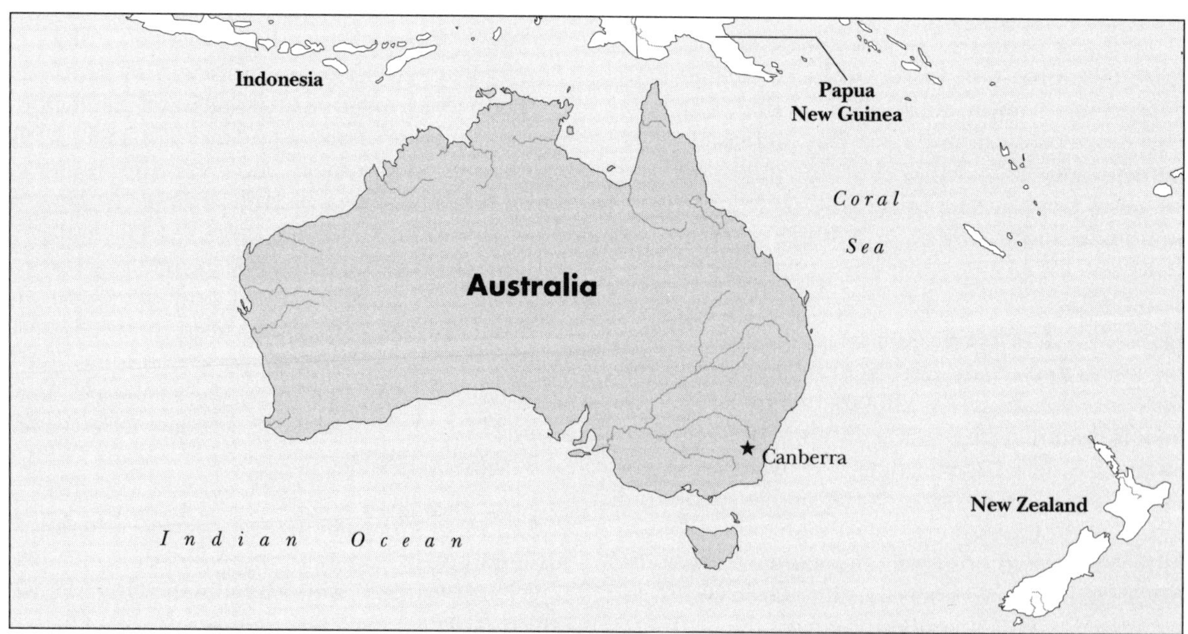

former are better able to pay the cost of emission cuts than are the latter.

In 2008, Australia participated in negotiations on reducing deforestation at the fourteenth Conference of the Parties (COP-14) in Poznan, Poland. (COP is the highest body of the U.N. Framework Convention on Climate Change and consists of environment ministers who meet once a year to discuss the convention's developments.) Australia, Canada, New Zealand, and the United States opposed provisions designed to protect the rights of indigenous peoples. As a result, major changes to the draft agreement on deforestation were required prior to the 2009 conference (COP-15) in Copenhagen, Denmark. The goal of the required changes was to preserve the rights of indigenous and local peoples and communities, promote biodiversity, and address the causes of deforestation in a manner acceptable to the objecting parties.

• Australia as a GHG Emitter

As of 2008, Australia was ranked as the world's sixteenth-highest GHG emitter. To reduce such emissions, the Kyoto Protocol established carbon quotas for member countries, which may develop new carbon sinks, such as reservoirs of foliage or forests, in order to offset their carbon emissions. These sinks are known as "Kyoto lands." The use of carbon sinks to mitigate the global warming effects of emissions may be useful to countries with large areas of forest or other vegetation that are otherwise struggling to comply with the protocol. Specific legally binding quotas for reduction of GHG emissions have been established for the developed, Annex I nations, including Australia. Developing, non-Annex I countries, such as Brazil annd Indonesia, are not compelled to restrict their GHG emissions. In such countries, emissions may come in large part from the cultivation of lands and the destruction of forests. However, for developed nations such as Australia, land use would have little effect in meeting Kyoto quotas, since most Australian land has already been cultivated.

While the Organization of Petroleum Exporting Countries (OPEC) Gulf States have the highest GHG emissions, data from 2000 show that—of the top twenty emitters—those with highest per capita emissions were the Annex I countries. Australia, the United States, and Canada ranked fifth, seventh, and ninth, respectively. Their per capita emissions (7.0, 6.6, and 6.1 metric tons per person) were approximately double the emissions of the highest-ranked developing country in the top twenty (South Korea, at 3.0 metric tons), and they were six times those of China (1.1 metric tons). Australia's high per capita emissions are due in part to the nation's large area and low population density (about 2.7 persons per square kilometer in 2006), as well as its dependence on energy-intensive fossil fuel to transport people and goods over large distances, its use of coal to generate power, and the energy it needs to expend on resource extraction generally.

• Summary and Foresight

Before 2007, the Howard government argued that ratifying the Kyoto Protocol would jeopardize Australian jobs and industry. It opposed Australia's commitment to the treaty, on the grounds that some major polluters, including developing, non-Annex I countries such as China and India, would not be compelled to cut their GHG emissions. Critics argued that by not ratifying the protocol, Australia would tarnish its image as an environmentally progressive nation. Australia has been a leader in opposing whaling and in advocating for conservation of Antarctica and the South Pacific Ocean.

The Rudd government, despite ratifying the Kyoto Protocol, was criticized as it entered the climate change debate at COP-14, because it did not set specific targets for cutting GHG emissions by 2020. Developing nations such as India questioned why they should commit to such targets if industrialized countries did not. Earlier, the Intergovernmental Panel on Climate Change (IPCC) had declared that global GHG emissions must peak by 2015 and then begin to decline if catastrophic environmental consequences are to be avoided. The European Union has already committed to reducing GHG emissions by 20 percent by 2020.

COP-14 was supposed to provide continuity between the negotiations begun at COP-13 in Bali (2007) and a finalized agreement to be reached at COP-15, where the successor treaty to the Kyoto Protocol would be completed. COP-14 was unsuccessful; developed (Annex I) nations such as the United States, Canada, and Australia did not submit promised proposals on GHG emission reductions, finance, and technology. At COP-13, Australia had been welcomed with enthusiasm because of its environmental expertise in the region, but its refusal to set 2020 emission targets undercut its reputation, because setting such targets is perceived to be a criterion for global leadership. Without making such committments, it is difficult for developed nations to convince developing nations that they are serious about fighting climate change. One way of trying to close the gap between the developed, Annex I nations and the developing, non-Annex I countries may be to focus on per capita emissions or GHG emission intensity (emissions measured per economic unit) in order to provide individual nations with more equitable shares of energy use. For Australia, as well as the United States and Canada, this may involve more constraints, but for the developing countries there may be greater opportunities for economic growth.

Cynthia F. Racer

• Further Reading

Lohmann, Larry, ed. *Carbon Trading: A Critical Conversation on Climate Change, Privatization, and Power.* Uppsala, Sweden: Dag Hammarskjöld Foundation, 2006. Tackles the controversial aspects of carbon trading.

Lowe, I. *Living in the Hothouse: How Global Warming Affects Australia.* Melbourne: Scribe, 2005. Details Australia's global warming challenges.

Macintyre, S. *A Concise History of Australia.* New York: Cambridge University Press, 2000. Complete history of Australia, from 1600 to 1999.

See also: Annex B of the Kyoto Protocol; Kyoto lands; Kyoto mechanisms; Kyoto Protocol; United Nations Framework Convention on Climate Change.

Automobile technology

• **Categories:** Transportation; economics, industries, and products; science and technology

Efforts to refine or replace automotive internal combustion engines are immensely important for preventing further damage to the atmosphere. Advances in construction materials, increased use of electronic sensors, and strategies for reducing size and weight are also important.

• Key concepts

electric cars: automobiles that run completely on electricity, without the need to burn fuels

flexible (flex) fuel: a technology that allows vehicles to burn more than one type of fuel in the same engine

hybrid vehicles: automobiles that run on more than one power source, such as electric batteries and fuel-burning engines

hydrogen engines: engines that run on hydrogen, based either on fuel cells that oxidize the hydrogen or on hydrogen combustion

internal combustion engines: engines that burn fuel in a chamber to generate force

• Background

Although cars with steam engines were common in the early twentieth century and a few electric cars were produced, the internal combustion engine became the dominant means of powering automobiles after ignition systems were refined. Large vehicles such as the Cadillac became iconic symbols of prosperity, but as early as the late 1950's, American consumers and manufacturers became interested in the advantages of smaller cars, leading to the success of the American Motors Corporation's Rambler and Ford's Mustang.

In the second half of the twentieth century, efforts to limit the negative consequences of internal combustion engines focused on controlling the amount of toxins in their emissions. After the oil crisis of 1973, Japanese and European automakers aggressively marketed small cars that required less fuel than larger automobiles. In the United States, the use of unleaded gasoline and catalytic converters was mandated in the last decades of the twentieth century in order to reduce air pollution.

Larger cars returned to popularity after oil prices stabilized, but tightening emissions standards, increasing concern about the environment, and further price fluctuations eventually led to even smaller cars being produced. Swiss watch manufacturer

Swatch originated the idea of the extremely compact "smart car." In India, the Tata Motor Company's Nano, introduced in 2009, economized by reducing both its size and the number of parts. It used only three nuts on the wheels and one windshield wiper, and it was glued together rather than welded.

• Flexible-Fuel and Alternative-Fuel Engines

The design of flexible-fuel engines has given more choices to consumers. In Brazil, in response to the government's ethanol requirements, the introduction of the Volkswagen Gol, GM's Chevrolet Celta, and others has helped transform that nation's consumption patterns. The number of flexible-fuel vehicles in Brazil, where ethanol is made from sugarcane, rose to 7 million by 2009. In the United States, where agricultural land is more scarce and ethanol is made from corn, many engines are not capable of running on mixtures with high ethanol content, and colder weather could lead to ignition problems. Modern flexible-fuel engines have electronic sensors to detect fuel content and are able to change valve timing, cylinder pressure, fuel injection, and other functions automatically.

The search for new engine technologies has also been a focus for entrepreneurs, such as Johnathan Goodwin, whose company H-Line Conversions specializes in converting the engines of large vehicles such as Hummers and vintage Cadillacs so they can run on biodiesel fuel. His expensive vehicles have been popular with celebrities. Manufacturers have also experimented with the use of hydrogen in internal combustion engines and in fuel cells, but because existing methods of processing hydrogen

Compact smart cars are designed to minimize environmental footprint while maximizing efficiency. (Getty Images)

consume so much energy, it is generally thought to be an impractical automobile fuel.

• Beyond the Combustion Engine

Although the technology is older than combustion engines and was initially more popular, electric cars were sidelined by gasoline-powered vehicles from the 1920's through the twentieth century. Electric cars were cleaner, quieter, and simpler, but they were also more expensive and slower. Their batteries required recharging, which limited the distances they could travel. In the 1990's, electric cars were revived in response to environmental demands. Between 1997 and 1999, the major automakers introduced several all-electric cars in California. These included GM's EV1 and S-10 electric pickup, Honda's EV Plus, a Ford Ranger pickup, and Toyota's RAV4 EV. However, these models were all discontinued. It was difficult to reverse the petroleum trend and restructure the interdependent automobile and fuel industries. The lack of a supportive infrastructure (such as sufficient recharging stations equivalent to gas stations) also remained a problem.

To circumvent these problems, the hybrid vehicle was developed. Such vehicles could supplement the electric power they drew from batteries with a gasoline-burning engine for long distances, thus achieving greater fuel economy and power with the combined technologies than either could deliver alone. The first hybrid electric automobile in the U.S. mass market was Honda's two-door Insight, introduced in 1999. In 2000, Toyota debuted the Toyota Prius, the first hybrid four-door sedan, and the Honda Civic Hybrid followed in 2002. All major automobile manufacturers have developed hybrid vehicles. Amid the high gas prices of 2008, hybrids were in demand, especially the industry-leading Toyota Prius. However, by April, 2009, lower gas prices and a worldwide sales slump escalated competition. Honda's new, lower-priced Insight challenged the Prius. Other new models included the Mercury Milan and Ford Fusion hybrid sedans.

Because all-electric and hybrid vehicles can be charged by solar and wind sources, as well as by plugging into power grids, they are valued for their potential to combat global warming. The most popular all-electric car at the beginning of the twenty-first century was the REVA, a three-door car produced in India. Additional manufacturers such as Tesla Motors have entered the market, and sales are increasing worldwide. Electric and hybrid cars are also taking advantage of new advances in chassis construction, introducing lighter but strong materials as an alternative to steel.

• Context

The rapid changes in the automobile industry and the diversity of automotive technologies at the beginning of the twenty-first century herald a period of intense competition and experimentation, in contrast to the massive standardization characteristic of the late twentieth century. The move away from fossil fuels and toward smaller vehicles contributes significantly to efforts in fighting anthropogenic global climate change.

Alice Myers

• Further Reading

Bethscheider-Kieser, Ulrich. *Green Designed Future Cars: Bio Fuel, Hybrid, Electrical, Hydrogen, Fuel Economy in All Sizes and Shapes.* Los Angeles: Fusion, 2008. This beautiful pictorial book features futuristic, eco-friendly automobiles.

Blume, David, and Michael Winks. *Alcohol Can Be a Gas! Fueling an Ethanol Revolution for the Twenty-first Century.* Santa Cruz, Calif.: International Institute for Ecological Agriculture, 2007. Over twenty-five years in the making, this is the definitive book on alcohol fuel. Illustrations, index, bibliography, glossary.

Carson, Lain, and Vijay Vaitheeswaran. *Zoom: The Global Race to Fuel the Car of the Future.* New York: Twelve, 2007. Award-winning journalists analyze the auto industry and alternative fuels and call for a grassroots rebellion demanding clean energy. Index and bibliography.

Clemens, Kevin. *The Crooked Mile: Through Peak Oil, Biofuels, Hybrid Cars, and Global Climate Change to Reach a Brighter Future.* Lake Elmo, Minn.: Demontreville Press, 2009. Presents an optimistic overview of trends. Illustrations, bibliography, index.

Fuhs, Allen. *Hybrid Vehicles and the Future of Personal Transportation.* Boca Raton, Fla.: CRC Press, 2009. This comprehensive guide to hybrid vehicles includes their history, environmental impact, and implications for energy efficiency. Illustrations, bibliography, index.

Hasegawa, Yozo, and Tony Kimm. *Clean Car Wars: How Honda and Toyota Are Winning the Battle of the Eco-friendly Autos.* Singapore: John Wiley & Sons, 2008. Explains how the leading Japanese automakers are using clean-car technology to expand global market share.

Sperling, Daniel, and Deborah Gordon. *Two Billion Cars: Driving Toward Sustainability.* New York: Oxford University Press, 2009. This scholarly study includes statistics on emissions and alternative technologies. Illustrated. Bibliography and index.

See also: Biofuels; Catalytic converters; Fossil fuels; Fuels, alternative; Hybrid automobiles; Motor vehicles; Transportation.

Average weather

• **Category:** Meteorology and atmospheric sciences

• Definition

The term "average," from a statistical point of view, denotes the arithmetic mean of a set of numbers taken from a sample or a representative population. That is, the average rainfall for the month of January for a specific place is the average of the actual January rainfalls for a period of time. In most cases, this use of "average" in weather and climate studies and applications makes sense, but in some cases it can give an erroneous impression of the average weather.

For example, to determine the average January rainfall for the city of Tauranga, New Zealand, one must first decide which January rainfalls to average. In Tauranga, rainfalls have been measured from several sites for the periods 1898-1903, 1905-1907, and 1910-2009. For a correct analysis, data for all the observation sites must first be carefully adjusted so that they apply to the current recording site. When this is done, the average for all of these January months is 88 millimeters. This average includes a rainfall of 268 millimeters in January, 1989 (the wettest January recorded), and a rainfall of only 1 millimeter in January, 1928 (the driest January recorded). The average in this case is therefore of some use and reveals something about the January rainfalls over a period of one hundred years in Tauranga, but it is not an exhaustive description, especially if one seeks to determine whether Tauranga is getting wetter or drier.

For other weather elements, such as temperature, sunshine, cloudiness, and wind, the situation is similar: The average for a particular period, such as a month, season, year, decade, or set of years, is simply the average of the values found in a series of observations.

• Significance for Climate Change

As noted, in most cases, this use of the average in weather and climate studies and applications makes sense, but in some cases it can give an erroneous and sometimes distorted view of the average weather. For example, in the case of temperatures, a problem has arisen in climate change discussions, because the average temperature is traditionally determined by climatologists by taking the average of the highest (maximum) and the lowest (minimum) temperatures for a particular day. While this practice produces useful information, especially when comparing the average temperatures of, say, Chicago with those of Bangkok, a difficulty arises when the daytime and nighttime temperatures are important. For example, in a continental climate, such as that of Moscow, the difference between the daytime and nighttime temperature is significant, whereas in a tropical climate, such as that of Singapore, the day-to-night temperature difference is relatively small.

Average weather is generally reflected by measurements of average rainfall, average day- and nighttime temperatures, average sunshine, and so on. However, when one considers changes in climate—small changes in the average weather over time—it becomes necessary to take particular note of the period involved and the specific weather element being measured. For example, climate change may cause a particular place to be wetter in the winter and drier in the summer. To measure such a change, one would need to assess the changes in the summer rainfall and the winter rainfall over a period of time of at least thirty years and ideally one hundred years. Such an analysis might show that during the first fifty of the last one hundred years, winter rainfalls were lower than those of the following fifty years, whereas summer rainfalls were higher. One must therefore treat all climate data with a degree of caution and be careful always to compare apples with apples.

For example, if one considers again the city of Tauranga, New Zealand, as a typical example, temperature observations have been taken there since 1913. Various observation sites have been used, and data from all sites have been adjusted to reflect what the temperature would have been if all observations had been taken from the same site. With these adjustments, the highest average monthly temperatures for each month in Tauranga have occurred in January, 1935; February, 1928; March, 1916; April, 1938; May, 1916; June, 2002; July, 1916; August, 1915; September, 1915; October, 1915; November, 1954; and December, 1940. The highest annual average daily maximum temperature was 20.4° Celsius, recorded in 1916, followed by 20.2° Celsius in 1928, 20.1° Celsius in 1914, and 20.0° Celsius in 1998. These observations show that, despite the indications of global warming in some parts of the world, not all areas are the same, and memories can be deceiving.

W. J. Maunder

See also: Extreme weather events; Meteorology; Rainfall patterns; Seasonal changes; Weather forecasting; Weather vs. climate.

Axelrod, Daniel

American paleoecologist

Born: July 16, 1910; New York, New York
Died: June 2, 1998; Davis, California

Axelrod showed how past climate change had caused the plants of an area to change. Extrapolating from his observations and methods, scientists may be able to predict the effects upon vegetation of future changes in climate.

• Life

Daniel Axelrod was the oldest of five children born to parents who immigrated to the United States from Russia. The family soon moved to Guam, then to Waikiki, Hawaii, and, when Daniel was a teenager, to Berkeley and finally to Oakland, California. Axelrod, a Boy Scout, spent a lot of time out in the open country looking at plants and animals. He claimed that during that time he became a naturalist. With a B.A. in botany from the University of California at Berkeley, he went to work for the California Forest Service to earn money to go to graduate school. Axelrod received an M.A. in 1936 and a Ph.D. in 1938, both in paleobotany from the University of California at Berkeley. His Ph.D. work was under Ralph Chaney, who required him to minor in geology and to study climatology, evolution, and genetics before he started his graduate work.

After graduate school, Axelrod worked at the U.S. National Museum and the Carnegie Institution in Washington, D.C., as a postdoctoral research fellow. During World War II, Axelrod served in the Army Air Forces as a photo analyst. His knowledge of vegetation in the Pacific allowed him to distinguish enemy sites and movements, as well as conditions that U.S. troops were likely to encounter. His work earned him the rank of major and a Bronze Star. After the war, Axelrod (or, as he liked to be called, Ax) joined the geology faculty of the University of California, Los Angeles (UCLA). From 1962 through 1967, he held a joint appointment in botany and geology. In 1967, he moved to the University of California at Davis to help start a new geology department. Retirement from Davis in 1976 provided Axelrod with more time to conduct research. He was preparing breakfast before going to work in 1998 when he experienced a fatal heart attack. He was survived by his widow, Marilyn Gaylor Axelrod, and a stepdaughter.

• Climate Work

Educated in both botany and geology, Axelrod was well trained to analyze botanical fossils and correlate changing flora with the geological indicators of climate change. From fossils of seeds, leaves, stems, and flowers, he was able to build a model of the vegetation and climate of a time period. His specialty was the western area of North America during the Tertiary period (65 million to 1.8 million years ago). In his long career, he produced more than 140 publications and collected more than eight thousand fossil specimens. While at the U.S. National Museum, he learned how to preserve, store, and catalog specimens. He was known for his carefully preserved fossils.

One of Axelrod's insights was into the distribution of angiosperms, plants whose seeds are inside a fruit, such as a tomato or cucumber. By looking at angiosperm fossils, Axelrod determined that they originated in tropical areas and, over millions of years, migrated toward both poles. Axelrod also introduced the concept of altitude as a factor that controlled which plants grew in which area. His deduction that the Rocky Mountains are young on a geological time scale came from the fact that the same type of plants grow on both sides of the mountains. There has not been enough time for the plants to evolve into different types. He published papers that show where the continents were connected based on the places where similar plants existed.

Axelrod's most important contribution to global warming studies is to be found in the many papers in which he demonstrated how climate controls the movement of vegetation. Madrean plants that were in the Mohave area 18-20 million years ago were in the San Francisco area by 7-8 million years ago. The plants had moved northward as the desert grew. Axelrod showed that forests were more diverse in the past than they are now. As climate has changed in a certain area, only trees that are adapted to the new climate survive, and today there are forests composed predominantly of one tree species. Axelrod even predicted the effects of a change to a drier climate over a half million years and a million years. His methods were straightforward enough to enable others to follow in his footsteps, predicting changes in vegetation that may result from changes in climate.

C. Alton Hassell

• Further Reading

Axelrod, Daniel I. *History of the Coniferous Forests: California and Nevada.* Berkeley: University of California Press, 1976. Provides insight into the evolution of western forests in response to changes in western climate.

_______. *A Miocene (10-12 Ma) Evergreen Laurel-Oak Forest from Carmel Valley, California.* Berkeley: University of California Press, 2000. Typical of the work of Axelrod, this book presents the connection between the botany and the geology of the Carmel Valley. Illustrations, maps, bibliography.

Barbour, M. G. "Dan Axelrod: Paleoecologist for the Ages." *Fremontia* 27, no. 1 (1999): 29-30. Among the various obituaries and tributes to Daniel Axelrod, this is the one most commonly cited by the people who knew him.

Erwin, Diane M., and Howard E. Schorn. "Daniel I. Axelrod, 1910-1998: In His Own Words." *Bibliography of American Paleobotany.* St. Louis, Mo.: Paleobotanical Section, Botanical Society of America, 1998. The bulk of this article is drawn from a 1994 taped interview with Axelrod. It includes a list of the scientist's publications.

Marriott, Gerard, and Ian Parker. *Biophotonics.* Part B. San Diego, Calif.: Academic Press, 2003. Although Axelrod thought modern paleobotanists spent too much time with instrumentation and not enough time outside, he made use of methods such as total internal reflection fluorescence microscopy, as described in this book. Illustrations, bibliography, index.

See also: Atmospheric structure and evolution; Earth history; Ecosystems; Geological Society of America.

Bangladesh

- **Category:** Nations and peoples
- **Key facts**

Population: 156,050,883 (July, 2009, estimate)
Area: 144,000 square kilometers
Gross domestic product (GDP): $206.7 billion (purchasing power parity, 2007 estimate)
Greenhouse gas (GHG) emissions in millions of metric tons of carbon dioxide equivalent (CO_2e): 33.7 in 2004
Kyoto Protocol status: Ratified October, 2001

- **Historical and Political Context**

Among the world's nations, Bangladesh has one of the largest populations at risk from coastal flooding due to sea-level rise and storm surges. It is located in South Asia and borders India, Myanmar, and the Bay of Bengal. The country became part of the new nation of Pakistan in 1947, after India and Pakistan became independent from England. It was known then as East Pakistan. In 1971, it became independent with the help of India after a brief civil war and changed its name to Bangladesh.

Bangladesh is about the size of Wisconsin, with a population of about 154 millon, which makes it one of the most densely populated countries in the world. Its government is a parliamentary democracy, and Islam is the state religion. The 1972 constitution has undergone fourteen amendments and creates three branches of governments, the executive, legislative, and judicial branches. The prime minister is the head of government (executive) and elected by the majority party in parliament. The unicameral parliament is known as Jatia Sangsad and is made up of 345 members, including 45 seats reserved for women. The highest judicial body is the supreme court, which is independent of the executive branch.

The nation has experienced political instability since its inception. It had fourteen governments between 1972 and 2008. The first two national leaders were assassinated, beginning with President Sheikh Mujibur in 1975. The country's limited resources can be blamed in part on this political violence. Being residents of a developing nation, a majority of Bangladesh's people depend on agriculture. However, flooding has continued to decrease the available farmland in the country. The government has been unable to address basic issues, such as protection of life and property.

In many poor countries, when governments fail to provide for the people, those people turn to religious extremism for an answer. There is a growing Islamic extremist movement in Bangladesh that threatens domestic stability. For example, in 2002, the government arrested seven Arab nationals on charges of providing arms and training to militant Islamic groups in Dhaka who were funded by Al-Haramain, a Saudi-based agency. As a result of this growing Islamic extremism, India decided to fence off Bangladesh in order to keep out Muslim terrorists.

At a cost of over $1 billion, India is building a 4,000-kilometer border fence that will run through five Indian states. The fence is designed to prevent infiltration by Islamic terrorists, smuggling, and illegal immigration from Bangladesh. On January 22, 2007, Bangladesh postponed an election and declared a state of national emergency as a result of continued political instability. Fakhruddin Ahmed, with the help of the military, established a caretaker government aimed at cracking down on corruption and preparing for free elections in late 2008.

On December 29, 2008, a general election was held, and the Awami League Party led by Sheikh Hasina Oajed won the election by defeating the Bangladesh National Party. As a result, the military-controlled caretaker government handed power to Sheikh Hasina as prime minister on January 5, 2009. The eldest daughter of Sheikh Mujibur Rahman, the founding father of Bangladesh, Hasina had already served as prime minister from 1996 to 2001.

• Impact of Bangladeshi Policies on Climate Change

Flood control measures in Bangladesh were limited mainly to building embankments (artificial levees), polders, and drainage canals. The Bangladesh Water Development Board has constructed about seventeen hundred flood-control structures along with several thousand kilometers of embankments and drainage canals. Most of these projects have created a false sense of security for the residents, even though many of the projects have experienced breaching and erosion since they were constructed. During the 1999 floods, the Gumtl embankment at Etbarpur was breached, creating substantial damage to properties and the environment. The government also adopted a World Bank-sponsored flood action plan after the 1988 flood. The plan calls for the construction of hundreds of kilometers of tall embankments along the major rivers of the country's delta, as well as huge drains and several compartments on the floodplains.

By 1992, the government began to shift its policy from a narrow focus on flood control to flood and water management. It produced several five-year plans with guidelines for development. One of the plans entailed involving all concerned government agencies, as well as local people, in implementing

A Bangladeshi man washes clothing in the polluted Buriganga River in Dhaka. (Andrew Biraj/Reuters/Landov)

future embankments and other flood-control and drainage programs.

After the 1998 floods, the government worked with several nongovernmental organizations (NGOs) and donor nations to set up both short-term and long-term projects aimed at controlling or managing floods. These projects included government distribution of free seeds to farmers to reduce food shortages, as well as construction of large flood-protection shelters raised above the ground to protect both people and animals. The government also constructed flood-proof storage sheds to hold grains and other food supplies, dams upstream of the capital city of Dhaka, and a major embankment around the city itself. Emergency flood warning systems were improved and contingency plans formulated for the deployment of rescue and relief services. Villages—particularly remote villages that are difficult to reach during flooding—were stocked with emergency medical stores. The government also implemented reforestation programs and animal grazing controls in an attempt to increase absorption and reduction of water runoff. These projects were bold and very costly for Bangladesh alone, but the government hoped that the United Nations, donor countries, and NGOs would come to its aid.

• Bangladesh as a GHG Emitter

According to a United Nations development report, Bangladesh in 2004 accounted for 0.1 percent of global emissions, an average of 0.3 metric ton of carbon dioxide (CO_2) per person. As a result, it is not bound by a specific target for greenhouse gas (GHG) emission reductions. The emission levels of Bangladesh and other developing countries are so low that they are not bound by the Kyoto treaty. The treaty commits only the industrialized countries that ratified it to reduce the amount of six GHGs by 5.2 percent of the 1990 levels during the five-year period from 2008 to 2012.

• Summary and Foresight

Bangladesh is situated in the delta of three major rivers, the Ganges, Brahmaputra, and Meghna, which eventually empty into the Bay of Bengal. These rivers have large volumes of water with large drainage basins that increase the flood risk. Moreover, Bangladesh is a very low-lying country: Almost 70 percent of its land area is less than 1 meter above sea level, and 80 percent of it is located in a floodplain. Thus, the country's location, climate, and geography make it susceptible to the effects of climate change and also extremely hard to protect from those effects. The courses of its rivers are constantly shifting, making it difficult to build up riverbanks to protect farmland. Bangladesh has responded to climate change with huge projects and programs, but it will need cooperation from its neighbors, especially India.

Femi Ferreira

• Further Reading

Cash, Benjamin A., Xavier Rodó, and James L. Kinter III. *Non-ENSO Variability and the Regional Climate of Bangladesh: Implications for Cholera Risk.* Calverton, Md.: Center for Ocean-Land-Atmosphere Studies, 2008. Brief, detailed examination of Bangladesh's climate and climate variability with a focus on epidemiology and climatic effects on disease.

Huq, Saleemul, et al. *Mainstreaming Adaptation to Climate Change in Least Developed Countries (LDCs).* London: International Institute for Environment and Development, 2003. Bangladesh and Mali are the two nations discussed as case studies in this treatise on proper responses to global warming in developing nations.

Orford, Margie. *Climate Change and the Kyoto Protocol's Clean Development Mechanism: Brazil, Bangladesh, Indonesia, South Africa.* London: ITDG, 2004. Detailed monograph examining the Kyoto Protocol-related obligations and activities of Bangladesh and three other nations.

Zedillo, Ernesto, ed. *The Future of Globalization: Explorations in Light of Recent Turbulence.* New York: Routledge, 2008. Collection of essays on global economic development, including one on Bangladesh's place in the world economy and two on the environmental implications of global development.

See also: Flood barriers, movable; Floods and flooding; India; Nongovernmental organizations; Water resources, global.

Barrier islands

• **Category:** Geology and geography

• **Definition**

Barrier islands are narrow strips of land that are oriented parallel to the mainland and that lie on average a few kilometers offshore. Barrier islands are largely composed of sand-sized grains of sediment plus some finer sedimentary material and organic debris. They are separated from the mainland by an intervening body of water called a lagoon or bay. Barrier islands therefore have two shorelines, one facing the open ocean and the other facing the mainland. The side facing the open ocean is typically a much more energetic shoreline, experiencing higher waves and storm surges. The ocean-facing side is typically much straighter than the mainland-facing side, which is a low-energy coastal area.

Barrier islands may have several origins, but the most common is thought to be related to the post-ice age rise in sea level that commenced about eighteen thousand years ago. This sea-level rise inundated the world's coastal areas, and many barrier islands probably formed from coastal sand ridges and dune fields that were surrounded by the rising sea. Once the barrier islands were established, sand that was eroded from land and washed along the shore by wave currents helped build and maintain the islands.

Barrier islands are nourished—that is, restocked with sand—by a process called longshore drift. Longshore drift is the movement of sand down the seaward coast of a barrier island, resulting from the effects of wind waves that approach the seaward coastline of the island at a slight angle. These wind waves wash up on the barrier island coast (beach face) at an angle, and their return flow is directly back to the sea but slightly displaced in the wave's direction. As a result, the waves move sand and other material down the beach.

• **Significance for Climate Change**

Barrier islands are significantly affected by climate change. As global sea levels rise, barrier islands are at risk of being inundated or eroded away, as has

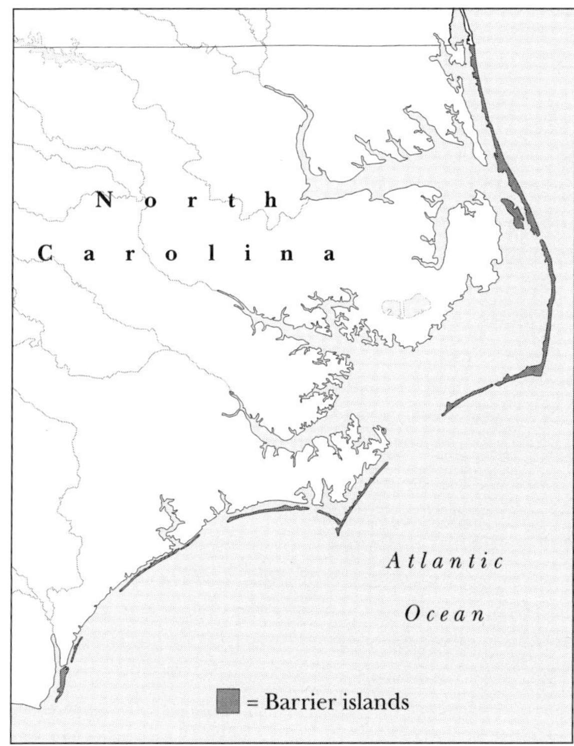

been seen in many areas of the world. The loss of barrier islands or a diminution in their width and continuity is a direct effect of sea-level change in many places. Increased storm activity, a result of climate change in general and of changes in sea surface temperatures in particular, affects barrier islands because they are at the front line when a sea storm comes ashore. Storm wave energy is spent on barrier islands, and the result can be dissection of the barrier islands or their complete destruction.

Barrier islands are integral parts of typically fragile coastal ecosystems, which are home to many plant and animal species. If climate change causes the loss of barrier-island habitats for coastal species, these organisms will need to adapt or perish.

Barrier islands require a continual sand supply to persist. This supply can be interrupted in many ways, including through climate change. Climatic changes that affect local runoff, for example, might reduce the flow of rivers that bring sand to the coastline and nourish the barrier islands there. There are other ways to lose sand supply, including human intervention.

The potential loss of barrier islands represents the possibility that areas of mainland currently sheltered by barrier islands will become primary coastlines. This would result in many changes in coastal geomorphology and ecosystems, as the protected mainland would become the focus of wave energy, altering the shape, population, and dynamics of shorelines.

Barrier islands on some coastlines of the world contain some of the most expensive real estate developments. These developments, some of which lie upon narrow sand islands less than 3 meters above sea level, are at imminent risk of loss as climatic conditions change. Such loss could entail huge economic losses for many countries and states, which depend upon tax revenues from entities built upon the islands.

In the past, barrier islands developed along the world's coastlines in much the same way that they had developed since the end of the last ice age. Studying ancient barrier islands, which are now part of the sedimentary rock record of the Earth, and comparing them to modern barrier islands helps scientists understand ancient climates and the history of climate change and its effects. There are lessons to be learned from the sedimentary record of barrier islands—particularly that conditions are always changing, especially at the oceanic coastline.

Barrier islands are analogous to living things in that they use energy to survive, they grow larger and smaller, they experience and recover from injuries, and they have a life cycle that ends in death. Climate change can bring on the death of a barrier island or present challenges to the continued survival of that island.

David T. King, Jr.

• Further Reading

Morton, Robert A. *Historical Changes in the Mississippi-Alabama Barrier Islands and the Roles of Extreme Storms, Sea Level, and Human Activities.* Reston, Va.: U.S. Geological Survey, 2007. Extensive study of natural and anthropogenic effects upon the Mississippi-Alabama barrier islands; invaluable resource for understanding the diversity of influences impinging upon these islands.

Pilkey, Orrin H., and Mary Edna Fraser. *A Celebration of the World's Barrier Islands.* New York: Columbia University Press, 2003. Very well illustrated popular-science book, coauthored by one of the world's foremost experts on barrier islands, Pilkey. Discusses the natural history of barrier islands and their modern peril.

Pilkey, Orrin H., Deborah Pilkey, and Craig A. Webb. *The North Carolina Shore and Its Barrier Islands.* Durham, N.C.: Duke University Press, 2000. Scientific and ecological guide to one of the classic barrier island coastlines of the world, North Carolina and its Outer Banks. Provides an outstanding introduction to barrier-island studies and issues in a changing global environment.

See also: Alliance of Small Island States; Coastal impacts of global climate change; Coastline changes; Islands; Ocean dynamics.

Basel Convention

• Categories: Laws, treaties, and protocols; pollution and waste

• Date: Opened for signature March 22, 1989; entered into force May 5, 1992

The Basel Convention is an international agreement to promote the environmentally sound management of hazardous waste and to reduce irresponsible disposal in less developed countries.

• Participating nations: *1989:* Jordan; *1990:* Hungary, Norway, Saudi Arabia, Switzerland; *1991:* Argentina, China, El Salvador, Finland, France, Mexico, Nigeria, Panama, Romania, Sweden, Uruguay; *1992:* Australia, Bahamas, Bahrain, Brazil, Canada, Chile, Cyprus, Estonia, India, Latvia, Liechtenstein, Maldives, Mauritius, Monaco, Poland, Senegal, Sri Lanka, Syrian Arab Republic, United Arab Emirates; *1993:* Antigua and Barbuda, Austria, Bangladesh, Belgium, Czech Republic, Ecuador, Egypt, Indonesia, Iran, Japan, Kuwait, Malaysia, Netherlands, Peru, Philippines, Saint Lucia,

Seychelles, Slovakia, Slovenia, Tanzania; *1994:* Comoros, Côte d'Ivoire, Croatia, Cuba, Democratic Republic of Congo, Denmark, European Community, Greece, Ireland, Israel, Italy, Lebanon, Luxembourg, Malawi, New Zealand, Pakistan, Portugal, Republic of Korea, Saint Kitts and Nevis, South Africa, Spain, Trinidad and Tobago, Turkey, United Kingdom, Zambia; *1995:* Barbados, Costa Rica, Germany, Guatemala, Guinea, Honduras, Iceland, Micronesia, Morocco, Namibia, Oman, Papua New Guinea, Paraguay, Qatar, Russian Federation, Tunisia, Vietnam; *1996:* Bolivia, Bulgaria, Colombia, Kyrgyzstan, Mauritania, Nepal, Saint Vincent and the Grenadines, Singapore, Turkmenistan, Uzbekistan, Yemen; *1997:* Belize, Benin, Burundi, Gambia, Mongolia, Mozambique, Nicaragua, Thailand, former Yugoslav Republic of Macedonia; *1998:* Algeria, Botswana, Dominica, Republic of Moldova, Niger, Venezuela; *1999:* Albania, Andorra, Armenia, Belarus, Burkina Faso, Cape Verde, Georgia, Lithuania, Madagascar, Uganda, Ukraine; *2000:* Dominican Republic, Ethiopia, Kenya, Kiribati, Lesotho, Mali, Malta, Serbia; *2001:* Azerbaijan, Bosnia and Herzegovina, Cambodia, Cameroon, Guyana, Libyan Arab Jamahiriya, Nauru; *2002:* Bhutan, Brunei Darussalam, Djibouti, Samoa; *2003:* Equatorial Guinea, Ghana, Jamaica, Kazakhstan, Marshall Islands; *2004:* Chad, Cook Islands, Liberia, Rwanda, Togo; *2005:* Eritrea, Guinea-Bissau, Swaziland; *2006:* Central African Republic, Montenegro, Sudan; *2007:* Congo (Republic of the); *2008:* Democratic People's Republic of Korea, Gabon

• Background

The Basel Convention is a treaty designed to promote environmentally sound management of hazardous waste while controlling and ultimately reducing the transboundary movement of hazardous waste for disposal to countries with less stringent regulations or enforcement. The convention requires countries to engage in more responsible management of hazardous waste, with a preference toward preventing pollution and minimizing the amount of waste produced. It promotes the domestic management of waste, rather than its disposal in other nations. In addition to protecting ecosystems from hazardous materials, the convention has a potential positive effect on Earth's climate: If less waste is produced and fewer raw materials and less energy are consumed, fewer greenhouse gases (GHGs) will be emitted into the atmosphere.

In the early and mid-1980's, the environmentally irresponsible management of hazardous waste drew international attention and outrage. Throughout the 1980's, industrialized nations began strengthening domestic hazardous-waste laws and regulations to foster more responsible hazardous-waste management to protect public health and the environment. An important component of this phase was the incorporation of the concept of waste minimization into the waste-management hierarchy. Waste minimization, the reduction in the quantity or toxicity of the hazardous waste generated, is the most desirable option under the waste-management hierarchy; it is followed by recycling, then treatment, and finally, the least preferable, responsible disposal. This model sought to eliminate or reduce significantly the disposal of untreated hazardous waste. However, the cost to treat and manage hazardous waste increased, resulting in escalated irresponsible dumping of untreated hazardous waste in less developed countries.

• Summary of Provisions

In response to this situation, the international community under the auspices of the United Nations Environment Programme drafted the Basel Convention in 1987. The convention was opened for signature in 1989 and went into force in 1992. It stipulates that if waste from one country is to be managed in another country, its transportation and management are to be conducted under conditions that do not endanger human health or the environment. The convention also prohibits the irresponsible disposal of hazardous waste in less developed countries. By 2008, 170 parties had ratified the convention. However, not all the major industrial countries are parties. For example, the United States became a signatory to the convention in 1990 but has never ratified it. The two other nonparty signatories to the convention are Afghanistan and Haiti.

Although the Basel Convention generally prohibits the shipment of hazardous waste between parties and nonparties, Article 11 of the convention allows the transboundary shipment of hazardous waste

Preamble to the Basel Convention

The Basel Convention's Preamble, excerpted below, sets out the context and international legal framework within which the convention was conceived.

The Parties to this Convention. . .

Affirming that States are responsible for the fulfilment of their international obligations concerning the protection of human health and protection and preservation of the environment, and are liable in accordance with international law,

Recognizing that in the case of a material breach of the provisions of this Convention or any protocol thereto the relevant international law of treaties shall apply,

Aware of the need to continue the development and implementation of environmentally sound low-waste technologies, recycling options, good house-keeping and management systems with a view to reducing to a minimum the generation of hazardous wastes and other wastes,

Aware also of the growing international concern about the need for stringent control of transboundary movement of hazardous wastes and other wastes, and of the need as far as possible to reduce such movement to a minimum,

Concerned about the problem of illegal transboundary traffic in hazardous wastes and other wastes,

Taking into account also the limited capabilities of the developing countries to manage hazardous wastes and other wastes . . .

Convinced also that the transboundary movement of hazardous wastes and other wastes should be permitted only when the transport and the ultimate disposal of such wastes is environmentally sound, and

Determined to protect, by strict control, human health and the environment against the adverse effects which may result from the generation and management of hazardous wastes and other wastes,

HAVE AGREED AS FOLLOWS. . . .

from parties to nonparties, provided such shipments are subject to separate agreements that are no less stringent that the requirements of the convention. For example, the United States, a nonparty country, has bilateral agreements for hazardous-waste management with Mexico and Canada, both party countries. In addition, the United States has a multilateral agreement addressing transboundary shipments of hazardous waste with the thirty member-countries of the Organization for Economic Cooperation and Development (OECD).

The Basel Convention controls the transboundary movements of hazardous and other wastes (those wastes listed in Annex I of the convention) under a "prior informed consent" procedure. Any shipment made without prior consent is deemed illegal. Under Article 11, shipments to and from nonparties are illegal unless there is a special agreement in place that does not undermine the Basel Convention.

For shipments between party countries, the Basel Convention requires the state of export to notify the state of intended disposal, as well as any states through which the shipment is intended to pass (transit states). This notification must contain detailed information describing the proposed shipment of hazardous waste. The shipment may commence only upon receipt of the written consent of the state of import, upon confirmation of the existence of a contract with a disposer specifying that the hazardous waste will be managed in an environmentally sound manner, and upon confirmation that the states of transit have consented to allow the waste to move across their territories. If the shipment is authorized, it must be accompanied at all times by a movement document that provides detailed information about the shipment and that must be signed by each person who takes charge of the waste. Finally, the disposer must confirm receipt of the waste and completion of disposal in accordance with the original notification documents.

The Basel Convention also requires parties to ensure that hazardous waste is managed and disposed of in an environmentally sound manner. In accordance with the convention, environmentally sound procedures include minimization of quantities moved across borders, treatment and disposal of wastes as close as possible to their place of generation, and prevention and minimization of waste generation in the first place. In addition, parties to the convention are expected to adopt controls applicable to the movement and manipulation of hazardous waste from its generation through its storage, transport, treatment, reuse, recycling, recovery, and final disposal.

In 1995, an amendment to the convention was offered as Decision III/1, known as the Basel Ban. If ratified, the ban would prohibit all hazardous-waste exports from the most industrialized countries of the OECD and the European Union to all non-OECD and non-European Union member countries. However, the Basel Ban amendment has not yet entered into force, as only about one-half the number of required countries has ratified it.

• Significance for Climate Change

One of the underlying goals of the Basel Convention is to recast waste as a resource, rather than as an undesirable residue. This recasting promotes the reuse and reclamation of valuable and finite resources, a principle of sustainable development. When waste is reduced, reused, or recycled, there is less demand for virgin, raw materials in the manufacturing and processing stages. This significantly reduces energy consumption throughout the life cycle. As a result of less waste and energy being consumed, the emission of GHGs is correspondingly reduced. However, some criticism remains that without ratification of the Basel Ban, the "effluent of the affluent" will continue, countering principles of sustainability. By forcing all countries to better manage their waste rather than allowing for controlled exports, the Basel Ban likely would generate a greater focus on waste minimization, resulting in further reductions in GHG emissions.

Travis Wagner

• Further Reading

Clapp, Jennifer. *Toxic Exports: The Transfer of Hazardous Wastes from Rich to Poor Countries.* Ithaca, N.Y.: Cornell University Press, 2001. Detailed discussion of the evolution of the Basel Convention, its weaknesses, and the challenges related to the international trade of hazardous waste.

Lipman, Zada. "A Dirty Dilemma." *Harvard International Review* 23, no. 4 (Winter, 2002): 67-72. Discusses some of the economic challenges and issues of sovereignty relating to countries that rely on exporting hazardous waste, as well as others dependent on its importation.

"The New Frontier." *Environmental Policy and Law* 37, no. 1 (2007): 22-24. Summarizes the most significant challenges facing the Basel Convention parties, including electronic waste and ship breaking.

United Nations Environment Programme. *The Basel Convention: A Global Solution for Controlling Hazardous Wastes.* New York: Author, 1997. Comprehensive overview and description of the Basel Convention.

See also: International agreements and cooperation; Sustainable development; United Nations Environment Programme.

Baseline emissions

• **Category:** Pollution and waste

• Definition

Baseline emissions are the greenhouse gas emissions that would take place in the absence of emission mitigation policies or projects. Sometimes referred to as the emissions in the "business-as-usual" scenario, baseline emissions are often compared with the actual emissions brought about by a project or policy in order to determine how effective the project or policy was at reducing emissions. This information is often used to award carbon offset credits to the project's sponsor.

- **Significance for Climate Change**

Determining baseline emissions is an important part of emissions-offsetting schemes, in which nations, corporations, or other entities receive tradable credits in return for reducing their emissions. Most notably, the "project-based mechanisms" under the United Nations Framework Convention on Climate Change (UNFCCC)—joint implementation (JI) and the clean development mechanism (CDM)—award carbon credits by subtracting the actual emissions of a project from the baseline emissions. The resulting quantity is the reduction in emissions generated by the project when compared to the business-as-usual scenario. Carbon credits may then be purchased by entities that expect to exceed their minimum negotiated emission levels. They may also be used by the recipient to offset high emission levels elsewhere in the country.

Baseline emissions cannot be measured. They are counterfactual, involving what would have happened had a project not taken place, and thus require expert judgment to be ascertained. In the CDM, the executive board, with the help of a methodologies panel, approves methodologies for determining baseline emissions for various types of projects.

Baseline emissions are directly tied to the profitability of specific emission-reduction projects and to the environmental integrity of emissions-offsetting mechanisms. If the baseline emissions of a given project are low, fewer credits will be awarded for the project sponsor to sell or use to offset emissions elsewhere. The project becomes less profitable, and fewer investors are drawn to invest in emission-reduction projects. If the baseline is high, more credits may be awarded to the project, allowing the project sponsor, or whoever buys the credits, to emit more elsewhere. The project becomes more profitable, but awarding more credits than were "actually" reduced threatens the environmental integrity of the mechanism. Because of the stakes involved, the entity awarding carbon credits generally implements procedures to help prevent "gaming" the baseline—that is, manipulating the baseline to change the number of credits awarded in order to benefit certain parties.

The cost, effort, and uncertainty associated with setting the baseline for emission reduction projects is sometimes so great that it makes projects unprofitable or unattractive. In response to this, some have suggested standardizing baselines to streamline the process. Another suggestion has been to move from project-based offsetting to "sectoral" offsetting, in which baseline emissions for an entire sector would be calculated and compared to actual emissions in order to award credits to mitigation strategies on a scale larger than individual projects.

Douglas Bushey

See also: Certified emissions reduction; Clean development mechanism; Emissions standards; Fossil fuel emissions; Industrial emission controls; Industrial greenhouse emissions; United Nations Framework Convention on Climate Change.

Bayesian method

- **Category:** Science and technology

- **Definition**

In statistics, there are two very different approaches to making inferences about unknown parameters: the frequentist, or non-Bayesian, and the Bayesian. One of the key differences between these two approaches is the notion of probability employed. The frequentist approach defines probability as the relative frequency of an event occurring in repeated trials; probability under this definition is also termed "objective probability." The Bayesian approach regards probability as a measure of the uncertainty inherent in a researcher's rational belief about the values of parameters or unknown quantities of concern. Another important difference is related to the specification of unknown parameters. The frequentist approach considers parameters as unknown but nonrandom values. By contrast, the Bayesian method regards parameters as random variables and uses probability distribution to specify possible values of those parameters.

In application, the Bayesian method is basically a way of learning from data. A Bayesian application refers to a three-step process employed to update a researcher's rational belief about unknown parameters or about the validity of a proposition, given the data observed. The first step concerns the formulation of a prior distribution for an unknown parameter in a statistical model of concern. The prior distribution reflects knowledge or results from past studies. Next, data or observations are collected to incorporate information about the parameter that generates those data. In the final step, the prior distributions are updated with the new data to create a new distribution. This method follows from a theorem formulated by and named after the Reverend Thomas Bayes, a British mathematician (1702-1761).

Applications of the Bayesian method and its implications for rational decisions are especially useful in studies for which the researcher has just a few data points available or for which uncertainty over some parameters needs to be resolved in the light of new data or observations. The method has become very popular over the past few decades in part because of the drastic growth of computer power, which renders much more feasible the calculations necessary to resolve simulations. There are various applications of the Bayesian method in such sciences as biostatistics, health outcomes, and global climate, to name just a few.

- **Significance for Climate Change**

Climate is the long-term average of weather events occurring in a region. The weather on a day in January in Chicago may be mild or sunny, but the winter climate in the city is on average cold, snowy, and rainy. Climate change reflects a change in long-term trends of the aggregate of these weather events. For example, annual precipitation can increase or decrease, and the climate can become warmer or colder. On a global scale, global warming refers to an increasing trend of Earth's temperature, which in turn causes changes in rainfall patterns, a rise in sea level, and a wide range of impacts on ecological systems and human life. The prediction of these changes and their impacts is difficult, because many uncertainties are associated with various relations and parameters present in the climate system. However, there has been significant study of these uncertainties under different methodologies.

As an approach to analyze uncertainty that allows incorporating expert knowledge and empirical observations into the analysis of updated data, the Bayesian method provides a powerful tool to study climate change. Recent studies have employed the Bayesian method to construct statistical models characterizing climate change. In those models, typical climate variables include, but are not limited to, surface air temperatures, precipitation, sea level, and ocean heat contents, on either a global or a regional scale. These models are built to serve many purposes, such as attribution, estimation, detection, and prediction of climate change.

In order to illustrate how the Bayesian approach can be applied to climate change, take the case of sea-level rise as an example. A recent study in this area tries to develop a Bayesian model for using evidence to update probability distributions for a climate model's parameters, which reflect the unknown states of nature, including sea-level rise. Once developed, the model and its updated probability distributions can be used to make projections of sea-level rise.

The steps to build such a model are as follows: First, define a prior probability distribution over the parameters of a model of sea-level rise as formulated based on expert knowledge or past studies. Second, draw a certain number of samples of those parameters at random from the prior distribution, then feed these samples into the model to calculate projected sea levels. Third, observe the actual sea levels and use the data to update the model's projections using the Bayes theorem. The updated projections are then translated into a new probability distribution. In the next cycle, with new data on sea levels obtained, the second and third steps are repeated, and the probability distribution is updated once more. Thus, each new observation of sea levels is incorporated into the model, providing better data and refining the probability distributions to increase the predictive accuracy of the system. The model parameters are partially resolved over time.

Although the Bayesian method provides an attractive approach for analyzing climate change, it is subject to some criticism. First, the idea of subjective judgments of prior probabilities, which influence the inferences drawn from models, is not accepted by many scientists. Critics of the Bayesian method argue that subjectivity prevents observers from viewing data objectively, so inferences should be based on observed data alone. Another problem is that people do not actually think like Bayesians. There is ample empirical evidence that people fail to update their prior beliefs using Bayes' law and that they act differently from the assumptions of Bayesian analysis would predict. It should be noted that there are alternative approaches to the Bayesian method to analyze uncertainty involving climate change, such as fuzzy set theory.

To N. Nguyen

• Further Reading

Bolstad, William M. *Introduction to Bayesian Statistics.* Hoboken, N.J.: John Wiley & Sons, 2007. Introduces the Bayesian approach to statistics. Covers the topics usually found in an introductory statistics book but from a Bayesian perspective.

Gelman, Andrew, et al. *Bayesian Data Analysis.* New York: Chapman & Hall, 2004. Includes an introduction to Bayesian inferences starting from basic principles, a text for graduate students to learn current approaches to Bayesian modeling and computation, and a handbook of the Bayesian method for researchers in the sciences.

Hobbs, Benjamin F. "Bayesian Methods for Analyzing Climate Change and Water Resource Uncertainties." *Journal of Environmental Management* 49, no. 1 (January, 1997): 53-72. Outlines the advantages of the Bayesian method when applied in analysis of uncertainty involving climate change, with an emphasis on the risks that changes pose to water resources systems.

See also: Climate prediction and projection; Meteorology; Weather forecasting.

Benefits of climate policy

• **Category:** Laws, treaties, and protocols

Climate policy decisions are particularly difficult to make because the accuracy of climate projections is difficult to evaluate. Even if the changes themselves were predictable, the consequences of those changes are not. Thus, the relative benefits and dangers of a particular course of action are difficult to determine.

• Key concepts

climate impact: the effects of climate and climate change on the socioeconomic well-being of an area

climate impact statement: empirical case study designed to help predict future impacts of climate on society

market benefits: positive effects of a given climate policy on production and trade, including prevention or mitigation of damages

nonmarket benefits: positive effects of a given climate policy on health, social and psychological welfare, and other attributes that are not primarily economic

• Background

In the 1980's, such books as W. J. Maunder's *The Uncertainty Business: Risks and Opportunities in Weather and Climate* (1986) emphasized the risks and opportunities presented by Earth's natural climate, subject to contingent human influences. Since then, mainly as a result of the works of the Intergovernmental Panel on Climate Change (IPCC), emphasis has moved markedly to considerations of the human impact on the climate system. However, irrespective of this change in emphasis, weather and climate have always given rise to risks and opportunities, and communities and individuals who can adapt to these challenges will always be in a position to lessen the costs of climate variations and climate changes. They will also be in a better position to increase the benefits and profits arising from climate variations and climate changes.

In Maunder's follow-up project, *The Human Impact of Climate Uncertainty: Weather Information, Economic Planning and Business Management* (1989), most of the emphasis was on variations in the natural climate. While these remain important, during the following twenty years a greater emphasis was placed on anthropogenic effects on the climate system. Nevertheless, good economic planning should take into account the best possible advice from climate experts on likely climatic changes during the relevant planning horizons, regardless of the causes of those changes.

• The Reliability of Climate Forecasts

To understand how a society might best respond to a change in its regional—as well as the global—climate, it is highly desirable to know how societies have been affected by, and how they have coped with, past climatic events, such as droughts, warm periods, cold periods, and wet periods. The climate of the future might not be exactly parallel to the climate of the past, particularly if anthropogenic factors become significantly more influential. Barring unforeseeable shocks to social and economic systems, however, socioeconomic institutions are likely to act in ways similar to their actions in the past. With better climate forecasts, societies can act in a much more informed manner to eradicate their weaknesses and capitalize on their strengths. Socioeconomic organizations would then be better prepared for future climate change, even if the nature of that change remains uncertain.

U.S. secretary of energy Stephen Chu, left, and U.S. secretary of commerce Gary Locke shake hands in Beijing, China, after announcing a joint project by the two nations to conduct energy-efficiency research. Climate policy, based in science, has profound effects upon trade and commerce. (Jason Lee/Reuters/Landov)

• The Formulation of Climate Policy

If organizations and individuals are going to take a positive attitude in dealing with climate change, again considering climate change from all causes, it is important that they understand the uncertainties of any climate forecast. The forecasts made by the various reports from the IPCC must be taken into account, but these forecasts have been subject to changes since the first IPCC report was published in the early 1990's, and they will continue to be modified as new information comes to light. Furthermore, while the majority of climate scientists generally agree with the forecasts made by IPCC scientists, there is a sizable group of climate scientists who have considerable concerns about the lack of emphasis being placed on the natural causes of climate change.

In particular, some scientists believe that the IPCC minimizes the role of variations in solar output, volcanic eruptions, the oceans, and other factors beyond human control. From a policy point of view, therefore, it is important for decision makers to take note not only of the average and extreme values forecast by the various IPCC reports but also of the possibility that some of these forecasts may prove to be wrong, particularly as the natural causes of climate change become better understood. Caution, therefore, should be a key concern, and decision makers concerned with climate change should be aware of the uncertainties involved in understanding and predicting that change.

• Context

Society would benefit considerably from correct policy decisions based on accurate climate forecasts. The world of climate forecasting and the world of decision making, however, are both far from perfect. The state of contemporary climate forecasting in particular is difficult to assess, because new computer models are making predictions about events decades in the future, and the accuracy of those predictions and their underlying methodology will not be known until decades have passed. Even accurate scientific predictions must be interpreted through the lens of policy and politics, adding a significant further complication.

For example, the 2007 IPCC report, states

> continued greenhouse gas emissions at or above current rates would cause further warming, and induce many changes in the global climate system during the twenty-first century that would "very likely" be larger than those observed during the 20th century.

Similarly, the 2007 IPCC report gives a best estimate, and a likely range of best estimates, for a global average temperature range for the last decade of the twenty-first century compared with the last two decades of the twentieth century.

Depending on the climatic greenhouse gas (GHG) scenario used, the best estimates of temperature increase range from 0.6° Celsius to 4.0° Celsius, and the likely range within these scenarios extends from 0.3° Celsius to 6.4° Celsius. Given these ranges, policy makers must determine which is the most appropriate forecast to use and how such a forecast will be used in planning activities in the future. Relevant activities vary widely, from constructing dams, to building roads to ski resorts, to planting new vineyards, and even to wholesale relocation of island inhabitants to avoid the consequences of sea-level rise.

The 2007 IPCC report projects many regional impacts of climate change during the next one hundred years. Under a range of climate scenarios, Africa's arid and semiarid land is projected to increase by 5-8 percent by 2080. In Europe, climate change is expected to magnify the regional differences in the distribution of natural resources and assets. Negative impacts will include the increased risk of inland flash floods, more frequent coastal flooding, and increased erosion due to storminess and sea-level rise. North American cities that currently experience heat waves are expected to be further challenged by an increased number, intensity, and duration of heat waves, with potential for adverse health impacts. Whether such forecasts are correct and whether society adapts itself to such forecasts remains to be seen.

W. J. Maunder

• Further Reading

Emanuel, K. "Increasing Destructiveness of Tropical Cyclones over the Past Thirty Years." *Nature* 436 (2005): 686-688. Written by a key research scientist in the field of tropical storms, particularly their destructiveness, frequency, and intensity.

Maunder, W. J. *The Uncertainty Business: Risks and Opportunities in Weather and Climate.* London: Methuen, 1986. Considers the atmosphere as a unique resource that may be tapped, modified, despoiled, or ignored. The "risks and opportunities" concept is particularly significant in any consideration of global warming.

Singer, S. Fred, and Dennis T. Avery. *Unstoppable Global Warming: Every Fifteen Hundred Years.* Rev. ed. Blue Ridge Summit, Pa.: Rowman & Littlefield, 2008. Describes a fifteen-hundred-year climate cycle that, according to the authors, offers the only explanation for modern global warming supported by physical science.

See also: Ecological impact of global climate change; Economics of global climate change; U.S. energy policy.

Bennett, Hugh Hammond

American soil conservationist

Born: April 15, 1881; near Wadesboro, Anson County, North Carolina
Died: July 7, 1960; Burlington, North Carolina

Bennett recognized that climate influences soil erosion, and he incorporated assessments of temperature, precipitation, and other climatic factors into soil-conservation techniques.

• Life

During childhood, Hugh Hammond Bennett observed soil damages on his family's 486-hectare North Carolina farm. He studied chemistry and geology at the University of North Carolina, earning a B.S. in 1903. Employed by the Bureau of Soils in the U.S. Department of Agriculture (USDA), Bennett traveled throughout the United States, documenting topsoil losses detrimental to agriculture and noting scientific causes of erosion. A pioneer soil conservationist, he stressed that soil resources were finite. In 1933, he became director of the Soil Erosion Service, a federal agency offering farmers conservation assistance. In 1935, Bennett encouraged U.S. legislators to establish the Soil Conservation Service in the USDA and was selected as its chief, promoting scientific conservation methods while holding that position and after his 1951 retirement.

• Climate Work

Bennett emphasized evaluating climate factors while planning soil conservation strategies to control erosion. He noted how climate affects crops in his book *The Soils and Agriculture of the Southern States* (1921), and he prepared reports discussing climate and conservation for scholarly journals. Bennett urged agriculturists to comprehend climate issues associated with lands they farmed in order to avoid such disasters as dust storms like those which had devastated the western United States in the 1930's. In public lectures and magazine articles, he promoted planting cover crops to collect precipitation, using contour tillage to stop runoff, and leaving stubble mulch in harvested fields to protect soil from wind.

Bennett devoted chapters to climate and its role in soil erosion in several books. In *Soil Conservation* (1939), he described how temperature, wind speed, humidity, and the amount and duration of precipitation affect soils. Bennett noticed that varying climates influence the speed and form of erosion, remarking that warmer temperatures accelerate chemical activity such as leaching in soil. He also stated that climates could be altered from moist to arid after winds blew topsoil away, exposing clays that could not absorb sufficient precipitation. Bennett reiterated this information in both editions of his classic text, *Elements of Soil Conservation* (1947 and 1955).

Bennett's climate insights helped soil conservation achieve recognition as a scientific field and inspired soil conservationists internationally, as they modeled their work on his techniques. Bennett's concepts retain value as global warming intensifies climate changes that impact twenty-first century landscapes.

Elizabeth D. Schafer

See also: Agriculture and agricultural land; Conservation and preservation; Soil erosion.

Berlin Mandate

- **Category:** Laws, treaties, and protocols
- **Date:** Negotiated March 28-April 7, 1995

The Berlin Mandate established the series of international meetings that led to the Kyoto Protocol.

• **Participating nations:** Albania, Algeria, Antigua and Barbuda, Argentina, Armenia, Australia, Austria, The Bahamas, Bahrain, Bangladesh, Barbados, Belgium, Belize, Benin, Bolivia, Botswana, Brazil, Burkina Faso, Cameroon, Canada, Chad, Chile, China, Comoros, Cook Islands, Costa Rica, Côte d'Ivoire, Cuba, Czech Republic, Democratic People's Republic of Korea, Denmark, Dominica, Ecuador, Egypt, Estonia, Ethiopia, Fiji, Finland, France, The Gambia, Georgia, Germany, Greece, Grenada, Guinea, Guyana, Hungary, Iceland, Italy, Jamaica, Japan, Jordan, Kenya, Kuwait, Lao People's Democratic Republic, Lebanon, Liechtenstein, Luxembourg, Malawi, Malaysia, Maldives, Mali, Malta, Marshall Islands, Mauritania, Mauritius, Mexico, Micronesia, Monaco, Mongolia, Myanmar, Nauru, Nepal, Netherlands, New Zealand, Nigeria, Norway, Pakistan, Papua New Guinea, Paraguay, Peru, Philippines, Poland, Portugal, Republic of Korea, Romania, Russian Federation, Saint Kitts and Nevis, Saint Lucia, Samoa, Saudi Arabia, Senegal, Seychelles, Slovak Republic, Solomon Islands, Spain, Sri Lanka, Sudan, Sweden, Switzerland, Thailand, Trinidad and Tobago, Tunisia, Tuvalu, Uganda, United Kingdom, United States of America, Uruguay, Uzbekistan, Vanuatu, Venezuela, Vietnam, Zambia, Zimbabwe

• Background

When the United Nations Conference on Environment and Development, also known as the Earth Summit, was held in Rio de Janeiro in 1992, the result was the international treaty known as the

United Nations Framework Convention on Climate Change (UNFCCC). The purpose of this treaty, which went into effect in 1994, was to prevent "dangerous anthropogenic interference with Earth's climate system." Instead of specific limits on emissions for individual nations, the treaty simply called for future protocols that would set these limits, but the intention was that industrialized nations, known as Annex I countries, would reduce their emissions to 1990 levels by the year 2000. In 1995, the first annual Conference of the Parties (COP-1) was held in Berlin, Germany, to analyze progress and make plans for the future. It was generally agreed that industrialized nations would not be able to honor their emission reduction commitments by 2000. The Berlin Mandate, an agreement reached at that meeting, was an attempt to establish goals that could reasonably be met.

• Summary of Provisions

Realizing that the 2000 targets would not be met, the framers of the Berlin Mandate gave the parties the ability to begin making plans and commitments "for the period beyond 2000." Following the understanding reached in the UNFCCC, industrialized nations would continue to bear most of the responsibility for emissions reductions, because they generated most of the emitted greenhouse gases (GHGs) and because they were most able to make reductions. However, it was agreed that

> the global nature of climate change calls for the widest possible cooperation by all countries and their participation in an effective and appropriate international response, in accordance with their common but differentiated responsibilities and respective capabilities and their social and economic conditions.

The Berlin Mandate called for a two-year analytical and assessment phase to be conducted by the Intergovernmental Panel on Climate Change (IPCC) and suggested that this research phase might lead to new targets

Berlin Mandate

The Berlin Mandate is an agreement to create a process to strengthen provisions of the United Nations Framework Convention on Climate Change. Article 2 of the mandate, reproduced below, establishes the objectives and protocols governing that process.

The process will, inter alia:

(a) Aim, as the priority in the process of strengthening the commitments in Article 4.2(a) and (b) of the Convention, for developed country/other Parties included in Annex I, both

- to elaborate policies and measures, as well as
- to set quantified limitation and reduction objectives within specified time-frames, such as 2005, 2010 and 2020, for their anthropogenic emissions by sources and removals by sinks of greenhouse gases not controlled by the Montreal Protocol, taking into account the differences in starting points and approaches, economic structures and resource bases, the need to maintain strong and sustainable economic growth, available technologies and other individual circumstances, as well as the need for equitable and appropriate contributions by each of these Parties to the global effort, and also the process of analysis and assessment referred to in section III, paragraph 4, below;

(b) Not introduce any new commitments for Parties not included in Annex I, but reaffirm existing commitments in Article 4.1 and continue to advance the implementation of these commitments in order to achieve sustainable development, taking into account Article 4.3, 4.5 and 4.7;

(c) Take into account any result from the review referred to in Article 4.2(f), if available, and any notification referred to in Article 4.2(g);

(d) Consider, as provided in Article 4.2(e), the coordination among Annex I Parties, as appropriate, of relevant economic and administrative instruments, taking into account Article 3.5;

(e) Provide for the exchange of experience on national activities in areas of interest, particularly those identified in the review and synthesis of available national communications; and

(f) Provide for a review mechanism.

for the years 2005, 2010, and 2020. The second Conference of the Parties (COP-2) was to present a progress report including "an analysis and assessment, to identify possible policies and measures for Annex I Parties which could contribute to limiting and reducing emissions by sources and protecting and enhancing sinks and reservoirs of greenhouse gases." The study, with recommendations for further binding agreements, would be completed in time to be presented by the Ad Hoc Group on the Berlin Mandate (AGBM) at COP-3.

• Significance for Climate Change

The AGBM met eight times between COP-1 and COP-3. At Kyoto, Japan, where COP-3 was held, it proposed what became known as the Kyoto Protocol. Based on the findings of the IPCC, the protocol created binding targets for Annex I countries. The United States, the largest emitter at the time, was to reduce its GHG emissions to about 7 percent below 1990 levels; European Union nations would reduce by 8 percent, and Japan by 6 percent. Although they had agreed in theory to the provisions of the Berlin Mandate, many industrialized nations protested that refusing to hold developing nations, or non-Annex I nations, to binding reduction targets was both unfair and unwise. These developing nations were undergoing rapid population growth and industrial development, and it was recognized that by 2010 they would be the largest emitters of GHGs. Citing this disparity and its own need for continued economic growth, the United States did not ratify the treaty, although by 2008, 183 other countries had ratified it.

Cynthia A. Bily

• Further Reading

Chasek, Pamela S., David Leonard Downie, and Janet Welsh Brown. *Global Environmental Politics.* 4th ed. Boulder, Colo.: Westview, 2006. Introductory textbook in environment-related international relations; begins with the emergence of global environmental politics and considers climate change, among other issues.

Montgomery, W. David. *Toward an Economically Rational Response to the Berlin Mandate.* Boston: Charles River Associates, 1995. Written immediately after the Berlin Mandate was approved, this brief book proposes economic considerations and tools important for drafting a U.S. response.

Niskanen, William L. *Reflections of a Political Economist: Selected Articles on Government Policies and Political Processes.* Washington, D.C.: Cato Institute, 2008. This collection by an advocate of free markets and limited government includes the essay "Too Much, Too Soon: Is a Global Warming Treaty a Rush to Judgment?"

Paterson, Matthew. *Global Warming and Global Politics.* New York: Routledge, 1996. A history of the early years of global warming as a political issue and of the negotiations leading to international agreements including the Berlin Mandate.

See also: Greenhouse gases; International agreements and cooperation; Kyoto Protocol; United Nations Framework Convention on Climate Change; United States.

Biodiversity

• **Categories:** Animals; plants and vegetation

Global warming is an emerging threat to biodiversity around the world. As temperatures rise, the geographic location of climatic envelopes will shift significantly, forcing species to migrate. Those that are not able to keep up with their respective climatic envelopes may become extinct.

• Key concepts

biodiversity: the total variation within and among all species of life in a given area, ecosystem, or context

climatic envelope: the range of temperatures, precipitation, and other climatic parameters to which a species has adapted

coral bleaching: whitening of coral that occurs when the coral expels a single-celled, symbiotic alga as a result of stress caused by environmental factors such as warm temperature and pollution

living planet index (LPI): a number representing the population sizes of vertebrate species representative of terrestrial, freshwater, and marine ecosystems around the world; a higher LPI is indicative of higher biodiversity

• Background

Biological diversity allows myriad species to work together to maintain the environment without costly human intervention. Thus, biodiversity is an irreplaceable natural resource crucial to human wellbeing. Habitat destruction, pollution, invasive species, overexploitation, and global climate change have led to an accelerated decline in biodiversity in recent decades. As a result, the current rate of species extinctions is estimated at one hundred to one thousand times greater than the natural rate. If the biodiversity decline is not slowed, Earth will be much less inhabitable for future generations.

• Past and Current Extinctions

There is no doubt that climate plays a central role in fluctuations of biodiversity. At least four out of five recognized mass extinction events on Earth are attributable to climate changes. The Triassic event (around 200 million years ago) was triggered by atmospheric carbon dioxide (CO_2) levels increasing to one hundred times the current level. The Permian-Triassic extinction (about 251 million years ago), the greatest mass extinction on record, resulted from global temperature rises to between 10° and 30° Celsius higher than today's average temperature. That single event is believed to have wiped out 95 percent of the life in Earth's oceans and nearly 75 percent of terrestrial species.

Climate change has also led to the emergence of new species. In fact, some experts believe that rapid climate changes in Africa might have created suitable conditions for the emergence of modern humans. In particular, rapid changes in water sources are believed to have forced primitive hominids rapidly to change and adapt. Under this theory, *Homo sapiens* evolved from its progenitor species as a result of these changes. In general, however, the birth of new species following a mass extinction usually takes millions of years. Recovery from severe losses of biodiversity does not occur on a human timescale.

Many experts believe that Earth is currently in the midst of a sixth mass extinction event. Although there are many debates raging over the extent of the current biodiversity loss, very few scientists dispute the fact that species extinction is occurring as a result of climate change and human activities. Comprehensive studies conclude that 15 to 37 percent of Earth's species will become extinct by 2050 as a result of projected climate changes. Such massive extinctions will wreak havoc on ecosystems around the world.

• Impact of Climate Change on Biodiversity

Climate changes can affect species in a number of ways. These include the expansion, contraction, and shift of habitats; changes in temperature, precipitation, and other environmental conditions; increased frequency of diseases; emergence of invasive species; and disrupted ecological relationships. With a rising temperature, extreme and severe weather events will become more frequent, including extreme high temperatures, extreme severe storms, large floods, a decrease in snow cover and ice caps, rising sea levels, and alteration to the distribution of infectious diseases and invasive species.

As temperatures rise, habitats for many plants and animals will be altered, eliminating the homes and niches to which they have adapted. It is estimated that up to 60 percent of northern latitude habitats could be affected by global warming. In response to global warming, plants and animals will migrate to more suitable climes. Specific effects have already been observed. The tree line near Olympic National Park has moved up in altitude by more than 30 meters since the 1980's. The red fox is spreading northward in Australia in response to the warming climate. Many fish species along the Pacific coast are shifting their habitats northward in search of cooler waters.

• Biodiversity Loss Due to Human Activities

Before the rapid explosion of human populations, many species may have responded to climate change by migrating northward or southward to the cooler regions. Contemporary human activities such as urbanization, road construction, agriculture, and tourism have fragmented, converted, and destroyed many habitats and potential migration routes and thus make it much more difficult for species to migrate. As a result, many species struggle to cope with climate change as they decline in population, often facing outright extinction.

The replacement of low-intensity farming systems with industrial agro-ecosystems has led to a significant decline in biodiversity. For example, the deforestation of tropical rain forests represents an alarming threat to biodiversity. In some regions, forest cover remains high, but intensive management has turned natural forests into stands of very few or single tree species, leading to the loss of many animal species as well. The disappearance of wetlands worldwide has been dramatic over the last century, ranging from 60 percent in China to 90 percent in Bulgaria. The living planet index declined by 37 percent between 1970 and 2000.

• Context

The loss of biodiversity is both real and accelerating, as Earth continues to warm. Approximately 25 percent of conifers, 52 percent of cycads, 12 percent of bird species, 23 percent of mammals, and 32 percent of amphibians are threatened with extinction. The irreversible loss of biodiversity poses serious threats to the well-being of present and future generations. The societal response to this threat has been slow, stemming from a lack of awareness of the vital role biodiversity plays. The key to the solution is education.

There is often a perception of conflict between the need to preserve biodiversity and the goal of economic development. Policy makers and businesses will need to work together to encourage and reward economic development that is friendly to biodiversity. Biodiversity is more complex than many other environmental concerns because it involves several levels of biological organization, including genes, individuals, species, populations, and ecosystems. It cannot easily be measured by a single indicator such as temperature or rainfall. Nevertheless, countries can work together to build consensus and achievable goals aiming at slowing down the biodiversity decline or even restoring habitats for the recovery of threatened species.

Ming Y. Zheng

A May, 2008, meeting of the United Nations Conference on Biological Diversity. From left: General Assembly president Srgjan Kerim, European Union Commission president Jose Barroso, German chancellor Angela Merkel, German environment minister Sigmar Gabriel, and Canadian prime minister Stephen Harper. (Ina Fassbender/Reuters/Landov)

• **Further Reading**

Braasch, Gary. *Earth Under Fire: How Global Warming Is Changing the World.* Afterword by Bill McKibben. Berkeley: University of California Press, 2007. Very important topic clearly presented and supported by photographs and firsthand data the author collected during his travels around the world. The author's research is meticulous, his reasoning is sound, and his conclusion is clear.

Emanuel, Kerry. *What We Know About Climate Change.* Cambridge, Mass.: MIT Press, 2007. Useful reference summarizing what is and is not known about global warming. Arguments are well substantiated.

Fotherqill, Alastair, et al. *Planet Earth: As You've Never Seen It Before.* Berkeley: University of California Press, 2006. As a companion to a television series on Discovery Channel and the British Broadcasting Corporation, this remarkable book depicts natural wonders and represents an eloquent rallying call to significantly strengthen efforts to preserve these wonders. Illustrated with spectacular color photographs.

Lovejoy, Thomas E., and Hannah Lee, eds. *Climate Change and Biodiversity.* New Haven, Conn.: Yale University Press, 2005. Series of essays, well edited and presented in a cohesive flow. A must read for all who love nature and wildlife and want to know the impacts of climate change upon them.

See also: Amazon deforestation; Amphibians; Dolphins and porpoises; Forests; Mangroves.

Bioethics

• **Category:** Ethics, human rights, and social justice

Climate change raises difficult moral issues because the need to allocate limited funds and resources requires some classes, peoples, and generations to be prioritized over others. Bioethics seeks to create rational frameworks to facilitate the resolution of these issues.

• Key concepts

ethics: the application of moral philosophy to real-world decision making

intergenerational equity: relative equality of treatment between present and future generations, including the obligation of present generations to preserve limited resources for the future

prioritization: using a rational principle to determine the order of precedence among individuals or groups when allocating limited resources

• Background

Bioethics is a discipline that seeks to determine the most ethical course of action when faced with a decision involving medical care, medical or biological research, or life and living processes generally. As such, it is implicated in many decisions dealing with the human response to climate change, because those decisions require balancing the interests of people in different locations, of different classes, and of different generations.

Sensitivity to temperature, and to change in temperature, is a central fact of almost all physical objects and processes, living or not. Extremes of cold and of heat disrupt physiological processes and thereby undermine ecological stability. Heat waves in summer in large population centers, for example, frequently result in illness and death. Such health threats create practical problems of predicting and planning for them, as well as training for and executing appropriate responses. They also raise many of the issues familiar in discussions of medical ethics.

• Prioritization and Prediction

A central question in preparing responses to possible threats related to global warming is that of prioritization. Finite resources must be allocated equitably or according to some other compelling moral principle. It may be necessary to sacrifice a group or individual to advance the well-being of others or to abrogate property rights to serve the general welfare. Moreover, because predicting the future is never a certainty, principles must be established to measure or evaluate the uncertainty of predictions and the acceptable level of risk. Often, a balance must be struck between the two: If an event seems unlikely and its consequences would be minor, preparing for it is less important than preparing for either a near certainty with minor consequences or an unlikely event with severe consequences.

A particularly thorny issue involves intergenerational equity, the comparative claims of present and future generations. The moral philosopher John Rawls spoke of the "just savings" a society is required to put aside for the future. Choices made today will often affect those who live decades or centuries from now. Insofar as these effects can be anticipated, the needs, preferences, and well-being of future generations must be taken into account. It is difficult, however, to weigh those interests against those of the present generations, not least because only one party is able to participate in making decisions. This means not only that the decision makers will be biased in favor of themselves but also that it is impossible to determine with certainty what the preferences of unborn future generations will be. Indeed, as utilitarian theorist Derek Parfit emphasizes, it is possible that the choices made by present generations will determine which individuals are and are not born to be members of future generations.

In order to make ethical decisions regarding the proper response to global warming, it is necessary to establish a framework for evaluating future harms and benefits. Some economists invoke a social "discount rate" to resolve this problem. This method entails calculating the current cost of preventing future harm and comparing it to the cost of that harm in the future.

• Competing Present Interests

Even restricting attention to the present, serious questions arise regarding the most ethical distribution among and within nations of the burdens of climate calamity and the costs of avoiding it. The Kyoto Protocol, for example, treats developing and industrialized nations differently and establishes more stringent standards for the latter. To hold every nation to the same standard (say, so much greenhouse gas emission per capita) would advantage those nations already well developed industrially, which would thereby benefit from their past pollution, and disadvantage developing nations that need to produce significant pollution in the present if they are to progress economically.

On the other hand, using different standards may disadvantage industrialized nations that are competing with developing nations in a global market. More stringent rules regulating labor and environmental impact, for example, can make production in developed nations more expensive than production of the same products in developing nations. However, the lax standards that make production cheaper in developing nations may impose indirect costs not only on companies but also on all members of society. As journalist Alexandra Harney remarks,

> pollution from Asia is believed to be affecting weather up and down the west coast of North America. . . . The sacrifices China makes to stay competitive in manufacturing affect the rest of the world.

• Context

Bioethics is a branch of applied ethics, a relatively recent subdiscipline of the ancient discipline of ethics. Bioethics seeks to ascertain consistent principles to guide decisions regarding such literal life-and-death issues as euthanasia, medical ethics, and proper responses to global warming. Because climate change is such a large-scale phenomenon, both spatially and temporally, the unique challenges it poses to bioethics involve the need to reconcile the disparate interests of all nations and their inhabitants, as well as present and future generations. The scale of the problems posed by climate change requires both individuals and governments to think globally while acting locally.

Edward Johnson

• Further Reading

DeVries, Raymond, and Janardan Subedi, eds. *Bioethics and Society: Constructing the Ethical Enterprise.* Upper Saddle River, N.J.: Prentice Hall, 1998. Sociologists discuss the emergence of the field of bioethics and its supplantation of law, literature, and religion among medical decision makers and their overseers.

Friedman, Thomas L. *Hot, Flat, and Crowded: Why We Need a Green Revolution, and How It Can Renew America.* New York: Farrar, Straus and Giroux, 2008. A well-known journalist and social commentator offers practical arguments about the proper responses to overpopulation, global warming, and other environmental problems.

Garvey, James. *The Ethics of Climate Change: Right and Wrong in a Warming World.* New York: Continuum, 2008. Broad philosophical survey of the ethical issues raised by global warming.

Harney, Alexandra. *The China Price: The True Cost of Chinese Competitive Advantage.* New York: Penguin, 2008. Account of the environmental and other sacrifices China has made to achieve its economic preeminence.

Lomborg, Bjørn. *Cool It: The Skeptical Environmentalist's Guide to Global Warming.* New York: Knopf, 2007. Lomborg approaches questions of how best to weigh the costs of future harms in the spirit of an economist.

Parfit, Derek. *Reasons and Persons.* New York: Oxford University Press, 1984. Subtle discussion of the intellectual complexities of moral thought from a staunchly utilitarian perspective, including analysis of the social "discount rate" as a problem in dealing with the claims of future persons.

Posner, Richard. *Catastrophe: Risk and Response.* New York: Oxford University Press, 2004. A prolific federal judge and public intellectual analyzes flaws in familiar approaches to catastrophic risk—cases in which the natural tendency is to underestimate impact because of low probability.

Rawls, John. *A Theory of Justice.* Cambridge, Mass.: Harvard University Press, 1971. Analysis from a leading nonutilitarian theorist, with implications for how justice is understood across generations.

Reich, W. T. "The Word 'Bioethics': Its Birth and the Legacies of Those Who 'Shaped It.'" *Kennedy Institute of Ethics Journal* 4 (1994): 319-335. Discusses the 1971 origin of the word and the field, and the consequences of their beginnings for later thinkers.

Singer, Peter. *One World: The Ethics of Globalization.* 2d ed. New Haven, Conn.: Yale University Press, 2004. One of the world's most influential experts on applied ethics applies his utilitarian perspective to international problems, including environmental pollution and global warming.

See also: Benefits of climate policy; Conservation and preservation; Intergenerational equity; Journalism and journalistic ethics; Polluter pays principle; Poverty; Precautionary approach.

Biofuels

- **Category:** Energy

- **Definition**

Biofuels are renewable fuels generated from organisms or by organisms. Biofuels are considered by many as a future substitute for fossil fuels. For millions of years, living organisms played a crucial role in the formation of fossil fuels such as oil, natural gas, and coal. As fossil fuels are being depleted, humankind is looking for alternative energy sources. Once again, living organisms can be used to generate such fuels, including ethanol, biodiesel, butanol, biohydrogen, and biogas (mostly methane). One of the applications for biofuels is to use them as gasoline and diesel substitutes. At present, two biofuels are used in vehicles globally: ethanol and biodiesel. Biogas is mainly used to generate electricity. These biofuels are made from plant biomass, including corn, soybean, sugarcane, rapeseed, and other plant material. Solar energy is converted and stored in plant cells in the form of carbohydrates or lipids, and this energy can be transferred into biofuel energy.

Ethanol (C_2H_5OH), or grain alcohol, the most common biofuel, is produced by yeast fermentation of sugars derived from sugarcane, corn starch, and grain. In the United States, most of ethanol is produced from corn starch. Biodiesel, another commonly used biofuel, is made mainly by transesterification of plant oils, such as soybean, canola, or rapeseed oil. Its chemical structure is that of the fatty acids alkyl esters. Biodiesel may also be produced from waste cooking oils, restaurant greases, soap stocks, animal fats, and even algae.

Biogas, which is 50-75 percent methane, is a product of anaerobic fermentation of biomass. In many countries, millions of small farmers maintain

Biofuel Energy Balances

The following table lists several crops that have been considered as viable biofuel sources and several types of ethanol, as well as each substance's energy input/output ratio (that is, the amount of energy released by burning biomass or ethanol, for each equivalent unit of energy expended to create the substance).

Biomass/Biofuel	*Energy Output per Unit Input*
Switchgrass	14.52
Wheat	12.88
Oilseed rape (with straw)	9.21
Cellulosic ethanol	1.98
Corn ethanol	~1.13-1.34

Data from the British Institute of Science in Society.

a simple digester at home to generate energy. There are more than five million household digesters in China, used by people mainly for cooking and lighting, and over a million biogas plants of various capacities in India.

Other types of biofuels are on the road to commercialization. Among them, the most promising is butanol. Butanol (C_4H_9OH) is an alcohol fuel, but, compared with ethanol, it has higher energy content (roughly 80 percent of gasoline energy content). It does not pick up water as ethanol does, nor is it as corrosive as ethanol, and it is more suitable for distribution through existing pipelines for gasoline. Butanol is produced as a result of fermentation by the bacterium *Clostridium acetobutylicum.* Substrates for fermentation utilized for butanol production—starch, molasses, cheese whey, and lignocellulosic materials—are the same as for ethanol.

• Significance for Climate Change

Use of fossil fuels releases carbon dioxide (CO_2), which contributes to global warming. Biofuel utilization can produce considerably less CO_2 compared to fossil fuel utilization. In general, burning biofuels releases only that CO_2 that was captured by plants during photosynthesis. They can be considered "CO_2 neutral," in that the CO_2 released by burning can be reassimilated by plants. However, significant amounts of CO_2 are also generated during biofuel production. Estimated CO_2 emission during production of biofuels greatly depends on the method of their manufacture.

Production of biofuels from crops such as corn under the current fossil-fuel-based agricultural system would significantly increase greenhouse emissions. Emissions result from growing feedstock, applying fertilizers, transporting the feedstock to factories, processing the feedstock into biofuels, and transporting biofuels to their point of use. Manufacturing biofuels from corn starch and plant vegetable oil requires burning considerable amounts of natural gas, diesel, or coal to provide energy.

In contrast, net emissions of CO_2 during biofuel production from lignocellulose can be nearly zero. Lignocellulose is a combination of lignin, cellulose, and hemicellulose that strengthens plant cell walls. Cellulose and hemicellulose are made from sugars that can be converted into biofuels. Electricity from burning the lignin could provide enough energy to replace coal or natural gas during the production of biofuels from lignocellulose. Burning lignin does not add any CO_2 to the atmosphere, because the plants that are used to make the biofuels absorb CO_2 during their growth. Most important, lignocellulose may be obtained from nonedible plants, such as switchgrass and poplar, or nonedible parts of plants, such as corn stalks and wood chips.

Lignocellulose is also a very attractive biofuel feedstock because of its abundant supplies. On a global scale, plants produce about 90 billion metric tons of cellulose per year, making it the most abundant organic material on Earth. In addition, cultivation of nonedible plants for biofuel production requires fewer nutrients, fertilizers, and herbicides and less cultivated land and, thus, fewer energy resources. However, current methods to process cellulosic parts of plants into simple sugars in order to ferment them into ethanol are costly. Major research and development efforts are under way to lower the cost of converting lignocellulose to biofuel.

There is also considerable interest in algae-based biofuels, especially biodiesel. Research conducted by the U.S. Department of Energy Aquatic Species Program from the 1970's to the 1990's demonstrated that many species of algae produce sufficient quantities of oil to become economical feedstock for biodiesel production. Oil productivity of many algae greatly exceeds oil productivity of the best-producing oil crops. Algal oil content can exceed 80 percent of cell dry weight, with oil levels commonly at about 20 to 50 percent of cell dry weight. In addition, cropland and potable water are not required to cultivate algae, because they can grow in wastewater. Although development of biodiesel from algae is a very promising approach, this technology is not yet ready for immediate commercial implementation and needs further research.

Sergei Arlenovich Markov

• Further Reading

Chisti, Yusuf. "Biodiesel from Microalgae." *Biotechnology Advances* 25 (2007): 294-306. Review of the current state of technology for generating biodiesel from microalgae.

Glazer, Alexander N., and Hiroshi Nikaido. *Microbial Biotechnology: Fundamentals of Applied Microbiology.* New York: Cambridge University Press, 2007. Analysis of the pros, cons, and pitfalls of generating and burning ethanol for fuel.

Wald, Matthew L. "Is Ethanol for the Long Haul?" *Scientific American*, January, 2007, 42-49. Excellent discussion about the future of ethanol fuel.

See also: Biotechnology; Clean energy; Energy efficiency; Energy from waste; Energy resources; Ethanol; Fossil fuel emissions; Fossil fuel reserves; Fossil fuels; Fuels, alternative; Pollen analysis.

Biotechnology

• **Category:** Science and technology

Although the application of traditional biotechnology can be linked to a number of detrimental effects on the environment, modern biotechnological methods appear poised to reduce the levels of GHGs being released into the atmosphere.

• Key concepts

biodiesel: plant-derived oil that can be used to power diesel engines

biofuels: renewable resources that originate from living organisms and that can be burned for no net gain in CO_2 emissions

biomass: plant and other organic materials that can be used as fuel

ethanol: ethyl alcohol, a product of fermentation that can be used as a biofuel

modern biotechnology: advanced biological methods, such as genetic manipulation and cloning, that alter organisms so they are better suited to achieve human goals

recombinant DNA technology: methods whereby DNA from one organism can be recombined with that of another; the basis of modern biotechnology

traditional biotechnology: selective breeding of living organisms

white biotechnology: environmental biotechnology, the application of biotechnological principles affect the environment to positively

• Background

Biotechnology, broadly defined as the use of living organisms to achieve human goals, has been practiced for thousands of years, beginning with the domestication of animals and plants. Common usage of the word, however, typically refers to the use of more modern biological methods to achieve many of the same purposes. The term "modern biotechnology" is sometimes used to differentiate contemporary techniques from "traditional biotechnology." Biotechnology is not a scientific discipline in itself but is intrinsically interdisciplinary in nature, involving mostly agricultural techniques at first but more recently combining principles from such fields as microbiology, cell and molecular biology, and engineering.

• Traditional Biotechnology

Biotechnology likely had its origins between 10,000 and 9,000 B.C.E. with the domestication of dogs in Mesopotamia and Canaan. The first crops, consisting of emmer wheat and barley, are thought to have been grown in this same area within the following millennium. At this time, human impact on the environment was minimal, but as more land was subsequently cleared for the growing of crops and the raising of livestock, the potential to affect the environment increased, albeit slowly. These changes accelerated following the Industrial Revolution, as technologies to modify Earth's landscape were developed. By the late twentieth century, deforestation had become a major contributor to atmospheric carbon dioxide (CO_2) levels, while livestock could be linked to the release of methane, another greenhouse gas (GHG), into the atmosphere.

• Modern Biotechnology

Around this time, biotechnology underwent a revolution, as scientific breakthroughs made it possible directly to change the genetic makeup of virtually any organism. Prior to this, desired changes in organisms had been achieved primarily by selective breeding, a slow and inexact process. Beginning in the 1970's, techniques were developed that allowed scientists to cut deoxyribonucleic acid (DNA) at specific sequences (using purified enzymes called restriction enzymes) and to "glue" these liberated fragments of DNA into a vector that allowed for their propagation in host organisms, thereby cloning a particular gene or DNA segment. This entire process, sometimes called recombinant DNA technology, greatly altered both the speed and the scope of the genetic changes that could be achieved in targeted organisms.

It was not long before recombinant techniques had led to such outcomes as the production of human insulin in the bacterium *Escherichia coli* (in 1982), the production of ethanol from sugar in the same microbe (in 1991), and the development of a tomato that instead of ripening on the vine, could be picked while green and artificially ripened following shipping (in 1992). These particular examples represent the first applications of modern techniques in three different categories of biotechnology: medicine ("red biotechnology"), environmental science ("white biotechnology"), and agriculture ("green biotechnology"). The first category quickly became dominant over the other two in terms of money invested in the science, accounting for nearly 90 percent of venture capital in the twenty-five years following its introduction. The other two categories split the remaining investments nearly equally during the same time period.

• White Biotechnology

Sometimes called "environmental biotechnology," white biotechnology has been utilized to clean up contaminated environments via bioremediation, prevent the discharge of pollutants from currently existing industries, and generate resources in the form of renewable chemicals and biofuels. While bioremediation comprises a large portion of white biotechnology, it is the latter two goals that are expected to have the greatest effects on alleviating global warming.

Recognizing that the burning of fossil fuels is not likely to disappear overnight, scientists have been focusing on the use of living organisms to remove a portion of the CO_2 found in fossil fuel emissions. One candidate for this removal is phytoplankton, microscopic aquatic algae. Phytoplankton are known to make up a large portion of the carbon fixation cycle, which converts CO_2 (or dissolved carbon) to sugars during photosynthesis, thereby removing it from the surrounding environment. In nature, phytoplankton eventually die and sink to the bottom of the ocean, removing the carbon from circulation, if only temporarily on a geological timescale.

The burning of fossil fuels has the undesirable consequence of releasing into the environment carbon that has been sequestered in this way for millions of years, adding to the "new" carbon that is released from the burning of biomass. Experiments have been performed in which effluents from power plants were passed through columns filled with algae to reduce their CO_2 emissions. One problem that remained, however, was how to dispose of the algae, since simply allowing them to decompose would return the sequestered carbon back to the atmosphere. One possible solution that was explored involved burning dried algal pellets as fuel. Although still resulting in the release of CO_2, this burning allowed more energy to be obtained for a given amount of emission.

The burning of biomass for fuel is preferable to the burning of fossil fuels in terms of its effects upon global warming, since this process can be thought of as "CO_2 neutral," in that it simply returns to the atmosphere CO_2 that was recently sequestered by the organism in question. Biomass made up the majority of fuel prior to the Industrial

Revolution, and it is still relied on by about 50 percent of the world's population to meet its daily fuel needs. The widespread use of wood for fuel, however, although technically CO_2 neutral, is less desirable than the use of other forms of biofuel, since it contributes to deforestation and utilizes a resource that takes a considerable amount of time to replenish, compared to agricultural products. Another drawback of solid biofuel such as wood or plant waste is that it does not represent a practical replacement for petroleum-based products in automobiles, a major contributor of CO_2 emissions, after power stations and industrial processes. Biomass is therefore typically converted into various alcohols, oils, or gases in order to be used in such applications as the powering of automobiles. Conversion to these forms eases its transportation and storage.

- **Biofuels**

The fermentation of crops to obtain ethyl alcohol, or ethanol, was first performed in Egypt around 4000 B.C.E., although it was not known at the time that the process was being carried out by microscopic yeast. The purpose of such fermentation, however, was the brewing of alcoholic beverages, not the production of fuel. It was not until the energy crunch of the 1970's that ethanol began being mass-produced for fuel from either corn or sugarcane, the former conversion being practiced primarily in the United States, with the latter being most prevalent in Brazil. Brazil subsequently emerged as one of the few success stories regarding biofuels, as a result of the somewhat unique situation of having its sugarcane-growing centers in close proximity to its main population centers. This served to reduce shipping costs, while processing costs were contained by using the residual cane waste, or bagasse, as fuel for the processing plants.

In the United States, where the price of ethanol is more closely linked with the price of oil by the increased costs of processing corn and shipping ethanol to major population centers, debate continues on whether its production as a fuel is economically viable. It has been argued that any fuel that directly competes with food crops will result in increased food prices and ultimately lead to the expansion of farming, so that CO_2 emissions could actually experience a net increase as a result of U.S. corn ethanol production. At current production efficiencies, it has been estimated that, even if all of the corn grown in America were converted to ethanol, it would be able to replace only about 20 percent of domestic gasoline consumption. In order to solve some of these drawbacks, focus has shifted to the use of genetically modified organisms to efficiently convert the cellulose found in corn stover, the waste equivalent of bagasse, into ethanol for fuel.

One alternative to the production of ethanol as fuel is the use of plant oils in diesel engines. The use of these oils actually dates from the origination,

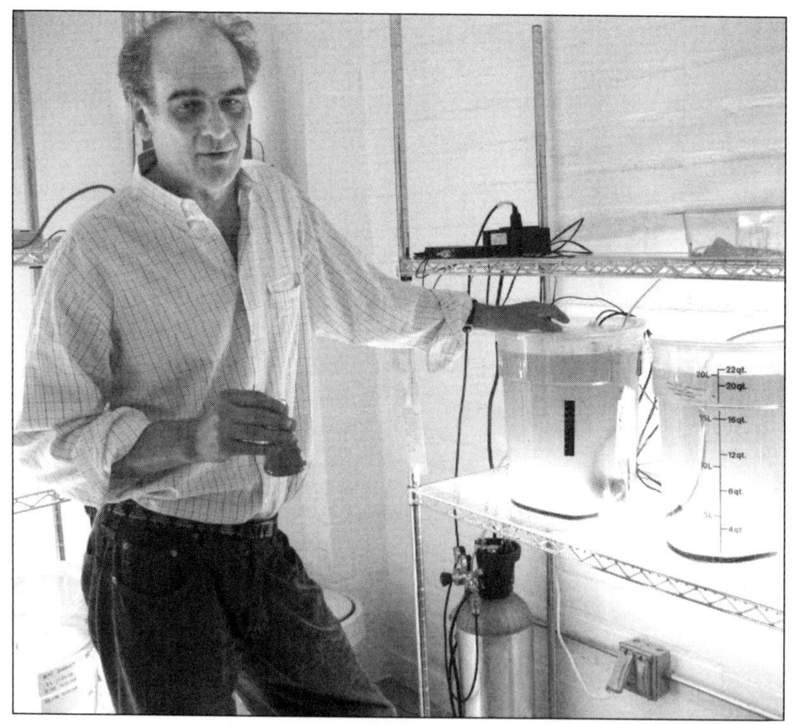

Nicholas Eckelberry, cofounder of OriginOil, displays containers of algae that his company is attempting to turn into an efficient biofuel. (Reuters/Landov)

in 1894, of the engine itself, which was designed by its German inventor, Rudolf Diesel, to burn a variety of fuels, including coal dust and peanut oil, in addition to petroleum products. The use of plant oils, or biodiesel, has seen its greatest adoption in Europe, often in conjunction with public transportation fleets, but is increasingly used in the United States as well. Rapeseed oil is typically used in the former case, while soybean oil is a more likely fuel in the latter. Unfortunately, biodiesel has shared many of the same problems that ethanol production from biomass has, including relatively high costs of production, as well as competition with the use of the same crops for food. One possible solution has been the genetic modification of algae so they accumulate excess oil, which can then be purified. Such algae could potentially be grown in aquatic environments that would not compete with the land normally used for food crops.

One biofuel that perhaps holds the most promise is hydrogen gas, H_2. While this gas can be easily adapted for use in automobiles or to generate electricity, it differs from the other fuels discussed in that the combustion of H_2 produces no CO_2 whatsoever, only water. One technical hurdle that must be overcome is that H_2 is normally released at a very low efficiency by the algae and bacteria that are known to produce it. These organisms typically undergo a process called photolysis, where an H_2O molecule is split using energy derived from sunlight. The enzyme responsible for creating the H_2 gas, however, is inhibited by the presence of the oxygen that is created during photolysis. Genetic engineering may be the key to improving the efficiency of H_2 production, so its use as a biofuel may soon be realized.

• Context

Throughout the years, humans have used the living things around them to meet their basic needs, as well as to achieve various other purposes, slowly changing these organisms through selective breeding in order to cause them to be better suited for their desired application. Achieving human purposes has not always had a positive effect on the environment, with the domestication of both plants and animals being responsible for steadily releasing large amounts of CO_2 and methane into the atmosphere. It has only been fairly recently that humans have acquired the motivation and the technology to begin to address some of these detrimental changes. The ability to change organisms rapidly via recombinant DNA technology may hold the promise of engineering organisms to clean up the environment and to reduce the emissions of greenhouse gases. Although still in its early stages, compared to biotechnology aimed at alleviating medical problems, environmental biotechnology is emerging as one possible solution to the threat of global warming.

James S. Godde

• Further Reading

Evans, Gareth M., and Judith C. Furlong. *Environmental Biotechnology: Theory and Application.* Hoboken, N.J.: John Wiley & Sons, 2003. Explains carbon cycles in phytoplankton, as well as detailing the organism's potential use in reducing the CO_2 emissions of power plants. Describes Rudolf Diesel's use of plant oil in his engine.

Kircher, Manfred. "White Biotechnology: Ready to Partner and Invest In." *Biotechnology Journal* 1 (2006): 787-794. Gives the history of different "colors" of biotechnology, along with the relative levels of venture capital invested in them.

Scragg, Alan. *Environmental Biotechnology.* New York: Oxford University Press, 2005. Defines biotechnology, including white biotechnology, giving a historical perspective. Explains the concept of CO_2 neutrality. Contains extensive discussion of various biofuels.

Tollefson, Jeff. "Not Your Father's Biofuels." *Nature* 452 (February, 2008): 880-883. Describes a patent for ethanol-producing *E. coli.* Details the competition between crops for food and crops for fuel, the cost of shipping ethanol in the United States, and the use of cellulose for ethanol production.

See also: Biofuels; Clean energy; Energy efficiency; Energy from waste; Energy resources; Ethanol; Fossil fuel emissions; Fossil fuel reserves; Fossil fuels; Fuels, alternative; Pollen analysis.

Bowen's ratio

• **Category:** Meteorology and atmospheric sciences

• **Definition**

The main process for transporting the daytime energy surplus at Earth's surface to the atmosphere is convection, which is heat transfer by turbulent air motion. There are two types of convective flows (fluxes), the sensible and latent heat fluxes. If the movement of energy from the surface can be sensed as a rise or fall in temperature, then it is referred to as the sensible heat flux (D). If the flow of energy involves moisture, as in evaporation or transpiration from plants, there will be no change in temperature. In this case, the energy is held dormant in the evaporated moisture for release should the water vapor revert to its liquid state via condensation. This is called the latent heat flux (E). Climate is characterized by the apportionment of energy at the Earth's surface between D and E, the ratio of which is known as Bowen's ratio (β), so that $\beta = D/E$. Negative β values indicate that one and only one of the two fluxes is negative. D is negative when the air is heating the surface. E is negative when condensation (dew) occurs.

• **Significance for Climate Change**

Bowen's ratio is a climate index. If β is greater than 1, D is greater than E. Since such values indicate that most of the heat being moved into the atmosphere is in sensible form, they also indicate that the climate is warmer than it is when β is less than 1. When β is less than 1, E is greater than D and most of the energy transfer does not contribute directly to warming the air. This transfer may increase atmospheric humidity, however, making the climate cool and humid. The size of Bowen's ratio is determined chiefly by the availability of water for evaporation. If water is available, the latent heat flux will dominate.

A rise in the concentration of greenhouse gases in the atmosphere means there is more available energy at the Earth's surface. This additional energy can be used either to heat the atmosphere by way of the sensible heat flux (increased warming) or to evaporate water via the latent heat flux. Since 71 percent of the Earth's surface is water, most of the additional energy would contribute to an enhanced latent heat flux. The resulting warming of the atmosphere would be less in this case than if all the additional available energy was accounted for by the sensible heat flux alone.

C R de Freitas

See also: Atmospheric dynamics; Dew point; Evapotranspiration; Greenhouse effect; Humidity; Latent heat flux.

Brazil

• **Category:** Nations and peoples
• **Key facts**

Population: 191,908,598 (May, 2008)
Area: 8,456,510 square kilometers
Gross domestic product (GDP): $1.838 trillion (purchasing power parity, 2007 estimate)
Greenhouse gas (GHG) emissions in millions of metric tons of carbon dioxide equivalent (CO_2e): 550 in 2004
Kyoto Protocol status: Ratified, 2002

• **Historical and Political Context**

Settled by Portugal in 1500, Brazil developed as a colony (1500-1822), then as an independent empire (1822-1889), and finally as a republic (1889-present). Of continental proportions, Brazil extends east across South America from the Andes Mountains and south from the equator to well below the Tropic of Capricorn. Slave labor, supporting mineral and agricultural plantation exports, forged the socioeconomic axis of the country for nearly four centuries (slavery was abolished in 1888). Several million Africans were forcibly transported to Brazil, one-third of all people shipped in the transatlantic slave trade. Brazil has the second-largest population of African descent in the world after Nigeria. The population is largely multiracial, and there is a vibrant Afro-Brazilian culture.

The abolition of slavery brought a wave of immigrants to Brazil from southern Europe. A nucleus of salaried labor emerged, producing a core of consumer demand that stimulated the country's industrial, capital, and urban development during the twentieth century. At the beginning of the twenty-first century, Brazil rose as the major economy of the region. However, its historical roots of inequality had produced a society with extreme divisions of wealth and poverty.

Subject to political instability and sporadic authoritarian regimes, Brazil achieved a democratic equilibrium around the turn of the twenty-first century. It is a nation rich in a vast array of mineral resources, including vast petroleum reserves, is endowed with abundant labor and land, and is widely favored by foreign investment capital. Brazil, together with Russia, India, and China (the so-called BRIC countries), stands out as among the most energetically and fervently developing nations in the world. With long-suppressed mass consumer demand, economic development requires extensive energy resources.

• Impact of Brazilian Policies on Climate Change

Although Brazil ranks among the leading emitters of greenhouse gases (GHGs) in the world, in no way does it contribute on the scale of the United States or China. However, Brazilian emissions are significant because of their unique mix. Brazil is the largest industrial economy in South America and is home to major automobile, steel, cement, electronics, communications, and aviation manufacturers. It has vast mineral-extraction operations, and its abundant reserves of iron ore are shipped to all parts of the world. Moreover, it is a major cattle producer: The national herd approaches nearly 200 million head and grows at about 2 percent per year. Furthermore, Brazil is the home of the largest tropical rain forest in the world, the Amazon, with vast stretches being burned or cut down for logging or for agricultural or pasture land.

Long an underdeveloped country, Brazil has resolutely and energetically engaged in development. Its populist, labor government has worked to bring the benefits of economic growth to all classes of society, with annual GDP growth averaging 4 percent in the early twenty-first century. Such development intensified GHG emissions through expanded manufacturing, construction, and transportation; enlarged farming and pasture areas; and deforestation.

In some ways, Brazil has been able to limit its GHG emissions. Its extensive hydroelectric resources allow it to economize on fossil fuels. With Paraguay, Brazil has built and operates the largest hydroelectric facility in the world, the Itaipú Dam. Brazil has also been a pioneer in the development of sugarcane ethanol. Virtually all vehicles manufactured in Brazil must use gasohol, a mix of fossil and vegetable fuel. However, as a result of Brazil's sophisticated development of offshore oil drilling, the country holds the promise of becoming a petroleum exporter. Satellite monitoring of the rain forest only moderately and unevenly has checked voracious deforestation.

• Brazil as a GHG Emitter

Although Brazil is a signatory to the Kyoto Protocol, as a developing country it is not among the Annex I nations. Thus, it is not required to provide regular, standardized accountings of its GHG emissions to the United Nations. The Brazilian Institute for Tropical Agriculture calculates Brazil's total GHG inventory as more than one-half billion metric tons. This amount places Brazil among the top

ten emitters in the world. However, such emissions account for less than 2 percent of the world total, ranking Brazil with countries such as the United Kingdom, South Korea, and Mexico.

The U.N. Statistics Division reports that one-third of Brazil's GHG emissions are carbon dioxide (CO_2) resulting from fossil fuel burning and deforestation. During the initial years of the twenty-first century, more than 20,000 square kilometers of rain forest (about the size of New York State) were being destroyed annually, contributing 200 million metric tons of CO_2. The nation's extraordinarily large livestock herd also produces significant methane emissions, and the fertilizers employed in ever-expanding agricultural areas emit nitrous oxide.

• Summary and Foresight

Brazil is a unique country of tropical and subtropical abundance and diversity. Its forests and mineral resources offer the promise of development, both for itself and for numerous other regions. The global growth of commodities markets has resulted in a singular phase of Brazilian prosperity. This growth came after years of political and economic instability. In 2002, a union leader who headed the Workers' Party, Luis Inácio da Silva, won the presidency, bringing an unprecedented populist yet market-friendly party to power. Da Silva helped drive a political will to exploit Brazil's natural resources for the benefit of all classes.

The energy requirements for such development, however, result in GHG emissions that contribute to environmental degradation. Brazil is acutely aware of the dilemma it faces and the responsibility it has for maintaining the integrity of its tropical environment, especially the Amazon. With its ethanol programs and satellite vigilance of the rain forest, it has made some guarded progress in controlling its GHG emissions. The Kyoto Protocol was prefigured by the United Nations Conference on Environment and Development, convened in Rio de Janeiro in 1992, which issued the Rio Declaration on the Environment and Development.

Edward A. Riedinger

• Further Reading

Andersen, Lykke E. *The Dynamics of Deforestation and Economic Growth in the Brazilian Amazon.* New York: Cambridge University Press, 2002. Provides comparative econometric models of deforestation in the Amazon River basin and economic growth as a means for formulating economic policy.

Barbosa, Luiz C. *The Brazilian Amazon Rainforest: Global Ecopolitics, Development, and Democracy.* Lanham, Md.: University Press of America, 2000. Examines the political issues in Brazil related to Amazon deforestation in order to formulate viable environmental policy for sustainable development.

Hudson, Rex A., ed. *Brazil: A Country Study.* 5th ed. Washington, D.C.: Federal Research Division, Library of Congress, 1998. The section on Brazil's economy includes data on resources, production, and other factors affecting the nation's contribution to global climate change.

Wood, Charles R., and Roberto Porro, eds. *Deforestation and Land Use in the Amazon.* Gainesville: University Press of Florida, 2002. Collection of articles reviewing historical phases of Amazon deforestation and contemporary patterns of economic and geographic use.

See also: Amazon deforestation; Deforestation; Forests; Kyoto Protocol; United Nations Conference on Environment and Development; United Nations Framework Convention on Climate Change.

Brundtland Commission and Report

• **Category:** Organizations and agencies
• **Date:** Commission established 1983; report published as *Our Common Future* in 1987
• **Web address:** http://www.un.org/documents/ga/res/38/a38r161.htm

• Mission

The secretary-general of the United Nations, heeding growing concerns about the global environment, convened the World Commission on Environment and Development in 1983. The commission took its name from its head, Gro Harlem Brundtland, a former prime minister of Norway. Representatives of the commission traveled the globe, gathering statements of mounting environmental concerns and possible solutions. It quickly became clear that environmental problems do not respect national boundaries and that only a collective, global effort could effect change. The commission's mandate arose from the "accelerating deterioration of the human environment and natural resources and the consequences of that deterioration for economic and social development."

The Brundtland Commission, in its report to the U.N. General Assembly, recommended the creation of a long-term environmental strategy to achieve sustainable development. Begun in 1983, the body produced a mission statement that anticipated a continuing effort to the year "2000 and beyond." The final report looked forward to change brought through cooperation among all countries, regardless of their socioeconomic condition or fundamental resources. The commission hoped to accomplish its goals by integrating environmental, economic, and social objectives to form a comprehensive strategy.

The Brundtland Commission, which published its report *Our Common Future* in 1987, recommended the creation of a universal declaration on environmental protection and sustainable development in the form of a new charter. The commission recognized the need for broad participation in environmental programs and solutions, so it required that at least half its members come from developing countries and that the voices of governmental and nongovernmental organizations, industry, scientists, and others with environmental concerns be heard and consulted. In an effort to embrace and consider a broad range of views, the commission stressed that there must be a continuing dialogue among not only the scientific community and environmentalists but also all sections of public opinion, particularly youth, as well as those concerned with the fragile balance between development and the environment.

The commission's message to the General Assembly was not an inflexible mandate but addressed varying environmental concerns for societies rich and poor, developed and undeveloped, resource abundant and resource scarce, industrial and rural. It expressed the hope for an inclusive and worldwide cooperative effort. Finally, the commission asked for "development that meets the needs of the present without compromising the ability of future generations to meet their own needs."

The findings of the World Commission on Environment and Development, as presented in *Our Common Future*, served as a framework for later environmental discussion. This was clearly in evidence during the 1992 U.N. Conference on Environment and Development in Rio de Janeiro, where the concept of "sustainable development" was firmly set in place as a "worthy universal goal."

Brundtland Report

The Brundtland Report provides the following summary statement of the status of the relationship between nature and humanity near the end of the twentieth century.

Over the course of this century, the relationship between the human world and the planet that sustains it has undergone a profound change. When the century began, neither human numbers nor technology had the power radically to alter planetary systems. As the century closes, not only do vastly increased human numbers and their activities have that power, but major, unintended changes are occurring in the atmosphere, in soils, in water, among plants and animals, and in the relationships among all of these. The rate of change is outstripping the ability of scientific disciplines and our current capabilities to assess and advise.

• Significance for Climate Change

The Brundtland Report raised awareness of the accelerating human impact on the environment and declared that solutions depended upon cooperative global effort. The paperback version of the report, published late in 1987, warned readers that "Most of today's decision makers will be dead before the planet suffers the full consequences of acid rain, global warming, ozone depletion, widespread desertification, and species loss." In response to that prophetic statement, in 2005, a U.N. Millennium Project report confirmed that sixteen out of twenty-five ecosystems were being critically degraded and demonstrated the negative impact of unsustainable development paths. Further, as if echoing the twenty-two-year-old concerns of the Brundtland Commission, the Intergovernmental Panel on Climate Change demonstrated the negative effects of unsustainable development paths. When *Our Common Future* was published, the fight against global warming was a distant consideration, and only four pages were devoted to the subject. It is noteworthy, however, that the report's central tenet is basic to the challenges of climate change: the need for global cooperation to achieve solutions to global environmental problems.

Richard S. Spira

• Further Reading

Bruno, Kenny, and Joshua Karliner. *Earthsummit.biz: The Corporate Takeover of Sustainable Development.* Oakland, Calif.: Food First Books, 2002. From Rio to Johannesburg, the book details corporate accountability to environmental interests and the need to achieve sustainable practices in the twenty-first century.

Park, Jacob. "*Our Common Future* Twenty Years Later: Rethinking the Assumptions of the World Commission on Environment and Development." Paper presented at the International Studies Association Forty-eighth Annual Convention. Chicago: International Studies Association, 2007. Comments on the practical applications of the Brundtland Report for business and on industrial environmental practices that have gained growing acceptance in the United States.

World Commission on Environment and Development. *Our Common Future.* New York: Oxford University Press, 1987. The Brundtland Report, detailing the need for global cooperation to address mounting environmental concerns in the late twentieth century.

See also: Agenda 21; Antarctic Treaty; Earth Summits; International agreements and cooperation; Kyoto Protocol; Stockholm Declaration; United Nations Climate Change Conference; United Nations Conference on Environment and Development; United Nations Conference on the Human Environment; United Nations Division for Sustainable Development; United Nations Environment Programme; United Nations Framework Convention on Climate Change.

Budongo Forest Project

- **Categories:** Organizations and agencies; environmentalism, conservation, and ecosystems
- **Date:** Established 1990
- **Web address:** http://www.budongo.org

• Mission

The Budongo Forest Project was founded in 1990 by chimpanzee specialist Vernon Reynolds in the Budongo Forest of Uganda, East Africa. Its mission is to blend "research and conservation to ensure sustainable management and utilisation of the Budongo Forest Reserve as a model for tropical rain forest management." Originally created to preserve the forest with the aim of protecting the native chimpanzees that inhabit it, the project has increased the scope of its research to include other species. The Budongo Forest, at more than 352 square kilometers, is home to over 300 bird species, 850 plant species, and 400 types of butterflies and moths. The project has built accommodations for staff and for visiting researchers and students from around the world, and dozens of articles based on research at Budongo have been published in scientific journals.

In 2007, the project was recognized by the Ugandan government as an official nongovernmental organization and was renamed the Budongo Conservation Field Station. The station receives much of its funding from the Royal Zoological Society of Scotland, the Oakland Zoo in California, and the United States Agency for International Development (USAID). The Edinburgh Zoo in Scotland operates a large primate enclosure named the Budongo Trail.

The Budongo Forest was used by loggers for six decades, and the governmental forestry service had maintained records and maps going back almost a century. These records have proven to be invaluable to scientists studying how tropical forests grow and how they respond to being harvested. With this information and with new research, the project hopes to protect against deforestation from hunting, logging, gathering, clearing for farming, and other human interference, including global warming. For example, one conservation effort encourages local inhabitants to grow the medicinal plant *Ocimum kilimandscharicum*, used for aromatherapy, in sustainable community farms, rather than gathering the plant in the wild. Other projects encourage beekeeping and other ways of earning a living without cutting down trees. While logging and gathering of seeds and other nontimber products is legal in the Budongo Forest, the project explores sustainable ways to manage these harvests.

• Significance for Climate Change

While the connection between tropical rain forests and global warming has never been an emphasis of the Budongo Forest Project, the world's attention to global warming has presented the project with a larger audience and potentially with a larger base of support. Tropical rain forests, in addition to providing habitat for a rich variety of plant and animal species, protect the Earth from some of the effects of global warming by acting as carbon sinks, absorbing carbon from the atmosphere. It has been estimated that the Budongo Forest sequesters more than 726,000 metric tons of carbon dioxide each year. As tropical forests shrink, their capacity to absorb carbon also decreases, increasing the threat of global warming.

Researchers at Budongo have initiated projects to help local people grow and make a living from sustainable tree crops. These projects are described in scientific articles whose titles include "Effectiveness of Forest Management Techniques in Budongo," "The Ecology of Long-Term Change in Logged and Non-logged Tropical Moist Forest," and "Understanding of National and Local Laws Among Villagers with Special Reference to Hunting." Unfortunately, these projects have had difficulty obtaining funding; although Uganda is a signatory to international conservation treaties, it has not given funding priority to efforts to preserve biodiversity, and the field station itself has only a small budget. In 2007, Uganda and its neighbors proposed that the carbon sequestered by natural forests should be included in the global carbon credit system created by the Kyoto Protocol, thus generating funds for conservation in forested nations. The resolution did not pass.

The success of conservation work at Budongo has led to another major effort in Uganda: In October, 2008, the National Forestry Authority and the poverty relief organization World Vision launched a national campaign to plant thirteen million trees by 2011. Community groups living at the edges of forest reserves planned to make formal agreements with the Forest Authority, which would provide incentives for conservation and nonconsumption projects, including tree planting, ecotourism, and the establishment of woodlots.

Cynthia A. Bily

The Budongo Forest Project was initially founded to protect Uganda's native chimpanzees. (©iStockphoto/Liz Leyden)

• **Further Reading**

Howard, P. C. *Nature Conservation in Uganda's Tropical Forest Reserves.* Gland, Switzerland: International Union for Conservation of Nature, 1991. Reports on the status of Uganda's tropical high forest reserves, describing native species and human disturbance and making recommendations for conservation and sustainable development.

Malhi, Yadvinder, and Oliver Phillips. *Tropical Forests and Global Atmospheric Change.* New York: Oxford University Press, 2005. Scientific analysis by two prominent researchers, demonstrating the effects of global warming on tropical forests and their implications for the future of carbon storage and exchange as the forests decline.

Reynolds, Vernon. *The Chimpanzees of the Budongo Forest: Ecology, Behaviour, and Conservation.* New York: Oxford University Press, 2005. Written by the founder of the Budongo project, this technical but accessible volume combines natural history of chimpanzees with the author's reflections on the importance of conservation.

Weber, William, et al., eds. *African Rain Forest Ecology and Conservation: An Interdisciplinary Perspective.* New Haven, Conn.: Yale University Press, 2001. Written by conservation scientists, this collection describes the ecology and human uses of African rain forests and argues that the forests must be conserved.

See also: Deforestation; Forestry and forest management; Forests; Intergovernmental Panel on Forests.

Building decay

• **Category:** Economics, industries, and products

• **Definition**

Human-made structures have evolved architecturally for the environmental conditions in which they have traditionally been constructed. They are fixed in their locations and constructed from materials that age at different rates and with different consequences. As a result, most human structures are ill suited to almost any significant environmental change. Their long-term viability will likely be hampered by any change in climate or biological environment. Any change to the global climatic regime thus presents significant dangers to the integrity of most human structures.

• **Significance for Climate Change**

If climate change continues to become more pronounced, rising temperatures and climate patterns will shift toward polar regions. As these shifts occur, ecosystems will also migrate, subjecting structures to new conditions. One likely and dramatic result of increased global warming is a rise in sea level. Even small changes in sea level would have spectacular localized effects, and, because a disproportionate share of the world's population lives within 50 kilometers of a sea shoreline, a large number of human structures would be at risk of inundation, in which case the residents would become environmental refugees.

Changes in sea level would have dramatic effects on storm patterns, stressing many structures not designed to sustain severe weather conditions. Changes in rainfall patterns generally result in changes in runoff: Many foundations and gutter systems would prove inadequate for significantly increased runoff and would suffer from turbulent flow patterns. Extreme fluctuations in temperature, wind, and rainfall coinciding with global warming would result in changes to humidity, subjecting building materials to wetter conditions than they were originally designed to repel. Expected results would include accelerated material disintegration, mold infestation, corrosion, and rot, which would degrade and weaken structures and their foundations.

Low soil moisture conditions preceding severe rainfall events increases the impact and magnitude of flooding. In areas expected to undergo decreased soil moisture, floods will cause increasing amounts of damage. Between floods, the same low soil moisture could lead to increased ground movement, causing building foundations to degrade. Moreover, changing climate patterns often cause species migration. If this trend continues, buildings in areas previously unaffected by certain pests will

This middle-school classroom in Lianyuan, China, collapsed as a result of termite damage in May, 2007. Increased global warming may increase the population and range of these pests. (Reuters/Landov)

need protection from an influx of destructive, invasive species.

Sea-level rise could cause changes to groundwater regimes, resulting in saltwater infiltration, subsidence, and drainage alterations, all potentially affecting the structural integrity of buildings. If shifts in global climatic conditions continue, buildings could be subjected to drastic changes in freeze-thaw cycles, oscillating between evaporation and condensation and between wet and dry conditions. Changes in a range of conditions, including temperature, humidity, light, wind turbulence, and vibration, would all create stresses on buildings that they were not constructed to withstand.

As the effects of global climate change broaden, three basic agents of structural deterioration will be prominent: biological agents, physical agents, and chemical agents. These agents may contribute to decomposing a structure independently or by co-association. Increased humidity, higher temperatures, and sufficient water availability may allow for both the spread of and longer seasonal life cycles among invasive plants, insects, animals, algae, fungi, lichens, bacteria, and boring worms. If so, biodegradation of buildings would increase significantly. Higher populations of climbing plants, termites, ants, molds, roosting birds, and tunneling animals would negatively affect the mechanical and physical integrity of many structures.

Physical decay includes discoloration of building materials from radiation damage due to increased sunlight, such as may result from changes in cloud cover and from ozone layer loss. Organic building materials, such as wood, straw, reeds, leafs, and grasses, suffer greater thermal decomposition when temperatures increase. Differential thermal expansion results in increased mechanical failures in porous building materials such as wood, stone, glass, and concrete. Mechanical stress due to increased sway and vibration from altered wind patterns also increases structural wear. Hygrometric stress due to changes in humidity can result in swelling and warping of organic and composite building materials.

Changes to weather patterns could subject many buildings to increased nonchemical erosion from windblown materials, such as dust and sand, and increase weathering due to water flow across their surfaces. Changes to global climate could result in changes to regional atmospheric chemistry, resulting in modifications to normal building-decay rates. Invasive windblown materials can increase static electrical charges, attracting dust, humidity, and soot to structural surfaces and enhancing chemical deterioration cycles.

As shifts in the climatic regime take place, chemical decay will likely increase in certain areas, as water and salts are introduced into porous structures, causing crystallization stress that may lead to mechanical failure. Stone and masonry buildings are particularly vulnerable to such stress. Organic materials may experience photodegradation by a combination of light, chemicals, and humidity, inducing destructive chemical changes. Humidity, chemicals, and temperature may also hydrate and make brittle structural materials such as glass, plastic, bronze, and copper.

Chemical dissolution and transformational stresses are also likely to increase in certain regions as global climate changes. Acidic precipitation—rain, snow, and fog carrying reactive solids, chemical pollutants, catalytic particles, trace gases, and corrosive compounds—is especially damaging to construction materials, including ferrous metals, granite, limestone, sandstone, marble, gypsum, dolomite, glass, and wood. Tarnished metals, rust, discoloration, exfoliation, acid etching, loss of surface details, decomposition, and enhanced weathering are all well-documented symptoms of atmosphere-induced chemical structural decay. Human behaviors release large quantities of reactive chemicals into the atmosphere as air pollution, resulting in increased acid deposition and enhanced structural decay. The chemicals responsible for these deteriorative effects are often the same chemicals implicated as anthropogenic causes of global warming and climatic change.

Randall L. Milstein

• Further Reading

Brimblecombe, P., ed. *The Effects of Air Pollution on the Built Environment.* London: Imperial College Press, 2004. Discusses ways in which structural materials are likely to be damaged by air pollutants.

Prykryl, R., and B. J. Smith, eds. *Building Stone Decay: From Diagnosis to Conservation.* London: Geological Society, 2007. Outlines the causes of structural decay and the environmental factors controlling them.

Samuels, R., and D. K. Prasad, eds. *Global Warming and the Built Environment.* London: E&FN Spon, 1994. A discussion of the interaction between built and natural environments relating to global climate change.

Smith, B. J., and P. A. Warke. *Stone Decay: Its Causes and Controls.* Shaftesbury, Dorset, England: Donhead, 2004. A descriptive volume on the environmental and biological causes of stone and masonry deterioration; discusses remediation measures.

See also: Air pollution and pollutants: anthropogenic; Flood barriers, movable; Floods and flooding; Levees; Rainfall patterns.

Byrd-Hagel Resolution

- **Category:** Laws, treaties, and protocols
- **Date:** Passed July 25, 1997

The Byrd-Hagel Resolution declared the "sense of the Senate" that the United States should not agree to limit its greenhouse gas emissions unless developing nations, including China and India, were also required to limit theirs.

• Background

When the United Nations Framework Convention on Climate Change (UNFCCC) was adopted in

May, 1992, it identified major industrialized countries, including the United States, as Annex I parties, and other countries as underdeveloped parties. This division became a long-standing source of controversy, as the United States and other industrialized nations came to resent the fact that they were subject to regulations under the treaty to which the developing nations were not subject.

In 1995, at the First Conference of the Parties (COP-1) in Berlin, the Berlin Mandate was adopted. It called for the creation of a new protocol that would require strict commitments on the part of Annex I parties to limit greenhouse gas (GHG) emissions but would exempt developing countries from new commitments. This provision increased the dissatisfaction on the part of the United States, particularly in the light of the fact that China, India, and Brazil—all heavy polluters—were among the developing nations that would not face strict regulation.

Through the 1990's and well into the first decade of the twenty-first century, the United States was the world's largest emitter of carbon dioxide (CO_2) because of its heavy use of fossil fuels. However, the GHG emissions of the developing country parties was rapidly increasing, and by some estimates they were expected to surpass those of the industrialized nations by 2015.

On July 25, 1997, Senate Resolution 98 was introduced before the 105th Congress. Its cosponsors were Democrat Robert Byrd of West Virginia and Republican Chuck Hagel of Nebraska. What became known as the Byrd-Hagel Resolution provided a chance for the Senate to express its opinion about ongoing international climate negotiations. Agreeing to the resolution by a vote of 95 to 0, the senators declared that the United States should not agree to any binding restrictions on its own GHG emissions if the developing country parties were excluded from such restrictions. Inconsistent restrictions, they asserted, would cause serious harm to the U.S. economy, because companies in developing nations would have an unfair advantage over their U.S. competitors in trade and in wages.

• Summary of Provisions

The resolution declared that it was the "sense of the Senate" that the United States should not sign any protocol of the UNFCCC mandating new commitments on the part of Annex I parties without plac-

The Byrd-Hagel Resolution

The Byrd-Hagel Resolution was a "sense of the Senate" resolution, expressing the majority opinion of the body without having the full weight of law. However, because the resolution passed on a 95-0 vote, it sent a clear message to President Bill Clinton that the Senate would not ratify the Kyoto Protocol unless it was significantly altered.

Resolved, That it is the sense of the Senate that—

(1) the United States should not be a signatory to any protocol to, or other agreement regarding, the United Nations Framework Convention on Climate Change of 1992, at negotiations in Kyoto in December 1997, or thereafter, which would—

(A) mandate new commitments to limit or reduce greenhouse gas emissions for the Annex I Parties, unless the protocol or other agreement also mandates new specific scheduled commitments to limit or reduce greenhouse gas emissions for Developing Country Parties within the same compliance period, or

(B) would result in serious harm to the economy of the United States; and

(2) any such protocol or other agreement which would require the advice and consent of the Senate to ratification should be accompanied by a detailed explanation of any legislation or regulatory actions that may be required to implement the protocol or other agreement and should also be accompanied by an analysis of the detailed financial costs and other impacts on the economy of the United States which would be incurred by the implementation of the protocol or other agreement.

ing limits on the emissions of developing country parties—nor should it sign any protocol that would harm the U.S. economy. In addition, the resolution reiterated the constitutional requirement that any treaty be submitted to the Senate for ratification and called for detailed explanations, including information about financial cost, to support any future climate change agreement.

• Significance for Climate Change

The vote on the Byrd-Hagel Resolution occurred when the Kyoto Protocol was nearly written but not yet finalized. The fact that the resolution was put forward by bipartisan cosponsors and that it passed unanimously seemed to indicate a strong consensus within the Senate. However, Byrd stated later that he had intended the resolution only to provide broad guidelines for negotiators and that he thought the Kyoto Protocol was a sensible compromise document. Hagel, on the other hand, opposed the Kyoto Protocol and continued to work against any agreement that went against the specifics of the resolution.

On December 11, 1997, less than five months after the passage of the Byrd-Hagel Resolution, the Kyoto Protocol was adopted at the Third Conference of the Parties (COP-3). The United States signed the treaty in 1998 but did not ratify it. During the Bill Clinton administration, which ended in 2001, the White House did not send the Kyoto Protocol to the Senate for ratification. The George W. Bush administration stated formally that it would not support the treaty, citing the Byrd-Hagel Resolution as evidence that the U.S. economy would be unfairly harmed by any agreement that did not include restrictions on developing nations, particularly China. By the end of the Bush administration, the United States had neither ratified nor officially withdrawn from the treaty. The language and the arguments of the Byrd-Hagel Resolution continued to inform negotiations and resolutions through Bush's two terms as president.

Cynthia A. Bily

• Further Reading

Lisowski, Michael. "Playing the Two-Level Game: U.S. President Bush's Decision to Repudiate the Kyoto Protocol." *Environmental Politics* 11, no. 14 (Winter, 2002): 101-119. Describes how President George W. Bush invoked the Byrd-Hagel Resolution as part of his rationale for rejecting the Kyoto Protocol.

Rabe, Barry George. *Statehouse and Greenhouse: The Emerging Politics of American Climate Change Policy.* Washington, D.C.: Brookings Institution Press, 2004. Examines the shortcomings of federal regulations regarding climate change and the emergence of sensible state-level policies.

Schneider, Stephen Henry, Armin Rosencranz, and John O. Niles. *Climate Change Policy: A Survey.* Washington, D.C.: Island Press, 2002. Textbook surveying in accessible language the scientific, economic, and policy contexts of the global warming debate.

Victor, David G. *Climate Change: Debating America's Policy Options.* New York: Brookings Institution Press, 2004. Analysis of three approaches to climate change policy, with an appendix containing transcripts of the Senate debate over the Byrd-Hagel Resolution.

See also: Berlin Mandate; International agreements and cooperation; Kyoto Protocol; United Nations Framework Convention on Climate Change; U.S. energy policy; U.S. legislation.

California

- **Category:** Nations and peoples

California has the largest population, most vehicles, most diverse economy, and most productive agricultural sector of any U.S. state. It produces about 1.4 percent of the world's GHGs and 6.2 percent of total U.S. GHGs.

• Key concepts

geothermal energy: energy generated from heat stored in the Earth

greenhouse gases (GHGs): atmospheric trace gases that absorb and emit infrared radiant energy and warm the atmosphere and the Earth's surface

ozone: a molecule of three oxygen atoms that can form in the lower atmosphere by photochemical reaction of sunlight with hydrocarbons and nitrogen oxides

runoff: precipitated water that flows into rivers, lakes, streams, oceans, and other water bodies

snowpack: accumulated snow in mountainous areas that melts during warmer months

• Background

California's climate, population, economy, and agriculture share a complex relationship with global warming. The state has a highly productive and diverse industrial complex, ranging from manufacturing and oil production to biotechnology and telecommunications. Population and industry are concentrated in the state's southern and coastal regions, where the water supply is limited. An expansive agricultural sector supports a food production capacity important for California, the United States, and the world, but it depends on irrigation water and a workforce of more than 1 million. California has led the United States in food and agricultural production for over fifty years and leads the nation in exporting agricultural products. Irrigated crops account for a majority of the $34 billion market value contribution agriculture makes toward California's total economic output of $1.8 trillion.

• Population and the Water Supply

California's 37 million people represent 12 percent of the nation's population, and the state's population is expected to approach 50 million by 2025. Nearly 60 percent of the populace is concentrated in Southern California, and 90 percent resides in metropolitan areas. Population growth is expected to be greater in inland counties, but coastal counties containing the major metropolitan areas will continue to account for over 60 percent of the population. Population growth has critical implications for almost every area of public policy, but issues particularly relevant for California are social services, education, transportation, the environment, the water supply, and loss of agricultural land.

Contrasting seasonal precipitation and temperature patterns, the influence of topography, and proximity to the coast create a variety of climates within California. However, the population favors the water-limited regions of southern and coastal California, while the mountains of northern and central California produce an abundant but variable water supply. The time and space mismatch between the water supply and the water needed by the population and agriculture is overcome by a statewide water storage and conveyance system. California's water system is a collection of constructed facilities and natural features blended by a mixture of overlapping agencies and jurisdictions from all levels of government.

• Vulnerability to Climate Change

California's climate is expected to warm by as much as 1.1° Celsius over thirty years, with greater summer than winter warming likely. Precipitation changes are uncertain, but the seasonal precipitation pattern is expected to continue. The warmer climate will produce a reduced winter snowpack; an earlier start of spring snowmelt; an increase in winter runoff as a fraction of total runoff; an increase in winter flooding frequency; decreases in total runoff; increased stress on drinking water and irrigation water supplies; increased droughts, floods, and wildfires; decreased agricultural productivity; and increased threats to human health. California's 2050 global warming impacts could approach total annual costs and revenue losses of $15 billion.

The impact of warmer temperatures on the win-

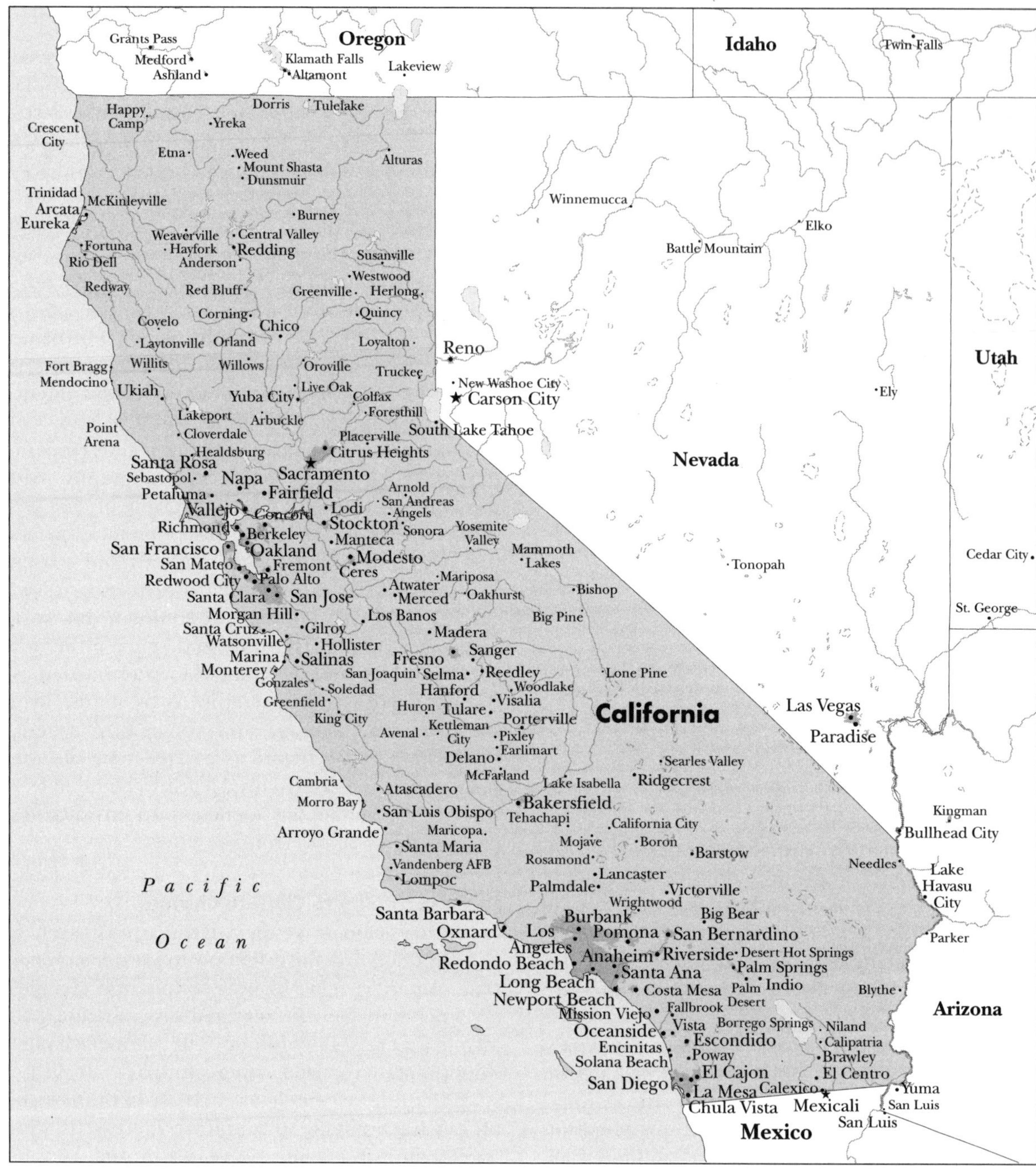

ter snowpack has major consequences. Snow accounts for up to 50 percent of California's stored water, but this figure may be reduced to 12 to 42 percent in a warmer climate. More precipitation occurring as rain rather than snow increases immediate runoff and heightens chances for more severe and frequent floods. Protecting against floods requires lower reservoir levels that reduce the water supply capacity of reservoirs, reduce available irrigation water, and reduce the ability of reservoirs to offset drought. Furthermore, summer hydropower generation is reduced when reservoir levels are lowered.

Most crops and forests will benefit from increased atmospheric carbon dioxide (CO_2) concentrations, but warmer temperatures may aggravate ozone pollution, which makes plants more susceptible to diseases and pests. Higher and more extreme temperatures increase the likelihood of air pollution episodes, which are a public health threat, and over 90 percent of California's current population lives in areas that violate the state's air quality standards for ozone or particulate matter. Warmer temperatures will increase the wildfire risk to homes and forests by 11 to 55 percent and will contribute to air quality problems.

• Reducing Climate Change Contributions

California is the world's twelfth-largest emitter of greenhouse gases (GHGs), producing 500 million metric tons, or 13.5 metric tons per capita. However, the state has reduced per capita GHG emissions by 10 percent. Transportation is California's largest source of GHG emissions, but the state has a long history of adopting rigorous vehicle GHG emission standards beyond federal Environmental Protection Agency requirements. Beginning in 2009, all new vehicles sold in California have an environmental performance sticker, rating the car on smog emissions and the amount of GHG emissions per mile, including emissions related to the production and distribution of the fuel.

The California Global Warming Solutions Act of 2006 is the United States' first comprehensive regulatory and market-mechanisms program. The legislation, along with related initiatives that include penalties for noncompliance, are designed to achieve quantifiable, cost-effective GHG emissions equal to 1990 levels by 2020. By 2050, GHG emissions are to be reduced to 80 percent of their 1990 levels. A 2008 Governor's Executive Order requires California utilities to derive 33 percent of their energy from renewable sources by 2020.

Electric power generation accounts for 28 percent of California's GHG emissions, and natural gas accounts for 56 percent of annual power generation. However, California has aggressively pursued renewable energy, which currently accounts for 14 percent of California's annual energy use. Geothermal is California's largest renewable source of electric power. Small hydroelectric, wind, and biomass contribute about equal shares, and solar provides about 0.2 percent. Nearly half of the nationwide venture capital investment in clean technology is in California.

• Context

California's role in global warming extends well beyond its position as an individual state. If California was a nation, it would have the thirty-third-largest population and the seventh-largest economy in the world. It is a world leader in developing energy-efficient technology, in implementing new technology, and in efforts to shift energy production to renewable resources. California's gross domestic product (GDP) is 11.5 percent of the U.S. GDP.

Projected temperature increases for California are smaller than estimated global temperature increases, but even small temperature increases are significant for California's population, agriculture, and economy, because they alter the character of a water supply that is already approaching its limits. The challenge of providing a reliable water supply is magnified by global warming and the uncertainty of its effects on California's climate.

Marlyn L. Shelton

• Further Reading

Godish, Thad. *Air Quality.* 4th ed. Boca Raton, Fla.: Lewis, 2004. Comprehensive overview of air-quality fundamentals and issues, including global warming, public health, and regulatory programs. Illustrations, figures, tables, bibliography, index.

National Assessment Synthesis Team. *Climate Change Impacts on the United States: The Potential Consequences of Climate Variability and Change.* New York: Cambridge University Press, 2000. This report by leading scientists combines national-scale analysis with an examination of potential climate change impacts at the regional scale and for five selected sectors. Illustrations, figures, maps.

Vanrheenen, Nathan T., et al. "Potential Implications of PCM Climate Change Scenarios for Sacramento-San Joaquin River Basin Hydrology and Water Resources." *Climatic Change* 62, nos. 1-3 (2004): 257-281. Impacts on California water resources are explored using several climate change and water management assumptions.

See also: Florida; Louisiana coast; New Orleans; United States.

Canada

- **Category:** Nations and peoples
- **Key facts**

Population: 33,487,208 (July, 2008, estimate)
Area: 9,984,670 square kilometers
Gross domestic product (GDP): $1.307 trillion (purchasing power parity, 2008 estimate)
Greenhouse gas (GHG) emissions in millions of metric tons of carbon dioxide equivalent (CO_2e): 600 in 1990; 650 in 1995; 758 in 2004
Kyoto Protocol status: Ratified, 2002

• Historical and Political Context

Canada was first explored by English and French explorers at the end of the fifteenth and beginning of the sixteenth century. Both England and France laid claim to Canada; England by John Cabot's landing at Newfoundland in 1497 and France by Jacques Cartier's discovery of the Saint Lawrence River. A long period of conflict between the two countries over ownership of Canada ensued. With the Treaty of Paris signed in 1763, France recognized Canada as belonging to England and relinquished all of its claims. Canada has continued to maintain a close relationship to England (today referred to as the United Kingdom). However, through a series of British parliamentary acts Canada has been granted legislative independence from the United Kingdom. This process began in 1867 with the British North American Act and continued into the twentieth century with the Statute of Westminster (1931) and the Canada Act (1982).

Canada has also enjoyed a long association with the United States, Britain's other former North American colony. Economically the two countries have had a long mutual dependency as each is the other's major trading partner. During their long trade relationship, Canada and the United States have often disagreed, imposed tariffs, and at times been almost isolationist in their attitudes. However, with the implementation of the North American Free Trade Agreement in 1994, their relationship has in general become one of free trade. Canada's economy has evolved from rural to industrialized, yet primary material, timber, oil, and gas, continue to play a very important role in the country's economic sector.

Domestically, Canada is a country of a varied geographic makeup and of a widely divergent population. Multiculturalism and diversity dominate Canada's social and cultural life. Both English and French are recognized as official languages. While British and French cultural heritage and traditions are certainly a major part of Canada's culture, those of aboriginal tribes and Irish, Scot, and other immigrant groups as well as American influences are very visible in Canadian society. Canada is an active participant in the global community as a member of many international organizations including the United Nations, the G-8, the North Atlantic Treaty Organization (NATO), and Francophonie.

• Impact of Canadian Policies on Climate Change

Canada is a vast country which possesses very large reserves of natural resources, both renewable and nonrenewable. Hydropower, a clean renewable source of energy, has been an important source of energy for Canada, especially for electricity. However, from 1990 to 2004, the demand for electricity in Canada increased some 23 percent and Canada increased its use of fossil fuels to meet this demand. Canada is in the process of developing the oil sands in the Athabasca Basin in the province of Alberta. The oil in the oil sands is bitumen, a thick viscous oil which requires considerable processing to be usable. The oil cannot be extracted by drilling but requires either open pit extraction or strip mining. The extraction of this oil emits in excess of 33 million metric tons of greenhouse gases (GHGs), accounting for approximately 5 percent of Canada's emissions. If Canada follows its present program in regard to the oil sands, the emissions from the oil sands production would reach 12 percent by 2020. The oil producers and the Canadian government

are involved in a heated battle with environmentalist groups over the extraction and use of this oil.

In addition, Canadians have changed their preferences in choice of motor vehicles. Sport utility vehicles (SUVs) and small pickup trucks now outnumber passenger automobiles on the roads of Canada. This has resulted in an increase in fuel consumption. Consequently, the majority of the increase in GHG emissions during the period has come from the energy sector, which is responsible for 81 percent of the GHGs emitted.

• Canada as a GHG Emitter

According to Environment Canada's report, Canada had GHG emissions totaling 600 million metric tons of carbon dioxide equivalent in 1990. The Kyoto Protocol, which Canada ratified in December of 2002, calls for Canada to reduce these base year (1990) emissions by 6 percent. The GHG emissions target established for Canada by the 2008-2012 period is 568 million metric tons. From 1990 to 2004, Canada's GHG emissions rose some 27 percent. Although the greenhouse emissions per unit of GDP fell by 2004 by 14 percent, the Canadian economy experienced considerable expansion that resulted in a net increase in the total emissions. In addition, the ratio of GHG emissions to the population rose by 10 percent. Canada was responsible for approximately 2.3 percent of the GHG emissions in the world. This amount of emissions ranked Canada seventh in the world in GHG emissions. The 2004 emissions of 758 million metric tons were 35 percent above the Kyoto Protocol target.

• Summary and Foresight

Canada's greenhouse emissions were 35 percent above the Kyoto Protocol target in 2004; by 2006 Canada had reduced its emissions to 21 percent above the Kyoto Protocol. In January of 2006, a conservative government which opposed Canada's participation in the Kyoto Protocol was in place. In April, the government announced that Canada could not possibly meet its Kyoto Protocol target for the 2008-2012 period. The government further stated that it was seeking an alternative to participation in the Kyoto Protocol and proposed the possibility of joining the Asian-Pacific Partnership on Clean Development and Climate. The federal government also proposed legislation setting mandatory emission limits for industry. Subsequently, a bill was introduced to force the government to take the necessary steps for Canada to achieve its Kyoto Protocol target. The bill passed but has been ignored by the government.

Shawncey Webb

• Further Reading

Charnovitz, Steve, and Gary Clyde Hufbauer. *Global Warming and the World Trading System.* Washington, D.C.: Peterson Institute for International Economics, 2009. Discusses reductions in GHGs and their relation to trade and to trade organizations, especially the World Trade Organization.

Details methods of reduction of GHGs without hurting domestic and global carbon-intensive industries. Appendix on using biofuel to save energy and reduce GHG emissions.

Dessler, Andrew E., and Edward A. Parson. *The Science and Politics of Global Change: A Guide to the Debate.* New York: Cambridge University Press, 2006. Excellent introduction to climate change from a scientific viewpoint, as well as a clear review of global politics and their effects on decision making. Written for the general reader. Well illustrated with graphs and tables. Excellent suggestions for further reading.

Lee, Hyun Young. "Sand Storm." *Wall Street Journal,* March 9, 2009, pp. R8-R9. Good article from an economics and world trade viewpoint. Also discusses how U.S. cap-and-trade policy could affect Canadian oil sands development and vice versa.

Marsden, William. *Stupid to the Last Drop: How Alberta Is Bringing Environmental Armageddon to Canada (And Doesn't Seem to Care).* Reprint. Toronto: Vintage Canada, 2008. Argues that development of the oil sands would destroy natural resources and habitat. Discusses the roles of various levels of government.

Shogren, Jason F. *The Benefits and Costs of the Kyoto Protocol.* Washington, D.C.: AEI Press, 1999. Excellent for understanding the Kyoto Protocol and the framework it creates for Canadian climate policy.

See also: Canadian Meteorological and Oceanographic Society; Kyoto Protocol; United Nations Framework Convention on Climate Change; United States.

Canadian Meteorological and Oceanographic Society

- **Categories:** Organizations and agencies; meteorology and atmospheric sciences
- **Date:** Established 1967
- **Web address:** http://www.cmos.ca

• Mission

The Canadian Meteorological and Oceanographic Society (CMOS) is Canada's national umbrella organization promoting research in weather and weather extremes, global warming, ozone depletion, and surface air quality. Among approximately eleven hundred members of CMOS are meteorologists, climatologists, oceanographers, limnologists, hydrologists, and cryospheric scientists. Governing the society are its national executive and council, the latter comprising the heads of fourteen centers located throughout Canada. The council appoints an executive director and members to serve on various committees, among them the Scientific, University, and Professional Education Committee; the School and Public Education Committee; the Prizes and Awards Committee; the Consultant Accreditation Committee; and the Media Weathercaster Endorsement Committee. CMOS is headquartered in Ottawa.

In addition to promoting professional and public education, CMOS accredits media weather forecasters and helps specialists study the meteorological aspects of hydrology, agriculture, and forestry. It publishes a scientific journal, *Atmosphere-Ocean,* as well as a bimonthly bulletin and *Annual Review.* It presents annual awards to scientists, graduate students, postdoctoral fellows, and volunteers and provides scholarships. Is also sponsors precollege teachers' participation in Project Atmosphere, run jointly with the American Meteorological Association and the U.S. National Oceanic and Atmospheric Administration.

• Significance for Climate Change

In March, 2007, CMOS issued its Comprehensive Position Statement on Climate Change. Noting that Canada's climate was changing rapidly, affecting its ecosystems and wildlife, the statement insisted that there was strong evidence (95 percent confidence) that atmospheric and oceanic warming in the previous fifty years were mostly the result of fossil-fuel burning and clearing of forest vegetation. Global warming, in fact, disproportionately impacts Canada and the Arctic. Accordingly, the CMOS issued a call to action.

Given the increasing significant load of greenhouse gases (GHGs) in the atmosphere, future warming seems inevitable. Because upcoming changes will occur rapidly, appropriate adaptation policies and programs must be designed to help increase humans' adaptive capacities. Climate-change protection is needed now. The global reduction of GHG emissions cannot be achieved by any single country, but each country must contribute its share toward accomplishing the global goal. The atmosphere, ocean, and biosphere together constitute a highly complex system. Climatologists' ability to make comprehensive climate projections is hindered by the lack of accurate characterization of important processes and feedbacks within this system. Further process studies, assessments of paleoclimate evidence, modeling, and observations are needed to understand and to improve predictions of climate change at the global and especially regional levels. Research is essential to humans' ability to mitigate and adapt to climate change.

Roger Smith

See also: Canada; Meteorology; Ocean-atmosphere coupling; Ocean dynamics; Weather forecasting; Weather vs. climate.

Canaries in the coal mine

• **Category:** Meteorology and atmospheric sciences

• Definition

Originally, canaries were used in coal mines, because small quantities of methane gas and carbon monoxide are deadly to them. Thus, a canary in a coal mine would die long before these substances became lethal to the miners carrying it. The term has come to be used to describe any such sensitive indicator of an emergent threat or problem. Several species and events act as "canaries" in the "coal mine" represented by global warming. These include the ecosystems and species most vulnerable to increases in extreme weather patterns (those most threatened by extreme heat, drought, flooding, hurricanes, and so on); ice sheets and other frozen formations in danger of melting; endangered animal species such as polar bears, amphibians, and butterflies; and plants vulnerable to increased or invasive exotic insect populations.

• Significance for Climate Change

Nineteen of the twenty hottest years since 1850 occurred after 1980, and the Earth was warmer in 2005 than at any time in the past two thousand years. Periods of extreme heat place strain on people, animals, and plants and can be responsible for hundreds of human deaths in a given area. A record European heat wave in 2003 is blamed for the loss of over fifty thousand human lives. In such heat waves, the most common causes of death are pollution or preexisting medical conditions that are aggravated by the heat. Because those with such conditions are more sensitive to heat waves than are healthy people, they may be considered to function as canaries in the coal mine.

Warm air holds more water than does cold air. Thus, with warmer temperatures, more moisture is removed from the ground, and droughts occur. The droughts, such as those in Africa, kill millions of people through starvation and increased vulnerability to disease. Warmer temperatures also mean that rains or snows fall harder in a shorter period of time, increasing the risks of flooding. Monsoons—patterns of heavy rainfall that particularly affect Asia—strengthen and weaken in relation to the Earth's temperature. Major flooding is responsible for loss of human lives and property, as well as the destruction of crops. The most vulnerable to this destruction are the poorest members of society, those with the fewest resources.

Tropical storms (also known as hurricanes, cyclones, or typhoons) are becoming stronger and more frequent. Oceans store a vast amount of heat in their depths, and warm ocean waters give rise to tropical storms and provide the fuel they need to grow. Because of the many factors involved with the storms, there is no clear evidence that they may occur more frequently as a result of global warming. The strength of the storms in the Atlantic Ocean, however, does appear to be affected somewhat by global warming. Three of the six 2005 North American hurricanes were the most powerful Atlantic storms ever observed. The year marked the first

Amphibians such as this gold frog are particularly sensitive to their environment, causing them to function as canaries in the coal mine of climate change. (Alberto Lowe/Reuters/Landov)

time four Category 5 hurricanes occurred in a single year since records began being kept in 1851.

The rate of change and level of warming are far above average for the Arctic region. The warm temperatures cause melting of the ice sheets near the shore, so life for subsistence whalers and seal hunters becomes more difficult. In addition, less ice near the shore means that violent storms are not kept offshore and crash into the coastal towns, eroding the coastline.

Polar bears are greatly affected by the decreasing ice sheet and warmer Arctic temperatures. The bears hunt for seals on sea ice in the springtime. When the ice is gone, the bears fast on land. With ice-free spells lengthening, cub birth weights are dropping, females are becoming thinner, and desperate bears are appearing in human settlements seeking food. The World Wildlife Foundation estimates that polar bears may not be fat enough to reproduce by 2012 if the lower weight trend continues. If polar bears cannot adapt to a land-based life, they may become extinct in the wild.

As glaciers in the Arctic and Antarctic melt, sea levels rise and the oceans become warmer and more acidic. As the sea level increases, millions of people living close to the ocean and in drought- or flood-prone areas will be greatly affected, as lower-lying areas are more prone to flooding and storm surges. Venice is already sinking in the rising waters and is threatened by storm surges. Minor flooding requires residents and visitors regularly to wade through ankle-deep water in the streets.

Amphibians and reptiles, because of their need for a particular range of air temperature and their slow mobility, cannot adapt to climate changes quickly. Frogs are among the most threatened species worldwide, and climate change is responsible for several cases of depletion or extinction. Climate change helps fungi attack amphibians more effectively, while drought allows light to penetrate shallow water sources and weaken the developing embryos of some toad species. In 1987, after an unusually warm, dry spring, twenty of fifty frog species vanished completely in Costa Rica.

Butterflies, which have been carefully studied for centuries, indicate climate change as well. Drought and flood can cause a population to crash. Slowly rising temperatures, as well as other pressures that restrict their range, threaten the populations. As temperatures rise and climate zones move, some species have expanded their habitat northward.

Insects are extremely adaptable to change, but their presence may not be beneficial. Insects may eat crops, and warming northern forests are being devastated as beetles that thrive in warmer climates migrate into them. As temperatures warm, parasitic diseases are more likely to spread and become more severe. Mosquitoes carrying malaria are expanding their territory, infecting more than four times as many people in 2005 as in 1990. Dengue fever, caused by four potentially fatal, mosquito-borne viruses, is also on the increase. Cases of the West Nile virus are showing up further north each year.

The frogs, butterflies, and trees suffering from climate change are canaries in the coal mine. They act as signals of the dangers to stronger and more resilient species that may develop if global warming continues. By the same token, those segments of human society most vulnerable to climate change now will not remain the only segments vulnerable in the future. These canaries demonstrate dangers that will only increase if the Earth's temperature continues to rise.

Virginia L. Salmon

• Further Reading

DiMento, Joseph F. C., and Pamela Doughman, eds. *Climate Change: What It Means for Us, Our Children, and Our Grandchildren.* Cambridge, Mass.: MIT Press, 2007. Explains various aspects of climate change, including different public and scientific responses. Illustrated, with glossary and index.

Henson, Robert. *The Rough Guide to Climate Change: The Symptoms, the Science, the Solutions.* New York: Penguin Putnam, 2006. Presents well-balanced, organized, and easy-to-follow information on climate change. Includes illustrations, charts, side articles, and bibliography.

Lomborg, Bjørn. *Cool It: The Skeptical Environmentalist's Guide to Global Warming.* New York: Alfred A. Knopf, 2007. Looks at global warming from the perspective of both environmental and human concerns. Includes notes, index, and extensive bibliography.

See also: Amphibians; Antarctica: threats and responses; Arctic; Arctic peoples; Coastal impacts of global climate change; Extreme weather events; Health impacts of global warming; Islands; Penguins; Polar bears; Poverty.

Carbon cycle

• Categories: Chemistry and geochemistry; plants and vegetation

The natural geochemical carbon cycle maintained an equilibrium of atmospheric and dissolved CO_2 for thousands of years. This equilibrium, however, has been disrupted by the anthropogenic emission of CO_2, which has increased CO_2 concentrations in the atmosphere and is associated with climate change.

• Key concepts

carbon fixation: conversion of CO_2 into organic molecules through such processes as photosynthesis or calcification

detritus: the residue left behind by decay or disintegration

limnetic zone: the upper layers of open ocean through which light penetrates and which thus supports photosynthesis by planktonic organisms

pH: a measure of the acidity of a solution, which is related to the activity of dissolved hydrogen ions in that solution

photosynthesis: the physiological process in plants that converts light energy into chemical energy

respiration: the physiological process that breaks down organic carbon-containing molecules to obtain energy and that releases CO_2 as a by-product

sedimentation: settling of suspended particulate material, for instance in response to gravity, to form a bottom layer

• Background

The biological carbon cycle primarily depends on two physiological processes of living things: photosynthesis and respiration. Photosynthesis converts light energy to chemical energy, which then is used to incorporate carbon dioxide (CO_2) into organic compounds, a process called carbon fixation. The organic compounds can be used by organisms as a source of energy through the process of respiration, which breaks chemical bonds to release energy for use. This breakdown of organic molecules provides energy to living cells and releases CO_2 back to the environment as a by-product. All living organisms respire, but plants do so at a much slower rate than they normally photosynthesize. As a result, overall global net photosynthesis is approximately equal to net respiration, and CO_2 levels are in balance.

- **Terrestrial Ecosystems**

In terrestrial ecosystems, the atmosphere is a major reservoir of carbon, holding about 750 billion metric tons (gigatons) of carbon in the form of CO_2. Nearly as much, about 610 gigatons of carbon, forms the biomass of living plants. Each year, plants convert about 120 gigatons of carbon from CO_2 into new organic tissue, with nearly that amount returned to the atmosphere through respiration of plants and animals, as well as through decomposition of dead organisms. A small amount of excess fixed carbon is added to the soil as humus. Over the millennia, this accumulation of organic material has formed soils that provide a reservoir of about 1,500 gigatons of carbon.

The depth and richness of soil are directly related to the photosynthetic productivity of the plants supported in a particular environment. The vegetation of Earth's temperate areas has produced deep, rich soils during the past ten to twenty thousand years that in turn promote high plant productivity in these regions. (Thus, although plants are predicted to migrate poleward in response to increasing temperatures associated with climate change during the twenty-first century, they will be moving into areas of poorer soils that will decrease plant productivity, even though other environmental factors may be positive.) Over a longer period of time during the past 3 to 3.5 million years, organic remains that did not decompose were converted into geological deposits of coal and oil. About 4 trillion metric tons of carbon are sequestered in these fossil fuels.

Scientists calculate that, at the start of the Industrial Revolution in the mid-eighteenth century, the atmospheric reservoir contained about 580 gigatons of carbon. This means that atmospheric CO_2 has increased by about 30 percent during the past

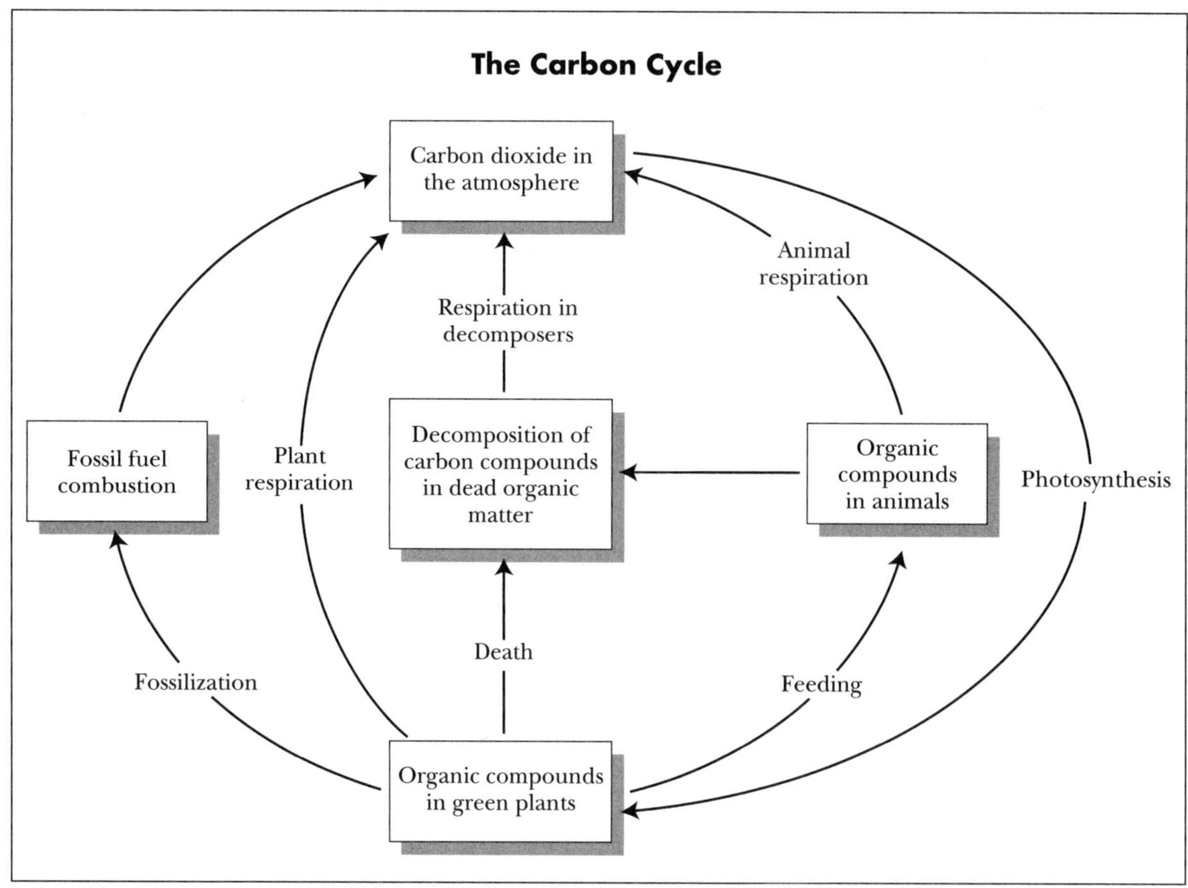

three hundred years, largely as a result of CO_2 being released to the atmosphere by burning fossil fuels. Fossil fuel consumption during the early twenty-first century adds about 5.5 gigatons of carbon to the atmosphere every year. In addition, changes in land use, particularly deforestation, contribute another 1.6 gigatons of carbon per year. While added CO_2 stimulates photosynthesis, particularly by carbon 3 plants, this activity fixes less than 1.9 gigatons of carbon per year. About 2 gigatons of carbon per year diffuse into the oceans, particularly in colder waters, but the majority remains in the atmosphere, increasing CO_2 concentrations and promoting global warming.

• Marine Ecosystems

A direct connection between the terrestrial and marine ecosystems exists in the equilibrium between atmospheric CO_2 and dissolved CO_2 in the oceans. This equilibrium shifts slightly toward the oceans as CO_2 reacts with water (H_2O) to form carbonate (CO_3^{-2}) or bicarbonate (HCO_3^-) ions along with hydrogen ions (H^+), which lowers pH, increasing the acidity of the oceans. Oceans absorb about 2 gigatons of carbon per year in this way.

The limnetic zone of the oceans supports a biotic carbon cycle similar to that of terrestrial ecosystems. About 1 trillion metric tons of carbon is dissolved in the near-surface waters that serve as a reservoir for marine organisms, which account for about 3 gigatons of carbon. These organisms convert CO_2 to calcium carbonate, a building block of shells and other portions of marine life. The dissolved organic compounds produced by marine organisms account for nearly 700 gigatons of carbon.

• Sedimentary Rock and Fossil Fuels

When marine organisms die, their remains form a detritus of organic materials that slowly sinks to the seafloor. A huge amount of carbon, about 38.1 trillion metric tons, is accounted for by this deep ocean detritus, which eventually forms bottom sediments.

Marine sediments are molded by the pressure of the ocean as well as by volcanic heat, eventually forming sedimentary rocks. By far the largest amount of carbon on Earth is found in the lithosphere, in carbon-containing sediments and sedimentary rocks such as chalk, limestone, and dolomite. Scientists estimate that these calcium-carbonate-rich deposits—formed by corals, shell-producing animals, coralline algae, and marine plankton millions of years ago—contain up to 100 quadrillion metric tons of carbon.

Similarly, coal and oil are the carbon remains of ancient organisms stored in the Earth. As is true of sedimentary rock, these remains are transformed over millions of years by the heat and pressure of the Earth, which compresses them into a new form. Contemporary burning of fossil fuels returns about 5.5 gigatons of carbon to the atmosphere each year.

• Context

Initially, it seemed that rising CO_2 levels in Earth's atmosphere would be partially compensated for by increased rates of plant photosynthesis and also by increased uptake of CO_2 by the oceans. The increased photosynthesis promoted by higher CO_2 levels, however, is countered by a decrease resulting from higher temperatures. Moreover, the rising acidity of the oceans resulting from higher levels of dissolved CO_2 counters the photosynthetic benefits of higher CO_2 levels. The global carbon cycle is thus much more complex than was imagined even a decade ago.

Marshall D. Sundberg

• Further Reading

Fagan, Brian. *Floods, Famines, and Emperors: El Niño and the Fate of Civilizations.* New York: Basic Books, 1999. Good introduction to climatology, integrating atmospheric and oceanographic interactions and the relationship between major climatological events and human history.

Field, Christopher B., and Michael R. Raupach, eds. *The Global Carbon Cycle: Integrating Humans, Climate, and the Natural World.* Washington, D.C.: Island Press, 2004. This introduction to the carbon cycle covers relevant biological, geochemical, and societal interactions.

Odum, Eugene. *Ecological Vignettes: Ecological Approaches to Dealing with Human Predicaments.* Amsterdam, the Netherlands: Harwood Academic, 1998. Examines human impacts on the environment and urges the adoption of an economic approach to ecology.

Wigley, T. M. L., and David Steven Schimel, eds. *The Carbon Cycle.* New York: Cambridge University Press, 2000. The multiauthored chapters of this comprehensive volume summarize results of the Intergovernmental Panel on Climate Change and address both the scientific and the political aspects of this complex relationship.

See also: Carbon dioxide; Carbon dioxide fertilization; Carbon 4 plants; Carbon isotopes; Carbon monoxide; Carbon 3 plants; Carbonaceous aerosols; Photosynthesis; Plankton; Sequestration.

Carbon dioxide

- **Category:** Chemistry and geochemistry

Carbon dioxide is the greenhouse gas that contributes the most to global warming. A naturally occurring gas, it is also a by-product of burning fossil fuels, which has become the most important determinant of its rate of increase in Earth's atmosphere.

• Key concepts

anthropogenic: created by human action

fossil fuel: fuels formed by pressure on plant or animal material over time, including oil, gas, and coal

greenhouse gases (GHGs): gases that trap heat within Earth's atmosphere, increasing global temperatures

• Background

Carbon is one of the building blocks of life on Earth; all plants and animals are composed largely of carbon. Carbon dioxide (CO_2) emission is a normal part of the respiration cycle. CO_2 is also emitted from the burning of plant or animal material. The burning of fossil fuels for energy causes CO_2 to be emitted into the atmosphere, and forest clearing also results in a net increase in atmospheric CO_2. Once in the atmosphere, CO_2 acts as a greenhouse gas (GHG), trapping heat in the atmosphere and contributing to global warming. Presently, CO_2 is the most common GHG in the atmosphere, and its atmospheric concentration is growing rapidly. The proper regulation of human activities that exacerbate this situation is therefore the focus of a good deal of attention.

• Fossil Fuels

Burning fossil fuels is the major contributor to the anthropogenic contribution of CO_2 in the atmosphere. Approximately 57 percent of GHGs are constituted by CO_2 generated by burning fossil fuels. Another 17 percent comes from the decay of biomass and deforestation. The Intergovernmental Panel on Climate Change (IPCC) has indicated with a high degree of confidence that human activities since 1750 have contributed to the addition of these and other GHGs to the atmosphere. When the scientist Charles Keeling began measuring the accumulation of CO_2 in the atmosphere from an observatory atop Mauna Loa, Hawaii, in 1958, his measurements indicated that the concentration of CO_2 in the atmosphere was 315 parts per million. By 2005, the measured concentration of CO_2 was 379 parts per million, which exceeded the natural range of atmospheric CO_2 levels over the last 650,000 years. The CO_2 content of the atmosphere, derived from ice-core data, has varied over time by 10 parts per million around a mean value of 280 parts per million.

Industrialization was initially fueled by coal in the eighteenth and nineteenth centuries. Coal continued to be a major source of energy for much of the twentieth century, but oil and natural gas also began to be used for energy in increasing amounts. Gasoline, derived from oil, fueled the growth of automobile culture in many parts of the world in the twentieth century. One thing that all of these energy sources have in common is that they are fossil fuels. Combustion of fossil fuels produces CO_2, along with smaller quantities of other GHGs. Much of industrial civilization in the early twenty-first century is powered by fossil fuels.

Some cleaner fuels that do not generate CO_2 are becoming available, and more efficient energy sources and automobiles are being built that generate less CO_2. However, CO_2 will continue to be a byproduct of industrial society for years to come. Controlling the impact of CO_2 on global climate will require different approaches to energy generation and consumption, as well as new technologies such as carbon sequestration that remove carbon from the atmosphere.

U.S. Anthropogenic GHG Emissions, 2006

Greenhouse Gas	*Amount (kilograms of CO_2 equivalent)*	*Percent of Total GWP*
Energy-related carbon dioxide	5,825.5	82.3
Other carbon dioxide	108.8	1.5
Methane	605.1	8.6
Nitrous oxide	378.6	5.4
All other Kyoto gases*	157.6	2.2

Data from U.S. Energy Information Administration.
*Hydrofluorocarbons, perfluorocarbons, and sulfur hexafluoride.

• Land Use

Human use of the land has led to the accumulation of CO_2 (and methane, another GHG) in the atmosphere for a period extending back several thousand years. The growth of population in the last two hundred years has magnified this impact. Clearing forested land for agriculture has generally led to the burning of much of the cleared vegetation. As noted above, combustion of carbon-based entities produces CO_2 as a by-product. The decay of biomass, often generated by agriculture, also produces CO_2. As societies have increased their use of metals over time, metal smelting—first using wood for its energy source, then coal—has also contributed CO_2 to the atmosphere.

Some of the CO_2 that has been anthropogenically generated has been fixed in the oceans and wetlands as peat, rather than in the atmosphere. As wetlands are drained for other uses, however, this carbon sink is diminished, so that more carbon enters the atmosphere. Population pressure and the demands for agricultural products drive land clearing and wetland degradation in many parts of the world. Land clearing continues to increase dramatically in some parts of the world, such as the Amazon basin in Brazil.

• Who Produces CO_2?

The industrial nations of the world have been the major producers of CO_2 over time. As each nation has industrialized, it has begun burning fossil fuels extensively, as well as clearing land for agriculture. Industrial countries continue to generate most of the world's CO_2, either directly through energy generation and automobile use or indirectly by their demand for agricultural products from other areas, which leads to further land clearing in those areas. Nations that are rapidly industrializing, such as China and India, are doing so largely through the use of fossil fuels. Although the United States is currently the largest consumer of fossil fuels, China is poised to move into second place and will probably surpass the United States by 2030 as a producer of GHGs. Even less-industrialized nations make extensive use of fossil fuels.

Dealing with CO_2 emissions in the future will require concerted efforts by industrial countries. Less industrialized nations cannot be expected to forgo the economic progress generated by industrialization, but they too will have to manage their carbon footprints over time.

• Context

CO_2 is the major driver of what is called the greenhouse effect. Some estimates of the amount of CO_2 that will be in the atmosphere by 2100 are as high as 1,000 parts per million if emissions grow unchecked. Such a concentration could produce a global temperature increase of 5° Celsius or more, a change not seen for several million years. Even if the growth of CO_2 in the atmosphere is brought somewhat under control, its concentration could still reach 440 parts per million, which could produce a temperature increase of as much as 3° Celsius. Controlling the rate of carbon emissions into the atmosphere without harming people's economic well-being will be a challenge. Failure to control the growth of carbon emissions will lead to what several authorities consider to be a much different and undesirable life for much of the planet. The production of CO_2, which has been the hallmark of economic progress, may lead to economic decay if it is not checked.

John M. Theilmann

- **Further Reading**

Broecker, Wallace S., and Robert Kunzig. *Fixing Climate: What Past Climate Changes Reveal About the Current Threat—and How to Counter It.* New York: Hill and Wang, 2008. Examination of the role of CO_2 in climate change and what can be done. Broecker is a leading climate scientist.

Houghton, John. *Global Warming: The Complete Briefing.* 4th ed. New York: Cambridge University Press, 2009. Excellent, comprehensive discussion of global warming and the role of CO_2 that includes extensive references. Houghton is the former chairman of the Scientific Assessment Working Group of the IPCC.

Ruddiman, William F. *Plows, Plagues, and Petroleum.* Princeton, N.J.: Princeton University Press, 2005. Especially good at demonstrating the long-term impact that agriculture has had on climate.

Smil, Vaclav. *Cycles of Life.* New York: Scientific American, 1997. Good description of the carbon cycle and how it relates to human life.

Volk, James. *CO_2 Rising.* Cambridge, Mass.: MIT Press, 2008. Excellent analysis of the carbon cycle, as well as energy use and CO_2 emissions.

See also: Carbon cycle; Carbon dioxide equivalent; Carbon dioxide fertilization; Carbon equivalent; Carbon footprint; Carbon 4 plants; Carbon isotopes; Carbon monoxide; Carbon taxes; Carbon 3 plants; Emissions standards; Greenhouse effect; Greenhouse gases.

Carbon dioxide equivalent

- **Category:** Pollution and waste

- **Definition**

Various greenhouse gases (GHGs) such as methane, nitrous oxide, chlorofluorocarbons, and carbon dioxide (CO_2) have different potential effects on global warming. In order to compare GHGs based on their global warming potential (GWP), CO_2 is commonly chosen as a reference gas. The carbon dioxide equivalent (CO_2e) for a given amount of GHG is defined as the amount of CO_2 that would have the same GWP measured over a specified time period, usually one hundred years. More formally, the CO_2e for emissions of a given gas is calculated by multiplying the amount of the gas emitted by its associated GWP.

For example, the GWP for methane over a one-hundred-year time horizon is 21. Thus, the CO_2e of one metric ton of methane is equal to 21 metric tons. In other words, one metric ton of methane released into the atmosphere would have the same effect on global warming over one hundred years as would 21 metric tons of CO_2.

- **Significance for Climate Change**

The CO_2e metric provides a universal standard measure by which to evaluate and compare the global warming effects of emissions of various GHGs, as well as to calculate the total effects of the GHGs present in Earth's atmosphere. Those uses of CO_2e have important implications for decisions about climate-change mitigation. On one hand, CO_2e measurements of GHG emissions make it easy to compare the various impacts of different plans to prevent such emissions. The fact that the CO_2e of methane is 21 times that of CO_2 helps determine whether to focus on methane or CO_2 reduction in a given context.

On the other hand, the calculation of total effect of GHGs expressed in CO_2e concentration can provide a way to compare existing concentrations of GHGs in the atmosphere with theoretical critical threshold levels. Recent studies have confirmed that one such threshold is represented by a global temperature rise of 2° Celsius during the twenty-first century. In order to prevent surpassing this threshold, the total concentration of all GHGs must be less than 450 parts per million in CO_2e. However, GHG concentration is already beyond that level.

In addition to facilitating global climate study, the CO_2e metric as a reference for evaluating GHGs also plays an important role in emissions trading. Various national and international regulatory structures allow polluters that reduce their emissions below a certain level to trade emissions credits (the right to pollute) to other entities whose emissions are still above the maximum permitted level. The existence of a universal standard of emission measurement, CO_2e, allows these credit systems to function much more effectively and efficiently.

To N. Nguyen

See also: Baseline emissions; Carbon dioxide; Carbon equivalent; Carbon footprint; Certified emissions reduction; Emissions standards; Fossil fuel emissions; Greenhouse gases; Industrial emission controls; Industrial greenhouse emissions.

Carbon dioxide fertilization

• **Categories:** Chemistry and geochemistry; plants and vegetation

• **Definition**

In the context of climate change, carbon dioxide (CO_2) fertilization is the stimulation of biospheric carbon uptake by rising atmospheric CO_2 concentrations. That is, as the CO_2 level increases, plants absorb more carbon and sequester it in biomass, removing it from the atmosphere and increasing their own size and productivity. This fertilization is realized by photosynthetic enzymes. It involves multi-scale processes, from the level of individual leaves and plants all the way up to regional and global ecosystems, and it is regulated by many other factors.

CO_2 and water (H_2O) are the two basic substrates of photosynthesis, which is driven by solar energy and regulated by an enzyme, Ribulose-1,5-bisphosphate carboxylase/oxygenase (RuBisCO). When all other factors are constant, photosynthesis increases as CO_2 concentration increases. Thus, photosynthesis is sensitive to CO_2 concentration.

The photosynthetic sensitivity to CO_2 concentration is universal to all carbon 3 plants, because all carbon 3 plants share the same enzyme, RuBisCO, to catalyze photosynthesis. The sensitivity is also virtually independent of light and nutrient levels, but it declines with CO_2 concentration. This sensitivity can be measured as the percentage increase in carbon fixation caused by a 1 part per million increase in CO_2 concentration. In other words, a CO_2 sensitivity of 0.2 percent means that if CO_2 concentration increases by 1 part per million, plants will fix 0.2 percent more carbon than they did at the previous concentration.

When atmospheric CO_2 concentration increases from 280 parts per million (the preindustrial level) to 440 parts per million (a level projected to be reached in the 2030's), photosynthetic CO_2 sensitivity declines from 0.183-0.352 percent to 0.077-0.183 percent. Assume that atmospheric CO_2 concentration increases by 2 parts per million per year and that global photosynthetic carbon fixation is 120 billion metric tons per year and is accomplished entirely by carbon 3 plants. Under such conditions, Earth's land ecosystems will fix roughly an additional 600 million metric tons of carbon per year as a result of photosynthetic sensitivity. Carbon 4 plants, however, are less sensitive to CO_2 concentration than are carbon 3 plants.

When plants grow under different CO_2 concentration, they adjust their leaf structures and biochemical properties, resulting in photosynthetic acclimation. Photosynthetic acclimation is regulated by sugar signals at the biochemical level and is usually related to nutrient supply. Acclimation may increase or decrease photosynthetic sensitivity. Many studies indicate that photosynthetic acclimation greatly varies with species and environmental conditions. However, on average across all studies conducted in natural ecosystems, photosynthetic acclimation is not very substantial.

Photosynthetically fixed carbon is stored in plant biomass and soil organic matter (SOM). Carbon storage may last for years, decades, or centuries in plant wood pools and for up to thousands of years in soil pools, but it is very ephemeral in leaf and fine root pools. Carbon storage in wood pools is regulated by plant allocation and species. Carbon storage in soil pools is regulated by nutrient availability. Many CO_2 experiments in natural ecosystems indicate that rising atmospheric CO_2 concentration results in carbon storage in plant and soil pools. However, the CO_2 fertilization may not occur under conditions in which some other growth factor is severely limiting, such as low temperature or low nutrient availability.

• Significance for Climate Change

CO_2 fertilization is a major mechanism of terrestrial carbon sequestration and a negative feedback mechanism protecting ecosystems from climate change. CO_2 is a major greenhouse gas (GHG). Buildup of CO_2 in the atmosphere results in climate warming. Since terrestrial ecosystems absorb approximately 120 billion metric tons of carbon from the atmosphere every year, a small stimulation of photosynthetic carbon uptake by CO_2 fertilization can substantially influence global carbon balance and reduce the likelihood or degree of climate change. Several global analyses indicate that nearly 30 percent of the CO_2 emitted by human activities is absorbed by land ecosystems. CO_2 fertilization is one of the major mechanisms responsible for land carbon sequestration.

Because photosynthetic enzymes are sensitive to CO_2 concentration, the CO_2 fertilization factor should gradually decline as atmospheric CO_2 concentration increases. However, as the world continues to consume fossil fuels, the yearly increase in atmospheric CO_2 concentration becomes larger over time. CO_2 fertilization will remain a major mechanism in the regulation of atmospheric CO_2 concentration.

Most experimental and modeling studies demonstrate that nitrogen deposition acts synergistically with atmospheric CO_2 concentration to stimulate carbon sequestration in land ecosystems. Nitrogen deposition is expected to increase by another two- to threefold in the future. This increase is likely to enhance the effects of CO_2 fertilization on plant growth and carbon sequestration. CO_2 fertilization effects are also regulated by other global change factors, such as climate warming and altered precipitation regimes.

In addition to regulation of CO_2 concentration in the atmosphere, stimulation of food production by CO_2 fertilization helps mitigate climate change impacts on developing countries. Although CO_2 fertilization's effects on food production are relatively small in comparison to those of nitrogen fertilization and genetic breeding, food production still can increase by 10-20 percent when atmospheric CO_2 concentration increases by 200-300 parts per million.

CO_2 fertilization-based increases in plant growth and carbon sequestration are fundamentally driven by rising atmospheric CO_2 concentrations. If those concentrations level off as a result of effective climate-mitigation activities, the effects of fertilization will also level off. If atmospheric CO_2 concentration declines, as hypothesized in some scenarios of the Intergovernmental Panel on Climate Change, CO_2 fertilization will also decrease, and land ecosystems will likely release some CO_2 from plants and soils into the atmosphere. Thus, any analysis of climate dynamics and climate change mitigation must take into account CO_2 fertilization effects, whether positive or negative.

Overall, CO_2 fertilization stimulates carbon storage in plants and soil, reduces buildup of GHGs in the atmosphere, and determines the airborne fraction of anthropogenic carbon emissions.

Yiqi Luo

• Further Reading

Curtis, P. S., and X. Z. Wang. "A Meta-analysis of Elevated CO_2 Effects on Woody Plant Mass, Form, and Physiology." *Oecologia* 113 (1998): 299-313. Summarizes more than five hundred reports of the effects of elevated CO_2 on key plant physiological processes. Determines that both total biomass and net CO_2 assimilation increase significantly when CO_2 is elevated.

Luo, Y., D. Hui, and D. Zhang. "Elevated Carbon Dioxide Stimulates Net Accumulations of Carbon and Nitrogen in Terrestrial Ecosystems: A Meta-Analysis." *Ecology* 87 (2006): 53-63. Synthesizes over one hundred studies of CO_2 effects on plant and soil carbon sequestration as regulated by nitrogen availability. On average, elevated CO_2 significantly increased plant growth, soil carbon storage, and accrual of nitrogen capital in ecosystems, potentially supporting long-term carbon sequestration.

Luo, Y., and H. A. Mooney. "Stimulation of Global Photosynthetic Carbon Influx by an Increase in Atmospheric Carbon Dioxide Concentration." In *Carbon Dioxide and Terrestrial Ecosystems*, edited by G. W. Koch and H. A. Mooney. San Diego, Calif.: Academic Press, 1996. Systematically analyzes CO_2 fertilization according to the CO_2 sensitivity of RuBisCO, which is an invariant function independent of species and major environmental factors. Photosynthetic sensitivity becomes a scalable parameter to estimate global plant carbon uptake as stimulated by rising atmospheric CO_2 concentration.

See also: Carbon dioxide; Carbon 4 plants; Carbon 3 plants; Forests; Photosynthesis; Reservoirs, pools, and stocks; Sequestration; Sinks.

Carbon equivalent

• **Categories:** Pollution and waste; meteorology and atmospheric sciences

• **Definition**

The carbon equivalent (CE), similar to carbon dioxide equivalent (CO_2e), is a metric measure used to quantify how much global warming a given quantity of gas can cause (global warming potential, or GWP), using the functionally equivalent concentration of carbon, converted to the amount of carbon dioxide (CO_2) that would have the same GWP as the reference. In contrast to the CE, the CO_2e uses the functionally equivalent CO_2 concentration as the reference (the amount of CO_2 that would have the same GWP). CE is usually given for a specified timescale, effectively expressing the time-integrated radiative forcing. This also differentiates CE from CO_2e, which describes the instantaneous radiative forcing.

• **Significance for Climate Change**

The CE of a gas or a mixture of gases is a metric measure that can be used to compare the GWP of greenhouse gases (GHGs) for the purposes of analysis, computer modeling, and reporting—as was done, for example, in such international treaties as the Kyoto Protocol and the Montreal Protocol. The CE of a gas is calculated by multiplying the mass (in metric tons) by the GWP of the gas and is expressed for a specified timescale. CE values can be converted to CO_2e units simply by multiplying the CO_2 by $^3/_{11}$ (the ratio of the molecular weight of carbon to that of CO_2). Time is an important factor when comparing CE or GWP values, as they are functions of the time periods over which these values are calculated.

The United Nations climate change panel, known as the Intergovernmental Panel on Climate Change (IPCC), an intergovernmental organization charged with evaluating the risk of climate change caused by human activity, expresses CE in billions of metric tons of CO_2e. In industry, CE units are often expressed in millions of metric tons of CO_2e; for vehicles, units are in grams of CO_2e per kilometer. For example, the GWP for methane gas over one hundred years is 25; the equivalent value for nitrous oxide is 298. This means that the emission of 1 million metric tons of methane or nitrous oxide has the equivalent ability to contribute to global warming as 25 million or 298 million metric tons of CO_2, respectively. Over the same timespan, the hydrofluorocarbons fluoroform (HFC-23) and 1,1,1,2-Tetrafluoroethane (HFC-134a)—shown to have large negative effects on the environment—have GWPs of 14,800 and 1,430, respectively.

Rena Christina Tabata

See also: Carbon dioxide; Carbon dioxide equivalent; Global warming potential; Greenhouse gases; Hydrofluorocarbons; Kyoto Protocol; Methane; Montreal Protocol; Nitrous oxide.

Carbon footprint

• **Category:** Pollution and waste

Measuring one's carbon footprint is a useful tool for evaluating the effect upon the climate of one's GHG emissions and for ascertaining the results of specific changes in laws, regulations, or behavior.

• **Key concepts**

carbon dioxide equivalent (CO_2e): measurement of the environmental effect of a given amount of greenhouse gas that specifies the amount of CO_2 that would have the same effect

carbon-neutral activity: an activity that has no net effect on the concentration of greenhouse gases in the atmosphere

emission credits: purchasable shares in an emission-reducing project that are used to measure attempts by an entity to offset its GHG emissions
greenhouse gases (GHGs): atmospheric trace gases that trap heat, preventing it from escaping into space
offsetting: the practice of supporting GHG-reducing efforts elsewhere in order to mitigate the effects of one's own emissions

• Background

The carbon footprint of an entity or an activity is its total annual contribution to the greenhouse effect. It is measured in units of mass of carbon dioxide (CO_2) per year, but it includes the entity's output of all greenhouse gases (GHGs) into the atmosphere. The effects of other GHGs are measured in terms of the equivalent amount of CO_2 necessary to produce the same effect, the gases' carbon dioxide equivalent (CO_2e).

The U.S. emission of CO_2 into the atmosphere increased at an average rate of 1.1 percent per year during the 1990's, but at an average of 3.1 percent per year in the early twenty-first century. This rate of increase is attributable to a higher intensity of energy use per unit of domestic product. The carbon footprint of a person is expressed as the CO_2e, in kilograms, of GHG emissions caused by that individual's lifestyle.

One portion of a person's carbon footprint is constituted directly by the CO_2 emitted by such activities as home heating, cooking, and driving an automobile. Other portions are less direct, as they comprise the CO_2 emissions of manufacturers whose products the person purchases or benefits from, vehicles used to ship those products, power plants whose electricity the person uses, and so forth. Carbon footprint measurements are being developed to educate individuals, organizations, and nations as to how to reduce the global warming effect of their activities. A given carbon footprint is a subset of the overall ecological footprint of the totality of human activities.

• Carbon Footprint by Individual Activity

When measuring the carbon footprint of an individual, one generally excludes breathing and other bodily functions and focuses on a person's discretionary lifestyle, although people will disagree as to what is and is not discretionary. The primary factors that contribute to an individual's carbon footprint are the amount of travel performed, the energy consumed in heating and cooling homes, and the amount of trash generated. Several "calculators" have been developed that take into account various factors.

As an example, 1 liter of gasoline contains 637 grams of carbon. Burning that gasoline in air generates 2,334 grams of CO_2. Thus, a car running 19,200 kilometers per year at an average of 10.5 kilometers per liter of gasoline emits about 4.26 metric tons of CO_2 per year. The carbon footprint of an average American family of four might include the emission of 12.7 metric tons of CO_2 from home energy use, 11.6 metric tons from driving two cars, 10.7 metric tons from food (including its growth or husbandry, processing, packaging, transportation, and preparation), and 3.2 metric tons from air travel, for a total of 38.2 metric tons of CO_2 emissions per year.

The carbon footprint of food is a complex and controversial issue. It depends strongly on where one lives (because of the cost and energy used by transportation and storage) and the type of food product. Some food products go through numerous processing steps that consume large amounts of energy. On average, one unit of energy gained from food consumed in the United States requires as many as seven to ten units of energy to produce. Some of this inefficiency is due to the macroeco-

Global Annual Carbon Emissions by Sector, 2000

Economic Sector	*Percent of Total CO_2 Emissions*
Power stations	29.5
Industrial processes	20.6
Transportation fuels	19.2
Residential, commercial, and other sources	12.9
Land use and biomass burning	9.1
Fossil fuel retrieval, processing, and distribution	8.4

Data from the Netherlands Environmental Assessment Agency.

Average Carbon Footprints Around the World

Nation	Average Carbon Footprint per Capita, 2004 (metric tons)
Kuwait	38.00
United States	20.40
Australia	16.30
Russia	10.50
Germany	9.79
United Kingdom	9.79
Japan	9.84
Italy	7.69
China	3.84
India	1.21

Data from U.S. Department of Energy.

nomics of agriculture in a nation like the United States. For example, it was estimated in the late 1990's that over 90 percent of all fresh vegetables consumed in the United States came from the San Joaquin Valley of California, implying large transportation costs. Breakfast cereal requires thirty-two times as much energy to produce as does an equivalent amount of blended flour.

The energy cost of packaging food is also very large. The aluminum container of a prepackaged frozen dinner, which has no nutritive value, requires 3 times as much energy to produce as does 1 kilogram of blended flour. Producing a 36-centiliter aluminum soda can requires 3.4 times as much energy. All of this energy comes with attendant costs in CO_2 emissions. Thus, several lifestyle choices made for convenience lead to the creation of much larger carbon footprints than are necessary.

People who live in some metropolitan areas may appear to have smaller carbon footprints if they use public transportation or enjoy shorter commutes rather than driving themselves long distances to work. The per capita carbon footprints of residents of many U.S. cities can be compared to one another, taking into account only the contributions from transportation and residential energy use. Among the cities with the smallest footprints is Honolulu, Hawaii, at 1.356 metric tons per year. Los Angeles; New York; Portland, Oregon; Seattle, and San Francisco all have per capita carbon footprints between 1.4 and 1.6 metric tons. At the other extreme, smaller but spread-out, semirural towns in the Snow Belt have footprints of around 3.4 metric tons.

The average per capita carbon footprint for the one hundred largest U.S. metropolitan areas is 2.235 metric tons. However, this calculation is based primarily on personal transportation and heating costs. It does not include the significant costs of transporting food and other supplies to these cities. For some people, a large carbon footprint is calculated simply because they have to take long airplane trips occasionally. Although the fuel mileage of modern airliners per person is considerably better than that of most cars, airline travel can easily exceed twice the automobile miles driven per year, and emissions of airplanes are more damaging than are those of automobiles, because they are released so much higher up in the atmosphere.

Food, air transportation, and the energy involved in making personal technology products contribute vastly to individuals' carbon footprints. When these are included, the average per capita footprint of a U.S. resident is 20 metric tons, compared to a worldwide average of 4.4 metric tons. Much of the difference lies in the energy intensity of the products that are used in developed nations, as is demonstrated by the previous discussion of food. Industrialized nations have built very efficient systems for minimizing the monetary cost of consumer goods, but these systems operate at a high cost in energy use. They are able to obtain and employ energy at low prices through efficient power plants and transmission grids. However, the cost in CO_2 emissions of this heavy energy use shows up as a large national carbon footprint. Because of these features of the U.S. economy, even the least privileged and the most conservation-minded Americans appear to have carbon footprints that are double that global average. Thus, measures to reduce U.S. carbon footprints must be accompanied by national policy decisions and major systemic changes in order to be successful.

• Carbon Footprints of Businesses and Organizations

Businesses and organizations have large carbon footprints, because they use fleets of vehicles and large buildings that have to be lit and climate-controlled every day. The emissions due to transportation and incidental energy use accrue in addition to direct emissions from manufacturing. On the other hand, larger organizations can implement dramatic reductions in their carbon footprints through means that are not yet available to individual homes. Examples include the installation of large areas of rooftop solar photovoltaic panels by companies such as Google and WalMart.

Facilities located in areas with predictable weather, such as California, are able to produce much of the electricity they require with solar technology that does not generate emissions. Producing the solar panels themselves requires a significant amount of energy and emissions, but once installed they operate with zero emissions for a long time. Likewise, many large organizations have switched to vehicles that are operated on natural gas (mostly methane) instead of gasoline or diesel fuel. Since methane released into the atmosphere has a global warming potential more than twenty times that of CO_2, burning methane achieves a net reduction in equivalent CO_2 emission. Large buildings are also able to use combined solar heating, power generation, and even air conditioning using appropriate building surface systems.

Many organizations are working to become carbon neutral. Such entities buy credits from low-emission projects such as wind farms and apply the credits toward their own emissions. It is often impossible to reduce actual emissions to zero, or in some cases at all. For example, air travel cannot be avoided completely. In such cases, trading emission credits to achieve carbon neutrality is a viable solution, as it supports the development of green technologies in the future.

• Measures to Reduce Carbon Footprints

Individuals' carbon footprints may be reduced by recycling plastics, glass, paper, and magazines; using more fuel efficient cars; reducing the number of miles driven per year per person by either car pooling, using public transportation, or simply reducing travel; changing to fluorescent lightbulbs; and turning thermostats up during summer and down during winter. National carbon footprints can be reduced through appropriate policies. The signatories to the Kyoto Protocol have adopted a complex system of certified emissions reduction credits (or carbon credits) that can be traded on the open market, allowing a large marketplace to develop for emission reduction schemes. In the United States, federal tax credits have offset the higher initial costs of installing modern, low-emission water heaters, home heaters, air conditioners, and energy-saving windows in homes, as well as the costs of hybrid and fuel-cell automobiles. Some nations are phasing out the use of incandescent lightbulbs, forcing their replacement with more efficient, longer-lasting compact fluorescent bulbs. Policies that encourage telecommuting at work also contribute strongly to emission reductions.

Rapid increases in the cost of fossil fuels have induced dramatic changes in the economics of energy use. For instance, as mass-produced food prices increase, locally grown food from smaller farms becomes competitive. When consumers purchase more local foods, the emissions from food shipments decrease.

Policies that improve transportation options for people and help them live closer to population centers have been shown to reduce carbon footprints. In the United States, individuals are moving farther away from cities in order to enjoy better living environments, and when jobs change, people may choose to remain in the same homes rather than move closer to their new workplace. This results in long-distance commuting. Moreover, more houses are built to meet the increasing demand for rural accommodation, and these houses require heating using mainly natural gas. Thus, the individual's desire for better living can have a negative effect on the environment with increased emission of CO_2. Other activities, such as the frivolous use of fire extinguishers, can have huge carbon footprints, because fire suppressants such as halon 1310 can have CO_2e's thousands of times greater than that of CO_2.

• Context

The carbon footprint has become an increasingly important concept as decisions at the personal, community, industry, state, and national levels are guided by the need to reduce GHG emissions. Changes in energy policy that reward utilities for the efficiency of their overall operation, rather than just for delivering ever-larger amounts of power, can greatly help reduce carbon footprints. Finally, shifting to increased use of "clean coal" power plant technology and to nuclear power can substantially reduce carbon footprints by rendering the power generation underlying much industrial human activity carbon neutral.

Padma Komerath

• Further Reading

Brown, M. A., F. Southworth, and A. Sarzynski. *Emission Facts: Average Carbon Dioxide Emissions Resulting from Gasoline and Diesel Fuel.* Washington, D.C.: Author, 2005. Short note showing examples of how to calculate the CO_2 emissions from motor fuels.

_______. *Shrinking the Carbon Footprint of Metropolitan America.* Washington, D.C.: Brookings Institution, 2008. This report gives data on the carbon footprints in U.S. metropolitan areas, with breakdowns by type of activity. It discusses policy issues in reducing carbon footprints and argues for improved coordination between different federal agencies.

Environmental Protection Agency. U.S. Office of Transportation and Air Quality. *Greenhouse Gas Emissions from the U.S. Transportation Sector, 1990-2003.* Washington, D.C.: Author, 2006. This sixty-eight-page report documents the emissions from different modes of transportation, including road, air, rail, and water travel. Appendix B gives data on different components of transportation.

See also: Anthropogenic climate change; Carbon dioxide; Carbon dioxide equivalent; Carbon equivalent; Carbon taxes; Energy efficiency; Greenhouse effect; Greenhouse Gas Protocol; Greenhouse gases; Human behavior change; Transportation.

Carbon 4 plants

• **Category:** Plants and vegetation

• Definition

During photosynthesis, plants absorb carbon dioxide (CO_2) from the atmosphere and convert, or fix, it into a variety of organic molecules that can be used to provide energy to cells or to provide the raw materials to build more cells and tissues. Carbon 4 fixation evolved over the past 30 million years, but it spread widely in tropical grasslands and savannas only about 5 million years ago as a result of its ability to concentrate CO_2 in photosynthetic plant tissues in a cooling climate without undergoing photorespiration. Carbon 4 plants have come to account for about 30 percent of worldwide photosynthesis, and they include important crops such as maize, sugarcane, millet, and sorghum. The carbon 4 pathway is particularly important in hot, dry climates, because this pathway is much more efficient than the carbon 3 pathway at fixing low levels of CO_2 without competition from oxygen.

Carbon 4 plants often have a tight bundle sheath of carbon 3 cells around the vascular bundles of leaves. This physically separates the cells where "normal" carbon 3 photosynthesis occurs from the leaf mesophyll cells where atmospheric CO_2 is fixed. In carbon 4 mesophyll cells, the enzyme Phosphoenolpyruvate carboxylase (PEP carboxylase) binds CO_2 to a three-carbon PEP molecule to form a four-carbon organic acid, malic acid. The malate diffuses from the mesophyll cells where it is produced into the bundle sheath cells where CO_2 is released to be used in the carbon 3 pathway.

• Significance for Climate Change

Carbon 4 evolved and spread rapidly in response to low levels of atmospheric CO_2 during past geological periods, because PEP carboxylase is much more efficient than is ribulose-1,5-bisphosphate carboxylase/oxygenase (RuBisCO) in binding CO_2, and it is not affected by oxygen levels. With rising CO_2 levels, it might seem that the advantage of carbon 4 over carbon 3 photosynthesis would decrease. However, hot, dry conditions strongly favor the carbon 4 pathway. Atmospheric gases must diffuse through microscopic openings in the surface of leaves and green stems called stomata. Warming causes plants to lose water vapor through their stomata, which leads to wilting and even death. Plants respond to water loss by closing their stomata, thus reducing gas exchange with the atmosphere. If the stomata remain closed for very long, CO_2 soon becomes depleted inside the plant tissues, and this favors carbon 4. Hot temperatures and dry conditions have a much greater influence on photosynthesis than do ambient CO_2 levels. As a result, with continued global warming, plants with carbon 4 metabolism will be increasingly important in agriculture and carbon 4 plants will become even more widespread in the environment.

Marshall D. Sundberg

See also: Carbon cycle; Carbon dioxide; Carbon 3 plants; Photosynthesis.

Carbon isotopes

- **Category:** Chemistry and geochemistry

- **Definition**

An atom consists of a small nucleus surrounded by a cloud of electrons. The nucleus is made of protons and neutrons, collectively called nucleons. All atoms of a specific chemical element have the same atomic number (number of protons or electrons) but differing numbers of neutrons. The total number of protons and neutrons in an atom is called its nucleon number. Forms of chemically identical atoms with differing nucleon numbers are called isotopes.

Carbon atoms have an atomic number of 6, but they may have a nucleon number of 12, 13, or 14. Carbon 12 (C^{12}) and carbon 13 (C^{13}) are stable isotopes; carbon 14 (C^{14}) is radioactive, decaying with a half-life of 5,730 years. Earth's supply of C^{14} is continuously replenished by cosmic-ray bombardment of individual nitrogen atoms, which can replace a proton of nitrogen 14 with a neutron, reducing its atomic number by one and changing the chemical identity of the atom from nitrogen to carbon. This process takes place only in the atmosphere, so any carbon that has been isolated from the atmosphere for more than sixty thousand years will be effectively free of C^{14}.

- **Significance for Climate Change**

Carbon dioxide (CO_2) is a greenhouse gas (GHG), contributing to global warming by slowing the escape into space of infrared radiation from the surface of the Earth. A change in the amount of CO_2 in the atmosphere is the net result of a disequilibrium between the processes that add atmospheric CO_2 (such as respiration and combustion) and the processes that subtract atmospheric CO_2 (such as photosynthesis and oceanic absorption). Estimates of the amount of carbon entering the atmosphere are substantially greater than the estimates of the amount removed. The difference is greater than the observed increase in atmospheric CO_2, however, implying that there is an unidentified carbon reservoir absorbing the remainder.

Carbon isotope ratios help constrain the type and location of processes collecting carbon. All atoms of carbon are chemically identical, but isotopes differ in mass and therefore in physical properties. In particular, atoms of C^{12} move faster at a given temperature than do atoms of C^{13} and more readily participate in chemical reactions. The ratio of the isotopes within a given sample of carbon is an important indicator of the chemical and physical history of that sample. Photosynthesis preferentially takes up C^{12}, so biomass is C^{13}-deficient compared to atmospheric CO_2. Respiration, combustion, and oceanic absorption of CO_2, however, show little discrimination among carbon isotopes.

Fossil fuels are of biological origin and are therefore deficient in C^{13}; because of their age, they are also completely free of C^{14}. Thus, the combustion of coal and oil emits a disproportionate amount of C^{12} into the atmosphere, thereby increasing the percentage of that isotope and reducing the percentage of C^{13} and C^{14} in the atmosphere. Analysis of the relative proportions of carbon isotopes can therefore provide a valuable clue as to the contribution of fossil fuels to the increase of total atmospheric CO_2.

Billy R. Smith, Jr.

See also: Carbon cycle; Carbon dioxide; Coal; Fossil fuels; Greenhouse gases; Ocean-atmosphere coupling; Photosynthesis.

Carbon monoxide

- **Category:** Chemistry and geochemistry

- **Definition**

Carbon monoxide (CO) is a colorless, odorless, tasteless, and highly toxic gas composed of an atom of carbon and an atom of oxygen chemically bonded together. It is useful in the production of a wide variety of chemicals in various industries, including the automotive, construction, agrochemical, cosmetics, pharmaceutical, plastics, and textile industries. Environmental CO is primarily produced from the incomplete combustion of carbon-containing materials.

Anthropogenic (human) sources of CO include incomplete combustion of fossil fuels in internal combustion engines, from which it is released in automobile exhaust; industrial plant exhaust, including exhaust from industry oxidation of hydrocarbons; cigarette smoke; burning of biomass; and various fuel-burning household appliances, including wood-burning stoves, water heaters, clothes dryers, furnaces, fireplaces, generators, and space heaters. CO released in automobile exhaust accounts for about 60 percent of all U.S. CO emissions. Moreover, such automobile exhaust can represent up to about 95 percent of all CO emissions in U.S. cities. Natural sources of CO include coal mines, forest fires, volcanoes, vegetation, soil (including water-saturated areas such as wetlands), the ocean, and atmospheric oxidation of hydrocarbons.

In the United States, CO is considered to be the leading cause of death from poisoning. CO is toxic in that it interferes with delivery of oxygen in the body. Normally, oxygen binds to a blood protein called hemoglobin, which then transports the oxygen throughout the body. CO has a higher affinity for hemoglobin than does oxygen. Therefore, when CO is inhaled, it binds to hemoglobin, displacing oxygen or preventing it from binding and thereby preventing the hemoglobin from delivering the oxygen to the cells that need it.

- **Significance for Climate Change**

CO affects global warming through its ability, either directly or indirectly, to increase the levels of other gases in the atmosphere. Such other gases, including carbon dioxide (CO_2), methane (CH_4),

The Conch Cement Company's Kiln Number 3, site of a carbon monoxide poisoning incident that left one person dead and twelve injured in January, 2009. (Li Jian/Xinhua/Landov)

and ozone (O_3), directly affect global warming. After the Earth is heated by the Sun, some terrestrial heat normally leaves the planet, escaping into outer space. This process allows the Earth to maintain a constant temperature, rather than growing steadily warmer as more solar energy impinges upon it. CO_2, methane, and ozone have the ability to trap the terrestrial heat attempting to leave the Earth, thereby preventing this heat from escaping into outer space. Although such gases, known as greenhouse gases (GHGs), play a role in maintaining a stable and moderate temperature on Earth, an excess of GHGs results in a terrestrial buildup of heat and an elevation of the Earth's temperature. The atmospheric concentrations of many GHGs have significantly increased since the advent of industrialization, around 1750.

Although CO is not a GHG, it acts directly or indirectly to increase the levels of Earth's GHGs—including methane, tropospheric ozone (the troposphere is the lowest portion of the Earth's atmosphere), and CO_2—by participating in various chemical reactions in the atmosphere. For example, CO indirectly affects the levels of methane and tropospheric ozone by reacting with the hydroxyl radical. The hydroxyl radical is a reactive molecule, consisting of an oxygen atom chemically bonded to a hydrogen atom, that is responsible for decreasing the levels of many atmospheric pollutants, including methane and tropospheric ozone. When atmospheric carbon monoxide reacts with the hydroxyl radical, it decreases the amount of this radical available to react with and remove methane and tropospheric ozone. As a result, methane and ozone build up in the atmosphere.

CO also indirectly affects the levels of tropospheric ozone by its involvement in reactions producing substances that can generate tropospheric ozone. For example, when CO reacts with the hydroxyl radical, one of the products formed in a series of reactions is the hydroperoxyl radical (a reactive molecule consisting of an atom of hydrogen and two atoms of oxygen). The hydroperoxyl radical can participate in reactions that form tropospheric ozone. Finally, CO directly affects the levels of atmospheric CO_2 through its reaction with the hydroxyl radical. When CO reacts with the hydroxyl radical in the atmosphere, it forms CO_2. CO_2 is one of the most potent GHGs.

Jason J. Schwartz

• Further Reading

Houghton, John. *Global Warming: The Complete Briefing.* 4th ed. New York: Cambridge University Press, 2009. Discusses the science of global warming, the effects of climate change on society, and ways these effects can be mitigated.

Intergovernmental Panel on Climate Change. *Climate Change, 2007—Synthesis Report: Contribution of Working Groups I, II, and III to the Fourth Assessment Report of the Intergovernmental Panel on Climate Change.* Edited by the Core Writing Team, Rajendra K. Pachauri, and Andy Reisinger. Geneva, Switzerland: Author, 2008. A multi-article scientific assessment of information related to climate change, including discussion of the role of greenhouse and other gases, including carbon monoxide, and an evaluation of the environmental and economic consequences of climate change.

Kroschwitz, Jacqueline I., and Arza Seidel, eds. *Kirk-Othmer Encyclopedia of Chemical Technology.* 5th ed. Hoboken, N.J.: Wiley-Interscience, 2004-2007. Provides detailed information on chemical compounds (including CO), including their physical and chemical properties, the reactions they participate in, their uses, and related information.

Van Ham, J., et al., eds. *Non-CO_2 Greenhouse Gases: Scientific Understanding, Control, and Implementation.* Amsterdam, the Netherlands: Kluwer Academic, 2000. Discusses the sources of non-CO_2 GHGs, including chemical reactions occurring in the troposphere and methods to control emission of such gases.

See also: Carbon dioxide; Greenhouse effect; Greenhouse gases; Methane; Oxygen, atmospheric; Ozone.

Carbon taxes

- **Category:** Laws, treaties, and protocols

- **Definition**

In the narrowest sense, carbon taxes are governmentally mandated fees levied on entities engaged in activities that cause carbon-containing greenhouse gases (GHGs), such as carbon dioxide (CO_2) and methane, to be emitted into the atmosphere. Such activities would include the combustion of fossil fuels, changes in land use that increase CO_2 emissions from soils, livestock propagation, solid-waste burial, and the production and disposal of goods that release carbon-containing GHGs to the environment when they decay. More broadly, as the phrase is used in political discourse, carbon taxes are fees levied on activities that lead to the emission of any of the major anthropogenic GHGs—CO_2, methane, nitrous oxide, sulfur hexafluoride, hydrofluorcarbons (HFCs), and perfluorocarbons (PFCs)—whether the emitted gases contain carbon or not.

The amount levied by a carbon tax is determined based upon how many units of carbon dioxide equivalent (CO_2e) are released into the atmosphere by a given activity. The CO_2e of a GHG is calculated using the gas's global warming potential (GWP), which is based on the potency and longevity of the gas. In theory, carbon taxes could also be levied on people for inactions, such as failing to suppress a fire in a tree farm that functions as a carbon reservoir or failing to recycle carbon-containing materials.

Carbon taxes may be thought of as a type of consumption tax, as they are based on the consumption or reduction of a finite good (the atmosphere's ability to tolerate GHG emissions). In a pollution-control context, carbon taxes represent taxes on the environmental damages inflicted by GHG emissions upon current and future generations by the production and sale of commodities. The difference between the market price of a good or activity and the overall cost that the good or activity inflicts on society is called the "externality cost" of the activity. If such externality costs are not "internalized" into the price of a good or activity, there is said to be a market failure, which many economists believe should be remedied by government action. The resulting taxes are often called Pigovian taxes in recognition of economist Arthur Pigou, who first developed the concept of taxing activities to compensate society for damages not priced into the activity in the free market.

- **Significance for Climate Change**

Carbon taxes have been proposed as a means to control the emission of GHGs. Economic theory predicts that increasing the cost of emitting GHGs will lead emitters to curtail their activities. The size of the resulting reduction would depend upon the elasticity of demand, or the relative value that people place on activities that emit GHGs compared to activities that emit less gases or none. Carbon taxes would have a broad variety of effects in addition to reducing GHG emissions.

By raising the price of goods and services throughout the economy, carbon taxes would reduce economic efficency. This would create what economists call "deadweight loss," a general consequence of all taxation. The result of accumulating deadweight losses is a reduction of wealth, productivity, and employment across the economy.

Like other taxes, carbon taxes redistribute wealth. The tax collected from GHG emitters is paid to the government, and the cost is passed on to consumers. The government can use collected revenues in many ways. For example, the government could use carbon tax revenues to administer government programs, compensate government employees, compensate contractors, or subsidize other forms of economic activity (such as the production of non-fossil-fuel-based energy). Alternatively, the government could rebate some or all of the tax to taxpayers, with or without regard for their income levels or for how much tax they have paid. Such revenue rebates are called "revenue recycling" in carbon tax discussions. Depending on the nature of the redistribution, carbon taxes could be proportional, progressive, or regressive. That is, they could place an economic burden on people in proportion to their incomes (proportional); increase the share of rebated revenue to poorer people (progressive); or impose a higher burden on poorer people than on wealthier people (regressive).

Carbon taxes are considered a price-based control measure, as their effectiveness is based on increasing the price of activities that lead to GHG emissions. Other forms of GHG control include quantity-based controls, such as carbon emission trading, and regulatory controls, such as vehicle emission standards. Among economists, both carbon taxes and carbon emission trading are considered to be more efficient than regulatory controls. Recent economic research suggests what then the ultimate cost of an externality is uncertain, as is the case for climate change, and when the ultimate cost of reducing GHG emissions is uncertain, carbon taxes are superior to both emission trading and regulatory approaches.

Kenneth P. Green

• Further Reading

Congressional Budget Office. *Policy Options for Reducing CO_2 Emissions.* Washington, D.C.: Author, 2008. This report summarizes the economic and environmental impacts of alternative carbon control regimes.

Green, Kenneth P., Steven F. Hayward, and Kevin A. Hassett. *Climate Change: Caps vs. Taxes.* Washington, D.C.: American Enterprise Institute, 2007. A short policy article that outlines the limitations of emissions trading and the potential benefits of carbon taxes.

Kaplow, Louis, and Steven Shavell. "On the Superiority of Corrective Taxes to Quantity Regulation." *American Law and Economics Review* 4, no. 1 (2002): 1-17. Makes the case for the superiority of carbon taxes in situations where there is uncertainty about control costs.

Nordhaus, William D. *Life After Kyoto: Alternative Approaches to Global Warming Policies.* Washington, D.C.: National Bureau of Economic Research, 2005. Nordhaus's economic modeling of the impacts of carbon taxes are prominent in academic discussions of carbon taxation.

Repetto, Robert, et al. *Green Fees: How a Tax Shift Can Work for the Environment and the Economy.* Washington, D.C.: World Resources Institute, 1992. An early publication that lays out the case for replacing taxes on goods (productivity) with taxes on bads (carbon dioxide emissions).

See also: Air quality standards and measurement; Baseline emissions; Certified emissions reduction; Ecotaxation; Polluter pays principle.

Carbon 3 plants

• **Category:** Plants and vegetation

• Definition

During photosynthesis, plants absorb carbon dioxide (CO_2) from the atmosphere and convert it into a variety of organic molecules that can be used to provide energy to cells or to provide the raw materials to build more cells and tissues. The process of converting CO_2 to organic molecules is called carbon fixation. The oldest and most common carbon fixation pathway is the carbon 3 pathway. Most trees and agricultural crops, including rice, wheat, soybeans, potatoes, and vegetables, use carbon 3 metabolism. In the carbon 3 pathway, the first organic molecule formed from CO_2, phosphogylcerate (PGA), contains three carbon atoms. Essential to this process is a five-carbon molecule called ribulose bisphosphate (RuBP) that has an affinity for CO_2. The enzyme ribulose-1,5-bisphosphate carboxylase/oxygenase (RuBisCO) binds water and a single CO_2 molecule to RuBP and immediately splits the six-carbon intermediate molecule into two PGA molecules, one of which includes the original CO_2 molecule from the atmosphere.

• Significance for Climate Change

Although RuBisCO has an affinity for CO_2, it also binds molecular oxygen (O_2), depending on the relative ratio of the concentrations of CO_2 to O_2. When CO_2 levels are low, O_2 begins to out-compete CO_2 for the binding site on RuBisCO, and the rate of photosynthetic carbon fixation decreases. In fact, as more O_2 is taken up by RuBisCO, a photosynthesizing plant begins to produce CO_2 through a process called photorespiration. This competition between CO_2 and O_2 is temperature dependent. At contemporary levels of atmospheric CO_2, about 380 parts per million, temperatures above

25° Celsius are unfavorable for carbon 3 plants, because CO_2 levels are suboptimal.

Some greenhouse growers speed up production of many crops by adding supplemental CO_2. This suggests that the rising levels of CO_2 associated with global warming should stimulate carbon 3 photosynthesis in most plants, which should help reduce CO_2 levels. However, it may not be so simple. For atmospheric gases to get into plants, they must diffuse through microscopic openings in the surfaces of leaves and green stems called stomata. Warming causes plants to lose water vapor through their stomata, a process called transpiration, which leads to wilting and even death. Plants respond to water loss by closing their stomata, thus reducing gas exchange with the atmosphere. If the stomata remain closed for very long, CO_2 contained in plant tissues' intercellular spaces soon is depleted enough that photorespiration begins to outpace photosynthesis.

Marshall D. Sundberg

See also: Agriculture and agricultural land; Carbon cycle; Carbon dioxide; Carbon dioxide fertilization; Carbon 4 plants; Photosynthesis.

Carbonaceous aerosols

• **Category:** Chemistry and geochemistry

• **Definition**

Aerosols are fine solid or liquid particles suspended in a gas. Carbonaceous aerosols are fine, solid carbon particles suspended in the atmosphere. They result from burning fossil fuels, which are not completely consumed in the combustion process. Sometimes, these aerosols are referred to as "soot." They can affect the global climate, as well as causing problems for people who breathe the air: They are associated with allergies and respiratory diseases, as they interfere with breathing by clogging the air sacs in a person's lungs. These aerosols are also a major cause of pollution-related mortality.

• **Significance for Climate Change**

Carbonaceous aerosols are made up of two parts: Organic carbon (OC), which scatters light, and black carbon (BC), which absorbs light. These particles can block radiation from the Sun and scatter light, so they can affect Earth's climate in several ways. They can scatter and absorb radiation from the Sun. They can reflect light back into space, increasing Earth's albedo directly, and they can also make clouds more reflective, increasing it indirectly. OC in particular is able to do this, offsetting the warming that greenhouse gases (GHGs) cause. Carbonaceous aerosols, particularly BC, can also heat the atmosphere by absorbing sunlight.

Carbonaceous aerosols can block light from reaching the Earth's surface. BC also does this, which can lead to cooling the Earth's surface. They can affect the amounts of trace gases in the atmosphere, which may affect warming or cooling of the atmosphere depending on which type of gas is affected. They can combine with each other and other particles to interact in different ways that lead to unusual, and sometimes perplexing, effects on the global climate.

Thus, carbonaceous aerosols can affect both the warming and cooling of the Earth and its atmosphere. Scientists are still trying to understand the complexities of how these aerosols affect global climate change, though some estimate that black carbon particles may be responsible for 15 to 30 percent of global warming.

Tihomir Novakov and his research group at Lawrence Berkeley National Laboratory have been leaders in performing significant research on carbonaceous aerosals since the 1970's, and it is mostly due to this work that these particles are now accepted as being common in the atmosphere. (Previously, scientists thought that fossil fuels burned completely and left no fine solid particles in the atmosphere.)

The effect that carbonaceous aerosols have on the environment can be influenced by the number of these particles contained in the total volume of air, the size of the particles, and the proportion of organic versus black carbon composing each particle. Studying aerosols can be difficult. Carbonaceous aerosols do not last long and do not mix in the same way in all areas across the Earth, which

A snowmobile trail cuts through the volcanic ash covering the snow in Skwentna, Alaska, following the eruption of Mount Redoubt in March, 2009. (Bonnie Dee Childs/MCT/Landov)

makes analyzing their effects difficult. The way these particles interact with water, particularly salt water, and over areas covered with snow and ice is not well understood and is a subject of scientific inquiry. Previous major studies, such as the Asian-Pacific Regional Aerosol Characterization Experiment (ACE-Asia) in 2001 and the Indian Ocean Experiment (INDOEX) in 1991, relied on large teams of scientists using aircraft, balloons, ships, and surface stations to help analyze these effects.

Carbonaceous aerosols can affect the hydrologic cycle by cutting down the amount of sunlight that is able to reach the ocean, affecting how quickly seawater evaporates into the air. They may also affect how clouds are formed. Both these actions can reduce the amount and frequency of rainfall. Carbonaceous aerosols can also affect plants by coating their leaves and affecting their ability to photosynthesize or use light to break down chemical compounds, thus also contributing to global climate change.

Marianne M. Madsen

• Further Reading

Gelencsér, András. *Carbonaceous Aerosol.* Dordrecht, the Netherlands: Springer, 2004. Contains thirty-one pages of references, a helpful index, and a list of abbreviations referring to carbonaceous aerosols.

Jacobson, Mark. "Strong Radiative Heating Due to the Mixing State of Black Carbon in Atmospheric Aerosols." *Nature* 409 (February 15, 2001): 695-697. Discusses studies of black carbon in aerosols and its variations in mixing in the atmosphere, describing external and internal mixing.

Novakov, T. "The Role of Soot and Primary Oxidants in Atmospheric Chemistry." *The Science of the Total Environment* 36 (1984): 1-10. Explains how soot in the atmosphere affects atmospheric chemistry and describes a methodology for estimating primary and secondary carbon particulate concentrations.

Ramanathan, V., and G. Carmichael. "Global and Regional Climate Changes Due to Black Carbon." *Nature Geoscience* 1 (2008): 221-227. Describes how black carbon mixes with other aerosols to form atmospheric brown clouds, especially over snow and ice surfaces, to affect the hydrologic cycle.

Rosen, H., T. Novakov, and B. Bodhaine. "Soot in the Arctic." *Atmospheric Environment* 15 (1981): 1371-1374. One of Novakov's early publications describing soot in Arctic ice. Describes how drilling into the Arctic ice and extracting cores led to the scientific acceptance of carbonaceous aerosols in the atmosphere and how their concentration has increased over time.

Wolff, G. T., and R. L. Klimisch, eds. *Particulate Carbon-Atmospheric Life Cycle.* New York: Plenum, 1982. This classic book on the carbon cycle includes Novakov's article "Soot in the Atmosphere."

See also: Aerosols; Air pollution and pollutants: anthropogenic; Air pollution and pollutants: natural; Albedo feedback; Carbon equivalent; Clouds and cloud feedback; Hydrologic cycle; Sulfate aerosols.

Carson, Rachel

American naturalist, ecologist, and writer

Born: May 27, 1907; Springdale, Pennsylvania
Died: April 14, 1964; Silver Spring, Maryland

Although at the time of her writing Rachel Carson did not attribute global warming to human causes, she was well aware of climate change and held the view that humans were the one part of nature distinguished largely by their power to alter it, perhaps irreversibly.

• Life

Rachel Carson, the youngest of three offspring of Robert Warden and Maria McLean Carson, was born near Springdale, Pennsylvania, and was brought up in a simple farmhouse there and in nearby Parnassus, where she graduated from high school. She graduated with honors in 1929 from the Pennsylvania College for Women (later Chatham College); a scholarship enabled her to earn an M.A. degree in zoology from Johns Hopkins University in 1932. As a child, she was encouraged by her mother to appreciate nature, and, although her goal was to become a writer, she changed her college major from English to biology after taking a required biology course. Increasingly, she developed a passion for the world of nature, especially for the ocean.

After teaching at Johns Hopkins and the University of Maryland, Carson did postgraduate work at the Marine Biological Laboratory in Woods Hole, Cape Cod, Massachusetts, then wrote a radio show, "Romance Under the Waters," for the Bureau of Fisheries in Washington, D.C. In 1936, after becoming the first woman to take and pass the civil service test, she became a junior aquatic biologist and moved up the ranks to become the chief editor of publications for the U.S. Fish and Wildlife Services. Carson began writing about her observations of life under the sea, and she published *Under the Sea-Wind: A Naturalist's Picture of Ocean Life,* the first book of a trilogy, in 1941. It was followed by *The Sea Around Us* (1951) and *The Edge of the Sea* (1955), a *New York Times* best seller for twenty weeks.

Carson's best-known book, *Silent Spring,* appeared in 1962. Documenting the dangers of pesticides and herbicides and thereby bringing on a storm of assault from the agricultural chemical industry, the book was ultimately instrumental in action being taken to regulate these substances at the national level. Carson has been called the mother of the modern environmental movement. *The Sense of Wonder,* encouraging the young to appreciate nature, was published posthumously in 1965. After an extended battle with cancer, Carson died on April 14, 1964, at her home in Silver Spring, Maryland.

• Climate Work

When *The Sea Around Us* was published, it brought Carson international acclaim and awards, including the Gold Medal of the New York Zoological Society, the John Burroughs Medal, the Gold Medal of the Geographical Society of Philadelphia, and the National Book Award. The book remained on the *New York Times* best-seller list for eighty-six weeks, thirty-nine at the number one position.

"A Fable for Tomorrow"

In 1962, Rachel Carson's Silent Spring *painted a bleak picture of a not-too-distant future in which unrestricted use of chemical pesticides had upset the balance of nature, resulting in a sterile, dead landscape:*

There was once a town in the heart of America where all life seemed to live in harmony with its surroundings. The town lay in the midst of a checkerboard of prosperous farms, with fields of grain and hillsides of orchards.

Along the roads laurel, viburnum and alder, great ferns and wildflowers delighted the traveller's eye through much of the year. Even in winter the roadsides were places of beauty, where countless birds came to feed on the berries and on the seed heads of the dried weeds rising above the snow. The streams flowed clear and cold out of the hills and contained shady pools where trout lay.

Then a strange blight crept over the area and everything began to change. . . . Everywhere was a shadow of death. . . .

There was a strange stillness. The birds, for example—where had they gone? Many people spoke of them, puzzled and disturbed. The feeding stations in the backyards were deserted. The few birds seen anywhere were moribund; they trembled violently and could not fly. It was a spring without voices. . . . The apple trees were coming into bloom but no bees droned among the blossoms, so there was no pollination and there would be no fruit. . . . Even the streams were now lifeless. Anglers no longer visited them, for all the fish had died.

In the gutters under the eaves and between the shingles of the roofs, a white granular powder still showed a few patches: some weeks before it had fallen like snow upon the roofs and the lawns, the fields and the streams.

No witchcraft, no enemy action had silenced the rebirth of new life in this stricken world. The people had done it themselves.

Source: Excerpted from Rachel Carson, "A Fable for Tomorrow," chapter 1 in *Silent Spring* (Boston: Houghton Mifflin, 1962).

Selling over 200,000 copies, the book and its reception enabled Carson to leave the Fish and Wildlife Service and devote herself full time to writing. Its popularity was due in part to Carson's ability to condense huge amounts of scientific information into terms that nonspecialists could understand.

Although Carson did not attribute global warming to human causes, she was well aware of a trend toward a warmer Earth. *The Sea Around Us* contains three sections. The first, "Mother Sea," focuses on the origin of Earth and the sea and analyzes the nature of the sea from its surface to the ocean floor. It explains the intricacy of the food chain in the ocean, the seasonal changes in surface waters, and the development and death of islands. Part 2, "The Restless Sea," explains the forces that affect the ocean: the wind, the Sun, the Moon, Earth's rotation, waves, tides, and currents, as well as the gravitational pull that affects every drop of water in the ocean.

The book's final section, "Man and the Sea About Him," is of particular importance. A chapter entitled "The Global Thermostat" explains how the ocean dominates the world's climate, poses the question of whether the ocean is an agent in bringing about the lengthy climate swings that have occurred throughout Earth's history, and cites a theory of Swedish oceanographer Otto Pettersson that supports an affirmative answer: The theory predicts that a tidal effect that prevailed about 500 C.E. will occur again about the year 2400, bringing global warming. Carson says that it is established beyond question that a change in arctic climate set in around 1900, became marked about 1930, and is still spreading. Acknowledging that there are undoubtedly other agents at work in bringing about climate changes in the Arctic regions, she concludes that, if the Pettersson theory is valid, there is no doubt that the pendulum is swinging toward a warmer Earth.

Victoria Price

• Further Reading

Brooks, Paul. *The House of Life: Rachel Carson at Work.* Boston: Houghton Mifflin, 1972. Provides selections of Carson's writing and reminiscences of Brooks and other friends of Carson to show how she achieved what she did.

Carson, Rachel. *The Sea Around Us.* New York: Oxford University Press, 1961. A special edition in which Carson's description of the ocean as part of the Earth's ecological system, which makes life on Earth possible, is supplemented by a chapter by Jeffrey Levinton, a leading marine ecologist, who updates the scientific side of her book.

Freeman, Martha, ed. *Always Rachel: The Letters of Rachel Carson and Dorothy Freeman, 1952-1964.* Boston: Beacon Press, 1995. The exchange of letters between Carson and her summer house neighbor provides a sense of Carson's philosophy comparable to that of a biography.

Lear, Linda. *Rachel Carson: Witness for Nature.* New York: Henry Holt, 1997. Considered the definitive biography of Rachel Carson. Herself an environmental historian, Lear brings to the account of Carson's life and work insightfulness and accuracy.

See also: Environmental economics; Environmental movement; Pesticides and pest management.

Catalytic converters

• **Categories:** Economics, industries, and products; transportation

• Definition

A catalytic converter is a device to reduce the amount of toxic emissions in the exhaust of an internal combustion engine. It was developed in response to efforts in the early 1970's by the Environmental Protection Agency (EPA) to tighten regulations on emissions from internal combustion engines. The 1975 model cars were produced with catalytic converters. The first converter, a two-way converter, oxidized carbon monoxide (CO) to carbon dioxide (CO_2) and oxidized unburnt hydrocarbons to CO_2 and water. The hydrocarbons were molecules of gasoline that escaped a car's pistons without being burned. Beginning in 1981, catalytic converters also had a third component that reduced nitrogen oxides to nitrogen and oxygen.

Catalytic converters are composed of three parts. There is a core or substrate honeycomb constructed of either ceramic material or stainless steel. The honeycomb design provides maximum surface area. A washout of silica and alumina is added to the substrate to increase the surface area. The catalyst is a precious metal attached to the washout. Platinum or rhodium is usually used as the reducing catalyst, and platinum or palladium is usually used as the oxidizing catalyst.

Exhaust gas reaches the reducing catalyst first. Nitrogen oxides attach to the catalyst, and the reaction to separate nitrogen from oxygen is aided by the nitrogen being bound. After the oxygen is removed, nitrogen atoms can combine to form diatomic nitrogen gas. Carbon from the CO in the exhaust is held in the oxidizing area, where oxygen reacts with it to form CO_2. Similarly, hydrocarbons are held to be oxidized to CO_2 and water.

• Significance for Climate Change

Catalytic converters drastically reduce the amount of nitrogen oxides, hydrocarbons, and carbon monoxide emitted into the atmosphere. Nitrogen oxides and hydrocarbons are two of the main reactants, along with sunlight, producing photochemical smog. The catalytic converter has been a major factor, along with reformed gasoline, in reducing smog in many places. For example, in 1975, the air

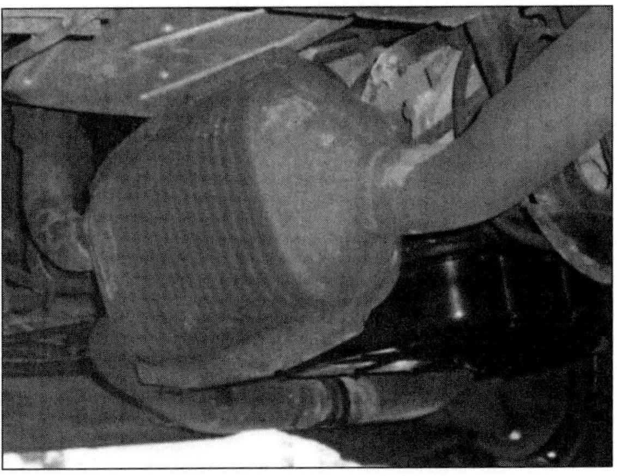

A catalytic converter on a 1996 Dodge Ram.

in Los Angeles exceeded the ozone standard 192 out of 365 days. By 2005, Los Angeles air exceeded ozone standards on only 27 days of the year. Los Angeles's air quality improved consistently over the thirty years between 1975 and 2005. Smog in the city was cut by two-thirds. In 1977, the city experienced 121 stage 1 smog alerts, the most severe designation denoting a day when air quality is particularly unhealthy. In 1980, there were 79 stage 1 smog alerts in Los Angeles, and there were only 7 such alerts in 1996.

The deployment of catalytic converters is not the only factor in achieving this significant improvement in air quality. It is, however, a major factor. By reducing the smog-related emissions from motor vehicles, the catalytic converter prevented a great deal of pollutants from being released into the atmosphere. Contemporary automobiles emit 90 percent less CO, hydrocarbons, and nitrogen oxides than did 1970 models.

The effects of catalytic converters have not all been positive, however. The devices oxidize CO and hydrocarbons, converting them to CO_2. CO_2 is a greenhouse gas (GHG) that may contribute to global warming. The transportation industry is one of the major sources of CO_2. In addition to transforming other pollutants into CO_2, catalytic converters often reduce fuel efficiency, increasing a vehicle's output of CO_2 per kilometer. Moreover, nitrogen oxides are not always reduced completely to nitrogen gas by catalytic converters. Instead, they may be reduced to nitrous oxide (N_2O), a GHG whose global warming potential is three hundred times that of CO_2. As the number of vehicles with catalytic converters has increased, the N_2O in the atmosphere has also increased. It now constitutes over 7 percent of the GHGs in the atmosphere.

Catalytic converters function well only after they are warmed up. Thus, they do not effectively reduce the emissions of the first gasoline burned after starting a motor vehicle. Catalytic converters can deteriorate if exposed to intense heat. Lead and some other elements will contaminate the catalyst. For this reason, it is illegal to use leaded gasoline in a vehicle with a catalytic converter. Contaminated and deteriorated converters may increase, rather than decrease, the pollution emitted from a car. Catalytic converters have helped clean the air of smog, but they have come to represent a new problem by generating GHGs.

C. Alton Hassell

• Further Reading

Bode, H. *Material Aspects in Automotive Catalytic Converters.* Weinheim, Germany: Wiley-VCH, 2002. Based on the second international conference on materials aspects in automotive catalytic converters, Munich, 2001. Illustrations, biography, index.

Kubsh, Joseph Edward. *Advanced Three-Way Catalysts.* Warrendale, Pa.: Society of Automotive Engineers, 2006. Describes the new generation of catalysts, both reducing and oxidizing. Illustrations, biography.

Society of Automotive Engineers. *Advanced Catalysts for Emission Control.* Warrendale, Pa.: Author, 2003. Covers the topics discussed at the Society of Automotive Engineers 2003 World Congress in Detroit. Illustrations, biography.

_______. *Emission: Advanced Catalysts and Substrates, Measurement and Testing, and Diesel Gaseous Emissions.* Warrendale, Pa.: Author, 2003. Covers the testing of catalytic converters, the emission of nitrogen oxides, and diesel motor exhaust gas. Illustrations, biography.

See also: Automobile technology; Greenhouse effect; Greenhouse gases; Motor vehicles.

Catastrophist-cornucopian debate

• **Category:** Environmentalism, conservation, and ecosystems

• Definition

Many people are pessimistic about Earth's future when they consider the myriad environmental problems facing today's world. In the early nineteenth century, Thomas Robert Malthus predicted a dismal future of overpopulation and mass starvation; neo-Malthusian environmentalists foresee a catastrophic future in which too many humans battle for ever-dwindling resources, leading to vice, misery, and the collapse of civilization.

On the other hand, optimists believe that technology is a cornucopia that, like the mythical horn of plenty, will provide an abundance of ingenious new cures for the world's environmental problems. This promethean environmentalism, named for the Greek titan who created humans and gave them the gift of fire, argues that past innovative technologies have repeatedly averted predicted disasters with new inventions. It asserts that this historic pattern of progress and abundance will continue indefinitely into the future.

• Significance for Climate Change

According to catastrophists, the disasters accompanying global warming are inevitable unless drastic changes in human society and behavior are implemented immediately. Even then, it may already be too late to avoid an impending doom. Although excessive dwelling on future disasters can become a self-fulfilling prophecy, if nothing is done, the feared consequences are more likely to occur. It is not unreasonable for environmentally concerned people to feel moral indignation over abundant excesses, abuses, and needless waste, but attempting to shock or shame people into altering such behaviors is often futile. Progressive, positive action is seldom motivated by fear alone. If a predicted disasters fail to materialize, the public may assume a false sense of complacency.

On the other hand, prometheans tend to emphasize historical precedents for the solution of environmental problems. Although this approach is comfortable, it can lead the public into a false sense of security by causing people to expect technology to fix all problems. Blind faith in technology then becomes an excuse for continuing behavior that exacerbates existing problems.

A balanced viewpoint also exists between these two extremes. This viewpoint recognizes that there are serious environmental problems facing the world but asserts that obstacles can be conquered when faced openly and creatively. Such a viewpoint embraces the cornucopian belief that solutions to all problems are possible, but it also embraces the catastrophist belief that the cooperation of nearly every human society and individual will be necessary to achieve those solutions. It rejects the inevitability of environmental doom, but it also rejects the belief that anonymous scientists or inventors will solve problems on humanity's behalf, a belief that excuses individuals from working to solve those problems themselves.

George R. Plitnik

See also: Ecocentrism vs. technocentrism; Environmental movement; Skeptics.

Cato Institute

- **Category:** Organizations and agencies
- **Date:** Established 1977
- **Web address:** http://www.cato.org/

• Mission

Classified as a nonprofit, tax-exempt 501(c)3 educational foundation, the Cato Institute deploys an extensive program of reports, books, press releases, and speeches. Since at least 1998, Cato has disputed the scientific evidence of climate change and its human causes, the urgency of taking action, and the efficacy of large expenditures to mitigate greenhouse gas (GHG) emissions.

The Cato Institute was cofounded by Edward H. Crane, a libertarian, and Charles Koch, the chief executive officer of Koch Enterprises, a vast, privately held company engaged in oil refining, forest production, and commodities trading, among other ventures. Cato's stated purpose is to broaden public policy debate by advocating for individual liberty, limited government, dynamic market capitalism, and peaceful relations among nations. An annual study of mainstream media citations of public policy think tanks conducted by Fairness and Accuracy In Reporting (FAIR) ranked Cato ninth in 2006, with 1,265 citations.

• Significance for Climate Change

In 2008, the Cato Institute employed 104 full-time employees, 72 adjunct scholars, and 23 fellows. Cato has released several books and reports on

What to Do About Climate Change

The following quotation, from Indur M. Goklany's 2008 policy analysis What to Do About Climate Change, *illustrates the Cato Institute's small-government philosophy.*

If future well-being is measured by per capita income adjusted for welfare losses due to climate change, the surprising conclusion using the Stern Review's own estimates is that future generations will be better off in the richest but warmest (A1FI) world. This suggests that, if protecting future well-being is the objective of public policy, governmental intervention to address climate change ought to be aimed at maximizing wealth creation, not minimizing CO_2 emissions.

global warming, including *Climate of Fear: Why We Shouldn't Worry About Global Warming* (1998), issued shortly after the Kyoto Protocol was adopted, and *The Improving State of the World: Why We're Living Longer, Healthier, More Comfortable Lives on a Cleaner Planet* (2007).

Patrick J. Michaels, Cato's senior fellow in environmental studies, is their most visible and prolific spokesperson on global warming. The institute's 2004 annual report states that Michaels's book *Meltdown* "formed the scientific basis" for Michael Crichton's 2004 best-selling mystery thriller, *State of Fear.* Michaels was a keynote speaker at the 2008 International Conference on Climate Change, giving a lecture entitled "Global Warming: Truth or Swindle" that was attended by around five hundred global warming skeptics. A Cato scholar, Jonathan Adler, wrote an amicus brief for the Supreme Court case *Massachusetts v. EPA* (2007) in late 2006, arguing on behalf of the U.S. Environmental Protection Agency that the Clean Air Acts (1963-1990) do not grant the agency the authority to regulate vehicular emissions of greenhouse gases.

Glenn Ellen Starr Stilling

See also: Cooler Heads Coalition; Libertarianism; Skeptics.

Cement and concrete

- **Category:** Chemistry and geochemistry

- **Definition**

Concrete, at its most basic, is composed of aggregates and a binding material (cement). Concrete aggregates are coarse, greater than 4.75 millimeters; fine aggregates are less than 4.75 millimeters in size. Aggregates—which are free of silt, organics, sugars, and oils—include sand, gravel, crushed stone, and iron blast-furnace slag. By volume, they make up about 75 percent of a concrete-cement mixture. The aggregates' size plays an important role in achieving maximum particle packing. Optimum packing reduces the amount of cement needed; with less cement, the durability and mechanical properties of the concrete are improved.

Compressive strength, the measured maximum resistance to axial loading, is one of the outstanding properties of cement. Tensile strength, a measure of resistance to stretching, is much lower for concrete, so it is often reinforced with steel bars to provide additional tensile strength. The durability of concrete is high, because it can be designed and manufactured for resistance to freeze-thaw cycles, seawater exposure, chemicals, and corrosion.

Cement, in the broadest sense, binds concrete elements together in the presence of water. Cement is instrumental in determining the quality of concrete. In properly manufactured concrete, every particle of aggregate must be surrounded by cement, and all voids must be filled with cement.

Early cements, known as soft lime cements, were prepared by burning slabs of limestone in a vertical kiln. After burning, the crumbly slabs were used immediately; slaked to produce a powder form that, when combined with sand and water, created soft lime mortars used for brickwork/masonry; or packed into barrels for later use.

There are three classes of hydraulic cement: Pozzolana (Pozzola or Trass), natural cement, and Portland cement. Pozzolana, a volcanic deposit, is finely ground then mixed with lime, sand, and water, creating strong cement that hardens underwater (hydraulic) and is impervious to salt water. Natural cement is produced by low-temperature burning of clay- or magnesium-rich limestone; upon completion of the burn, the limestone is crushed into smaller fragments, then pulverized, producing very strong cement.

Portland cement, patented by Joseph Aspdin in England in 1824, combines limestone and clay, then grinds them with water into fine slurry. The dried slurry is burned in a kiln and the calcined material is again ground to a fine powder. By the 1850's, the strength and setting qualities of Portland cement were improved by burning the mixture at very high temperatures—close to the fusion point within the kiln. This improvement and the ability to chemically analyze successful cement products allowed the Portland cement industry to grow. Portland cement began production in the United States between 1875 and 1890, with mills in Texas, Oregon, Michigan, New York, Maine, and the Lehigh District of Pennsylvania.

• Significance for Climate Change

The basic makeup of Portland cement is lime (CaO) from limestone and cement rock, silica (SiO_2) from clay and fly ash, alumina ($A1_2O_3$) from aluminum ore refuse, and iron oxide (Fe_2O_3) from iron ore. The proportions of these crushed elements are closely defined by industry standards. A mix or slurry of limestone and shale or clay is prepared for burning and final cooling in an inclined, rotating kiln. The dry cement mix slowly heats to 1,260° Celsius, and the carbonates (limestone and cement rock) burn and lose carbon dioxide (CO_2). Lime, alumina, and iron oxide fuse between 1,427 and 1,482° Celsius to complete the cement. Approximately 60 percent of all CO_2 emissions are from the lime-burning process; the remaining 40 percent of CO_2 emissions originate from fossil fuels used for combustion.

Originally, oil was used to heat kilns; the use of pulverized coal began in the late 1890's and has continued. Electricity, used in plant operation, is often generated by coal, and diesel or gasoline is used for quarrying raw materials. All are fossil fuels—all release CO_2 into the atmosphere. U.S. CO_2 emission data from the Energy Information Agency (EIA) for cement manufacture show that atmospheric CO_2 has risen from 33 million metric tons in 1990 to 46.1 metric tons in 2005, an increase of

Cement factories such as this one emit carbon dioxide and other pollutants into the atmosphere. (Reuters/Landov)

13.1 percent in fifteen years. Projected data beyond 2005 continue the increasing CO_2 emissions trend. It is estimated that 5 percent of all global atmospheric CO_2 is derived from cement manufacturing.

According to most climate scientists, elevated levels of CO_2 increase the Earth's temperature. CO_2 and other greenhouse gases absorb and prevent the longer wavelength heat radiation from leaving Earth's surface. This heat builds up and warms Earth's surface, atmosphere, and climate. This warmer climate may substantially melt glaciers and polar ice sheets, causing sea-level rise, increased evaporation, drought, flooding, and heat waves.

Mariana L. Rhoades

• Further Reading

"Concrete as a CO_2 Sink?" *Environmental Building News* 4, no. 5 (September/October, 1995): 5. Addresses CO_2 emission of cement manufacture, explains why the process releases CO_2, and details CO_2 sequestering during the curing process and after concrete is fully cured.

Energy Information Administration. *Emissions of Greenhouse Gases in the U.S.* Washington, D.C.: U.S. Department of Energy, 2006. Full set of industry-wide data on CO_2 emissions and fossil fuels consumed. Part of a series that is updated yearly and includes future projections.

Kosmatka, S. H., B. Kerkhoff, and W. C. Panares. *Design and Control of Concrete Mixtures.* Skokie, Ill.: Portland Cement Association, 2002. Complete manual of concrete and cement manufacture. Includes easily understood descriptions with useful graphics and photographs, a large section on concrete applications, a list of regulations, and research resources.

Long, Douglas. *Global Warming.* New York: Facts On File, 2004. Confronts global warming on all levels, including science community input (addressing skeptics) and international efforts and protocols to understand and mitigate global warming. Provides a large research resource base, including periodicals, books, and Web sites.

See also: Carbon dioxide; Greenhouse gases; Industrial ecology; Industrial emission controls; Industrial greenhouse emissions.

Center for the Study of Carbon Dioxide and Global Change

- **Category:** Organizations and agencies
- **Date:** Established 1998
- **Web address:** http://www.co2science.org/

• Mission

The Center for the Study of Carbon Dioxide and Global Change is an independent, nonprofit, science-based educational organization that provides regular reviews and commentary on new research findings about the climatic and biological significance of the continuing rise atmospheric carbon dioxide (CO_2) concentrations. The center reviews peer-reviewed scientific journal articles, original research, and other educational materials relevant to the debate over CO_2 and global change. Material is published weekly on the center's Web site.

The center's main aim is to separate research findings based on solid science and empirical data from the rhetoric in the emotionally charged debate over global climate change. Its stated commitment is to empirical evidence and real-world solutions, as opposed to speculation and information from untested hypothetical numerical models. Its position on global warming may be summarized as follows: There is little doubt that the CO_2 concentration in the atmosphere has risen significantly over the past 100 to 150 years as a result of humanity's use of fossil fuels and that the Earth has warmed slightly over the same period; however, there is no compelling reason to believe that the rise in temperature was caused primarily by the rise in CO_2. Moreover, real-world data provide no compelling evidence to suggest that the ongoing rise in the CO_2 concentration of the atmosphere will lead to harmful changes in Earth's climate. The center was founded in 1998 and is run by a father and two sons, all of whom have Ph.D's in fields directly related to the study of CO_2, climate, and global warming.

• Significance for Climate Change

Since its creation in 1998, the center has published over three thousand reviews of scientific journal articles on both the biological and climatological effects of atmospheric CO_2 enrichment. Accompanying each review is the full journal reference for the article reviewed, so users may access the articles and assess the information themselves. Reviews are archived in one or more topical categories inside a large subject index that includes more than eight hundred topics and subtopics relative to CO_2 and global change. The material is listed in detailed subject index summaries, which are continually updated as newer material is added. Web site users may use a keyword search engine for locating reviewed articles or summaries.

The center also provides Web site users with access to various air-temperature and precipitation databases, from which they may calculate and plot trends for the entire globe or for selected regions of the globe. Output includes a graphical representation of temperature or precipitation anomalies over a user-selected time interval, as well as access to these data in tabular form. A linear regression line of the temperature or precipitation trend over time is also displayed, along with its associated statistics. The site includes four global data sets and 1,221 individual station locations in the conterminous United States where users may calculate and plot climate data.

Two other important services provided by the center may be found in its plant-growth database and Medieval Warm Period Project. The plant-growth database is an ongoing project to build an archive of the results of peer-reviewed scientific studies that report the growth responses of plants to atmospheric CO_2 enrichment. Results are updated weekly and posted according to two types of growth response (dry weight and photosynthesis). The data are listed alphabetically according to plant names (both scientific and common) in individual tables. Each table begins with an abbreviated reference, followed by a brief description of the experimental growing conditions and the percentage increase in plant growth due to a 300, 600, or 900 part-per-million increase in the atmospheric CO_2 concentration. Full reference citations for each experiment are also available in linked files. The center has archived thousands of CO_2-enrichment studies.

The Medieval Warm Period Project is an ongoing project to document the magnitude and spatial and temporal extent of a significant period of warmth that occurred in Europe approximately one thousand years ago, when the atmosphere's CO_2 concentration was approximately 30 percent lower than it is currently. The purpose of this project is to show that Earth's near-surface air temperature was equally as high as, or even higher than, it is today during a period of lower CO_2 concentration. Thereby, the center reasons that current air temperatures are not unusual and need not be due to the recent rise in the CO_2 content of the atmosphere. Updates of new scientific studies documenting the climate of the Medieval Warm Period are provided weekly, and Web site users may graphically view individual study locations and attributes on an interactive map.

Students and teachers may use material found in the center's Global Change Laboratory, where instructions are given on how to conduct simple experiments that illustrate the effects of atmospheric CO_2 enrichment and depletion on vegetative growth and development. Utilizing the so-called Poor Man's Biosphere experimental technique pioneered by center president Sherwood B. Idso, plants are grown inside sealed aquariums or other containers maintained at different CO_2 concentrations, from which students can watch and plot growth progression data over the course of time. Several experiments have been developed, and these experiments can be performed by nearly anyone anywhere in the world at very little expense.

The center expanded its activities in 2008 to include the production of documentary digital video discs (DVDs) covering a wide range of topics in the global warming debate, including *Carbon Dioxide and the Climate Crisis: Reality or Illusion?*; *Carbon Dioxide and the Climate Crisis: Avoiding Plant and Animal Extinctions*; and *Carbon Dioxide and the Climate Crisis: Doing the Right Thing*. In addition, the center produces short two- to four-minute video segments on YouTube that highlight the findings of important research papers that have appeared in international science journals. Since 2001, the center has provided a professional service to assist U.S. companies in filing their greenhouse gas emission reports under the U.S. Voluntary Reporting of Greenhouse Gases Program—section 1605(b) of the Energy Policy Act of 1992.

C R de Freitas

• Further Reading

Essex, Christopher, and Ross McKitrick. *Taken by Storm: The Troubled Science, Policy, and Politics of Global Warming.* Rev. ed. Toronto: Key Porter Books, 2008. Thought-provoking skeptical analysis of aspects of key global warming issues that are widely misunderstood.

Idso, Sherwood B. *Carbon Dioxide and Global Change: Earth in Transition.* Tempe, Ariz.: IBR Press, 1989. Analyzes the climatic and biological implications of the rise in the atmosphere's CO_2 concentration. The author is the center's founder.

_______. *Carbon Dioxide—Friend or Foe? An Inquiry into the Climatic and Agricultural Consequences of the Rapidly Rising CO_2 Content of Earth's Atmosphere.* Tempe, Ariz.: IBR Press, 1982. The first substantial critique of anthropogenic global warming.

Kininmonth, William. *Climate Change: A Natural Hazard.* Brentwood, Essex, England: Multi-Science, 2004. Argues that global climate models represented by the Intergovenmental Panel on Climate Change are deficient as a basis for future planning for global warming.

Michaels, Patrick J., ed. *Shattered Consensus: The True State of Global Warming.* Lanham, Md.: Rowman & Littlefield, 2005. Collection of ten critical accounts of important global warming themes by eleven internationally known climate scientists.

See also: Scientific credentials; Scientific proof; Skeptics.

Certified emissions reduction

• **Category:** Laws, treaties, and protocols

• Definition

A certified emissions reduction (CER) is a unit of greenhouse gas (GHG) emissions reduced by sources or removed by sinks and achieved under a clean development mechanism (CDM) project. Under the CDM, developed countries can invest in or finance projects in developing countries. These projects must result in a reduction of emission or removal of GHGs from the atmosphere additional to any that would have occurred in the absence of the projects. The difference between the emissions that would have been released without the project (referred to as the baseline emissions) and what was actually released is validated and certified by an independent body and confirmation is issued by the CDM Executive Boards as a CER. Developed countries that have quantified emission limitation and reduction commitments (QELRCs) under the Kyoto Protocol can use CERs to contribute to meeting their targets. These QELRCs are set out in Annex B of the protocol, and developed countries are allowed to use CERs to achieve part, but not all, of their commitments. CERs are also known as carbon credits.

A CER represents 1 metric ton of carbon dioxide equivalent (CO_2e) GHG emissions, either removed from the atmosphere or prevented from being released into the atmosphere. The amount of CERs issued for 1 metric ton of GHG reduction or removal depends on the type of gas and its global warming potential (GWP). The GWP of a gas refers to an estimation of its contribution to climate change or the effect it has on the climate. All gases are defined relative to carbon dioxide (CO_2), whose GWP is 1, so for 1 metric ton of CO_2 reduced or removed, one CER or carbon credit can be issued. The GWP of other gases may vary depending on the time horizon over which the impact of the gas is being determined—for most gases, the impact reduces as the time horizon increases. The GWP of methane, for example, ranges from 56 for a time horizon of 20 years, to 21 for 100 years, to 6.5 for 500 years. This means that for 1 metric ton of methane reduced or removed from the atmosphere, 56, 21, or 6.5 CERs can be generated, depending on the relevant time horizon.

• Significance for Climate Change

As of October, 2008, about 227 million metric tons of CERs were being generated annually from registered CDM projects. The total CERs from all roughly 1,184 registered CDM projects are expected to amount to about 1.3 billion metric tons by the end of 2012. By the end of 2012, it is expected that the CDM will have generated approximately 3 billion metric tons of CERs. This represents 3 billion metric tons of CO_2e removed or reduced from the atmosphere. The purpose of the CDM includes assisting developing countries in their efforts to stabilize the amount of GHGs in the atmosphere at a safe level and assisting developed countries to meet their commitments to reduce their GHG emissions. The aim of the CDM therefore is to support the mitigation of climate change.

The quantity of CERs generated from CDM projects represents how successful the CDM has been in achieving this goal. In particular, the CER system is the only streamlined means for developing countries to participate in the CDM, because under the Kyoto Protocol, developing countries have no set emission-reduction targets. Since all CERs are generated from projects implemented in developing countries, they represent emission reductions achieved in developing countries, in addition to those achieved in developed countries. In addition, the CDM is expected to assist developing countries in achieving sustainable development, helping them move to a more climate-friendly development path.

CERs can also be traded on the carbon market. Some developed country entities invest in CDM projects in order to use the CERs generated to meet their reduction targets under the Kyoto Protocol. Other entities invest in projects in order to obtain CERs from the projects and sell them on the carbon market. In addition, developing-country entities also finance or invest in projects and trade the CERs generated from them in the carbon market. Trade in CERs has helped create a market for trade in carbon, thereby increasing the number of entities, especially private entities, interested in carbon credits, and potentially in efforts to reduce GHG emissions.

The issuance of CERs could also potentially have a negative impact on climate change. Developed countries with reduction targets can use CERs to meet these targets. If the emission reductions generated by a CDM project are overestimated and too many CERs issued for the project, this would result in no decrease, or even a net increase, in GHG emissions, as these CERs would count toward developed country targets, and these countries could in fact theoretically increase their emissions by this amount.

Tomi Akanle

- **Further Reading**

Ali, Paul A. U., and Kanako Yano. *Eco-Finance: The Legal Design and Regulation of Market-Based Environmental Instruments.* New York: Kluwer Law International, 2004. Chapter 3 of this book provides a legal analysis of the carbon market, carbon finance, and carbon trading.

Lecocq, Franck. *State and Trends of the Carbon Market, 2004.* Washington, D.C.: International Bank for Reconstruction and Development/World Bank, 2005. Uses case studies of carbon market transactions to examine the status and trends of the carbon market.

Yamin, Farhana, ed. *Climate Change and Carbon Markets: A Handbook of Emissions Reductions Mechanisms.* London: Earthscan, 2005. Provides a comprehensive description of the carbon market, including its rules and operation.

See also: Annex B of the Kyoto Protocol; Baseline emissions; Carbon dioxide; Carbon dioxide equivalent; Clean development mechanism; Emission scenario; Emissions standards; Kyoto mechanisms; Kyoto Protocol.

Charcoal

- **Category:** Chemistry and geochemistry

- **Definition**

Charcoal is the residue of biomass charring, or the burning of animal or vegetable substances such as wood (usually beech, birch, oak, or willow), peat, nut shells, bark, or bones into a hard material used for fuel. Chemically, it is mainly carbon (85 to 98 percent), with small amounts of oxygen and hydrogen. Microscopically, it retains some of the texture typical of plant structures.

Charcoal is formed when trees or other organic matter are burned in the absence of oxygen (usually in a kiln-type structure), preventing the plant material from turning into ash. The burned material forms charcoal, a hardened material that burns with little or no flame or smoke. Charcoal provides a greater amount of heat in proportion to its volume than the wood or other biomass material from which it is formed. This quality makes it a much sought after type of fuel, particularly in nonindustrialized countries.

• Significance for Climate Change

Charcoal and Carbon Dioxide Production. Charcoal historically has been a major source of fuel for cooking. It is still used in many countries every day in this way. When charcoal is burned for fuel, the carbon stored in it is released in the form of carbon dioxide (CO_2) back into the atmosphere. If this occurs at a greater rate than the rate at which the oceans or plants and other trees are able to convert the CO_2 back into oxygen and energy, more CO_2 is left in the atmosphere than is removed. CO_2 acts as an insulator; it keeps heat from evaporating into the atmosphere, causing the Earth to be warmer. Thus, the release of CO_2 from burning charcoal can affect global climate change if it changes the composition of the atmosphere. In addition, creating charcoal involves burning wood in a kiln very slowly for long periods of time, which also releases CO_2 into the air.

Charcoal and Deforestation. Cutting down trees and other plants to produce charcoal may also affect the global climate. As wood is removed and burned to create charcoal, fewer trees are left to grow, and deforestation occurs. Deforestation affects the global climate, and nonforested areas have a degraded environmental structure and result in less biodiversity in an area that was previously covered in forest.

In addition, denuding the land of forest materials decreases the amount of oxygen released into the atmosphere, as trees and other plants that remove CO_2 from the air during photosynthesis are removed. Deforestation also affects the amount of moisture in the atmosphere (trees take water up through their roots and release it into the atmosphere), which affects clouds and how much protection they are able to give the Earth from the Sun, as well as the amount of water that is retained in soil (trees, especially in their roots, keep water in the soil and discourage erosion, keeping an area cooler and increasing soil fertility).

Currently, Central America, South Asia, and Southeast Asia lead the world in deforestation rates. Some alternatives that may help reduce the amounts of charcoal burned for fuel in these countries include providing people with stoves that retain heat better than an open fire and creating briquettes that will fully combust out of other biomass materials. These technologies are often expensive or unavailable to the full population that would need to use them to reduce the amount of deforestation and CO_2 production often found with the use of charcoal as fuel. Ultimately, charcoal pro-

Pyrolysis, Charcoal, and Climate Change

Water-saturated charcoal can be used as a soil supplement, increasing fertility; it can be produced in pyrolyzers, which rapidly heat biomass, charring rather than burning it. This process produces three substances, each of which can be put to sustainable use:

- Bio-oil (60 percent): Can be used as an alternative fuel. 1 megagram of biomass will produce enough bio-oil to replace about 1.7 barrels of oil.
- Synthesis gas, or syngas (20 percent): Can be used to power the pyrolyzer, obviating the need for any outside energy source to drive production.
- Charcoal (20 percent): Can be buried in agricultural fields, increasing crop yields and sequestering carbon. Carbon buried as charcoal will remain sequestered for thousands of years.

duction and burning may lead to a downward spiral of producing much more CO_2 that escapes into the atmosphere, while leaving fewer trees to take CO_2 out of the atmosphere.

Marianne M. Madsen

• Further Reading

Anderson, L. E., et al. *The Dynamics of Deforestation and Economic Growth in the Brazilian Amazon.* New York: Cambridge University Press, 2003. Discusses the economics of deforestation in the Amazon Basin, including the sources and agents of deforestation and carbon emissions.

Johns, A. G., and J. Burley. *Timber Production and Biodiversity Conservation in Tropical Rain Forests.* New York: Cambridge University Press, 2004. Gives an overview of how deforestation affects the environment in rain forests, including how charcoal production affects deforestation and how charcoal burning affects air quality.

Lomborg, Bjørn. *The Skeptical Environmentalist: Measuring the Real State of the World.* New York: Cambridge University Press, 2001. Discusses traditional fuels, including charcoal and wood, and how they contribute to energy production and consumption.

Raven, P. H., and L. R. Berg. *Environment.* New York: Wiley, 2006. Includes discussions on energy, including burning charcoal for cooking fuel, and resources, including how deforestation occurs when wood is burned for charcoal.

Rosen, L. *Climate Change and Energy Policy: Proceedings of the International Conference on Global Climate Change—Its Mitigation Through Improved Production.* Edited by R. Glasser. New York: American Institute of Physics Press, 1992. These conference proceedings discuss charcoal burning in underdeveloped countries and give an overview of better production methods.

Thomas, P., and J. Packham. *Ecology of Woodlands and Forests: Description, Dynamics, and Diversity.* New York: Cambridge University Press, 2007. Provides an overview of forest structures and the dynamic changes in wooded areas. Includes an index, references, and illustrations.

See also: Carbon cycle; Carbon dioxide; Deforestation; Fossil fuels.

Chemical industry

• **Category:** Economics, industries, and products

As a major user of fossil fuels, the chemical industry contributes to global warming, and its processes and products that generate GHGs exacerbate this problem. Some chemical companies are developing environmentally friendly products and technologies that reduce or eliminate GHG emissions.

• Key concepts

fossil fuels: hydrocarbon materials formed from carboniferous organic life in geologic deposits of natural gas, petroleum, and coal

greenhouse effect: an atmospheric warming phenomenon by which certain gases act like glass in a greenhouse, allowing the transmission of ultraviolet solar radiation but trapping infrared terrestrial radiation

greenhouse gases (GHGs): gases such as carbon dioxide, chlorofluorocarbons (CFCs), methane, water vapor, and nitrous oxide that cause the greenhouse effect

ozone layer: a region of the stratosphere with high concentrations of triatomic oxygen, which absorbs much solar ultraviolet radiation that would be harmful to life if it reached Earth's surface

petrochemical industries: businesses that refine crude oil to create raw materials needed in the manufacture of most chemicals, from fertilizers and pesticides to plastics, synthetic fibers, and medicines

• Background

Before the nineteenth century, chemical industries were relatively modest in size and their releases of carbon dioxide (CO_2) and other pollutants tended to have local, rather than global, effects. As these industries expanded, processes in their factories and the uses of their products led to escalating emissions of CO_2, methane, and nitrous oxide that began influencing global temperatures. However, because of annual temperature fluctuations, the resulting rise in temperature was not clear until the second half of the twentieth century.

• How Chemical Industries Contribute to Climate Change

Chemical industries have been vital to the economies of developed and developing nations, but, particularly in the late twentieth and early twenty-first centuries, environmentalists have criticized these industries for their unsustainable use of raw materials and for their polluting products and by-products that, among other things, have contributed to global warming. In the late twentieth century, some scientists estimated that about one-third of the world's total energy consumption could be attributed to the chemical industry. The American chemical industry, the world's largest, manufactures over seventy thousand different products and employs over one million people, but in doing so it generates about 3.5 percent of U.S. CO_2 emissions.

In both developed and developing countries, chemical industries produce greenhouse gases (GHGs) not only directly through the manufacture of various products but also indirectly when these products are used. For example, a major product of the petrochemical industry is gasoline, whose use in automobiles is a significant source of CO_2 emissions. In the United States, automobiles discharge more of this gas than the total output of all but four countries. Chlorofluorocarbons (CFCs) are more potent GHGs than CO_2, and, because of their effect on the ozone layer, they have a more complex relationship to global warming than does CO_2. CFCs' phenomenal success as refrigerants and in aerosols led to a 500 percent increase in their manufacture by chemical companies from 1960 to 1985. By the late 1980's, scientists knew that CFCs depleted the ozone layer, resulting in dangerous ultraviolet radiation reaching Earth's surface, but CFCs also accelerated stratospheric cooling. Therefore, it was the radiation effect rather than global warming that, in 1988, led DuPont, the largest manufacturer of CFCs, to stop making them.

By this time, computerized general-circulation models of the atmosphere had become sufficiently sophisticated that the role of GHGs in global warming had been accepted by most scientists. Environmental activists criticized chemical companies for accelerating climate change. In his writings, Barry Commoner contended that the only reason the chemical industry was so successful was that it had failed to pay its "environmental bill." He also acknowledged that, if the chemical industry were to eliminate all pollutants, including GHGs, from their discharges into the environment, the costs would probably make these companies unprofitable. Environmental organizations such as Greenpeace attacked various companies such as ExxonMobil for their intransigence in accepting responsibility for adding GHGs to the atmosphere. Some of these chemical companies responded by claiming that circulation models failed precisely to predict specific rises in global temperatures, so chief executive officers should postpone policy decisions on global warming until scientific studies were deemed sufficiently mature to assign responsibility correctly.

• National and International Regulations

When, in the late twentieth century, consensus among climatologists developed that anthropogenic GHG emissions from various industries and the use of certain products influenced climate change, national and international actions were taken to reduce concentrations of these molecular culprits. Following the pattern set by laws designed to force chemical companies to reduce or eliminate toxic chemicals from their discharges, several nations passed laws whose purpose was to reduce GHG emissions. In developed countries, many chemical companies claimed that this "overregulation" and "increased environmental costs" would inevitably lead to the decline of their global competitiveness and the loss of many jobs. On the other hand, certain developed nations, such as the United States and the United Kingdom, were leaders in enacting international regulations that resulted in the gradual phase-out of CFCs. In 1987, at a meeting of industrialized nations in Montreal, Canada, an agreement known as the Montreal Protocol was achieved, despite protests from some chemical companies. This treaty was later amended, and eventually over 175 nations became signatories.

Because of controversies over global warming, consensus among climatologists about GHG emissions came later than the CFC consensus, but in 1997 more than twenty-two hundred delegates from 161 nations met in Kyoto, Japan, to formulate a treaty to help slow global warming. A central provision of the Kyoto Protocol required thirty-nine

Louisana Petrochemical plants line the Mississippi River in St. Charles Parish, Louisiana. (Brett Duke/The Times-Picayune/Landov)

developed countries to reduce GHG emissions by about 5 percent below 1990 levels by 2012. By 2008, over 180 nations had ratified the Kyoto Protocol, but the United States had neither ratified nor withdrawn from the treaty. President George W. Bush argued that the treaty was basically unfair, since it exempted such developing countries as China and India from restrictions on their GHG emissions, which were becoming significant as these populous countries industrialized. Bush's stance had the support of many American chemical companies, which had dramatically slowed the construction of U.S. chemical facilities and relocated many of their enterprises in developing countries.

- **How Chemical Industries Attempt to Mitigate Global Warming**

Despite criticisms of the Kyoto Protocol and the recognition that new and better international climate treaties were needed, an increasing number of chemical companies came to realize that it was in their best interest to reduce GHG emissions. In the United States, companies such as Dow Chemical and DuPont understood that they could increase their profits by making their operations more energy-efficient, with the concomitant benefit of a reduction in GHG emissions. By 2004, DuPont had cut its GHG emissions by 65 percent from their 1990 levels and simultaneously saved hundreds of millions of dollars. Dow Chemical, using ideas developed by its subsidiaries, invested $1.7 million in GHG reductions and received a 173 percent return on its investment. Similarly, chemical industries in Europe made significant improvements by reducing energy use per unit of product by 40 percent from 1990 to 2004. In Asia, several Japanese chemical companies formed the Global Warming Countermeasure Council to decrease substantially their GHG emissions. These and other countries used a variety of methods to achieve these gains in energy efficiency, including emissions-trading arrangements, combining heat and power plants, and the use of powerful catalysts to make products more efficiently and reduce harmful by-products.

Besides reducing their dependence on fossil fuels, many chemical companies have promoted climate protection by developing environmentally friendly products and technologies, and a number of these have even started their own environmental businesses. Some chemical companies have developed improved insulation for buildings, while others have created renewable products based on new biopolymers and enzymes. Some factories began making more efficient materials for solar and fuel cells, new ceramics for power plants, and new chemicals for removing GHGs before their discharge into the atmosphere. DuPont has created more than ten environmental businesses, and those that provide environmental remediation services are expected to grow by $4 billion over ten years. Many chemists working for various industries are engaged in research on sequestration, using nanotechnology, mineralization, and other methods to remove CO_2 from the atmosphere. Optimists believe that what the chemical industry has done, with its modest impact on global warming, can be multiplied and intensified, whereas pessimists claim that what has been done is "too little, too late."

• Context

Because of its influence on so many fields, chemistry has often been called the "keystone science," and, similarly, chemical industries play such vital roles for so many other industries that they have become essential to the success of the global economy. These industries have helped lift impoverished nations to prosperity and prosperous nations to even greater levels of wealth. These gains have not come without costs, particularly to the environment, and these negative environmental effects have resulted in the chemical industry becoming one of the world's most regulated. Chemical companies are required to spend billions of dollars in operating costs to meet national and international regulations, with the lion's share of the costs borne by those companies that convert such raw materials as petroleum into basic chemicals for commercial uses.

Compliance with the growing regulations on GHG emissions will entail heavy financial burdens, and chemical companies in the developed world have lobbied for flexibility in how these goals are achieved, to avoid not only business failures but also the collapse of national economies. Because the stakes are so high, between the Scylla of world economic collapse and the Charybdis of the flooding of major coastal cities, the scientifically correct diagnosis of and solution to the problem of global warming are imperative. Just as the chemical industry has had an important part in creating this problem, so, too, will it have to be deeply involved in its solution.

Robert J. Paradowski

• Further Reading

Begg, Kathryn, Frans van der Woerd, and David Levy, eds. *The Business of Climate Change: Corporate Responses to Kyoto.* Sheffield, South Yorkshire, United Kingdom: Greenleaf, 2005. Analyzes how industries have responded to the challenge of climate change. Chapter 12 in the "Sector Analysis" section deals with the chemical industry's response to climate change.

Fineman, Stephen, ed. *The Business of Greening.* New York: Routledge, 2000. Collection of articles written by scholars in several fields who investigate the relationship between businesses, including chemical companies, and environmental problems, including global warming. Bibliographic references, index.

Hoffman, Andrew J. *From Heresy to Dogma: An Institutional History of Corporate Environmentalism.* 1997. Expanded ed. Palo Alto, Calif.: Stanford Business Books, 2001. Emphasizes the dramatic changes that have taken place in American companies, including chemical industries, as a result of ever-increasing requirements for environmental protection.

Matlack, Albert S. *Introduction to Green Chemistry.* New York: Marcel Dekker, 2001. With its several thousand references, this book is more a treatise than a textbook, though it can serve as both an introduction to the field and a reference source. Index.

Spitz, Peter H., ed. *The Chemical Industry at the Millennium: Maturity, Restructuring, and Globaliza-*

tion. Philadelphia: Chemical Heritage Foundation, 2003. Some scholars in this compendium analyze the influence of environmental regulations on the chemical industry, including the future challenges posed by global warming.

See also: Aerosols; Chlorofluorocarbons and related compounds; Clean energy; Emissions standards; Fossil fuels; Greenhouse gases; Industrial ecology; Industrial emission controls; Kyoto Protocol; Montreal Protocol; Ozone.

China, People's Republic of

- **Category:** Nations and peoples
- **Key facts**

Population: 1,330,000,000 (2009 estimate)

Area: 9,826,630 square kilometers (several border areas are in dispute)

Gross domestic product (GDP): $7.8 trillion (purchasing power parity, 2008 estimate)

Greenhouse gas (GHG) emissions in millions of metric tons of carbon dioxide equivalent (CO_2e): 6,100 in 2004

Kyoto Protocol status: Ratified, 2002

- **Historical and Political Context**

The major factor in the global production of greenhouse gases (GHGs) and in general environmental degradation is the massive growth in the world human population, and China is home to a substantial portion of that population. From 400 million people before World War II, China's population has grown to over 1.3 billion. Following the establishment of the People's Republic of China in 1949, the new government focused on rebuilding the nation and growing the population. In the face of Mao Zedong's philosophy of controlling and shaping the environment to serve a new generation, it was difficult at first for population scientists to convince the government that there was a need to constrain growth.

It soon became obvious, however, that progress in literacy, food production, and modernization was undermined by uncontrolled population growth. Mao agreed first to a two-child policy and then to a one-child policy, limiting the number of offspring allowed to each family. These policies are enforced only for the Han Chinese; ethnic minorities have never been limited in their number of children. It is estimated that without the one-child policy, China's population would be greater by over 300 million—a number approximately equal to the population of the United States: The Chinese standard of living would be dramatically lower, and severe stresses would be placed on food supplies, living conditions, and energy. The nation would have added over 1 billion metric tons of carbon dioxide (CO_2) to its annual emissions.

Management of environmental issues in China has modified and expanded since the first National Environmental Protection Meeting in 1973. In 1982, the Ministry of Urban and Rural Construction and Environmental Protection was established. The State Environmental Protection Administration (SEPA) was established in 1988 and upgraded to ministry level in 1998. The Ministry of Environmental Protection (MEP) was established in 2008. On March 28, 2008, the MEP established five regional inspection offices with a total staff numbering under twenty-fix hundred, compared with the U.S. Environmental Protection Agency's seventeen thousand personnel.

The State Council of the People's Republic of China issued the *China National Environmental Protection Plan in the Eleventh Five-Years, 2006-2010.* That five-year plan set forth a goal of achieving 20 percent greater energy efficiency and a 10 percent decrease in pollutants. China's officials acknowledged that the need for sustaining economic growth could take precedence over environmental concerns and weaken enforcement of environmental regulations. Lu Xuedu, deputy director of the Chinese Office of Global Environmental Affairs, explained that "You cannot tell people who are struggling to earn enough to eat that they need to reduce their emissions." Economic arguments are not restricted to China. Because coal is much cheaper than oil and natural gas, Japan and Europe also make greater use of coal.

While the economic health of the population takes priority, China's officials are not hesitant to

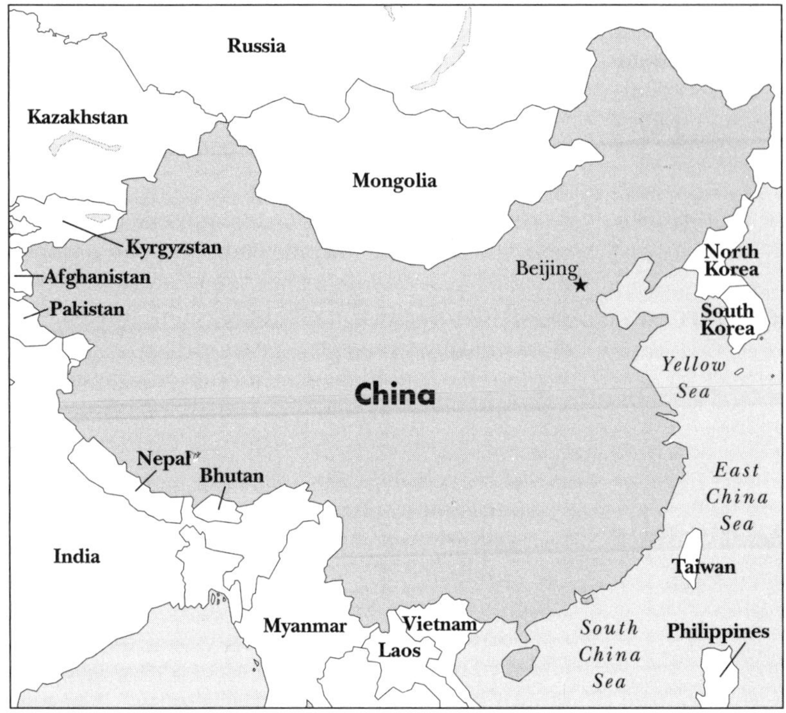

have been addressed before and directly blames ongoing problems on the lack of observation of laws, minimal punishments for lawbreakers, and poor enforcement of environmental laws. The most successful controls have been those cutting sulfur emissions, which ironically reflected sunlight back into space and therefore counteracted global warming. Thus, by decreasing acid rain, China has increased the greenhouse effect.

In contrast to American trends, China has avoided the rush to develop corn ethanol and other biofuels that could in any way displace foodstuffs for human consumption. China has concerns over food security originating in the nation's history of hunger and famine. Biofuels from nonfood plants, oilseeds, and an experimental *Jatropha curcas* plant are being considered, but great caution is being exercised to avoid soil erosion or reducing the number of food crops.

China recently completed the world's largest hydroelectric dam, a source of clean energy. Its twenty-six generators produce 700 megawatts of energy each. The dam's total 18,200 megawatt output is equal to that of fifteen of the largest nuclear power plants. It was designed to generate up to 10 percent of the country's power needs at the time of construction, but China's energy needs are rapidly increasing. The World Bank refused to fund the dam, and the United States led a boycott of bank funding for the project. China nevertheless built the massive dam based on the need to control devastating floods, generate clean power, and also bring ocean freighters to the interior industrial city of Chongqing.

China has provided subsidies to companies producing solar photovoltaic systems. Photovoltaics, or solar panels, are carbon neutral once they have been produced. However, as a source of electrical power, their costs have not yet dropped to a level

address the problems of climate change. *China's National Assessment Report on Climate Change (I)* (2007) notes that sea levels have risen 2.5 millimeters annually over fifty years. Glaciers have retreated, and air temperatures have increased 0.5°-0.8 ° Celsius over the last century, mostly in the last fifty years. The report predicts rural agricultural output will drop by 5 to 10 percent by 2030.

China's Scientific and Technological Actions on Climate Change further details that from 1986 to 2006, "China experienced 21 warm winters nationwide in succession." China is putting together large-scale observation networks and has laid out targets for very aggressive research in emission controls.

• Impact of Chinese Policies on Climate Change

The *China National Environmental Protection Plan in the Eleventh Five-Years, 2006-2010* by the State Council of the People's Republic of China has directed over one percent of gross domestic product (GDP) to environmental protection but considers the environmental situation "still grave." It laments that there were no breakthroughs in areas that should

where they can compete with natural gas or coal, a point called "grid parity" in China, and many solar units are shipped to Germany and other countries. General estimates are that solar power would have to drop to 14 cents per kilowatt hour to be economical in China. Contemporary costs run near 40 cents per kilowatt hour. The 2008 surge in oil prices provided the expectation that the cost of such alternative fuels would soon come close to competing with fossil fuels. However the subsequent global economic downturn also dramatically reduced the price of fossil fuels. The point at which solar power would become as cheap as fossil fuels, once optimistically thought to be as early as 2012, was deferred.

Many companies that make photovoltaics are located in China in order to take advantage not only of cheaper labor but also of cheaper land and materials. New buildings in Guangdong and other developed areas are being designed to use solar panels to provide their complete energy needs. The main market for Chinese solar panel production is in Europe, where regulations and subsidies promote the use of this more expensive power source. China itself remains cost conscious and is not ready to substitute more expensive power sources for cheaper coal plants. The need to serve the poorer population in the less developed countryside takes precedence.

Silicon, a central ingredient in solar cells, is also critical to the semiconductor industry. The solar industry has exceeded the semiconductor industry in its use of silicon. China has provided various electrical engineers with millions of dollars in start-up funds to establish state-of-the-art solar photovoltaic system factories in Wuxi and several other cities. Research has reduced the amount of silicon needed to produce solar cells. The economic downturn has also decreased the costs of silicon. Improved technology, much developed in China, is increasing the efficiency of silicon electricity production.

• China as a GHG Emitter

China was the first developing country to establish a national policy for addressing global warming, releasing its National Action Plan on Climate Change in June, 2007. According to the International Energy Agency, China surpassed the United States in CO_2 emissions in 2009. China's dramatic economic expansion since 1980 has pulled 400 million of its citizens out of poverty, but it has likewise increased per capita use of energy, especially visible in the nation's rapid adoption of automobiles. The increase in energy demand in developed regions has required China to enter the global market as a major player, negotiating the purchase of major shares of oil from Kazakhstan and other nearby fields.

According to the *World Energy Outlook, 2008*, produced by the Organization for Economic Cooperation and Development and the International Energy Agency, China's energy demand is expected to grow to such an extent that China will produce double the emissions of India and more than triple the emissions of other developed economic regions by 2030. This expected growth is a function not only of China's productive potential but also of the size of its markets.

While China is the second-largest consumer of oil at 7 million barrels per day, the United States leads at 21 million barrels per day. As is true of the United States, China has limited and rapidly declining domestic oil reserves. Oil imports account for 40 percent of Chinese consumption. By comparison, 60 percent of U.S. consumption is of imported oil.

With substantial coal reserves, China has used many smaller power plants located near population centers to deliver power locally with less lost to transmission. The nation has began requiring, however, that larger plants be built that are more efficient and use less coal per kilowatt-hour produced. China's main strategic energy reserve is coal. National energy consumption is so great that it also imports coal from Canada and Australia.

While scrubber technology reduces some emissions, burning coal is still a major CO_2 generator. China has begun thirty large-scale coal-to-liquid (CTL) projects using the Fischer-Tropsch process, which produces methanol for end users. Methanol is added to gasoline to produce a cleaner-burning fuel, and the country has committed to substantial use of methanol in 2011-2013. When oil prices are above $35 per barrel, coal-derived methanol is cost-effective for automobile fuel. China seeks to produce the methanol equivalent of 20 percent of its current oil consumption, although because its consumption is expected to double, this production level would amount to about 10 percent of the country's future needs.

China's largest joint venture with Royal Dutch/ Shell is a project in Ningxia that would produce 70,000 oil barrels equivalent of methanol per day. The National Development and Reform Commission established standards for CTL methanol, allowing the fuel to be used and projects such as the one in Ningxia to go forward. China was the first country in the world to develop methanol as an alternative fuel. The Chinese employ oxygenated gasification, a process developed with some U.S. government funding in earlier years, to isolate CO_2. This allows China to either sequester the CO_2 or use it to increase oil production from older wells.

A growing source of Chinese carbon emissions is the increase in motor vehicles. In the last few decades, government policy has promoted automobile production and encouraged the growing Chinese middle class to buy cars. The initial rationale was that the production, repair, and maintenance of cars would drive economic growth and reduce poverty; pollution control was deferred until after economic benefits had been realized. However, China has since established fuel-economy requirements for new car production that are more rigorous than those of the United States.

Compared to over-the-road trucking, rail is far more efficient for moving products on cost-per-mile and carbon emission bases. China has 72,500 kilometers of rail network compared to 228,500 kilometers still operable in the United States. China plans to invest $242 billion by the year 2020 in modernizing its railroad infrastructure, expanding the rail network and separating passenger and freight transport.

• Summary and Foresight

China's rapid economic growth has generated concern that the energy demands of such a huge population, if it grew to the per capita usage of the United States, would produce sufficient pollution and GHG emissions to represent an ecological disaster. However, the lifestyle of the new Chinese middle class retains a conservation ethic that may prevent runaway consumption. China combines economy and conservation of energy by using passive solar water heaters and has a long history of employing public transportation, bicycles, and electric bicycles to travel. In addition, the crowding of a large population requires severe limitations on pollution and provides a public awareness of the need to conserve and protect the environment. As a result, the government has begun aggressively recognizing and responding to the problems of global warming. China's ability to implement such policies is tempered by the need to pull half of its population, primarily in the countryside, out of poverty.

At the international level, under the Kyoto Protocol, the People's Republic of China is considered a developing country and therefore not required to meet emission timetables and targets. However, it can earn emission-reduction credits under the clean development mechanism. China's admission into the World Trade Organization and its leading

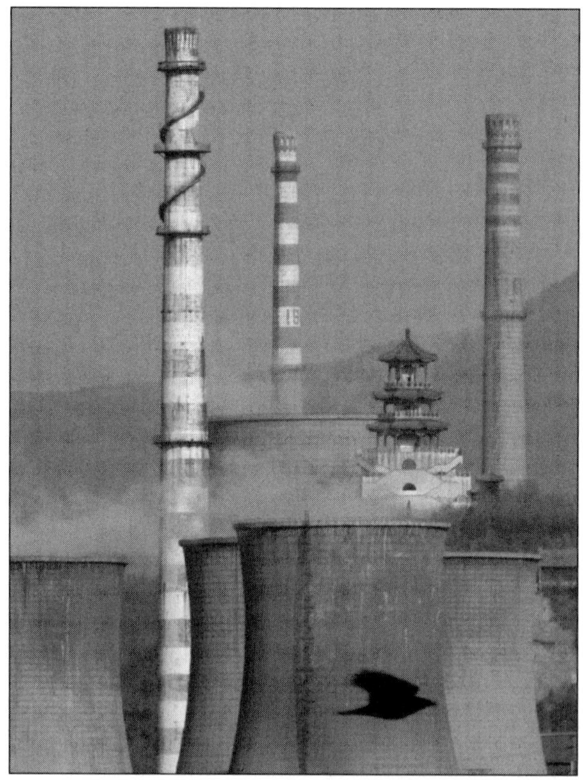

Smokestacks and cooling towers frame a pagoda outside the Chinese capital of Beijing. (Reinhard Krause/Reuters/ Landov)

role as a major nation have moved it to the position of needing to provide leadership in controlling GHGs.

John Richard Schrock

- **Further Reading**

Ding Yihui, et al. *Advances in Climate Change Research.* Supplement 3 to Vol. 1 of *China's National Assessment Report on Climate Change.* Beijing: National Climate Commission of China, 2007. Details the state of Chinese climate research in the early twenty-first century.

Economy, Elizabeth C. *The River Runs Black: The Environmental Challenge to China's Future.* Ithaca, N.Y.: Cornell University Press, 2004. This Council on Foreign Relations book examines the extent to which China's spectacular commercial advancement has depleted resources and challenged governmental regulatory agencies at all levels. Discussion extends to health problems, migration, economics, and social stability.

People's Republic of China. Ministry of Science and Technology, et al. *China's Scientific and Technological Actions on Climate Change.* Beijing: Author, 2007. Official state overview; candidly addresses the severity of climate change in China and the need to set strict targets.

People's Republic of China. State Council. *China National Environmental Protection Plan in the Eleventh Five-Years, 2006-2010.* State Council Document 39. Beijing: Author, 2005. China sets goals for all governmental issues in five-year plans. This official report details the failure to achieve major targets under the previous Tenth Five-Year Plan for National Environmental Protection and the necessity to emphasize environmental protection equally with economic goals.

Shapiro, Judith. *Mao's War Against Nature: Politics and the Environment in Revolutionary China.* New York: Cambridge University Press, 2001. Overview of the early years of the People's Republic under Mao; takes a negative view of the environmental degradation that accompanied the restructuring of Chinese society immediately after 1949.

See also: Coal; Fossil fuel emissions; Fossil fuel reserves; Fossil fuels; Fuels, alternative; Hydroelectricity; Kyoto Protocol; Motor vehicles; United Nations Framework Convention on Climate Change.

Chlorofluorocarbons and related compounds

- **Category:** Chemistry and geochemistry

CFCs are useful as refrigerants, solvents, and aerosol propellants, but they can act as catalysts for the destruction of stratospheric ozone. These compounds also contribute directly to global warming by acting as GHGs.

- **Key concepts**

Freon: trade name for CFCs made by DuPont chemical company

greenhouse effect: a phenomenon in which gases trap solar heat within Earth's atmosphere, preventing it from radiating away into space

halocarbons: the general family of compounds that includes CFCs, HCFCs, HFCs, and other molecules in which carbon bonds with halogen atoms

halon: a compound containing bromine, carbon, chlorine, and fluorine

Kyoto Protocol: a 1997 international agreement to limit greenhouse gas emissions

Montreal Protocol: a 1987 international agreement to phase out the manufacture and use of ozone-depleting chemicals

ozone layer: the portion of the Earth's stratosphere (10-50 kilometers high) where ozone has formed and absorbs dangerous ultraviolet radiation from the Sun

- **Background**

Developed in the 1930's as refrigerants, chlorofluorocarbons (CFCs) gained rapid acceptance in the 1940's. New uses were found for them as aerosol propellants, blowing agents, solvents, fire suppressants, and inhalation anesthetics. Production climbed, reaching as high as 566,591 metric tons in the United States by 1988. In 1971, it was shown that CFCs had accumulated in the atmosphere, and by 1974 a relationship was demonstrated between

Aerosol spray cans release a fine mist of chemicals suspended in liquid. CFCs used to propel this mist out of the can cause depletion of Earth's ozone layer. (©iStockphoto.com/Picsfive)

atmospheric CFCs and depletion of the ozone layer. Over the next twenty years, manufacture and use of CFCs were drastically reduced to protect the ozone layer from further harm.

- **Fate of Halocarbons in the Atmosphere**

CFCs, although denser than air, mix throughout the atmosphere and eventually reach the stratosphere (10-50 kilometers in altitude). Although CFCs have low chemical reactivity (and hence long lifetimes) in the lower atmosphere, in the stratosphere they encounter and absorb energetic ultraviolet radiation, resulting in their chlorine atoms being set free. These chlorine atoms can act as catalysts for the destruction of stratospheric ozone molecules. Because of the catalytic process, each chlorine atom can destroy thousands of ozone molecules. Other volatile chlorine compounds such as methyl chloroform and carbon tetrachloride can also form destructive chlorine atoms. Bromine atoms are also destructive of ozone, and bromine-containing compounds include the halons, used as fire suppressants and inhalation anesthetics, and methyl bromide, a soil fumigant and natural product of sea organisms.

In 1971, James E. Lovelock, using a sensitive detector, found traces of CFCs in air samples from different parts of the world. F. Sherwood Rowland and Mario Molina realized the potential ozone destructiveness of CFCs and stirred up concern among industrialists and politicians that led to the signing of the Montreal Protocol in 1987. Because of the long lifetimes of CFCs, however, even in the absence of further production it will take years for the existing pollutants to dissipate and allow the ozone layer to recover.

- **Substitutes for CFCs**

Phasing out CFCs meant that substitutes were needed for applications in refrigeration, air conditioning, and aerosols. The ideal substitute would be nontoxic, nonflammable, noncorrosive (like a CFC), and of suitable physical properties (such as boiling point and heat of vaporization), but without the ozone destructiveness of the CFCs. Attention naturally focused on the related compounds hydrochlorofluorocarbons (HCFCs) and hydrofluorocarbons (HFCs). The HCFCs, although they contain chlorine, tend to be destroyed by chemical reactions in the lower atmosphere before they reach the ozone layer. HFCs, which contain no chlorine, have negligible ozone destructiveness even should they reach the stratosphere. In addition to adopting these new compounds, the existing stocks of CFCs in abandoned equipment had to be trapped and either recycled or disposed of in an environmentally acceptable manner, rather than being vented into the atmosphere.

- **Context**

The impact of CFCs and their related compounds is mainly on the ozone layer, because of their catalytic effect. CFCs are also potent greenhouse gases but are present in the atmosphere at such low levels (0.1-0.5 parts per billion) that their contribution to the total greenhouse effect is relatively small. Loss of ozone in the stratosphere also affects temperature in complex ways, warming some parts of the atmosphere and cooling others. Nevertheless, the sheer numbers of individual halocarbons, even at low levels individually, add up to a warming potential that is worth controlling. Atmospheric levels of substances controlled by the Montreal Protocol have declined appreciably, but, not surprising, the levels of their substitutes have risen. This trade-off is good for the ozone layer, but it leaves much to be accomplished on the global warming front.

John R. Phillips

- **Further Reading**

Benedick, Richard E. *Ozone Diplomacy: New Directions in Safeguarding the Planet.* Enlarged ed. Cambridge, Mass.: Harvard University Press, 1998. Detailed discussion of the negotiations that led to the Montreal Protocol and its London Amendments.

Bily, C. A., ed. *Global Warming: Opposing Viewpoints.* Farmington Hills Mich.: Greenhaven Press, 2006. Collection of individual short essays arguing either pro or con regarding some aspect of global warming science and environmentalism.

Finlayson-Pitts, B. J., and J. N. Pitts, Jr. *Chemistry of the Upper and Lower Atmosphere.* San Diego, Calif.: Academic Press, 2000. Very thorough technical discussion of global warming, ozone depletion, and other topics. Hundreds of references to peer-reviewed literature. A graduate-level text.

Hiyama, T. *Organofluorine Compounds.* New York: Springer, 2000. A survey of the chemistry of organofluorine compounds, with a chapter on CFCs detailing their manufacture, uses, and fate in the environment.

Howe-Grant, M. *Fluorine Chemistry: A Comprehensive Treatment.* New York: Wiley, 1995. Contains basic information on the properties, manufacture, and uses of chlorofluorocarbons, hydrofluorocarbons, and perfluorocarbons.

Lomborg, B. *Cool It.* New York: Knopf, 2007. Lomborg, a social scientist, feels that some of the dire predictions about the consequences of global warming have been exaggerated. He concludes that huge sums spent for reducing carbon emissions are not cost-effective and would be better spent on clean water, treating AIDS, and other problems.

McNeill, J. R. *Something New Under the Sun.* New York: Norton, 2000. A comprehensive environmental history of the twentieth century. Considers global warming, ozone depletion, and many other topics with regard to various world regions.

Malmström, Bo G., ed. *Nobel Lectures: Chemistry, 1991-1995.* Singapore: World Scientific, 1997. Includes a lecture by F. Sherwood Rowland, who shared the 1995 Nobel Prize in Chemistry, in which he gives a technical discussion of atmospheric chemistry involving CFCs and ozone.

Midgley, Thomas, IV. *From the Periodic Table to Production: The Biography of Thomas Midgley, Jr., the Inventor of Ethyl Gasoline and Freon Refrigerants.* Corona, Calif.: Stargazer, 2001. Biography of Midgley (1889-1944), a chemist working for General Motors who pioneered the use of chlorofluorocarbons as refrigerants in the 1930's. Written by his grandson.

Walker, G., and D. King. *The Hot Topic: What We Can Do About Global Warming.* Orlando, Fla.: Houghton-Mifflin Harcourt, 2008. States the evidence for global warming in layperson's terms and proposes technological and political steps that should be taken. Includes a chapter on myths and misconceptions.

See also: Aerosols; Air conditioning and air-cooling technology; Carbonaceous aerosols; Chemical industry; Halocarbons; Halons; Montreal Protocol; Ozone; Sulfate aerosols.

Cities for Climate Protection

- **Category:** Organizations and agencies
- **Date:** Established 2005
- **Web address:** http://www.icleiusa.org/

- **Mission**

Greg Nickels, Seattle's mayor, launched the U.S. Conference of Mayors Climate Protection Agreement in 2005, after a season of unusually mild winter weather ruined the Pacific Northwest's skiing season. By mid-2008, 850 U.S. cities from Boston to Portland, Oregon, representing a total population of more than 85 million, had pledged to meet Kyoto Protocol standards. Many municipal leaders took action, because they believed the federal government during the administration of George W. Bush had failed to do so.

At the Seattle Climate Protection Summit of November, 2007, more than one hundred U.S. mayors called for a federal partnership to address energy dependence and global warming. Nickels said,

> We are showing what is possible in light of climate change at the local level, but to reach our goal of 80 percent reductions in greenhouse gases by 2050, we need strong support from the federal government.

The call for federal action occurred alongside the commitments of a large number of U.S. cities to reduce their residents' greenhouse gas (GHG) emissions.

Cities for Climate Protection organizes and coordinates efforts to combat global warming in U.S. towns and cities, coordinates action on a national level, and encourages efforts undertaken by individual urban areas. There is no limit on the size of a municipal district that may take part. The organization also works with cities and towns to improve air quality and enhance urban livability, coordinating with groups that have similar missions around the world.

• Significance for Climate Change

Cities for Climate Protection's street-level impact can be measured by programs in several affiliated urban areas. For example, in Los Angeles, Mayor Antonio Villaraigosa, in partnership with the Los Angeles City Council and environmental leaders,

The Five Milestones

Cities joining the Cities for Climate Protection are given five milestones to measure their progress toward implementing the organization's program.

Milestone 1. ***Conduct a baseline emissions inventory and forecast.*** Based on energy consumption and waste generation, the city calculates greenhouse gas emissions for a base year (e.g. 2000) and for a forecast year (e.g. 2015). The inventory and forecast provide a benchmark against which the city can measure progress.

Milestone 2. ***Adopt an emissions reduction target for the forecast year.*** The city establishes an emission reduction target for the city. The target both fosters political will and creates a framework to guide the planning and implementation of measures.

Milestone 3. ***Develop a Local Action Plan.*** Through a multi-stakeholder process, the city develops a Local Action Plan that describes the policies and measures that the local government will take to reduce greenhouse gas emissions and achieve its emissions reduction target. Most plans include a timeline, a description of financing mechanisms, and an assignment of responsibility to departments and staff. In addition to direct greenhouse gas reduction measures, most plans also incorporate public awareness and education efforts.

Milestone 4. ***Implement policies and measures.*** The city implements the policies and measures contained in their Local Action Plan. Typical policies and measures implemented by CCP participants include energy efficiency improvements to municipal buildings and water treatment facilities, streetlight retrofits, public transit improvements, installation of renewable power applications, and methane recovery from waste management.

Milestone 5. ***Monitor and verify results.*** Monitoring and verifying progress on the implementation of measures to reduce or avoid greenhouse gas emissions is an ongoing process. Monitoring begins once measures are implemented and continues for the life of the measures, providing important feedback that can be used to improve the measures over time.

Source: ICLEI—Local Governments for Sustainability.

has unveiled GREEN L.A., described as "an action plan to lead the nation in fighting global warming." Villaraigosa pledged to reduce his city's carbon footprint to 65 percent of its 1990 level by 2035, the most ambitious goal set by a major American city. Los Angeles also planned to increase its use of renewable energy to 35 percent of total energy usage by 2020, much of it through changes to its municipal electrical utility, the largest in the country.

In Austin, Texas, energy efficiency standards were raised for homes, requiring a 60 percent reduction in energy use by 2015. Chicago has attempted waterless urinals and planted several thousand trees. Philadelphia has been replacing black tarpaper roofs atop old row houses with snow-white, highly reflective composites. Keene, New Hampshire, requires parents waiting to pick up their children at schools to turn off their car engines. In Portland, Oregon, carbon emissions had been reduced to 1990 levels by 2007. Water flowing through Portland's drinking-water system also generates hydroelectricity. Mayors of at least 134 U.S. cities by 2007 were using more energy efficient lighting in public buildings, streetlights, parks, traffic signals, and other places. Many city governments' auto fleets had converted to alternative fuels or hybrid-electric technology.

The Chicago Climate Action Plan, announced in September, 2008, aimed to cut GHG emissions to 25 percent below 1990 levels by 2020. The plan required retrofitting of commercial and industrial buildings, increased energy efficiency in residences, and more use of renewable sources of electricity to reduce Chicago's emissions to 80 percent of 1990 levels by 2050. Buildings, which emit 70 percent of Chicago's carbon dioxide, are the major target of the Climate Action Plan. Chicago City Hall already has a green roof, designed as a model for as many as six thousand buildings citywide. Chicago's Smart Bulb Program by 2007 had distributed 500,000 free compact fluorescent light bulbs to residents.

The Dallas, Texas, municipal government decided early in 2008 to purchase 40 percent of its power from renewable energy sources, primarily wind power, which has been expanding rapidly in Texas. The city government also was reducing its energy use 5 percent per year by using lighting upgrades, solar panels, highly efficient heating and air-conditioning systems, and automated building temperature controls.

Using the AlbuquerqueGreen program, that city reduced natural-gas consumption 42 percent and cut GHG emissions 67 percent between 2000 and 2006. AlbuquerqueGreen promotes growth of green-tech companies, bicycle use, and pedestrian-friendly streets. Albuquerque requires that all new buildings be designed to be carbon neutral, with architecture suitable for 100 percent renewable energy use by 2030.

Several cities have targeted poor neighborhoods with subsidies and grants for insulation of older homes that often leak heat in winter. Such programs also allow some people to acquire insulation and energy-efficient compact fluorescent light bulbs as they replace older, inefficient basic electrical appliances, such as refrigerators, washers, and dryers.

With encouragement from Cities for Climate Protection, several major U.S. cities have launched sizable tree-planting programs, including Washington, D.C., Baltimore, Minneapolis, Chicago, Denver, and Los Angeles. Even so, an ongoing decline in urban tree cover has been accelerating since the 1970's in the United States, especially on private property and new development, according to American Forests, an environmental group that uses satellite imagery to document tree cover across the United States. Washington, D.C., is among the cities with the largest reduction in dense tree cover, with a 64 percent decline from 1973 to 1997, according to American Forests.

Bruce E. Johansen

• Further Reading

Cohen, Stuart, et al. *Cities for Climate Protection Policy and Practice Manual—Green Fleets: A Guide to Increasing Efficiency for Reducing Emissions from Municipal Fleets.* Toronto: International Council for Local Environmental Initiatives, [1996?]. Practical guide to city-level initiatives that can be undertaken to reduce motor-vehicle emissions.

U.S. Communities Acting to Protect the Climate: Achievements of ICLEI's Cities for Climate Protection—U.S., 2000. Berkeley, Calif.: International Council for Local Environmental Initiatives, 2000. Thirty-two-page pamphlet, published by Cities for Climate Protection's parent organization, detailing the program's accomplishments.

Young, Abby. "Forming Networks, Enabling Leaders, Financing Action: The Cities for Climate Protection Campaign." In *Creating a Climate for Change: Communicating Climate Change and Facilitating Social Change,* edited by Susanne C. Moser and Lisa Dilling. New York: Cambridge University Press, 2007. Detailed case study of the formation and activities of Cities for Climate Protection.

See also: Megacities; Urban heat island; U.S. energy policy.

Civilization and the environment

- **Category:** Nations and peoples

In a world where the map of global climate change is always shifting, the conditions that once made a civilization flourish can never be taken for granted. There is a constant ebb and flow of migration and return-migration, as changes in weather patterns modify a people's environment and sources of sustenance. These changes have often proven so drastic as to contribute to the demise of previously stable civilizations.

• Key concepts

deforestation: the process by which areas are stripped of forests and tree cover, exposing the underlying soil to erosive forces

desertification: the process by which once semi-arable lands are converted into desert

El Niño-Southern Oscillation: a climatic phenomenon wherein the normally cold ocean currents off the Pacific Coast of South America reverse, affecting global weather patterns

Little Ice Age: time period from around 1250 to 1850 characterized by cooling global temperatures

Medieval Warm Period: time period from around 800 to 1250 when European and possibly global temperatures were generally warming

• Background

The extent to which climatic change has affected the genesis, development, location, nature, rise, and fall of civilizations throughout history has been a matter of long-standing debate. The issue has gained a sharper edge and assumed a greater urgency, however, in an era of increasing environmental awareness. Any agreement among scholars seems to be on a broad scale, rather than on specifics. The wider view is that the globe has been in a predominantly warming trend for the last seventeen thousand years. This thaw in the wake of the last major ice age has been proposed as the main factor that started, propelled, and then sustained the spectacular advance of human civilization, notably from around 3000 B.C.E. to the present. However, when historians attempt to determine causal factors contributing to the success and failure of particular cultures, multiple factors other than climate have to be taken into account.

• Indus River Valley Culture

A long-standing debate has raged over what might have caused the demise of the Indus River Valley culture. Sprawled along the alluvial plain of the Indus River in what later became Pakistan, it was the largest in terms of land area of the four earliest cradles of civilization. Dependent on the agricultural productivity of the fertile soil along the riverbanks, it was highly urbanized and had strong trade links with Mesopotamia. The cities were painstakingly planned; buildings were constructed of oven-baked brick; spacious streets were almost perfectly laid out in modern grid fashion; and the cities employed advanced drainage and sewage disposal mechanisms.

The Indus culture flourished beginning around 2500 B.C.E. then declined and became depopulated by no later than 1200 to 1100 B.C.E. Different theories have been advanced to explain its downfall, none of which has proven conclusive. Increasing evidence indicates, however, that climatic and environmental factors brought about the downturn in the culture's fortunes, rather than theorized attacks by Aryan invaders. According to one scenario, a climatic change seems to have triggered increased rainfall, leading to massive flooding. The demand for wood to fire the kilns used to produce bricks had denuded the forests, and this deforestation had eroded the soil, rendering it vulnerable to the flooding. Another climate change theory has it that drought, warmer temperatures, and anthropogenic deforestation resulted in desertification, which caused the large-scale abandonment of the Indus cities.

• Egypt, Mesopotamia, and Greece

The Old Kingdom civilization of Egypt, 3100-2200 B.C.E., was renowned as the era of pyramid construction and exemplary of strength and prosperity. However, there is some evidence that global climate change possibly caused the disruption of the annual flooding of the Nile River—perhaps augmented by El Niño-Southern Oscillation (ENSO) effects—and thus might have brought about a devastating drought cycle that led to the termination of the Old Kingdom. The kingdom's demise was followed by a chaotic 150 years known as the First Intermediate Period.

The climatic vulnerability of Mesopotamia has been amply attested to in the cuneiform texts of its various civilizations, as well as being evidenced by the archaeological record. The Akkadian Empire (2400-2200 B.C.E.), which encompassed Mesopotamia and the Fertile Crescent up to the Levantine Coast, was apparently brought low by a series of very severe winters combined with drought conditions that were perhaps occasioned by volcanic eruptions. Another such massive drought cycle around 1200 B.C.E. is also believed to have contributed to the ending of the Hittite Empire and Mycenaean Greece.

• The Americas: Anasazi and Maya

Lack of documentation makes it exceptionally difficult to gauge the scale of the impact that climate may have had on pre-Columbian Native American cultures. The Maya of Mexico and Central America, whose civilization was the most advanced and probably the best documented, suffered the nearly wholesale destruction of their books by Spanish missionaries. The Anasazi, who dwelled in present-day portions of the southwestern United States from around 700 to 1150, had no such documentation to begin with, so their abrupt collapse has proven even more puzzling.

The Navajo term "Anasazi," or "Ancient Ones," has been used to denote the Pueblo cultures known variously in different regions as the Hohokam, Mogollon, and Patayan. These peoples inhabited the area around the modern Four Corners—where Utah, Colorado, Arizona, and New Mexico converge—and southward into the greater part of the modern state of Chihuahua, Mexico. They constructed the spectacular cliff dwellings that still dot the regions they once inhabited, including Mesa Verde, Chaco Canyon, and Canyon de Chelly. Around 1150, a particularly severe and prolonged period of drought drove the Anasazi to nearby areas where water was slightly more accessible, and the descendents of the Anasazi can be found among the present-day Pueblo.

Deriving, most likely, from the more ancient cultures of the Olmec and Teotihuacan in southern Mexico, the people known as the Maya flourished in the region of the Mexican Yucatán Peninsula and the modern countries of Guatemala, Belize, El Salvador, and Honduras around 200 to 900 C.E. Massive and sophisticated in their architectural designs and elaborate pyramids, Mayan city-states burgeoned into mini-empires under autocratic kings who claimed to serve as conduits to the gods to ensure their people's prosperity. The once-baffling collapse of these city-states is now thought to have been caused by a series of droughts; the resulting crop failures drove the Maya to abandon both their faith in these kings and the city-states themselves, which became isolated ruins.

• The Andes and Oceania

The enigmatic Moche culture of the Andes lasted from about 100 to 750 C.E. along the northern coastline of Peru and coincided with the Nazca culture to the south. Both ended abruptly, leaving little more than their major artifacts (Moche pottery and the Nazca lines in the Nazca Desert). There is some disputed evidence that extensive ENSO episodes set off alternating patterns of flooding and drought, causing both civilizations to crumble. The more extensive Tiahuanaco culture, centered around an urban site in Bolivia, flourished from about 700 to 1150, left giant stonework ruins, and seems to have perished from the effects of the same massive drought period that destroyed the Anasazi.

The more isolated islands of Pacific Oceania could be expected to offer more examples of vulnerability: Because of their remoteness, the scarcity of land and resources, and the potential capriciousness of nature, they have been home to some of the most frail of human societies. These societies, however, met with different outcomes. For example, the native culture of Easter Island seems to have come to an end in part through a murderous civil war brought on by deforestation and social unrest. It is possible that an ENSO episode exacerbated conditions of drought and scarcity, thus bringing about the culture's final destruction.

• Norse Greenland

The fate of the Norse settlements in Greenland offers the most readily documented example of the negative impact of climate change. In 968, the Viking warlord Eric the Red sailed from Iceland to establish a Scandinavian colony in Greenland. During the Medieval Warm Period, the colony thrived and expanded, concentrating into two major settlements. An eastern settlement was located on the island's extreme southern coast, and a less significant settlement was founded 320 kilometers up the western coast. In its heyday, each settlement may have had a population of five thousand. Their economic mainstays were cattle and sheep grazing, as well as trade, mainly with the Icelanders and Norwegians.

Around 1250, as the Medieval Warm Period gave way to the Little Ice Age, glaciers advanced dramatically across Greenland. (The term "Little Ice Age" is used differently by different writers. Many use it to refer to the climate cooling from about 1300 to 1850, while others use it for the latter half of that interval, when cooling was greatest, beginning around 1550 or 1600.) Growing seasons shortened; temperatures plummeted; and hunting, grazing, and agricultural land became scarce. The Greenlanders, always in an isolated position, became even more cut off as ice packs began blocking the passage routes to and from Iceland and Norway. Increasingly harsh conditions also drove the Inuit (Eskimos) south and west, and almost immediately, armed clashes broke out between them and the Norse, as both groups competed for dwindling sources of sustenance. The Inuit proved to be the more readily adaptable to change, while the Norse were reluctant to alter their ways. Thus, by 1400, the western and eastern settlements lay abandoned as the Vikings either starved, died at the hands of the Inuit, or fled.

• Context

Though there now exists little doubt that climate change and environmental factors have affected the course of civilization, there remains an element of debate over their relationship to other causal factors such as outside encroachment and internal dissension. It may well be different in each case. The ability of climate to influence history may affect contemporary politics and public policy decision making. If the correlation between climatic or environmental events and the decline of civilizations could be conclusively established, then it would logically follow that governments should work to prevent or to alleviate the effects of such events in the interests of self-preservation.

Raymond Pierre Hylton

• Further Reading

Avari, Burjor. *India—The Ancient Past: A History of the Indian Subcontinent from c. 7000 B.C. to A.D. 1200.* New York: Routledge, 2007. Evenhanded account that examines both major theories for the collapse of the Indus Valley culture, with the logical preponderance of evidence supporting environmental or climatic calamity, rather than outside invasion.

Diamond, Jared. *Collapse: How Societies Choose to Fail or Succeed.* New York: Penguin Books, 2006. Though the author concedes that climatic change is a significant factor in the demise of several civilizations, he emphasizes the impact of ill-advised decisions by the leaders and populaces as necessary contributing elements.

Fagan, Brian. *The Long Summer: How Climate Changed Civilization.* New York: Basic Books, 2005. Advances the theory that all the major developments of human civilization may be seen as coinciding with the lengthy warming period that followed the last global ice age.

Henderson, John S. *The World of the Ancient Maya.* Ithaca, N.Y.: Cornell University Press, 1997. Sets out unhesitatingly the probable role of climate-related subsistence crisis in triggering the abandonment of Mayan city-states.

Jones, Gwyn. *A History of the Vikings.* New York: Oxford University Press, 1969. Seminal study that establishes the pivotal role of Greenland's isolation and the onset of extreme global cooling to the subsistence crisis that overtook the westernmost Viking outposts.

Pike, Donald G. *Anasazi: Ancient People of the Rock.* Illustrated by David Muench. Palo Alto, Calif.: American West, 1974. Blames the Anasazi's decline in part on drought, but asserts that the depredations of nomads were a more significant factor.

Schele, Linda, and David Freidel. *A Forest of Kings: The Untold Story of the Ancient Maya.* New York: William Morrow, 1990. Extremely detailed description of the complexity of Mayan culture, including its language and religion. Scant and inconclusive discussion of the culture's decline.

Wheeler, Sir Mortimer. *The Indus Civilization.* 3d ed. New York: Cambridge University Press, 1968. Downplays the idea of climate change and human environmental blunders as causes for the decline of the Indus culture; strongly advocates the view that Aryan aggression was more decisive.

See also: Arctic peoples; Deforestation; Desertification; Easter Island; El Niño-Southern Oscillation; Greenland ice cap; Little Ice Age; Medieval Warm Period; Paleoclimates and paleoclimate change.

Clathrates

- **Category:** Chemistry and geochemistry

- **Definition**

A clathrate is a chemically constructed lattice structure in which one type of molecule traps and contains another type of molecule without forming a chemical bond. Clathrates may also be called gas hydrates (sometimes using the name of the gas, such as "methane clathrate"), host-guest complexes, or molecular compounds. They are sometimes referred to as "marsh gas."

The most common types of clathrates are those involving water trapping a gas, such as methane, ethane, propane, isobutene, butane, nitrogen, carbon dioxide (CO_2), or hydrogen sulfide. Methane trapped in water is the most common type of naturally occurring clathrate. Methane clathrates usually occur when organic matter in a low-oxygen environment is degraded by bacteria. This situation usually occurs under ice sheets in the ocean; usually, these structures are found in sediments. A clump of methane hydrate looks like a snowball, but at room temperature, the ball dissolves. It can even be lit on fire, as the methane inside the lattice structure burns.

- **Significance for Climate Change**

Because methane is one of the greenhouse gases (GHGs), gases that trap warmth on Earth's surface, the release of large amounts of methane could cause the Earth to become warmer. In the geologic past, sudden releases of large amounts of methane may have triggered abrupt climatic events. Geologists believe that about 635 million years ago, a large quantity of methane was released into the atmosphere from ice sheets that covered Earth. This release caused an abrupt, global warming, taking Earth from a cold, relatively stable climate to a very warm, still relatively stable climate. Scientists think this event quickly warmed the Earth by tens of degrees Celsius.

Sudden releases of methane could occur if either the temperature rises or the pressure falls on the sediment above the methane clathrate structure. Often, sediment continues to accumulate over the methane clathrate until the weight of the sediment causes the clathrate structure to become unstable, when it disintegrates into water and gas. The covering sediments then become unstable and release methane into the atmosphere. The methane reacts with oxygen to form CO_2, which can kill any animals that need oxygen.

The U.S. Geological Survey estimates that up to 8.5 million cubic kilometers of methane clathrate exist in the seafloor worldwide, more than in all other fossil energy resources combined, so methane clathrates are being investigated as a potential source of natural gas energy. However, getting to this energy source is dangerous, as the accidental release of methane into the atmosphere could be deadly in and of itself. In addition, a release of this sort could affect the global temperature or even cause tsunamis if landslides under the ocean were triggered. Safe methods of harvesting this resource will need to be developed, as methane can be very explosive. Deadly methane explosions have occurred in the past in mines in Harlan County, Kentucky, in 2006, and in Ulyanovskaya, Russia, in 2007. Methane can also cause an "exploding lake," such as Lake Kivu in Africa. There, the lake water interacts with volcanic gases to produce methane and CO_2, which are potentially deadly when released.

Harvesting methane from under ice sheets for fuel requires two different approaches, both of which are potentially dangerous. With the depressurization method, holes are drilled into a layer of methane hydrate just under the ice, which releases pressure and causes the methane to flow up a pipe. This approach uses less energy than the other approach, that of thermal injection. Thermal injection involves pumping hot water into a deposit of methane, which destabilizes the structure of the methane clathrate. A promising future approach involves injecting CO_2 into the hydrate formation, displacing the methane in the structure.

Marianne M. Madsen

• Further Reading

"Clathrate Hydrates: Icy Fuel Thawed with Antifreeze Proteins." *Canadian Chemical News*, May 1, 2006. Explains in detail what clathrate hydrates are and how they affect the environment. Includes illustrations and photos.

Kennedy, M., D. Mrofka, and C. von der Borch. "Snowball Earth Termination by Destabilization of Equatorial Permafrost Methane Clathrate." *Nature* 453 (May 29, 2008): 642-645. Describes how a large methane release in the past affected climate; explains how another such release could affect climate change now.

Kennett, J. P., ed. *Methane Hydrates in Quaternary Climate Change: The Clathrate Gun Hypothesis.* Washington, D.C.: American Geophysical Union, 2002. Uses varied sciences, such as paleoceanography, climate dynamics, paleobotany, biogeochemical cycling, analysis of methane hydrates (clathrates), and marine geology, to explain abrupt global warming.

Maynard, B. "Methane Hydrates: Energy Source of the Future?" *Popular Mechanics*, April, 2006. Explains methods of extracting methane as fuel. Includes maps and graphs.

Sloan, E. Dendy, Jr. *Clathrate Hydrates of Natural Gases.* Boca Raton, Fla.: CRC Press, 2008. Provides a historical overview of clathrate hydrates and discusses their structure and properties. Includes tables and illustrations.

Weber, E. *Molecular Inclusion and Molecular Recognition: Clathrates I.* New York: Springer, 1987. Classic textbook discussing clathrates and their chemical structure.

Weitemeyer, K. A., and B. A. Buffett. "Accumulation and Release of Methane from Clathrates Below the Laurentide and Cordilleran Ice Sheets." In *Global and Planetary Change.* Amsterdam, the Netherlands: Elsevier, 2006. Describes how ice-age cycles are associated with large fluctuations of methane, possibly due to clathrates, and how methane clathrates accumulate below ice sheets.

See also: Fossil fuel reserves; Fossil fuels; Greenhouse gases; Ice shelves; Methane.

Clean Air Acts, U.S.

- **Category:** Laws, treaties, and protocols
- **Date:** Enacted 1963, 1970, 1977, and 1990

The Clean Air Act of 1963 and its subsequent amendments have mandated reductions in various atmospheric pollutants, with Congress imposing increasingly stringent standards over time. Some of the provisions of the laws have limited anthropogenic contributions to climate change as well.

• Background

Air pollution has long been recognized as a problem in the United States. The passage of the Clean

Air Act (CAA) in 1963 provided federal support for efforts to reverse and control air pollution. In 1970, Congress passed the Clean Air Act Extension, amending the CAA to provide for strict national ambient air quality standards (NAAQSs) and their enforcement. The 1970 and later amendments have sought to reduce atmospheric ozone, carbon monoxide (CO), sulfur dioxide (SO_2), nitrogen oxides (NO_x), hydrocarbons, lead, and particulate matter. The 1990 amendments also set provisions for toxic air pollutants. The legislation has been primarily aimed at protecting human health and preventing or reducing acid rain, rather than mitigating climate change.

• Summary of Provisions

The CAA of 1963 provided for federal help in dealing with air pollution but indicated few specifics and no standards for enforcement. The 1970 and 1990 CAA amendments have been the most important in establishing specific air pollution policies. Dating from the 1970 amendments, the Environmental Protection Agency (EPA) identified six air pollutants in the United States (and, by extension, worldwide): ozone, sulfur dioxide, nitrogen oxides, carbon monoxide, lead, and particulates. The EPA divided the United States into regions and established acceptable levels for each pollutant in each region in order to reduce pollutant levels nationwide. Some of these pollutants are generated by industries, such as the electric power industry, while others are generated by automobiles. Some of the pollutants, such as SO_2 and NO_x, lead to the production of acid rain, which harms human health, bodies of water into which the rain falls, buildings, and crops. The 1970 and 1977 amendments to the CAA led to reductions in electric power plants' production of SO_2 and NO_x, as the plants turned to low-sulfur coal or adopted technological fixes, such as scrubbers, that removed these gases from their emissions. Automobile makers modified their engines to burn unleaded gasoline, reducing lead emissions. Decreasing ozone and CO levels proved to be more difficult.

Because of the high cost of reducing emissions, the EPA adopted standards that enabled utilities with low SO_2 emissions to trade some of their polluting capacity to utilities that were unable to meet EPA standards codified in the 1977 amendments. The 1990 CAA went farther and allowed utilities to sell some of their excess capacity on the open market. This approach has led to overall reductions in SO_2 in the United States, although levels remain high in some areas.

Congress tightened standards for automobile emissions of the various pollutants with the 1977

Time Line of U.S. Clean Air Acts

Year	*Law*	*Provisions*
1955	Air Pollution Control Act	First U.S. law addressing air pollution and funding research into pollution prevention.
1963	Clean Air Act of 1963	First U.S. law providing for monitoring and control of air pollution.
1967	Air Quality Act	Established enforcement provisions to reduce interstate air pollution transport.
1970	Clean Air Act Extension of 1970	Established first comprehensive emission regulatory structure, including the National Ambient Air Quality Standards (NAAQS).
1977	Clean Air Act Amendment of 1977	Provided for the prevention of deterioration in air quality in areas that were in compliance with the NAAQS.
1990	Clean Air Act Amendment of 1990	Established programs to control acid precipitation, as well as 189 specific toxic pollutants.

Source: U.S. Environmental Protection Agency.

and 1990 amendments and directed that the EPA should continue to develop even more stringent standards if ambient air pollution was not reduced. The 1990 amendments to the CAA provided for new enforcement standards for 189 toxic air pollutants such as asbestos, benzene, mercury, or cadmium compounds, as well as providing for the phasing out of stratospheric ozone-depleting substances by 1996. These standards affected various aspects of American life, from oil refineries that emitted numerous toxins to building demolition.

The CAA has been directed toward improving the health of the American people through removing or reducing various air pollutants that have been judged to be harmful to human health and by extension harmful to the environment. It has been a top-down approach that mandated ambient air quality standards and that left it up to industry to achieve the technological means to meet these standards. The record of the CAA has been mixed, but there have been some notable successes, such as the reduction of lead in the atmosphere.

• Significance for Climate Change

The CAA in its various iterations has been directed toward improving human health, not dealing with climate change. Nonetheless, some of the standards of the CAA have also had an impact on the climate. In particular two greenhouse gases (GHGs), nitrous oxide (N_2O) and chlorofluorocarbons (CFCs)—which deplete stratospheric ozone—have been regulated by the CAA.

With the 1970 amendments to the CAA, Congress mandated a reduction in NO_x in the atmosphere, which helps reduce nitrous oxide. Automobile exhaust and fertilizer containing nitrogen are major contributors of nitrous oxide to the atmosphere, and coal-burning power plants contribute other nitrogen oxides. Control of NO_x has been more difficult to achieve than has control of SO_2. The 1970, 1977, and 1990 CAA amendments have provided for progressively tougher standards for the emission of NO_x. The 1990 amendments, for example, provided for cutting nitrogen oxide emissions by an additional 1.8 million metric tons below the 1980 levels. Although N_2O has a lower presence in the atmosphere than do other GHGs such as carbon dioxide or methane, it is estimated to contribute slightly over 6 percent of the greenhouse effect worldwide, so reductions are important.

Nitrogen oxide compounds have long been products of industrial society, but production of CFCs has been of more recent vintage. CFCs are composed of chlorine, fluorine, and carbon atoms and are chemically almost inert. Because of this property, they came to be used as refrigerants, as well as propellants for aerosols such as deodorants. CFCs are also long-lived in the atmosphere, so once produced they will remain for a long time. When exposed to ultraviolet radiation in the upper atmosphere, CFC molecules break apart and free up chlorine atoms that, combined with oxygen in the stratosphere, help deplete the ozone layer.

Dating from the late 1970's, one CFC, CFC-11, was banned from use as a propellant. Because of the attractiveness of CFCs as refrigerants and solvents and the difficulty in obtaining substitutes, CFC-11 and two other CFC compounds (CFC-12 and CFC-113) continued in use in refrigerants, air conditioners, and industrial mechanisms. As the severity of the damage to the ozone layer became recognized, pressure built to ban or curtail the use of all CFCs. The 1990 amendments to the CAA mandated the elimination of new CFCs in the United States and provided for careful handling of CFCs in existing refrigerants and air conditioners so as to limit their escape into the atmosphere. Although some countries continue to make use of CFCs, the position of the United States as a major user made American reduction quite important. Although it will be quite some time before the holes in the Arctic and Antarctic ozone layers no longer exist, the reduction of additional CFC production has decreased further damage to the ozone layer.

In sum, the Clean Air Act and its various amendments have helped in dealing with climate change, although that generally was not their intent. The CAA has regulated two GHGs, limiting their production and use. In addition, the CAA has provided a precedent for congressional action dealing with other GHGs. Because the United States is a major producer of GHGs, any reduction in the production of these gases by that nation will have a large impact on climate change globally.

John M. Theilmann

• Further Reading

Blatt, Harvey. *America's Environmental Report Card.* Cambridge, Mass.: MIT Press, 2005. Provides a summary of the impact of U.S. policy on reducing air pollution and global warming.

Rosenbaum, Walter A. *Environmental Politics and Policy.* 7th ed. Washington, D.C.: CQ Press, 2008. Includes a chapter that places the CAA in the context of command-and-control environmental proctection.

Somerville, Richard C. J. *The Forgiving Air.* Berkeley: University of California Press, 1996. Useful discussion of the impact of the CAA on air pollution and climate change.

See also: Air pollution and pollutants: anthropogenic; Air pollution and pollutants: natural; Air pollution history; Air quality standards and measurement; Chlorofluorocarbons and related compounds; Environmental law; Nitrous oxide; U.S. legislation.

Clean development mechanism

• **Category:** Laws, treaties, and protocols

The CDM allows industrialized nations to fund climate-friendly development in developing nations. It fosters cooperation between the industrialized and developing world, but by giving industrialized nations credit for their actions abroad, it allows them to do less to reduce emissions at home.

• Key concepts

Annex I parties: industrialized nations listed in Annex I of the UNFCCC

CDM project cycle: the registration and issuance process all CDM projects must go through in order to generate CERs

certified emissions reductions (CERs): credits for contributing to reduced emissions in developing nations that Annex I nations can substitute for domestic reductions

quantified emission limitation and reduction commitments (QELRCs): Annex I parties' reduction targets, as set out in Annex B of the Kyoto Protocol

• Background

Article 3 of the Kyoto Protocol commits developed countries identified in Annex I of the United Nations Framework Convention on Climate Change (UNFCCC) to reduce their emissions of six greenhouse gases (GHGs). These reduction commitments, termed quantified emission limitation and reduction commitments (QELRCs), are spelled out in Annex B of the protocol. To help Annex I parties meet their commitments, three mechanisms were established: joint implementation, emissions trading, and the clean development mechanism (CDM). Under the CDM, developed countries can invest in or finance projects in developing countries that reduce or remove GHG emissions and use the reductions achieved toward meeting their own emissions targets.

• The CDM Regime

The CDM is a project-based mechanism under which projects in developing countries that result in lowered GHG emissions can generate certified emissions reductions (CERs), which developed countries can use to contribute to meeting their QELRCs. The mechanism was established by Article 12 of the Kyoto Protocol. That article sets out the aims of the CDM, which are to assist developing countries to achieve sustainable development and to contribute to the ultimate goal of the UNFCCC, as well as to assist developed countries to meet their QELRCs. The CDM is therefore intended to provide a dual benefit to the climate change regime: sustainable development benefits to developing countries and cost-effective emission reductions to developed countries. In addition to these, 2 percent of the proceeds of CDM projects are to be put into the Kyoto Protocol Adaptation Fund, to finance adaptation activities in developing countries.

Both public and private entities are allowed to participate in the CDM, but to do so they must be authorized by parties to the protocol. The CDM is overseen by an executive board, and under the authority and guidance of the Conference of the Parties to the Framework Convention on Climate

Change serving as the meeting of the Parties to the Kyoto Protocol (COP/MOP, or CMP).

• Operation of the CDM

Under the CDM, developed countries can invest in or finance projects in developing countries. These projects must result in a reduction or removal of emissions from the atmosphere additional to any that would have occurred in the absence of the projects. The difference between the emissions that would have been released (referred to as the baseline emissions) and those actually released, that is, the reductions or removals achieved by the project, are validated and certified by an independent body and issued by the CDM executive boards as CERs. Each CER represents 1 metric ton of carbon dioxide equivalent (CO_2e) reduction or removal achieved. That is, it represents either the reduction of 1 metric ton of atmospheric CO_2 or an amount of another GHG that has the same global warming potential (GWP) as 1 metric ton of CO_2.

CERs, also referred to as carbon credits, can then be used by Annex I parties to contribute toward achieving their QELRCs. A CDM project can also be implemented unilaterally. In that case, the investment or financing for the project is provided by an entity in the host country itself, without the involvement of a developed country. The CERs generated by the project can then be sold on the carbon market by the investing entity. Again, Annex I parties can buy and use these CERs to contribute to meeting their targets.

In a clean development mechanism project, these generators in China's Jiujiang Power Plant were made more efficient, allowing them to burn less coal to generate the same amount of electricity. (Hu Guolin/Xinhua/Landov)

• An Example of a CDM Project

An example of a registered CDM project is the West Nile Electrification Project, hosted by Uganda, with Finland, the Netherlands, and the International Bank for Reconstruction and Development as the investor entities. The project inter alia involves the installation and operation of a hydroelectric power plant to provide energy and displace the use of diesel-based electricity, as well as the installation and operation of a high-efficiency generator to displace low-efficiency diesel generators at isolated diesel stations and privately owned diesel generator sets. The project is expected to eliminate 36,210 metric tons of CO_2e annually.

• Operation and Provisions of the CDM

To participate in the CDM, countries must be parties to the Kyoto Protocol. In addition to this basic requirement, Article 12 of the protocol and Decision 3/CMP.1 (the third decision of the First Conference of the Parties serving as the meeting of the Parties, 2005) provide more detailed participation requirements that parties and projects must fulfill to be eligible for participation. Countries must designate national authorities, who serve within those countries as points of contact for information on the CDM. Developed countries have further requirements, including the need to comply with their methodological and reporting obligations under Articles 5 and 7 of the protocol. In addition, a CDM project must provide real, measurable, and long-term benefits related to climate change mitigation, including emission reductions or removals that exceed any that would occur in the absence of the project. For a project to be registered as a CDM project and generate CERs, it must go through a registration and issuance process, referred to as the CDM project cycle, to ensure that it satisfies these and other criteria.

• Participation in the CDM

There are 182 parties to the Kyoto Protocol. Of this number, around 135 have designated their national authorities, thus fulfilling the basic participation requirements of the CDM. As of October, 2008, 51 developing countries were hosting CDM projects and 16 developed countries had invested in the CDM, bringing the total participation to 67 parties. There were a total of 1,184 registered CDM projects.

• Status and Trends

Most of the 1,184 registered projects were hosted in Asia and the Pacific region, which had a total share of 65 percent of registered projects. This was followed by Latin America and the Caribbean region, with a total of 32 percent, and Africa, with less than 3 percent of the total. On a country-by-country basis, this distribution sees four countries accounting for about 70 percent of all registered projects. These are, in order of portfolio size, India, China, Brazil, and Mexico. This distribution of projects has been termed inequitable, and discussions are ongoing on ways of addressing this inequity. Several actions have been instituted to this end, such as the launch of the Nairobi Framework to catalyze the CDM in Africa.

• Context

The CDM arose from a proposal made by the government of Brazil for a clean development fund. This fund was to be replenished through compulsory contributions by developed countries that were in noncompliance with their emission reduction targets under the Kyoto Protocol. Countries welcomed this aspect of the fund for the flexibility it provided to developed countries unable to meet their targets and the potential financial benefits it provided to developing countries. Further negotiations led to agreement on the CDM, and its two core elements remain flexibility and cost-effectiveness for developed countries and financial investment and sustainable development benefits for developing countries.

Tomi Akanle

• Further Reading

Streck, Charlotte, and Jolene Lin. "Making Markets Work: A Review of CDM Performance and the Need for Reform." *European Journal of International Law* 19, no. 2 (2008): 409-442. Analyzes the operation of the CDM, highlighting some of its problems and suggesting ways of overcoming these problems.

United Nations Development Programme. *The Clean Development Mechanism: A User's Guide.* New York: Author, 2003. Examines the CDM and its rules and modalities; constitutes a guide on how to implement the CDM effectively.

Werksman, Jacob. "The Clean Development Mechanism: Unwrapping the Kyoto Surprise." *RECIEL* 7, no. 2 (1998). Describes the negotiations and circumstances leading up to the establishment of the CDM and examines the CDM's conceptual roots and basic provisions.

See also: Annex B of the Kyoto Protocol; Basel Convention; Baseline emissions; Certified emissions reduction; Emission scenario; Industrial emission controls; Kyoto lands; Kyoto mechanisms; Kyoto Protocol.

Clean energy

- **Category:** Energy

- **Definition**

Clean energy refers to any energy source whose use does not emit carbon dioxide (CO_2), or other pollutants or greenhouse gases (GHGs) that contribute to global warming. Examples of clean energy include solar energy, wind power, and hydroelectric power. Nuclear power is considered by some to be clean energy (it does not emit CO_2), but opponents point to the risk of radioactive contamination that it carries. Natural gas is essentially nonpolluting but emits CO_2 when burned. (Because methane has a much greater global warming potential than does CO_2, burning it reduces the total amount of CO_2 equivalent GHG in the world, despite generating a GHG.)

Solar power (or solar energy) is energy given off by the Sun. Humans harness only a tiny fraction of this energy (less than 0.0001 percent), as it is diffuse, is very difficult to trap, and can vary with seasonal and weather conditions. Photovoltaic and solar thermal technologies are the solar technologies most widely used for contemporary power generation. In a photovoltaic process, solar energy is used to generate electricity. Photovoltaic cells, or solar batteries, can be found in many home appliances, including calculators and watches. Each photovoltaic cell consists of two layers. When sunlight strikes the solar cell, electrons from the lower layer move toward the upper surface, creating an electrical potential between the layers. This potential provides an electrical current. Electrons from the upper layer flow through an electrical device (for example, a small motor) back to the lower layer, thus providing energy for the device. These cells can be arranged into panels that generate enough electricity to power an entire family house.

Another solar technology, known as a solar thermal system, uses sunlight to generate heat. These systems are very popular in warm, sunny climates and have been used to provide hot water for homes and factories for decades. A typical solar collector for heating water consists of a shallow box with a plastic top and a black bottom. Glass or plastic tubes filled with water run through the inside of the box. The black bottom absorbs light and conveys the heat to the water within the tubes; the clear top of the box prevents heat from escaping. This technology can also be used to generate electricity, using a technology called concentrating solar power. This technology deploys thousands of dish-shaped solar concentrators attached to an engine that converts heat to electricity.

Wind power is the conversion of wind energy into useful forms of energy, mainly electricity. The design of wind turbines is based on wind-driven propeller blades. Turbine design has recently improved, allowing much more electricity to be generated at competitive prices. Europe is the world's number one user of wind power, followed by the United States. Hydropower is the most widely used form of clean energy. In hydropower dams, water under high pressure flows through turbines and generates electricity.

- **Significance for Climate Change**

In order to reduce global warming, the United States and other countries are making a major commitment to develop clean energy sources that generate little or no CO_2. For decades, clean energy technologies were too expensive to compete with fossil fuels. Concerns about global warming, coupled with the high prices of oil, have pushed forward the use of clean (or green) energy technologies. In the first years of the twenty-first century, clean energy supplied less than 7 percent of the world's energy consumption, but the potential of such technologies was enormous. The amount of solar energy reaching Earth exceeds by six thousand times humanity's global energy consumption. In the United States, the potential for the expansion of wind power is enormous as well, especially in the windy Great Plains area. Hydropower already

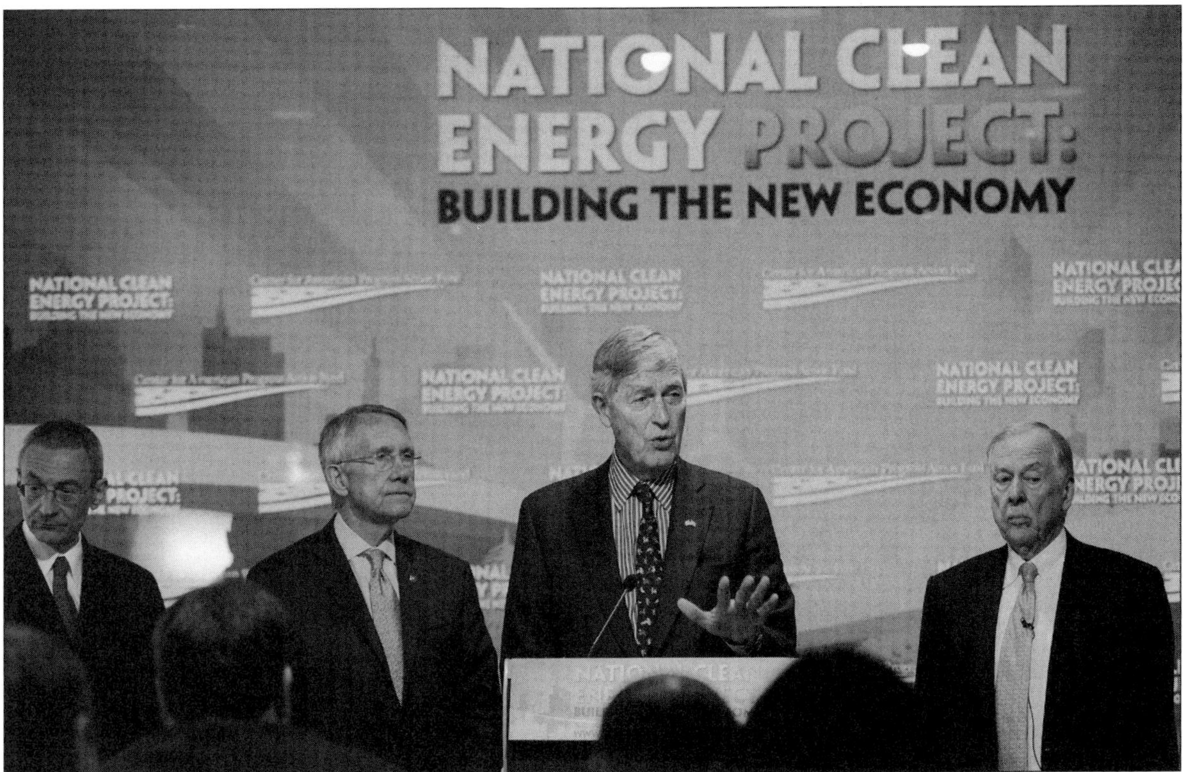

From left: Former White House chief of staff John Podesta, Senate Majority Leader Harry Reid, former senator Timothy Wirth, and energy executive T. Boone Pickens speak to the press about the need to develop clean energy technologies. (Jonathan Ernst/Reuters/Landov)

generates about 17 percent of the electricity used around the world.

The operation of wind turbines and solar power stations does not consume fossil fuels and does not produce GHG emissions. However, wind turbines and solar panels require the use of some fossil energy during construction and transportation. These initial CO_2 emissions are not significant compared to the environmental benefits of using clean wind and solar power. Wind turbines and solar panels do not require deforestation of land, so they do not interfere with forests' ability to sequester CO_2. In addition, land beneath wind turbines can be used for farming. Turbines may also be placed offshore.

Solar photovoltaic cells and collectors can be installed on the roofs and walls of homes and offices, as well as in the desert. Wind power is one success story involving the use of clean energy to reduce CO_2 emissions; the number of wind turbines used for the generation of electricity increased dramatically in the first decade of the twenty-first century. Prices for wind power decreased until they were very close to the price of electricity generated by burning coal, a major CO_2 producer. This decrease in price may allow replacing coal power with wind-generated power in the future. Some data indicate microclimate change around big wind farms, but these findings require additional studies.

Hydropower is also ideal for clean electricity generation. It does not produce CO_2 directly. However, hydropower stations are criticized, because they change the environment indirectly, in a way that may produce substantial amounts of both CO_2 and methane, a very powerful GHG. The water trapped in flooded areas directly upstream of hydropower dams contains a large amount of decaying plant material. This plant material is metabolized by microorganisms, leading to the formation of CO_2 and methane. However, hydropower technologies such as tidal energy exist that do not generate GHG emissions during use. Tidal power uses the energy generated by the daily ebb and flow of ocean tides. This energy is derived directly from the gravitational pull of the Moon and, to a lesser extent, the Sun. Large tidal power plants exist in France, Canada, and the United States.

Sergei Arlenovich Markov

- **Further Reading**

Kammen, Daniel M. "The Rise of Renewable Energy." *Scientific American* 295, no. 3 (September, 2006): 84-93. Describes different clean energy technologies and their future.

Nebel, Bernard J., and Richard T. Wright. *Environmental Science: Towards a Sustainable Future.* Englewood Cliffs, N.J.: Prentice Hall, 2008. Several chapters portray clean energy, its real-world applications, and its future.

Touryan, Ken J. "Renewable Energy: Rapidly Maturing Technology for the Twenty-first Century." *Journal of Propulsion and Power* 15, no. 2 (March/April, 1999): 163-174. Comprehensive review article that explains different types of clean energy technologies.

See also: Biofuels; Coal; Energy from waste; Energy resources; Ethanol; Fossil fuels; Fuels, alternative; Geothermal energy; Hydroelectricity; Hydrogen power; Nuclear energy; Renewable energy; Solar energy; Tidal power; Wind power.

Climate and the climate system

- **Category:** Meteorology and atmospheric sciences

- **Definition**

Climate is an aggregation of near-surface atmospheric conditions and weather phenomena over an extended period in a given area. It is characterized by statistical means and such variables as air temperature, precipitation, winds, humidity, and frequency of weather extremes. The time period is typically thirty years, as described by the World Meteorological Organization (WMO).

World climate is classified by either the empirical method, focusing on the effects of climate, or the genetic method, emphasizing the causes of climate. The empirical Köppen system, based on annual mean temperature and precipitation combined with vegetation distribution, divides world climate into five groups: tropical, dry, temperate, continental, and polar. Each group contains subgroups, depending on moisture and geographical location.

The genetic Bergeron, or air-mass, classification system is more widely accepted among atmospheric scientists, as it directly relates to climate formation and origin. Air-mass classification uses two fundamental attributes—moisture and thermal properties of air masses. Air masses are classified into dry continental (C) or moist maritime (M) categories. A second letter is assigned to each mass to describe the thermal characteristic of its source region: P for polar, T for tropical, and (less widely used) A for Arctic or Antarctic. For example, the dry cold CP air mass originates from a continental polar region. Sometimes, a third letter is used to indicate the air mass being cold (K) or warm (W) relative to the underlying surface, implying its vertical stability.

- **Climate System**

In a broad sense, climate often refers to an intricate system consisting of five major components: the atmosphere, hydrosphere, cryosphere, land surface (a portion of the lithosphere), and biosphere, all of which are influenced by various external forces such as Earth-Sun orbit variations and human activities. The atmosphere, where weather events occur and most climate variables are measured, is the most unstable and rapidly changing part of the system. The Earth's atmosphere is composed of 99 percent permanent gases (nitrogen and oxygen) and 1 percent trace gases, such as carbon dioxide (CO_2) and water vapor. All weather and climate phenomena are associated with the trace gases called greenhouse gases (GHG). Long-term increases in GHG concentration warm the climate, while day-to-day variations in atmospheric thermal and dynamic structures are responsible for daily weather events.

The hydrosphere comprises all fresh and saline waters. Freshwater runoff from land returning to the ocean influences the ocean's composition and circulations, while transporting a large amount of chemicals and energy. Because of their great thermal inertia and huge moisture source, oceans regulate the Earth's climate. The cryosphere consists of those parts of the Earth's surface covered by permanent ice in polar regions, alpine snow, sea ice, and permafrost. It has a high reflectivity (albedo), reflecting solar radiation back into space, and is critical in driving deep-ocean circulations.

Land surfaces and the terrestrial biosphere control how energy received at the surface from the Sun is returned to the atmosphere, in terms of heat and moisture. The partitioning between heating and moistening the atmosphere has profound implications for the initiation and maintenance of convection and thus for precipitation and temperature. Marine and terrestrial biospheres have major impacts on the atmosphere's composition through the uptake and release of GHG during photosynthesis and organic material decomposition.

- **Interactions Among Climate System Components**

The individual components of the climate system are linked by physical, chemical, and biological interactions over a wide range of space and time scales. The atmosphere and oceans are strongly coupled by moisture and heat exchange. This coupling is responsible for El Niño, the North Atlantic Oscillation, and the Pacific Decadal Oscillation, resulting in climate swings on interannual to interdecadal scales. The terrestrial biosphere and atmosphere exchange gases and energy through transpiration, photosynthesis, and radiation reflection, absorption, and emission.

These interactions form the global water, energy, and carbon cycles. The hydrologic cycle leads to clouds, precipitation, and runoff, redistributing water among climate components. Oceans and land surfaces absorb solar radiation and release it into the atmosphere by diffusion and convection. Global carbon and other gas cycles are completed by photosynthesis fixing CO_2 from the atmosphere and depositing it into the biosphere, soil, and oceans as organic materials, which are then decomposed by microorganisms and released back into the atmosphere.

Any change or disturbance to the climate system can lead to chain reactions that may reinforce or suppress initial perturbation through interactive feedbacks. If the climate warms, melting of glaciers and sea ice will ac-

A composite satellite image depicting Earth's interrelated climate systems. (NASA)

celerate, and the surface will absorb more solar radiation, further enhancing warming. On the other hand, warmer air temperatures result in more moisture in the atmosphere, increasing cloud cover, which increases albedo and reduces the absorbed solar radiation. This leads to cooling, compensating for the initial warming. There exist many such positive and negative feedback mechanisms, which makes the causality of climate change complex.

Zaitao Pan

• Further Reading

Ahrens, Donald C. *Meteorology Today: An Introduction to Weather, Climate, and the Environment.* 9th ed. Belmont, Calif.: Brooks/Cole, 2009. Textbook with many colorful illustrations of how climate and weather systems work. The last two chapters present concise summaries of the climate system and changes.

Bridgman, Howard A., and John E. Oliver. *The Global Climate System: Patterns, Processes, and Teleconnections.* New York: Cambridge University Press, 2006. Covers regional climate anomalies, global teleconnections, recent climate change, and human impacts upon the climate system.

Hartmann, Dennis L. *Global Physical Climatology.* San Diego, Calif.: Academic Press, 1994. Rigorous presentation of the physical principles that govern the energetics and gas exchanges within the climate system; focuses on the atmosphere.

McKnight, Tom L., and Darrel Hess. *Physical Geography: A Landscape Appreciation.* 9th ed. Upper Saddle River, N.J.: Prentice Hall, 2008. Covers all five components of the climate system; chapter 8 is devoted to climatic zones and types.

Oliver, John E., and John J. Hidore. *Climatology: An Atmospheric Science.* 2d ed. Upper Saddle River, N.J.: Prentice Hall, 2002. A descriptive textbook mainly for nonscientists who have a basic working understanding of the climate system.

See also: Abrupt climate change; Climate change; Climate feedback; Climate lag; Climate models and modeling; Climate prediction and projection; Climate sensitivity; Climate variability; Climate zones; Climatology; Continental climate; Global climate.

Climate change

• Category: Meteorology and atmospheric sciences

While Earth's climate is constantly changing in various ways, the planet tends to experience long-term trends toward either warming or cooling. The potential or actual contribution of postindustrial human activity to climate change, the consequences of that contribution, and the proper response to those consequences remain matters of crucial importance and signficant controversy.

• Key concepts

anthropogenic: caused by human activity

climate: long-term, average, regional or global weather patterns

emission scenario: a set of posited conditions and events, involving climatic conditions and pollutant emissions, used to project future climate change

greenhouse gases (GHGs): trace atmospheric gases that trap heat on Earth, preventing it from escaping into space

proxy: an indirect indicator of past climate conditions

weather: the set of atmospheric conditions obtaining at a given time and place

• Background

Climate is characterized by mean air temperature, humidity, winds, precipitation, and frequency of extreme weather events over a lengthy period of time, at least thirty years. Global warming is an example of climate change, and so are increases in the magnitude or frequency of floods and droughts experienced in many parts of the world during the past several decades. Climate change includes both natural variability and anthropogenic changes.

Although climate changes on longer than millennial timescales are natural, the global warming of the past 150 years or so is likely anthropogenic, according to the Fourth Assessment Report of the Intergovernmental Panel on Climate Change (IPCC) of the United Nations. The United Nations is concerned primarily with anthropogenic climate change, both because it poses a threat to global security and because it can be altered by altering human and governmental behavior. For this reason, the United Nations Framework Convention on Climate Change (UNFCCC) defines climate change as

> a change of climate which is attributed directly or indirectly to human activity that alters the composition of the global atmosphere and which is in addition to natural climate variability observed over comparable time periods.

• Climate Change Detection

Earth's atmosphere is chaotic, and weather can change dramatically in a matter of days or even hours. Temperature in some places may rise or fall by 20° Celsius or more in one day. On the other hand, climate, as the average of years of weather conditions, changes on a much smaller scale. For example, the global mean surface air temperature increased by only 0.6° Celsius during the twentieth century. By the same token, such a seemingly small increase can have extremely significant effects.

The climatic increase in mean surface air temperature is computed from tens of thousands of weather station records spanning decades. The difficulty of ensuring data continuity in time, uniformity in space, and constancy in observational methods poses serious challenges to climatologists. To discern slight trends amid diverging data, scientists use advanced mathematical tools to synchronize all observations, adjust discontinuities, and filter out local influences such as heat island effects.

Modern climate change has generally been observed with in situ thermometers and, later, with remote sensing devices. Paleoclimate change (change before about 1850) is inferred from proxy climate data. Tree rings can provide evidence of temperature and precipitation history for two to three thousand years, while tiny air bubbles trapped in the Antarctic ice deposits provide data on ice ages hundreds of thousands of years in the past. Pollen and zooplankton cells in river and sea sediments also contain useful proxy climate data.

Detecting climate change depends on individual variables. Temperature change is the most reliable such variable, because its internal variability is small and it is more widely observed than other variables. Long-term precipitation changes are more difficult to discern, because rain- and snowfall vary so greatly from one year to the next. The intensity and frequency of extreme weather events such as hundred-year floods are even more difficult to detect, because these events are rare, so a significant data set must cover many years.

Instrument records from land stations and ships indicate that the global annual mean surface air temperature rose during the twentieth century. The warming occurred more quickly in high latitudes than it did in the tropics. It was also faster over land than it was over the ocean and faster in the Northern Hemisphere than in the Southern Hemisphere. Winters warmed more than did summer, and nights warmed more than did days. Contemporary daily temperature ranges have narrowed, precisely because nights have warmed more than have days.

Extensive heat waves and intense floods have become more frequent in recent decades. Globally, the average number of tropical storms (about ninety per year) changed little during the twentieth century, although historical data are poor for some regions. In the North Atlantic, where the best records are available, there has been a clear increase in the number and intensity of tropical storms and major hurricanes. From 1997 to 2006, there were about fourteen tropical storms per year, including about eight hurricanes in the North Atlantic, compared to about ten storms and five hurricanes between 1850 and 1990.

On timescales of thousands of years or greater, the Earth's climate has been both warmer and much colder than it is today, although temperatures around the turn of the twenty-first century were the warmest in the past two thousand years. Based on ice-core proxy data, four major global glaciations occurred in past 450,000 years, about one every 100,000 years, correlating well with the cyclical variations in Earth's orbit known as the Milanković cycles. Various ice ages occurred, with the most recent one ending about 11,500 years ago. Before that, much of North America was covered in permanent ice. Over the course of Earth's history, its temperature has swung more than 10° Celsius between cold and warm modes.

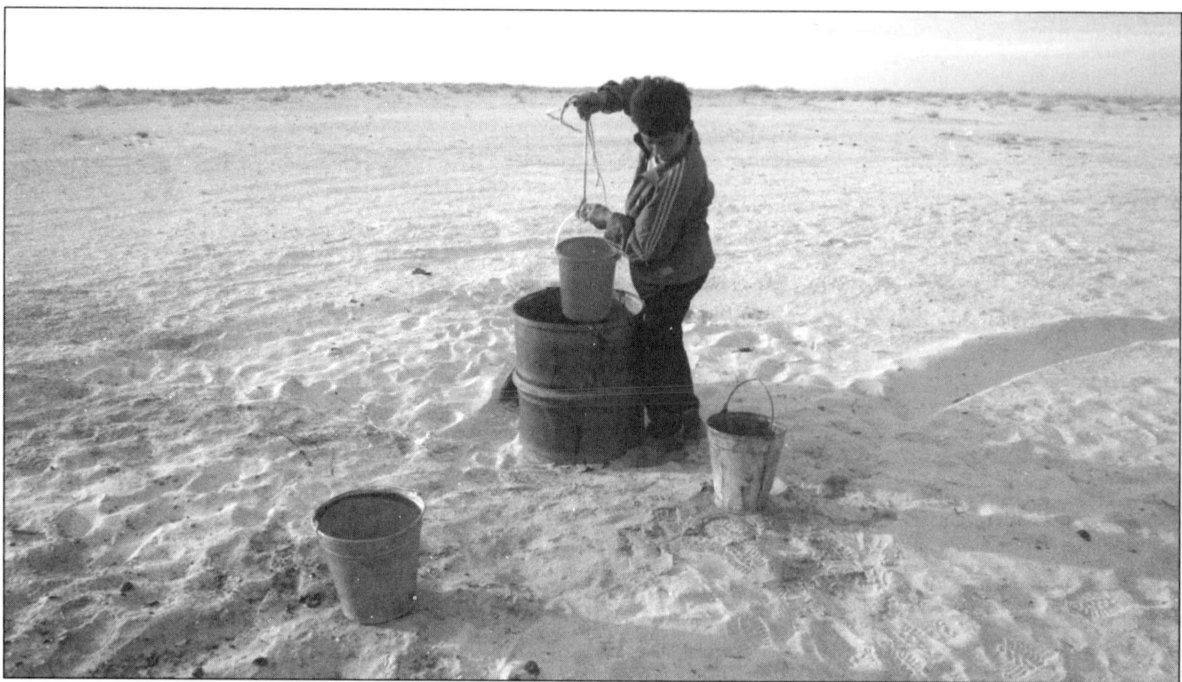

In the bed of what was once the Aral Sea, a Kazakh villager pulls water from a well sunk into the sandy, desert ground. (Shamil Zhumatov/Reuters/Landov)

• Climate Change Scenario

Future climate changes are predicted by climate models based on assumed greenhouse gas (GHG) emission scenarios. The scenarios range from high fossil fuel consumption, resulting in atmospheric carbon dioxide (CO_2) concentration of 800 parts per million, to low consumption, with CO_2 concentration reaching 550 parts per million. The reliability of these predictions depends on future global environmental, energy, and climate policy, as well as the accuracy of the models.

Most models project that climate change will accelerate during the twenty-first century and that the global average temperature will increase by between 1.8° Celsius and 4.0° Celsius by 2100. As in the past, warming will be more pronounced in the polar Northern Hemisphere during winter. Precipitation amounts are likely to increase in high latitudes and to decrease in most subtropical lands. Heat waves and heavy precipitation events will very likely increase in frequency. With warmer oceans, future tropical storms will become more intense, with greater peak wind speeds and heavier precipitation.

• Context

Climate change may be attributed to natural processes or to human activity. Natural factors include the Earth's internal processes, such as volcanic eruptions, as well as external parameters, such as solar luminosity and Earth's orbital pattern around the Sun. Anthropogenic activity includes GHG and aerosol emission and, to a lesser degree, changes in land use. Separating natural and anthropogenic causes of climate change is challenging, if it is possible at all. Since no controlled laboratory setting exists in which to conduct climate change experiments, climate scientists have developed computer models based on the laws governing climate systems. By altering model settings, one can simulate natural and anthropogenic effects on climate, separately or in combination, thereby tracing the causes of climate change. In general, on scales of a decade to a century, climate change is attributable to atmosphere-ocean interaction and to human activity. On scales of millennia to hundreds of thousands of years, the variations in Earth's orbit directly controls the planet's climate. This orbit is described by the Milanković cycles, which repeat every 20,000 to 100,000 years. Beyond the million-year timescale, tectonic drift is likely the main driver of climate change.

Zaitao Pan

• Further Reading

Easterling, David R., et al. "Maximum and Minimum Temperature Trends for the Globe." *Science* 277 (July 18, 1997): 364-367. Pioneering work on asymmetric warming between day and night.

Intergovernmental Panel on Climate Change. *Climate Change, 2007—The Physical Science Basis: Contribution of Working Group I to the Fourth Assessment Report of the Intergovernmental Panel on Climate Change.* Edited by Susan Solomon et al. New York: Cambridge University Press, 2007. Pools and synthesizes the most authoritative opinions on climate change from thousands of scientists worldwide. Reviews extensively past observations and future scenarios published in the early twenty-first century.

Mann, Michael E., Raymond S. Bradley, and Malcolm K. Hughes. "Global-Scale Temperature Patterns and Climate Forcing over the Past Six Centuries." *Nature* 392 (April 23, 1998): 779-787. The source of the famous and controversial "hockey stick" plot of the global warming trajectory, arguing for a sudden increase in temperature in the twentieth century.

Nakicenovic, N., et al., eds. *Special Report on Emission Scenarios: A Special Report of Working Group III of the Intergovernmental Panel on Climate Change.* New York: Cambridge University Press, 2000. Provides the baselines for GHG emission scenarios that are used globally to conduct climate change simulations.

National Research Council. *Abrupt Climate Change: Inevitable Surprises.* Washington, D.C.: National Academies Press, 2002. Comprehensive review of scientific theory and evidence for abrupt climate change.

United Nations Framework Convention on Climate Change, Article 1: Definitions. Geneva, Switzerland: United Nations, 1992. Gives specific definitions of climate-change-related terminology.

See also: Abrupt climate change; Climate and the climate system; Climate feedback; Climate models and modeling; Climatology; Global climate; Global dimming; Greenhouse effect; Hockey stick graph.

Climate Change Science Program, U.S.

- **Category:** Organizations and agencies
- **Date:** Established 2002
- **Web address:** http://www.climatescience.gov/

• Mission

The U.S. Climate Change Science Program (CCSP) is an interagency U.S. government program established by President George W. Bush that coordinates and publishes thirteen federal agencies' research related to the Earth's climate and to human impacts on climate change. The CCSP facilitates communication among the thirteen agencies, as well as coordinating their publications. The CCSP was established in 2002 to bring together scientific research from agencies including the Environmental Protection Agency, the National Science Foundation, the Smithsonian Institution, the Department of Energy, and the Department of Agriculture.

• Significance for Climate Change

The CCSP has been criticized as an effort by the Bush administration to delay practical responses to global warming. The program's integrity was questioned when, in 2005, a CCSP scientist accused White House officials of editing draft reports before they were made public in order to downplay references to global warming. However, in 2006, the CCSP published a final report that resolved inconsistencies created by scientific research that seemed to disprove global warming. The report confirmed that the Earth's temperature was measurably increasing and that human activity contributed to climate change.

The 1990 Global Change Research Act requires

the White House to provide Congress with an assessment of climate change every four years. Through the CCSP, the Bush administration opted instead to produce twenty-one synthesis and assessment reports. In November, 2006, environmental groups sued the administration for failing to comply with the 1990 law, and White House officials agreed to publish all twenty-one reports by May, 2008. By that date, however, only six reports had been completed. These reports provided scientific analyses of temperature measurements, the possible effects of increased or stabilized amounts of greenhouse gases in the atmosphere, North American production of carbon dioxide, methods for protecting ecosystems from the effects of climate change, anticipated extreme weather conditions, and the effects of climate change on natural resources and energy production.

The CCSP confirmed that global warming was a scientifically provable phenomenon, documented trends in global warming, and predicted its large-scale effects. In 2008, the Senate called for the CCSP to shift its focus away from global trends, data collection, and scholarly communication. Instead, the program was tasked with developing practical ways local governments and policy makers across various climatic regions of the United States could plan for and respond to climate change.

Maureen Puffer-Rothenberg

See also: Center for the Study of Carbon Dioxide and Global Change; Energy Policy Act of 1992; National Climate Program Act; National Research Council; U.S. energy policy; U.S. legislation.

Climate engineering

- **Category:** Science and technology

- **Definition**

Climate engineering is a species of geoengineering, an expansive field that ranges over a wide variety of subjects, combining the study of Earth's atmosphere, lithosphere, and biospheres with practical engineering principles. Also termed planetary engineering or macro engineering, the practice was suggested in the 1970's by Cesare Marchetti, who envisioned mitigating climatic impacts from burning fossil fuels by injecting carbon dioxide (CO_2) deep into the ocean. The term geoengineering then started to appear in publications by the National Academies of Science and entered the conventional climate change debate.

Practitioners of geoengineering or climate engineering develop technologies for the large-scale, intentional manipulation of the global environment. The threat represented by the greenhouse effect has generated international concern among politicians, policy makers, scientists, and engineers, who are continuously trying to fight global climate change and control warming. Climate engineering has gained a great deal of attention worldwide as a result of its potential to combat global warming. Goals of such engineering are to reduce the amount of solar radiation absorbed by the Earth, control the atmosphereic concentration of CO_2, and maneuver the ocean-atmospheric system by redirecting heat.

- **Methods of Climate Engineering**

Important climate engineering technologies and goals include use of atmospheric aerosols, CO_2 sequestration, re-icefication of the Arctic, use of ocean-cooling pipes, cloud seeding, genetically modified CO_2-eating trees, space mirrors, and glacier blankets. Use of atmospheric aerosols and CO_2 sequestration have gained the greatest reputation in the scientific community.

Aerosols are microscopic particles floating in the atmosphere. The chemical composition of aerosols can vary widely; however, the sulfate aerosols have aroused particular concern. There are two main types of atmospheric aerosols that influence the Earth's climate. Natural aerosols result from volcanoes, wildfires, desert dust, and terrestrial and marine biogenic activity. Anthropogenic aerosols include smoke particulates from burning fossil fuels and tropical forests, as well as by-products from industrial activity. Sulfate aerosols influence the global climate by scattering and absorbing solar radiation, increasing the albedo, and modifying the size and duration of clouds, which in turn produce a global cooling effect.

Research suggests that it might be possible to increase Earth's albedo and compensate for global warming by adding aerosols to the atmosphere. Mikhail I. Budyko for the first time proposed injecting sulfate aerosols into the atmosphere to create an artificial volcano effect. David W. Keith's analysis has indicated that about 1.5 to 10.0 teragrams of sulfate per year would balance the effect of a doubled atmospheric CO_2 concentration. The most serious potential side effects of Budyko's proposal would be alteration of atmospheric chemistry, ozone layer depletion, and whitening of the daytime sky, which would all destabilize Earths ecosystems in ways both predictable and unpredictable.

The enhanced greenhouse effect due to fossil fuel combustion can be reduced by capturing CO_2 emissions from industrial power plants and sequestering them. Contemporary climate engineering studies focus on technologies that will extract and compress CO_2 from power plants and store them in carbon reservoirs or carbon sinks. The CO_2 extraction and capture technologies, which involve precombustion, postcombustion, oxyfuel combustion, and industrial separation (such as natural gas processing and ammonia production), are expensive, require substantial energy, and remain mostly in the research and development phase.

Carbon sinks can be classified as biological, geologic, or oceanic. Carbon from fossil fuel combustion can be sequestered in geologic sinks such as coal, oil, and gas fields and saline aquifers. The Intergovernmental Panel on Climate Change (IPCC) estimates that there is enough capacity worldwide permanently to store as much as 1.1 trillion metric tons of CO_2 underground in geological formations. CO_2 can be removed biologically from the atmosphere via photosynthesis, afforestation, or changes in farming practices, as well as restoration of phyto-

In a modest example of climate engineering, Chinese farmers plant wheat straw at the edge of the Mu Us Desert. Since the 1970's, farmers in this area have transformed more than 470,000 hectares of desert. (Xinhua/Landov)

plankton by seeding the ocean surface with micronutrients such as nitrates, phosphates, silica, and iron. Conversely, such increased photosynthesis could increase CO_2 emissions, deplete oxygen, and trigger climate warming.

The third and greatest carbon sink on Earth is the ocean, which balances atmospheric CO_2 levels. About 80 percent of atmospheric CO_2 is absorbed by the oceans on an exponential timescale of about three hundred years. This timescale can be reduced by direct disposal of CO_2 in the deep ocean (via pipelines or through dumping from ships) or in geological formations beneath the ocean bed. Accomplishing such sequestration requires a thorough understanding of the feasibility, efficiency, and environmental consequences of such a project.

• Significance for Climate Change

Climate engineering appears to be a potential solution to the global warming crisis. However, great caution needs to be taken before attempting any large-scale manipulation of the global climate, as the climate system is extremely complex and chaotic. Moreover, the nations or powers that engage in climate engineering are not necessarily the ones that will suffer the greatest damage should that engineering go awry. Thus, various political and ethical debates arise over whether geoengineering should be considered as a viable option to ameliorate the effects of climate change.

Arpita Nandi

• Further Reading

Brovkin, Victor, et al. "Geoengineering Climate by Stratospheric Sulfur Injections: Earth System Vulnerability to Technological Failure." *Climatic Change*, September 17, 2008. Uses a climate-carbon-cycle model of intermediate complexity to investigate situations in which injected stratospheric sulfur was used as a measure to balance CO_2-induced global warming.

Intergovernmental Panel on Climate Change. *Carbon Dioxide Capture and Storage: A Special Report of Working Group III of the Intergovernmental Panel on Climate Change.* Edited by Paul Freund et al. Geneva, Switzerland: Author, 2005. This extensive report discusses the options for capturing and storing anthropogenic CO_2 emissions.

Keith, David W. "Geoengineering and Carbon Management: Is There a Meaningful Distinction?" In *Greenhouse Gas Control Technologies: Proceedings of the Fifth International Conference*, edited by D. Williams et al. Collingwood, Vic.: CSIRO, 2001. Discusses attempts to manage and mitigate CO_2 production from a geoengineering perspective.

Marchetti, Cesare. "On Geoengineering and the CO_2 Problem." *Climatic Change* 1, no. 1 (March, 1977): 59-68. Prevention of CO_2 emissions through physical capture of CO_2 from fuel power plants was first proposed by Marchetti. He argues here that CO_2 can be sequestrated into suitable sinking thermohaline currents that would carry and spread it into the deep ocean.

Rasch, Philip J., Paul J. Crutzen, and Danielle B. Coleman. "Exploring the Geoengineering of Climate Using Stratospheric Sulfate Aerosols: The Role of Particle Size." *Geophysical Research Letters* 35 (2008). Investigates how the amount and size of atmospheric sulfur aerosols affect the climate system.

See also: Anthropogenic climate change; Human behavior change; Reservoirs, pools, and stocks; Sequestration; Sinks; Sulfate aerosols; Technological change.

Climate feedback

• **Category:** Meteorology and atmospheric sciences

Negative climate feedback helps preserve the Earth's climate in its existing state, whereas positive feedback accelerates changes, creating potential tipping points beyond which global warming or cooling becomes extremely difficult to reverse. Understanding both types of feedback is necessary to develop adequate climate policies.

• Key concepts

aerosols: tiny particles or liquid droplets suspended in the atmosphere; some, such as sea salt, are natural, while others, such as soot from power plants, are of human origin

albedo: the extent to which an object reflects radiation; the reflectivity of objects with regard to incoming solar radiation

climate forcing: factors that alter the radiative balance of the atmosphere (the ratio of incoming to outgoing radiation)

greenhouse gases (GHGs): gases (or vapors) that trap heat in the atmosphere by preventing or delaying the outward passage of long-wavelength infrared radiation from the Earth's surface out to space

infrared radiation: radiation with wavelengths longer than those of visible light, felt by humans and animals as heat

lapse rate: the change in a variable with height, often used to discuss changes in temperature with altitude in a context of climate change

radiative balance: the balance between incoming and outgoing radiation of a body in space, such as Earth

• Background

Like all planets, the Earth is bombarded by radiation from the Sun, stars, and other space-based sources of energy. Eventually, that energy is returned to space. Incoming and outgoing radiation must ultimately balance out, in conformity with the first law of thermodynamics, the law of conservation of energy. In the case of the Earth, which has a significant atmosphere and varied surface geography, the pathways by which radiation reaches the surface and is radiated back out to space are somewhat convoluted. Some incoming radiation is reflected back to space by clouds and particles in the atmosphere before it reaches the ground, some is reflected back toward space at ground level, and some is absorbed and then reradiated away in the form of long-wave, or infrared, radiation. Some of this reradiated infrared radiation is bounced back toward the ground by atmospheric gases or water vapor before it eventually makes its way back into space. Still, over time, the total Earth-atmosphere system is said to be in radiative balance.

Factors that can alter Earth's radiative balance are called "climate forcings" and "climate feedbacks." Understanding climate forcings and feedbacks is at the heart of understanding how greenhouse gases emitted into the atmosphere might affect future temperature trends. Climate feedbacks are particularly important to understand because computer models projecting future warming incorporate assumptions about such feedbacks (especially water vapor) that significantly elevate predicted temperature increases due to greenhouse gas emissions.

• Climate Forcing

One cannot understand climate feedbacks without understanding climate forcings. A climate forcing (technically a "radiative forcing") is something that exerts a direct effect on the radiative balance of the Earth's atmosphere—that is, something that changes the balance of incoming versus outgoing radiation either permanently or transiently. Forcings include incoming solar radiation, the heat-retaining ability of greenhouse gases present naturally in the atmosphere, human greenhouse gas emissions and conventional air pollutants, changes in land use that might alter the reflectivity of the Earth's surface, and more. The Intergovernmental Panel on Climate Change (IPCC) identifies nine major radiative forcing components, some of which are considered well understood, and others less so. Forcings identified by the IPCC include the greenhouse gases, ozone, stratospheric water vapor, surface albedo, aerosols, contrails, and solar irradiance.

• Climate Feedback

Climate feedbacks are secondary changes to the radiative balance of the climate stemming from the influence of one or another climate forcing. Such climate feedbacks may be either positive or negative; a positive feedback would amplify the effect of a change in a given climate forcing, while a negative feedback would damp down the effect of a change in a given climate forcing. Thus, a change in the atmosphere's water vapor content could cause greater cloudiness, which could, depending on the type of clouds, constitute a positive or negative feedback.

According to the National Research Council (part of the U.S. National Academies of Science), climate feedbacks that primarily affect the magnitude of climate change include clouds; atmospheric water vapor; the lapse rate of the atmosphere (defined as the change in temperature with altitude); the reflectivity, or albedo, of ice masses; biological, geological, and chemical cycles; and the carbon cycle. Feedbacks that primarily affect temporary responses of the climate include ocean heat uptake and circulation feedbacks. Finally, feedbacks that mostly influence the spatial distribution of climate change include land hydrology and vegetation feedbacks, as well as natural climate system variability.

When used in projecting future temperatures stemming from greenhouse gas emissions, estimates of some climate feedbacks are incorporated into computerized models of the climate system. These climate feedbacks—clouds, water vapor, surface albedo, and the lapse rate—are expected to contribute as much (or more) warming to the atmosphere as changes in the greenhouse gases do by themselves.

• Context

The extent to which climate feedbacks might increase or decrease the heat-trapping effects of humanity's greenhouse gas emissions is an important factor in public policy development. If computer models understate the extent of positive feedbacks, future warming could be worse than projected, and actions undertaken to combat climate change might be insufficient to the challenge. By contrast, if computer models overstate positive feedbacks, or underestimate negative feedbacks, projected future warming scenarios could be too high. In this case, massive resources spent on controlling greenhouse gas emissions could be wasted, leaving society less able to deal with other challenges, environmental or otherwise.

Kenneth P. Green

An iceberg melts in the Jokulsarlon glacier lake in Iceland. Melting icebergs decrease Earth's albedo, increasing warming in the region, which causes further melting of icebergs. (©Dreamstime.com/36clicks)

• Further Reading

Houghton, John. *Global Warming: The Complete Briefing*. New York: Cambridge University Press, 2009. A leading textbook on climate change, updated to reflect findings of the IPCC's most recent assessment reports.

Intergovernmental Panel on Climate Change. *Climate Change, 2007—The Physical Science Basis: Contribution of Working Group I to the Fourth Assessment Report of the Intergovernmental Panel on Climate Change*. Edited by Susan Solomon et al. New York: Cambridge University Press, 2007. The first volume of the IPCC Fourth Assessment Report, focusing on the quantification of climate change, atmospheric greenhouse gas concentrations, climate forcings and feedbacks, and the fundamental science underlying the theory of climate change. This volume includes a summary for policy makers, a technical summary, and a "frequently asked questions" section accessible to the interested lay audience.

National Research Council. *Understanding Climate Change Feedbacks*. Washington, D.C.: National Academies Press, 2003. A publication examining the question of climate feedbacks and their roles in understanding and predicting climate change.

North, Gerald, and Tatiana Erukhimova. *Atmospheric Thermodynamics: Elementary Physics and Chemistry*. New York: Cambridge University Press, 2009. A textbook coauthored with a climate modeler deeply steeped in the study of climate forcings and feedbacks. Suitable for an undergraduate audience.

Spencer, Roy. *Climate Confusion: How Global Warming Hysteria Leads to Bad Science, Pandering Politicians, and Misguided Policies That Hurt the Poor*. New York: Encounter Books, 2006. Spencer, an expert in the satellite measurement of the climate, argues that a misunderstanding of climate feedbacks, particularly water vapor feedbacks, has led to an overestimation of potential climate change due to greenhouse gas emissions.

See also: Albedo feedback; Amazon deforestation; Carbon cycle; Climate lag; Climate sensitivity; Clouds and cloud feedback; Ecosystems; Forcing mechanisms; Industrial Revolution; Paleoclimates and paleoclimate change; Radiative forcing.

Climate lag

• **Category:** Meteorology and atmospheric sciences

• Definition

Climate lag denotes a delay in the climate change prompted by a particular factor. This can occur when the system is influenced by another, slower-acting, factor. For example, when carbon dioxide (CO_2) is released into the atmosphere, its full effect may not be recognized immediately, because some of the CO_2 may be partially absorbed, and much later released, by the oceans. The considerable amount of lag in geophysical systems can be seen in the delay between actions that increase or decrease climate forcings (changes that affect the energy balance of Earth) and their consequent impacts on the climate. Lag can be accounted for in several ways: Some occurs because of the length of time it takes for certain chemicals to cycle out of the atmosphere, some results from the effects of warming upon natural cycles, and some comes from the slow pace of oceanic temperature change. Many climate scientists estimate climate lag to be between twenty and thirty years; thus, even if all additional carbon emissions were to cease, Earth would still experience two to three decades of warming before the cessation took effect.

• Significance for Climate Change

If, during a two- to three-decade lag, rising temperatures begin to trigger climate feedback effects—such as a reduction of the polar ice cap, allowing more heat to be absorbed by the dark water, or large emissions of methane from melting permafrost—the lag time could be extended. The longer some kinds of climate disruption are delayed, and the more climate commitment is built up, the more likely it is that feedback effects will be seen. "Climate commitment" refers to the fact that climate reacts with a delay to influencing factors. For example, an increase in the concentration of greenhouse gases (GHGs) will influence Earth's climate over time, rather than all at once. There are at least three responses to this situation: prevention, mitigation, and remediation. The potential for dangerous feedback effects drives a prevention response, that is, action that reduces the global warming risk. Since it is not yet known how damaging to the environment feedback effects could be, it seems prudent to do everything possible to start eliminating the anthropogenic sources of GHGs and thereby keep the level of committed warming to a minimum. A mitigation response takes a practical approach: Climate disaster is already imminent, and the best prevention efforts may be too little, too late, but there is still the need to reduce the worst of the threats. A remediation approach would not look at ways to change greenhouse emissions and their consequences; instead, efforts would be made to use geoengineering, that is, to alter the core geophysical processes that relate to global warming.

Victoria Price

See also: Abrupt climate change; Climate change; Climate feedback; Climatology; Paleoclimates and paleoclimate change.

Climate models and modeling

• **Category:** Meteorology and atmospheric sciences

Earth's climate is a system of vast complexity, and modeling that system in useful ways poses significant challenges. Understanding the hierarchy of climate models, however, allows scientists to create models using the minimum complexity necessary for the specific climatic problem under investigation.

• Key concepts

albedo: the fraction of incident light reflected from a body such as Earth

convection: vertical transfer of heat by movement of warm parcels of air

emissivity: a measure of the ability to radiate absorbed energy

greenhouse gases (GHGs): atmospheric trace gases that trap heat, preventing it from escaping into space

Navier-Stokes equation: an equation describing the flow of air (or any other fluid)

parameterization: the approximation of processes

• Background

Climate models are the most important tools for quantitatively estimating how natural and anthropogenic forces affect different aspects of the climate system. A climate model is a numerical representation of the climate system, including the physical, chemical, and biological properties of its components and the interactions between those components. It consists of a large number of mathematical equations that quantitatively describe the processes occurring within the climate system. Climate models span the entire range from very simple models that can be solved on paper to very complex models that require large supercomputers.

• Hierarchy of Climate Models

Climate model equations describe the conservation and transport of atmospheric heat, moisture, and momentum along three directions: latitude, longitude, and vertical altitude. For simplicity, climate models are often averaged along one or more of these directions. Depending on the nature of averaging, the following hierarchy of climate models develops.

Zero-dimensional energy-balance models. The simplest form of climate model is averaged along all three directions. It consists of a single equation describing the balance of incoming and outgoing energy at the top of the atmosphere. These models calculate the temperature of the Earth as a function of incoming solar radiation, Earth's albedo, and emissivity.

The advantage of zero-dimensional models is their simplicity. These are the only type of climate models that can be solved by hand using a simple calculator. They have been widely used to study how global temperature may change in response to changes in solar radiation (during the eleven-year and longer solar cycles), in Earth's albedo (resulting from changes in cloud cover or sea-ice extent), and in emissivity.

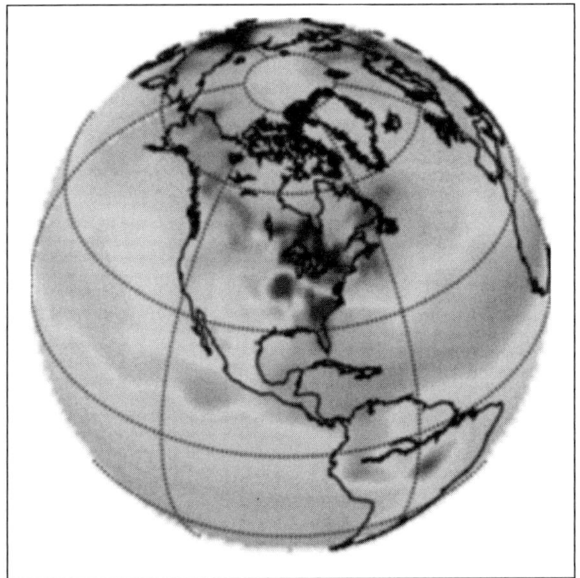

A computer model of global carbon dioxide emissions. (NOAA)

One-dimensional models. Radiative-convective (RC) models are averaged along latitude and longitude, but not along the vertical direction. They include the two most important processes of vertical energy transport in the atmosphere: upward and downward transport of radiation and upward transport of heat from Earth's surface by convection. The inclusion of vertical transport is important for climate modeling. Because gases in the atmosphere are unevenly distributed vertically, the impact of climate change varies with altitude. RC models can accurately simulate the vertical profile of temperature and temperature change in the atmosphere.

In contrast with RC models, energy balance (EB) models are averaged along the longitudinal and vertical directions. Thus, they may be used to account for equator-pole heat transport that arises because the equator receives more solar energy than do the poles. Unlike RC models, which use very sophisticated and realistic heat transport equations, the transport in EB models is described through relatively simple parameterizations that have limited applicability. EB models have been used to study the mechanisms of poleward heat transport, ice age climates, and ocean-atmosphere interactions.

Two-dimensional models. Two-dimensional RC models are averaged along the longitudinal direction. They thus combine one-dimensional RC and EB models. They account for the climate system's two most important heat transport processes, vertical and poleward. These models have been widely used to study atmospheric general circulation patterns such as the Hadley cell. Because of their relative simplicity, these models have also been coupled with chemistry and radiation models to study stratospheric chemical-radiative-dynamical interactions.

Two-dimensional EB models average atmospheric properties and processes over the height of the atmosphere to describe the energy balance over Earth's entire surface. These models have limited applicability and have been used to study ice-age climates.

• Three-Dimensional General Circulation Models

General circulation models (GCMs) are the most complex climate models. They solve the full set of Navier-Stokes equations for atmospheric flow and thermodynamic and microphysical equations for conservation of energy and moisture along all three cardinal directions. The three-dimensional nature of the models allows them realistically to handle the transport of energy, moisture, and momentum in the horizontal as well as vertical dimensions. Hence, GCMs are very useful for investigating regional aspects of climate and climate change.

Sometimes these models are coupled with dynamic ocean models, atmospheric chemistry models, or ecosystem dynamics models. Such multi-model coupling allows scientists to study the interactions between all three components of the climate system (atmosphere, biosphere, and hydrosphere).

The major disadvantage of GCMs, especially coupled atmosphere-ocean GCMs (AOGCMs), is that they are computationally very expensive. Only well-funded research groups have access to the supercomputers needed to run simulations with GCMs. Because of their size and complexity, it takes a long time, often weeks, to complete each simulation. Another drawback of these models is their coarse resolution. Each grid cell in a GCM is at least 1° latitude by 1° longitude. GCMs therefore cannot explicitly resolve fine-scale processes such as boundary-layer turbulence or cumulus clouds. The models do not completely ignore such fine-scale processes, however. Rather, these processes are approximated by subgrid parameterizations.

Earth systems models of intermediate complexity (EMICs) are another very powerful tool in climate studies. These models bridge the gap between complex GCMs and simpler models. EMICs simulate all the physical and dynamical processes contained in GCMs, but they use simpler parameterizations and coarser resolution, making them much faster to run than GCMs. Because of their computational efficiency, EMICs are becoming very popular in the climate policy field. They are widely used to study the feasibility of different adaptive and mitigation policies on climate change.

• Context

Climate models in general form the basis for nearly all rational argumentation as to the causes and severity of climate changes of the past and present, as well as the effects of human actions on climate in the future. They are thus crucial to both climate science and climate policy. However, different climate modeling systems—and different deployments of the same systems—have yielded different results, rendering it difficult thus far to come to definitive conclusions as to the optimal course of action in either the near term or the long term. Hierarchizing climate models can help reduce some of the uncertainty by deploying the most appropriate model for a given purpose, but precise, definitive predictions of climate change remain elusive.

Somnath Baidya Roy

• Further Reading

Harvey, D., et al. "An Introduction to Simple Climate Models Used in the IPCC Second Assessment Report." IPCC Technical Paper 2. Edited by J. T. Houghton, L. G. Meira Filho, D. J. Griggs, and K. Maskell. Geneva, Switzerland: Intergovernmental Panel on Climate Change, 1997. Surveys low-order climate models and modeling methods used by the foremost international scientific body working on climate change.

Held, Issac. "The Gap Between Simulation and Understanding in Climate Modeling." *Bulletin of American Meteorological Society* 86, no. 11 (November, 2005): 1609-1614. Held is a well-known climate modeler based in the NOAA Geophysical Fluid Dynamics Laboratory at Princeton, New Jersey. This article describes why even very simple models are essential for climate studies.

Intergovernmental Panel on Climate Change. *Climate Change, 2001—The Scientific Basis: Contribution of Working Group I to the Third Assessment Report of the Intergovernmental Panel on Climate Change.* Edited by J. T. Houghton et al. New York: Cambridge University Press, 2001. Provides a scientific assessment of past, present, and future climate change. Chapter 8 contains a thorough discussion of climate models and modeling hierarchy. Illustrations, references, index.

_______. *Climate Change, 2007—The Physical Science Basis: Contribution of Working Group I to the Fourth Assessment Report of the Intergovernmental Panel on Climate Change.* Edited by Susan Solomon et al. New York: Cambridge University Press, 2007. Revisits and expands the modeling discussion of the third report.

Kiehl, Jeffrey T. "Atmospheric General Circulation Modeling." In *Climate System Modeling*, edited by K. Trenberth. New York: Cambridge University Press, 1992. Discusses the different kinds of averaging of atmospheric dynamic and thermodynamic equations and the different kinds of climate models they produce.

See also: Bayesian method; Climate prediction and projection; General circulation models; Parameterization; Slab-ocean model.

Climate prediction and projection

• **Category:** Meteorology and atmospheric sciences

The methodology and databases accumulated over several decades to monitor and predict normal cyclical variations in climate provide a base against which to assess whether, and to what extent, current global warming is anthropogenic and to what extent it may be self-correcting.

• Key concepts

anthropogenic climate change: changes in overall long-term weather patterns in a region due to human activity

extreme weather events: natural disasters caused by weather, including floods, hurricanes, drought, and prolonged severe hot and cold spells

proxies: measurable parameters, correlated with climate, that are preserved in the geologic record (for example, oxygen isotope ratios and fossil pollen)

• Background

Weather is the sum of the atmospheric conditions we experience on a daily basis; climate is what produces those conditions in a given geographical area. Climatology has been described as "geographical meteorology." The principal purpose of studying a region's climate is to discover measurable factors that accurately predict, months or years in advance, what the general weather regime will be like in a given area. In making their predictions, climatologists first look at established historical patterns, both recorded in writing and documented through proxies. Increasingly they also investigate whether there are systematic perturbations in established historical patterns, and they also incorporate the effects of events whose probability is low or unknown.

• History

Attempts by humans to forecast climate date back several millennia, at least to the days of ancient Babylon, when astrologers used the motions of the Sun, Moon, and planets to predict whether the coming season would be favorable for agriculture. The practice may be even older. Anthropologist Johannes Wilbert recorded a religiously based system of climate prediction among the Warao Indians of Venezuela, a group of primitive Stone Age agriculturalists. The biblical story of Joseph (c. 1800 B.C.E.) relates how Joseph interpreted Pharaoh's dream of seven lean cattle devouring seven fat cattle as a prediction of impending drought and famine, enabling the Egyptians to prepare in advance.

Climate prediction for agricultural purposes remained a major function of astrology, and later astronomy, from Babylon to the pioneer European astronomer Tycho Brahe in the sixteenth century. Even today, many successful American farmers follow the *Farmers' Almanac*, with its astrologically based recommendations for planting crops. The practice persists in part because it has some basis in fact: The phases of the Moon, and to a lesser extent the orbits of Jupiter and Saturn, do affect climate.

Interest in a more scientific and systematic approach to climate prediction gained impetus in the late nineteenth century in response to expansion of Europeans into regions that experience more extreme and variable climatic conditions than Europe. The eleventh edition of the *Encyclopedia Britannica* (1911) divides the globe into climatic zones, describes the variability of each, and discusses evidence for a general warming trend, which the author of the encyclopedia article was inclined to dismiss as unproven.

A decade later, Sir Gilbert Walker began publishing his pioneering work on fluctuations in the Indian monsoons, their relationship to periodic famines, and the correlation between them and oscillation of high- and low-pressure areas between the Indian Ocean and the tropical Pacific. This Southern Oscillation, later shown to be linked to the El Niño phenomenon in the eastern Pacific, is the most important determinant of cyclical global weather patterns.

• Building a Global Climate Prediction Network

Climate prediction depends upon having large numbers of accurate measurements of many different variables, which can then be correlated mathematically. Correlation is an uncertain process at best, rarely yielding unequivocal results. For example, all of the complex hurricane-predicting machinery of the United States' National Oceanic and Atmospheric Administration (NOAA) produced, as of the beginning of June, 2009, the prediction that the upcoming August-October Atlantic hurricane season would have a 50 percent chance of being average and a 25 percent chance of being above or below average. Vague predictions such as this are one reason that skeptics such as Marcel Leroux,

In February, 2007, members of the Intergovernmental Panel on Climate Change sit in front of a projection of future changes in precipitation patterns caused by global warming. (AP/Wide World Photos)

of the University of Adelaide, can plausibly question that scientists have demonstrated any general global warming effect.

Global meteorological monitoring, coordinated through national weather services and the United Nations, involves a network of satellites capable of measuring physical parameters including surface temperatures, wind speeds, cloud cover, and barometric pressure, at points 50 kilometers apart, at hourly intervals. National weather centers are well apprised, for example, of the exact status of El Niño on any given day and how it has been developing, but unless it has recently exhibited extraordinary features, these data give only a general picture of what the climate will do in affected regions.

The main thrust of global climate prediction was, and to a large extent still is, extreme weather events, including droughts, floods, and cyclonic storms. Predictions impact disaster preparedness and help nations minimize mortality. Knowing in advance that El Niño is likely to produce drought in Australia and South Africa in a given year helps those countries stockpile grain, devote more acreage to drought-resistant crops, and prepare for wildfires. In the United States and elsewhere, projections for hurricane and tornado activity affect insurance policies and land-use decisions. For extreme weather projections, a decadal time frame is sufficient. Beyond that, only a few cyclical phenomena can be projected with yearly accuracy, and the number of unknown variables becomes too large to allow useful prediction.

• Climate Projection

All of the models used in predicting decadal climate variability were developed using historical data and assume that variables are constant, oscillate in a regular manner around a mean, or are increasing or decreasing at a constant rate. With respect to the carbon dioxide content of the atmosphere, none of these conditions is currently met. However, once a model is developed, climatologists can use a computer simulation to project, for example, how atmospheric carbon dioxide and global temperatures would respond if there were a large increase or decrease in emissions. Because of complexities, uncertainties, and unknown variables, such projections often prove to be far off track.

• Context

With respect to the controversy over global warming—whether is is occurring and to what it is attributable—input from climatologists associated with NOAA and other agencies suffers from distortion between laboratory and the media. When scientists correctly project from their models that a massive eruption of the Yellowstone supervolcano such as occurred 200,000 years ago would produce abrupt catastrophic cooling, completely dwarfing any anthropogenic warming, it implies neither that such an eruption is expected in the near future nor that efforts to curtail emissions and environmental degradation are futile in the face of overwhelming nature—yet that is the lesson many people would derive from their findings. The high resolution, global coverage, and international cooperation among climatology centers ensure that no event or trend of significance escapes attention. If media attention does not translate into action, it is not the fault of the climatologists.

Martha A. Sherwood

• Further Reading

Leroux, Marcel. *Global Warming: Myth or Reality? The Erring Ways of Climatology.* New York: Springer, 2005. A skeptical view focusing on the limitations of climate prediction; good nontechnical description of data and issues.

Mayewsky, Paul, et al. "Holocene Climate Variability." *Quaternary Research* 62 (2004): 243-255. A technical summary of high-resolution proxy records from many parts of the globe; offers some discussion of correlation with human history.

Saltzman, Barry. *Dynamical Paleoclimatology: Generalized Theory of Global Climate Change.* San Diego, Calif.: Academic Press, 2002. Treats interactions between abiotic, biotic, and anthropogenic variables; discusses controversies about the magnitude of human impact.

Thompson, Russell D., and Allen Perry, eds. *Applied Climatology: Principles and Practice.* New York: Routledge, 1997. This volume, aimed at professionals, offers a collection of papers emphasizing decadal-scale disaster prediction.

See also: Climate change; Climate feedback; Climate models and modeling; Extreme weather events; Famine; General circulation models; Global monitoring; Hockey stick graph; Parameterization; Seasonal changes; Skeptics; Weather forecasting; WGII LESS scenarios.

Climate Project

- **Category:** Organizations and agencies
- **Date:** Established 2006
- **Web address:** http://www.theclimate project.org

• Mission

The mission of the Climate Project is to increase "public awareness of the climate crisis at a grassroots level in the United States and abroad." It was founded in June, 2006, by Nobel laureate Al Gore, a former vice president of the United States (1993-2001) and an environmental activist and author.

The project grew out of a slide-show presentation that Gore, an early activist on the issue of preventing global warming, had been giving around the United States and internationally. In the slide show, Gore summarized the scientific arguments supporting the causes and dangers of global warming and described the political and economic consequences of ignoring the issue. The presentation was filmed, becoming the Academy Award-winning 2006 documentary *An Inconvenient Truth,* and a companion book became a *New York Times* best seller. Requests for showings of the film and presentations about global warming, and questions about Gore's next step in spreading his message, reached such a high level that Gore formed the Climate Project to help "educate, encourage, and promote dialogue about climate change as well as potential solutions." On the film's Web site, he sent out a call for one thousand "Climate Change Messengers," whom he would train to help deliver his message about global warming.

In Nashville, Gore himself began training hundreds of volunteers to deliver a version of the popular slide show. Volunteers in Australia, Canada, India, Spain, and the United Kingdom were also trained. The project expanded to include a special faith community training session, helping members of churches and other faith communities deliver the presentation, organize other faith-based activities to educate about global warming, and talk about global warming in the context of their faiths. Through the Climate Project Web site, anyone interested can request a presentation in the local community, delivered by a volunteer at no charge. Volunteers pay their own expenses to travel to training sessions, but donations to the Climate Project help support their travel as they make presentations, and pay for the brochures and other materials they distribute.

• Significance for Climate Change

Presenters, who commit to making at least ten presentations in a year, hope to alert citizens to the dangers of global warming, and to affect local and national policy decisions. For their programs, volunteers typically combine slides from Gore's presentation with new slides showing the local effects of global warming. In Australia, those who are uncomfortable with the idea of speaking at length in public can instead become "connectors," who commit to showing a digital video disc (DVD) called *Telling the Truth*, leading a discussion about global warming, and asking members of the audience to sign a statement that supports the Climate Project.

Volunteers, who must apply and be selected by the Climate Project, come from a variety of age groups and professions. They include a former Australian professional rugby player, a Canadian railroad executive, a French fashion model (who urged celebrity friends to design T-shirts to raise money for the project), a former member of the Canadian national women's hockey team, a mother and massage therapist, an astrophysicist, an eleven-year-old girl, and a woman in her nineties. Critics have objected that many of the presenters are not scientifically knowledgeable enough to make intelligent arguments about an issue as technically complex as global climate change. Others have praised the Climate Project for encouraging so many people to think seriously about an important issue. Public presentations, which run less than an hour, typically draw audiences of twenty to thirty people, but some draw a hundred or more.

Gore himself inspires and attracts volunteers. Most of the volunteers are trained by Gore, who has conducted two- and three-day training sessions in Nashville, Tennessee; Montreal, Canada; and Melbourne, Australia. After he spoke in Montreal, Canada, in April, 2008, the number of Canadian volunteers grew from 20 to more than 250 in six months. As of October, 2008, the Climate Project had trained more than 2,500 volunteer presenters, whose programs had reached audiences totaling more than four million people.

Cynthia A. Bily

• Further Reading

Adams, David. "Global Warming, Meet Your New Adversary." *St. Petersburg Times,* January 23, 2007, p. 1A. A history of Gore's founding of the Climate Project, focusing on Florida volunteer Roberta Fernandez.

Della Caca, Marco R. "Al Gore Trains a Global Army: Soldiers March Forth with Environmental Message." *USA Today,* April 25, 2007, p. 1D. A history of Gore's founding of the Climate Project, focusing on Texas volunteer Gary Dunham.

Gore, Al. *An Inconvenient Truth: The Planetary Emergency of Global Warming and What We Can Do About It.* Emmaus, Pa.: Rodale, 2006. A clearly written book based on Gore's slide show lectures on global warming. These presentations were captured in the film *An Inconvenient Truth* and are the basis of the Climate Project's presentations.

Meacham, Steve. "Standing Up for the Planet: The Messengers." *Sydney Morning Herald,* December 10, 2008, p. 19. Describes the activities of soccer player Al Kanaar, one of the first Australian Climate Project presenters.

See also: Gore, Al; *Inconvenient Truth, An.*

Climate reconstruction

• **Category:** Meteorology and atmospheric sciences

• **Definition**

Scientists use a combination of techniques to reconstruct and describe aspects of past climates, such as temperature, precipitation, and atmospheric carbon dioxide (CO_2) concentration, using historical accounts and proxies. These methods include analysis of preserved pollen, tree rings, and ice-core oxygen isotope ratios and CO_2 concentrations, as well as paleobotanical methods.

• **Significance for Climate Change**

To develop accurate climate-prediction models, scientists require data that predate modern science, forcing them to rely on historical records and a variety of climate proxies to reconstruct the climate of hundreds, thousands, and even millions of years ago. Climate factors that can be reconstructed include local and global temperature, precipitation, sea level and salinity, atmospheric pressure, atmospheric CO_2 concentration, ice volume, and ocean circulation.

Many studies use a combination of proxies in an attempt to minimize error. Data from these proxies can be combined with modern climate data to create models that infer past climate as well as predict future climate. Paleoclimatology and climate reconstruction are important for explaining current ecosystems and understanding factors that affect climate change. Climate reconstruction data also are important for improving models that help predict the effects of possible climate change scenarios, such as the potential effect of increased atmospheric CO_2.

Two of the most crucial sources of paleoclimatic data are ice and sediment cores. Deep ice cores preserve atmospheric gases, water, and pollen. Scientists analyze isotope ratios, CO_2 concentrations, and pollen assemblages to infer information about past climates. Perhaps the most famous ice core was taken from Lake Vostok in Antarctica. Data from this core has shown that East Antarctica was colder and drier and atmospheric circulation was more vigorous during glacial periods than they are now. Scientists can also correlate atmospheric CO_2 and methane with temperature.

Deep-sea sediment cores provide similar information about past climate through marine microorganisms such as diatoms and foraminifera, which preserve isotope ratios in their shells. These isotopes allow scientists to infer past water temperature, while community makeup of microorganisms can be used to make other inferences about their environments. Scientists also take sediment cores from lakes. Charcoal layers in sediment can indicate fires, and pollen can also provide climate information.

Pollen is extremely tough and holds up well for millions of years in the fossil record. Pollen assemblages from cores can be used to infer climate by comparison with modern plants and their climate tolerances, although scientists must be careful not to assume that plants today live exactly as their ancestors did. Pollen records are often correlated with records from other sources, such as marine plankton and ice cores, to minimize error.

Pollen records can have very high resolution, on the order of a single year when taken from annually deposited lake sediments. For example, researchers at the Faculte de St. Jerome in France were able to estimate climatic range and variability in the Eemian interglacial period, approximately 130,000 to 120,000 years ago. They found that the warmest winter temperatures occurred in the first three millennia of the period, followed by a rapid shift to cooler winter temperatures between 4,000 and

Sources of Climate Reconstruction Data

Years Ago	*Available Data Sources*
1,000	Written records, tree rings
10,000	Varved and lake sediments
50,000	Lake levels, mountain glaciers
250,000	Polar ice cores
1,000,000+	Ocean sediments, cave deposits

Source: National Ice Core Laboratory, U.S. Geological Survey.

5,000 years after the beginning of the Eemian. After that, annual variations of temperature and precipitation were slight, only 2-4° Celsius and 200-400 millimeters per year.

Tree-ring analysis is usually employed within the timespan of the historical record, although it can be employed on fossilized trees. The thickness of tree rings is affected by temperature, precipitation, and other environmental factors—trees grow thicker rings in years with optimal conditions. Scars and burn marks can also be used to identify fires and other events. These events can often be correlated with historical records to establish precise dates. Slices of different trees can also be correlated with one another to construct records stretching back hundreds and even thousands of years.

In areas with good tree records, such as the dry American Southwest, tree-ring analysis has extremely fine resolution. In the White Mountains, the bristlecone pine tree chronology goes back ten thousand years, to 7,000 B.C.E., almost to the end of the last ice age. Bristlecone pine chronologies have been used to recalibrate the carbon 14 dating process. Tree-ring analysis can also provide information about the effects of pollution. Using similar methods with coral, scientists have reconstructed sea surface temperatures and salinity levels for the last few centuries.

Several other methods of climate reconstruction are used to infer temperature and precipitation from millions of years ago. These methods often rely on the fossil record, particularly that of plants. Leaf physiognomy methods rely on physical characteristics of leaves thought to be independent of species in order to estimate precipitation and temperature. For example, leaf-margin analysis compares the ratio of leaves with smooth margins to leaves with toothed margins. Tropical environments have a higher percentage of smooth-margined leaves than do temperate environments. The stomatal index—the ratio of the tiny holes in a given area of a leaf to the overall number—can provide information about atmospheric CO_2. Scientists have used many other proxies to reconstruct aspects of past climates, and they develop new methods every year, further refining their understanding of the past and improving their models of the future.

Melissa A. Barton

• **Further Reading**

Alley, Richard B. *The Two-Mile Time Machine: Ice Cores, Abrupt Climate Change, and Our Future.* Princeton, N.J.: Princeton University Press, 2000. Author and geoscientist Alley, who has studied ice cores from Greenland and Antarctica, covers what ice cores reveal about past climate and about the climate challenges faced today. Illustrated.

Levenson, Thomas. *Ice Time: Climate, Science, and Life on Earth.* New York: HarperCollins, 1990. Good introduction to the causes of weather and climate and the history of climate science. Discusses the last ice age.

Redfern, Ron. *Origins: The Evolution of Continents, Oceans, and Life.* Norman: University of Oklahoma Press, 2001. Profusely illustrated history of the Earth, including the movement of continents, the rise and fall of oceans, and the fluctuations of temperature. Illustrated.

See also: Carbon isotopes; Climate models and modeling; Climate prediction and projection; Dating methods; Ice cores; Oxygen isotopes; Pollen analysis; Sea sediments; Tree rings.

Climate sensitivity

• **Category:** Meteorology and atmospheric sciences

• **Definition**

Global climate is a complex system that reacts to changes in its components, such as atmospheric carbon dioxide (CO_2) concentration. A change in the atmospheric concentration of CO_2 may cause a change in the radiation balance of the Earth; such a change is called "radiative forcing." Many changes can cause radiative forcing, including changes in greenhouse gas (GHG) concentration, the output of the Sun, ice cover, and aerosol concentration. In response to a change in the Earth's radiation balance, the planet's temperature will change until global energy balance is restored. How much the global temperature changes depends on internal feedbacks in the Earth's climate system that cause net amplification of the initial radiative forcing. Internal climate feedbacks include changes in water vapor, lapse rate, albedo, and clouds.

Equilibrium climate sensitivity (ECS) is a useful summary statistic of the behavior of the Earth's climate system. ECS is defined as the change in equilibrium of global average surface temperature in response to a doubling of the atmospheric concentration of CO_2 from preindustrial levels (from 280 parts per million to 560 parts per million). A doubling of atmospheric CO_2 causes radiative forcing of about 3.7 watts per square meter by increasing long-wave radiative absorption. ECS is not a simple measure of the amount of thermal energy added to Earth's climate system by the CO_2 alone, because climate changes cause feedback loops. For example, melting ice decreases Earth's albedo, causing the planet's surface to absorb more heat and to further increase thermal energy. If there were no internal climate feedbacks, then ECS would be about 1.2° Celsius. However, because of internal climate feedbacks, ECS is likely between 2° Celsius and 4.5° Celsius.

ECS allows scientists to assess the change in average global temperature after the Earth has reached equilibrium over several thousand years. However, it is computationally intensive to run complex global climate models to equilibrium. Thus, a modified concept, effective climate sensitivity, has been developed as an approximation of ECS. Effective climate sensitivity is calculated by estimating the climate feedback parameter at a specific point in time during transient climate conditions (not at equilibrium) using estimates of ocean heat storage, radiative forcing, and surface temperature change. Some studies find that effective climate sensitivity calculations underestimate the true ECS of a given model.

• Significance for Climate Change

ECS is used to summarize and compare different climate models, as well as to combine information from models, historical records, and paleoclimate reconstructions. It is immensely important for making, understanding, and reacting to projections of future climate change. Every 1° Celsius difference in ECS can imply vastly different impacts on the planet over the long term.

A large number of studies have estimated ECS using a variety of methods, models, and data. There are four main categories of strategies used to estimate ECS. One method is to estimate ECS directly from observations of past climate changes. Another is to compile expert opinions. A third is to take a single climate model, create multiple versions by varying its parameters, and then compare the climate simulated by each version with climate observations to determine which is most likely. A final strategy is to combine the results of multiple methods into a single probability distribution. There are two main sources of climate observations used in this type of research: modern instrumental observations (after 1850), and paleoclimate reconstructions over the past thousands or millions of years.

After considering all the available research, the Fourth Assessment Report of the Intergovernmental Panel on Climate Change (IPCC) concluded that ECS is likely (with a probability of greater than 66 percent) to be between 2° Celsius and 4.5° Celsius, very likely (greater than 90 percent probability) to be larger than 1.5° Celsius, and most likely to have a value of about 3° Celsius. The estimated range of ECS has been relatively stable over a thirty-year period: A range of 1.5° Celsius to 4.5° Celsius was proposed in 1979 in a report by the National Academy of Sciences. Scientists have improved the certainty of their estimates of the lower limit of climate sensitivity and of the transient climate response. The upper limit of ECS, however, remains difficult to quantify as a result of nonlinearities that cause a skewed probability distribution. The persistence of such large uncertainty in the value of ECS is a significant barrier to narrowing the range of projections of future climate change.

There are several important limitations to the concept of ECS. It is potentially dependent upon the state of the climate system and upon the rate and magnitude of radiative forcing: A doubling of CO_2 versus a halving of CO_2 may not cause the same magnitude of temperature change. Additionally, different forcing mechanisms can have different sensitivities to radiative forcing, so ECS values may be specific to changes in CO_2. Lastly, ECS quantifies equilibrium temperature change over thousands of years, so it does not give direct projections for future climate changes over periods of hundreds of years. A separate summary statistic, transient climate response (TCR), was developed to compare the transient responses of climate models and provide shorter-term projections.

Carolyn P. Snyder

• **Further Reading**

Edwards, T. L., et al. "Using the Past to Constrain the Future: How the Palaeorecord Can Improve Estimates of Global Warming." *Progress in Physical Geography* 31, no. 5 (October, 2007): 481-500. Provides an accessible overview of ECS research, with useful tables categorizing studies and a focus on the role of paleoclimate reconstructions in ECS estimations.

Intergovernmental Panel on Climate Change. *Climate Change, 2007—The Physical Science Basis: Contribution of Working Group I to the Fourth Assessment Report of the Intergovernmental Panel on Climate Change.* Edited by Susan Solomon et al. New York: Cambridge University Press, 2007. The IPCC provides an extensive overview of the scientific literature on ECS. Chapters 6, 8, 9, and 10 are most relevant.

Knutti, R., and G. C. Hegerl. "The Equilibrium Sensitivity of the Earth's Temperature to Radiation Changes." *Nature Geoscience* 1, no. 11 (November, 2008): 735-743. Provides an updated technical overview of the scientific literature on ECS, with useful figures comparing different estimates.

See also: Climate and the climate system; Climate feedback; Climate models and modeling; Climate prediction and projection; Climate reconstruction; Climate variability; Paleoclimates and paleoclimate change.

Climate variability

• **Category:** Meteorology and atmospheric sciences

A stable climate is not necessarily a static climate: Climates can vary within limits over time, and it is important to distinguish between this climate variability and genuine climate change in evaluating climatic history and making future projections.

• **Key concepts**

climate change: alterations in long-term meteorological averages in a given region or globally

climate fluctuations: changes in the statistical distributions used to describe climate states

climate normals: averages of a climatic variable for a uniform period of thirty years

climatic oscillation: a fluctuation of a climatic variable in which the variable tends to move gradually and smoothly between successive maxima and minima

climatic trend: a climatic change characterized by a smooth monotonic increase or decrease of the average value in the period of record

• **Background**

Climate is an abstraction, a synthesis of the day-to-day weather conditions in a given area over a long period of time. The main climate elements are precipitation, temperature, humidity, sunshine, radiation, wind speed, wind direction, and phenomena such as fog, frost, thunder, gales, cloudiness, evaporation, and grass and soil temperatures. In addition, meteorological elements observed in the upper air may be included where appropriate. The climate of any area may also be described as a statistical analysis of weather and atmospheric patterns, such as the frequency or infrequency of specific events.

In the most general sense, the term "climate variability" denotes the inherent tendency of the climate of a specific area to change over time. The time period considered would normally be at least fifty years, but a period of at least one hundred years is usually more appropriate. Instrumental climatic observations have been taken in most areas of the world for at least one hundred years, and in some areas for more than two hundred years, and any analysis of climate variability over time should utilize the full record of those observations. Longer time periods, from one thousand years to a geological period, may also be studied using proxies.

• Magnitude of Climate Variability

The degree or magnitude of climate variability can best be described through the statistical differences between long-term measurements of meteorological elements calculated for different periods. In this sense, the measure of climate variability is essentially the same as the measure of climate change. The term climate variability is also used to describe deviations of the climate statistics over a period of time (such as a month, season, or year) from the long-term statistics relating to the same calendar period. In this sense, the measure of climate variability is generally termed a climate anomaly.

• Climate Properties

Three basic properties characterize the climate of an area. Thermal properties include surface air temperatures above water, land, and ice. Kinetic properties include wind and ocean currents, which are affected by vertical motions and the motions of air masses, aqueous humidity, cloudiness, cloud water content, groundwater, lake lands, and the water content of snow on land and sea ice. Finally, static properties include pressure and density of the atmosphere and oceans, composition of the dry air, salinity of the oceans, and the geometric boundaries and physical constants of the system. These three types of properties are interconnected by various physical processes, such as precipitation, evaporation, infrared radiation, convection, advection, and turbulence. The climate is a complex system, and any consideration of climate variations, especially in terms of global warming, must be carefully evaluated.

• Climate Variability Over the Last Thousand Years

One thousand years ago, some—and possibly many—parts of the Earth were warm and dry. The Atlantic Ocean and the North Sea were almost free of storms. This was the time of the great Viking voyages. Vineyards flourished in England; in contrast, some frosts occurred in the Mediterranean area, and rivers such as the Tiber in Rome and the Nile in Cairo occasionally froze. This suggests that a shift occurred in the pattern of large-scale European weather systems. However, by about 1200, the benign climate in Western Europe began to deteriorate, and climate extremes characterized the next two centuries. From about 1400 to 1550, the climate grew colder again, and about 1550 a three-hundred-year cold spell known as the Little Ice Age began. (The term "Little Ice Age" is used differently by different writers. Many use it to refer to the climate cooling from about 1300 to 1850, while others use it for the latter half of that interval, when cooling was greatest, beginning around 1550 or 1600.) Around 1850, the cold temperatures began to moderate, and from about 1900 a relatively steady warming trend (with a few intervening cold periods) occurred in many areas of the world. However, since the peak warmth of the year 1998 for the world as a whole, global temperatures have remained relatively stable, with some cooling trends in a few areas.

• Context

The climate has always varied and will continue to vary, but it is important to differentiate internal variations, or changes that do not imply instability, with external variations, which are attributable to forcing. Changes in the intensity of seasons or of rainfall may presage global warming, or they may simply be temporary oscillations within an existing system. The longer the time period under discussion, moreover, the more difficult it is to tell where unidirectional climate change begins or ends.

Climate variability, when seen as an inherent characteristic of Earth's atmosphere, can be treated as a reason to accept global changes as natural and beyond human control. When seen as a result of human activity, the same variability can be treated as a call to action to reverse global warming. Evaluations of the limits of natural variability are therefore crucial precursors of evaluations of climate change itself. What will prove to be the correct argument remains for the future, but irrespective of the truth of the two arguments, it is evident that people and societies must adapt to climate changes, and those communities that adapt to the varying climate will be in a better position to withstand the climatic variability of the future.

W. J. Maunder

• Further Reading

Broecker, W. S. "Does the Trigger for Abrupt Climate Change Reside in the Ocean or the Atmosphere?" *Science* 300 (2003): 1519-1522. Good example of a discussion on some aspects of the natural causes of climate change.

Kininmonth, William. *Climate Change: A Natural Hazard.* Brentwood, Essex, England: Multi-Science, 2004. Kininmonth's career in meteorological and climatological science and policy spans more than forty-five years. His suspicions that the science and predictions of anthropogenic global warming have extended beyond sound theory and evidence were crystallized following the release of the 2001 IPCC report.

Maunder, W. J. *The Human Impact of Climate Uncertainty: Weather Information, Economic Planning, and Business Management.* New York: Routledge, 1989. Provides an overview of the economic dimensions of climate and human activities; explains how the variable nature of the atmosphere must be accepted as an integral part of the management package.

See also: Abrupt climate change; Climate change; Climate models and modeling; Climate prediction and projection; Climate sensitivity; Climatology; Intergovernmental Panel on Climate Change; Weather vs. climate.

Climate zones

• **Category:** Meteorology and atmospheric sciences

• Definition

Climate, the average weather conditions over a period of at least thirty years, is determined by various factors, the most important of which are the amount of precipitation and the temperature of the air. Climate controls the major ecological community types, or biomes; that is, the climate in a given region determines the flora and fauna that will thrive in that region. In 1900, Wladimir Köppen, a German climatologist, developed what has become the most widely used system for classifying world climates. The Köppen system identifies five major climate zones: tropical moist climates (A zone), dry climates (B zone), humid middle latitude climates (C zone), continental climates (D zone), and cold climates (E zone). Köppen also used two subgroups to more specifically describe the zones.

Tropical moist climates are characterized by year-round high temperatures and large amounts of rain. Rainfall is adequate all year round, and there is no dry season. This zone is typical of northern parts of South America, central Africa, Malaysia, Indonesia, and Papua New Guinea. The dry climate zone has little rain and a wide range of daily temperatures. There is a dry season in the summer and winter, with a mean annual temperature over or under 18° Celsius, as in the western United States, northern and extreme southern Africa, parts of central Asia, and most of Australia.

The humid middle latitude climate, or temperate zone, has hot-to-warm, dry summers and cool, wet winters, but no dry season as such. Southeastern sections of the United States and South America, westernmost Europe, and the southeast corner of China fit this category. The continental climate zone, in interior regions of large land masses such as Canada and northern Europe and Asia, experience varied seasonal temperatures and moderate rainfall. The cold climate zone, characterized by permanent ice and ever-present tundra, occupies Greenland and the most northerly parts of Asia.

• Significance for Climate Change

While acknowledging some unknowns and uncertainty, researchers predicts that, if global warming caused by carbon dioxide (CO_2) and other greenhouse gas (GHG) emissions continue at the current rate, some of the climate zones recognized in the early twenty-first century could disappear entirely by the end of the century, giving way to new climate zones on up to 39 percent of the world's land surface. Major areas that could be affected are tropical highlands and polar regions. Broad strips of areas labeled tropics and subtropics at the beginning of the twenty-first century could develop new climates that do not resemble any of the zones in categories assigned in the Köppen Climate Classification System. Researchers predict that heavily populated areas such as the southeastern United

An Adelie penguin leaps over a crack in the thinning sea ice off Ross Island. The polar climate zone is among the most susceptible to global warming. (Chris Walker/MCT/Landov)

States, southeastern Asia, parts of Africa such as its mountain ranges, the Amazonian rain forest, and South American mountain ranges are likely to be the most severely affected.

Climate change patterns could affect ecosystems on a global scale. For example, major changes in the forests of North America could result: Four species of tree—the yellow birch, the sugar maple, the hemlock, and the beech—are expected to move northward up to 1,000 kilometers while abandoning entirely their present-day locations. Animals could also be affected. Temperature and rainfall patterns could change breeding and migration patterns.

For humans, a grave concern is global food production. One model predicts that the corn belt in North America will move northward, possibly as far as Canada; winter wheat may replace corn in parts of the present corn belt. Within several decades, the Swiss Alps could become a Mediterranean climate, with wet winters and long, dry, warm summers. Within one hundred years, the climate zone in southern Switzerland may move northward by as much as 500 kilometers. In the western Alps, the climate may come to resemble that found in southern France in the early twenty-first century.

Biodiversity in South Africa is expected to be affected substantially by shifting climate zones: Species will experience extinction on a wide scale; up to half the country will see a climate not known before; succulent karoo, a globally important arid-climate hotspot, and biomes in the *fynbos* (a Mediterranean-climate thicket) will suffer. While the degree of this change remains speculative, availability of food could affect sub-Saharan West Africa, as vegetation zones move southward.

Deforestation will continue in the Amazonian forest areas of South America, and new climates are expected to be created near the equator. Some researchers predict that mountainous areas such as those found in Peru and the Colombian Andes, as well as regions in Siberia and southern Australia, could experience the disappearance of climates completely. Devastation of critical ecosystems and changes in agricultural patterns could severely affect Australia, New Zealand, and the developing island nations of the Pacific. With so many factors still undetermined or speculative, however, it remains to be seen how climate zone changes will play out in the future.

Victoria Price

• Further Reading

Burroughs, William James. *Climate Change: A Multidisciplinary Approach.* 2d ed. New York: Cambridge University Press, 2007. Discusses climate change's natural and anthropogenic causes, effects and consequences, methods of measurement, and contemporary predictions.

Diaz, Henry F., and Richard J. Murnane, eds. *Climate Extremes and Society.* New York: Cambridge University Press, 2008. Two parts discuss definitions and models of weather extremes and the impact of weather and climate extremes on the insurance industry.

Ennis, Christine A., and Nancy H. Marcus. *Biological Consequences of Global Climate Change.* Sausalito, Calif.: University Science Books, 1996. Explains the links between living organisms and climate; examines direct and indirect effects of GHGs on physical climate. Stresses uncertainties and challenges that affect future research and suggests that the most challenging aspect of climate change is that predicted changes will be occurring at the same time.

See also: Continental climate; Global climate; Holocene climate; Maritime climate; Mediterranean climate; Polar climate; Regional, local, and microclimates; Tropical climate.

Climatology

• **Category:** Meteorology and atmospheric sciences

• **Definition**

Climatology can be defined as the synthesis of weather studies over a time interval long enough to determine statistical properties. In other words, meteorology plus time equals climatology. Climatology differs from meteorology in that meteorology describes the atmospheric conditions at present, what are most often called "the weather." Meteorology also involves forecasting the weather for the very near future. Climatology involves studying the weather and its trends over a long period of time. Usually, the study of climate is limited to a particular area and, rather than providing a short-term forecast, can give an idea of the broad meteorological parameters in the area at any given time. The climate of a specific area can vary from year to year, decade to decade, or even century to century, and climatologists track these trends. Climatologists often use a thirty-year average to determine the normal climate for an area based on average humidity, precipitation, sunshine, and temperature. Meteorologists concentrate on the short term, studying and predicting weather systems that last only a few weeks.

• **Significance for Climate Change**

Climatologists study the climate, how it changes, and how those changes over time may affect the Earth, humans, and other forms of life. There are several different branches of climatology, such as paleoclimatology (reconstructing past climates by examining evidence such as ice cores or tree rings), paleotempestology (determining the frequency of hurricanes over thousands of years in order to understand their patterns), and historical climatology (focusing on climate changes that occurred on the Earth after humans appeared).

Climatologists study the way five different climate systems interact with one another. These systems include:

- *Atmosphere:* the gases that surround the Earth, including water vapor
- *Biosphere:* the Earth's ecosystems and all living organisms and dead organic matter on land and in water
- *Cryosphere:* any frozen water, such as floating ice, glaciers, permafrost, or snow
- *Hydrosphere:* all liquid surfaces, such as lakes, oceans, rivers, as well as underground water
- *Lithosphere:* the solid parts of the Earth

Based on the interactions between and among these systems and the way weather is created by them, climatologists are able to predict the changes in these systems that will affect the climate and, to a certain extent, what kinds of changes in the climate may occur as a result of these interactions and disturbances to them.

Climatologists study both broad issues (such as the circulation of ocean currents and how they affect the Earth) and seemingly small issues (such as how minor differences in the amounts of sunlight between urban and rural areas affect the overall climate of each area). Both these types of phenomena may affect the climate of a certain area; for example, the amount of heat retained in concrete may cause an area's climate to change dramatically if the area is urban.

Climatology can be a difficult science because of the long time periods involved and the complexity of the processes and interactions that must be dissected to arrive at any type of conclusion. Climatologists break these complexities down into mathematical differential equations that can be used to integrate different climate observations and help piece them together. Climatologists often use statistical or mathematical models to test how their hypotheses about climate will play out. These models simulate the interactions of climate systems in order to project future climate changes.

In the past, climatology was seen as a static field of study, one that changed little over time and involved compiling statistics about weather conditions. Concerns about global climate change beginning in the 1960's helped highlight this science as a field that could help explain the changes in climate that occurred in the distant past and predict how climate would affect the Earth and its living organisms in the future.

Marianne M. Madsen

• Further Reading

Barry, R. G. *Synoptic and Dynamic Climatology.* New York: Routledge, 2001. Discusses climate dynamics and mechanisms and synoptic-scale weather systems.

Bonan, G. *Ecological Climatology: Concepts and Applications.* New York: Cambridge University Press, 2008. Focuses on interactions between the climate and ecosystems; reviews basic meteorological, hydrological, and ecological concepts.

Critchfield, H. J. *General Climatology.* Englewood Cliffs, N.J.: Prentice Hall, 1998. Discusses all aspects of weather and climate as well as past and possible future climate changes. Includes charts, graphs, and black-and-white photos.

Hartmann, D. L. *Global Physical Climatology.* San Diego, Calif.: Academic Press, 1994. A textbook approach to physical climatology including the climate system. Includes a comprehensive index.

Oliver, J. E., ed. *The Encyclopedia of World Climatology.* New York: Springer, 2005. Discusses the subfields of climatology. Includes graphs, charts, drawings, and citations to other important climatology works.

Oliver, J. E., and J. J. Hidore. *Climatology: An Atmospheric Science.* 2d ed. Upper Saddle River, N.J.: Prentice Hall, 2002. Discusses why recent climate changes have occurred and explains the science behind those changes. Describes how basic processes operate in the atmosphere.

Perry, A. *Applied Climatology: Principles and Practice.* New York: Routledge, 1997. Emphasizes the effects of climate changes on life-forms.

Rayner, John N. *Dynamic Climatology: Basis in Mathematics and Physics.* Malden, Mass.: Blackwell, 2001. Discusses how statistics, dynamics, and thermodynamics affect the study of climate. Includes historical accounts.

Robinson, P., and A. Henderson-Sellers. *Contemporary Climatology.* Englewood Cliffs, N.J.: Prentice Hall, 1999. Covers topics such as the water cycle and use of satellite observations to track climate changes.

Rohli, R. V., and A. J. Vega. *Climatology.* Sudbury, Mass.: Jones & Bartlett, 2007. Provides a basic foundation in climatic processes. Discusses climatology versus meteorology.

See also: Atmospheric dynamics; Climate and the climate system; Climate models and modeling; Climate prediction and projection; Climate reconstruction; Ocean-atmosphere coupling; Weather forecasting; Weather vs. climate.

Clouds and cloud feedback

• **Category:** Meteorology and atmospheric sciences

Clouds are important components of Earth's life-sustaining greenhouse effect. Alterations in cloud density, prevalence, or altitude caused by global temperature increases may have far-reaching consequences for Earth's temperature equilibrium.

• Key concepts

aerosols: tiny particles suspended in Earth's atmosphere

enhanced greenhouse effect: increased retention of heat in the atmosphere resulting from anthropogenic atmospheric gases

global radiative equilibrium: the maintenance of Earth's average temperature through a balance between energy transmitted by the Sun and energy returned to space

greenhouse gases (GHGs): atmospheric gases that trap heat, preventing it from escaping to space

• Background

Clouds have a profound influence on local and global climate. Scientists have known for many years that clouds are a major component of the greenhouse effect, which makes life possible on Earth. What is not well understood is what effect clouds may have on global warming and cooling trends. Of the many variables that factor into global warming, clouds present the greatest uncertainty in predicting climate change.

• Clouds

Clouds are composed of droplets or ice crystals that form around aerosols in the atmosphere. These droplets accumulate until they become visible. The many shapes and sizes of clouds are divided into three basic types: cirriform (wispy and transparent), stratiform (layered), and cummuliform (mounded and fluffy). Three altitude divisions are also used for classification: high (more than 5,000 meters above sea level), middle (between 2,000 and 5,000 meters above sea level), and low (less than 2,000 meters above sea level). The terms are combined, with the upper and middle altitudes given the prefixes cirro- and alto-; a midlevel stratus cloud is called altostratus, for example.

Clouds reflect solar radiation back to space, and they reflect Earth's radiation back to the surface. Thus, all other factors being equal, nights are warmer and days are cooler when clouds are present. The amount of reflection back to space depends on cloud thickness, with a low of approximately 20 percent for cirrus clouds and a high of 90 percent for cumulonimbus clouds. Clouds contribute significantly to global radiative balance, or radiative equilibrium, an overall balance between solar radiation received by the Earth and heat reflected back to space. If the Earth absorbed all the radiation it receives, it would be much too hot to sustain life.

• Greenhouse Effect and Feedback

However, the Earth does not absorb all the solar energy it receives, but instead reflects visible radiation and emits infrared radiation. Without an atmosphere, the total return of energy to space would make the Earth much colder than it is—the average surface temperature would be approximately −18° Celsius. Clouds and other greenhouse gases (GHGs) absorb some of the energy, trapping it and sending some of it back to the Earth's surface. This is the greenhouse effect.

"Feedback" is a term that applies to any multipart system in which a change in one part produces a change in the other part, which then affects the original part, and so on. A simple example is a thermostat, which responds to a drop in temperature by turning on a furnace. The furnace raises the air temperature, which then causes the thermostat to turn off the furnace. Feedback systems can affect changes in cloud thickness and prevalence as a response to climate changes, particularly temperature. The greenhouse effect creates a state of equilibrium, and cloud feedback disturbs that equilibrium.

Cloud feedback can be either positive or negative. Positive feedback increases the enhanced greenhouse effect, while negative feedback lowers global temperatures. Most clouds provide both positive feedback (transmitting energy down toward the Earth) and negative feedback (transmitting solar radiation back to space). Determining whether the net feedback is positive or negative is a complicated process.

• Clouds and Climate Change

Climate change can increase both positive and negative feedback. Studies have shown that cold water

Types of Clouds

Name	*Altitude (km)*
Altocumulus	2-7
Altostratus	2-7
Cirrocumulus	5-13.75
Cirrostratus	5-13.75
Cirrus	5-13.75
Cumulonimbus	to 2
Cumulus	to 2
Nimbostratus	2-7
Stratocumulus	to 2
Stratus	to 2

Source: National Oceanic and Atmospheric Administration.

produced by the melting polar ice cap causes phytoplankton to release chemicals that produce more and brighter clouds, thereby increasing negative feedback. Aerosols may produce either positive or negative feedback, depending on the source of the aerosols. Volcanic eruptions and pollution from technologically advanced countries, consisting of sulfates and nitrates, generate clouds that produce negative feedback. However, the developing world produces pollution that contains these substances as well as large amounts of black carbon, the by-product of incomplete combustion of carbon-based fuels. Black carbon aerosols generate positive feedback. The net feedback of pollution is very difficult to determine.

A long-held belief is that if the Earth's climate warms, water vapor amounts in the atmosphere will increase, creating more low-level thick clouds that will generate negative feedback. That belief is being called into question by recent studies that have shown that turbulence created by rising warm air currents will actually lead to fewer clouds being formed overall.

• A Source of Uncertainty

The impact of clouds on climate change is extremely difficult to model. Most of the important climate modeling systems represent clouds with a small number of variables, masking the subtleties of cloud dynamics. This is due to the extremely complicated mathematics required to model clouds realistically. While analyzing the results of some large-scale climate studies, scientists have concluded that neither the magnitude nor the sign of cloud feedback can be relied upon.

When weather data are entered into climate models, the resulting pictures of cloud cover and thickness often do not match actual conditions. Further complicating matters, studies done with live data collection also show contradictory results. For example, some studies indicate that decreasing cloud cover over China (resulting from large amounts of pollution) may be responsible for increasing temperatures there. However, other studies performed in other parts of the world indicate that temperatures have increased as cloud cover has increased.

• Context

The impact of clouds on global climate cannot be overstated. Clouds are extremely sensitive to fluctuations in solar radiation, Earth's surface temperature, and many other environmental factors, including pollution levels. It is this sensitivity, combined with nearly infinite variations in cloud size, make-up, and altitude, that make predicting cloud feedback so difficult. Since clouds may either mitigate or increase global warming, it is imperative that scientists intensify their efforts to produce better predictive climate modeling systems, as well as promote studies that analyze data collected in the field. Throughout the world, decisions are being made using climate predictions that may be flawed as a result of the uncertainty presented by cloud behavior. As these decisions will have a considerable impact both economically and sociologically, minimizing the uncertainty presented by clouds in a warming climate may become a priority in environmental agendas.

Kathryn Rowberg and Gail Rampke

• Further Reading

Combs, Peter. "Clouds and Climate Change." *Focus* 46, no. 1 (Spring, 2000): 35-36. Examines ground-level data collection of actual clouds, rather than modeling.

Mlynczak, M. G., et al. "A Detailed Evaluation of the Statospheric Heat Budget and Global Radiation Balance and Diabatic Circulations." *Journal of Geophysics Research* 104 (1999): 6039-6066. This study of global radiative balance uses satellite data from two time periods (1978-1979 and 1991-1993), the second of which includes the Pinatubo eruption. The conditions existing during the two time periods make this worthwhile reading for those interested in cloud feedback.

Soden, Brian J., and Issac M. Held. "An Assessment of Climate Feedbacks in Coupled Ocean-Amospheric Models." *Journal of Climate* 19 (2006): 3354-3360. Examines global climate modeling systems, focusing on cloud feedback results and their interpretation.

Stephens, G. L. "Cloud Feedbacks in the Climate System: A Critical Review." *Journal of Climate* 18 (2005): 237-273. Excellent overview of cloud feedback systems and the difficulties they present when one is trying properly to parameterize them in climate modeling.

See also: Albedo feedback; Climate feedback; Noctilucent clouds; Polar stratospheric clouds.

Coal

- **Categories:** Fossil fuels; energy

Coal burning accounts for a large share of global anthropogenic CO_2 output. Net carbon emissions per unit energy produced by coal exceed those of any other major energy source. The relative cheapness of coal, enormous reserves, and the concentration of those reserves in countries that are among the top energy consumers make continued growth likely—and make reducing carbon output from coal an urgent priority.

• Key concepts

carbon budget: the total amount of CO_2 emissions that can be sustained, given a target atmospheric CO_2 concentration, and the way in which those emissions are apportioned

carbon-capture-and-storage (CCS) systems: processes that remove CO_2 from combustion products and sequester it—typically in underground storage

fossil fuels: combustible products of ancient photosynthesis

integrated gasification combined cycle (IGCC) systems: electrical generation technology that converts coal to combustible gas before burning it

• Background

Coal was the first fossil fuel to be used by humans, and it will likely be the last. Known world coal reserves are projected to last for another 250 years. Unlike petroleum and uranium, which must be transported great distances to reach oil burners and nuclear reactors, the largest concentrations of coal deposits exist within populous countries with high energy requirements.

Most coal is used in contemporary society for generating electricity. Historically, solid coal also fueled transportation and industry. Inefficient energy conversion, the weight of coal-fired steam engines, and massive pollution all encouraged conversion to petroleum products for most purposes other than electrical power generation in the twentieth century. These factors also constitute powerful arguments against reviving the older technologies. However, both fuel gas and oil can be produced from coal, eliminating much of the bulk and pollution associated with coal-based energy—but not the problem of carbon dioxide (CO_2) emissions. According to one estimate, if early twenty-first century trends continue, the CO_2 emissions from burning coal in 2030 alone will equal about 50 percent of the total emissions from 1750 through 2000.

Most of the world's coal reserves accumulated during the Carboniferous ("coal-bearing") age. This age, ocurring between 354 and 290 million years ago, was a very warm period when land-plant productivity greatly exceeded herbivore consumption and decomposition. The Chinese were already burning coal in the thirteenth century. The use of coal for cooking and heating in cities was well established by the middle of the eighteenth century in England, when a series of inventions enabled manufacturers to replace human and animal labor with increasingly sophisticated coal-fired steam engines. This Industrial Revolution enabled an individual worker greatly to increase productivity, leading to an overall increase in per capita consumption of the Earth's renewable and nonrenewable resources at a time when population was also beginning to increase at an exponential rate.

• History of Coal Consumption

Coal consumption in Great Britain increased from 2.45 million metric tons per year in 1700 to 45 million metric tons per year in 1850 to 227 million metric tons per year in 1900. Annual consumption in the United States was 36 million metric tons in 1870 and 469 million metric tons in 1913. Total world consumption of coal on the eve of World War I was roughly one-tenth of its 2006 level of 5.44 billion metric tons.

Global warming as an environmental concern of coal utilization is a relatively recent phenomenon. Pollution and environmental degradation due to unsound mining practices dominated discussions of coal until the 1980's. At first, efforts to curb rampant urban air pollution consisted of moving heavy industry away from population centers, building taller smokestacks to divert pollutants elsewhere, and removing soot and other particulates from emissions. These brought sulfur dioxide (SO_2) and acid rain to the forefront of the environmental consciousness.

To address the SO_2 problem, industry developed the integrated gasification combined cycle (IGCC) system, which subjects superheated pulverized coal to steam to produce coal gas, which is then used to fuel electrical generation or other industrial functions. Coal gasification converts sulfur to hydrogen sulfide, which is removed before combustion. Startup costs for an IGCC plant are high, but some of the operating expenses are recouped from industrial use of hydrogen sulfide.

In terms of carbon output, IGCC systems exacerbate coal's already unfortunate profile. The process itself requires large energy inputs, which increase the fuel's carbon footprint, and the gas produced includes a high proportion of carbon monoxide, a carbon-intensive fuel source. While in many respects equivalent to natural gas, coal gas is inferior to that gas in terms of its carbon emissions. Its toxicity is also a drawback for domestic use.

There is evidence that efforts to combat pollution have actually increased the greenhouse effect caused by coal burning, because the dirtier and more polluting a coal-fired plant is, the more soot and SO_2 it spews into the atmosphere. Both pollutants block incoming solar radiation directly and serve as nuclei for cloud formation. In contrast, soot on surfaces, especially snow and ice, has a positive warming effect. Coal mining also contributes to greenhouse gas (GHG) buildup by releasing methane stored in coal seams. This was once a significant source of atmospheric methane, but most mines now have recovery systems, using the gas for operations.

Coal Consumption Increase During the Industrial Revolution

Year	*British Consumption (mmt)**	*U.S. Consumption (mmt)*	*Global Consumption (mmt)*
1700	2.45		
1850	45		
1870		36	
1900	227		
1913		469	
1914			~550
2006			5,440

*mmt = millions of metric tons.

• Carbon Capture and Storage

In order to continue using coal as an energy source while avoiding escalating global warming, some practical method must be devised to remove CO_2 at the source, either storing it or converting it into nongaseous compounds. In practice, the energy costs of building and operating a system for sequestering the amounts of carbon emitted by a modern electrical generating plant are daunting. A very effective mechanism for using solar energy to convert CO_2 to organic compounds has existed for several billion years: photosynthesis. There have been proposals to couple CO_2 emissions with intensive algae farms, but it would unfortunately require thousands of square kilometers of surface area and hundreds of years to convert a large power plant's annual emissions into biomass fuel.

The most promising plans for carbon-capture-and-storage (CCS) systems call for using pure oxygen rather than air in IGCC plants, eliminating a large admixture of environmentally neutral nitrogen. The gas would then be injected into depleted petroleum reservoirs. Geologic formations suitable for CO_2 sequestration are typically far removed from coal-producing areas, requiring a delivery network for the waste gas comparable in cost and extent to that used for distribution of natural gas. The costs of retrofitting an IGCC plant for CCS, delivering the waste product to its destination, and maintaining ongoing operating costs make CCS commercially unviable in the absence of government subsidies and tax incentives. There have been many proposals and a few pieces of legislation addressing this dilemma, but none so far has stimulated investment in this technology.

The Fiddlers Ferry coal-fired power station in Liverpool, England. (Phil Noble/Reuters/Landov)

• Context

Electrical power plants have huge startup costs, very long half-lives, and limited capacity for adaptation. Many U.S. power plants date from the 1950's and even before, supplying cheap electricity in competition with costlier, newer plants built to more stringent environmental standards. Until these older plants are retired, the potential of the new technology cannot be realized. Reducing carbon emissions from coal-fired power plants will require massive public investment, either from government subsidies or sharply higher electricity costs. Neither alternative is likely to find an effective advocate in a shaky economy. Depending on the extent to which global markets contract, a deep recession could also reduce electrical demand and halt efforts to convert coal to other uses to reduce petroleum dependence. Such a reduction would temporarily dampen the escalating curve of coal consumption and CO_2 output.

Martha A. Sherwood

• Further Reading

Flatow, Ira. *Present at the Future: From Evolution to Nanotechnology—Candid and Controversial Conversations on Science and Nature.* New York: Collins, 2007. Chapter 14, "Is Coal Still King," discusses the merits of some new technologies.

Freese, Barbara. *Coal: A Human History.* Cambridge, Mass.: Perseus, 2003. Nontechnical; provides a narrative of coal use through history, a discussion of current environmental issues including global warming, and an assessment of future prospects.

Krupp, Fred. *Earth, the Sequel: The Race to Reinvent Energy and Stop Global Warming.* New York: W. W. Norton, 2008. Provides an extensive discussion of new, environmentally friendly technologies, most in the early experimental phase.

See also: Carbon cycle; Carbon equivalent; Clean energy; Fossil fuel emissions; Fossil fuel reserves; Fossil fuels; Greenhouse gases.

Coastal impacts of global climate change

• **Category:** Environmentalism, conservation, and ecosystems

Coastal environments include three of the four habitats on Earth that are most productive of biomass. Each of these habitats is directly affected by sea level and sea-level rise. Increasing concentrations of dissolved gases in the oceans will affect these ecosystems as well.

• Key concepts

brackish: having an intermediate salt concentration between freshwater (less than 5 parts per thousand) and seawater (35 parts per thousand)

primary production: the conversion of CO_2 into organic compounds, primarily by photosynthesis

salinity: the concentration of salts in water

• Background

The biomass productivity of marshes, estuaries, and coral reefs is comparable to that of tropical rain forests. This high productivity is due in part to the extreme range of environmental conditions at transitions between major habitats: terrestrial to marine, freshwater to seawater, and from the surface to considerable depths. These ecosystems not only are critical for the organisms whose entire lives are restricted to that ecosystem but also serve as nurseries for many of the fish and invertebrate species commercially harvested for human consumption.

• Coastal Wetlands

Coastal wetlands usually consist of a fringing border of marshlands grading through four general types, from open water toward land: salt marshes, brackish marshes, intermediate marshes, and freshwater marshes. The principal determinants of each grade are soil salinity and elevation. Soil salinity is slightly higher than the salinity of the surrounding water, so salt marsh plants, which are flooded by high tide, must be very salt tolerant. Such plants include salt grass and mangrove. Moving inland, soil elevation increases slightly with a consequent decrease in seawater influence versus freshwater input. The wetlands are integral to the coastal ecosystem. They filter freshwater runoff, trapping nutrients and sediments and thus extending the land seaward. Marshes protect the shore from wave action and storm damage by binding the soil and reducing wave energy. They serve as a nursery for many commercially important species such as shrimp, oysters, and crabs.

• Estuaries

Estuaries are semi-enclosed bodies of water where freshwater runoff from land mixes with seawater in a shifting gradient of brackish water with freshwater layered over denser seawater. Depending on the rate of freshwater flow, a wedge of salt water can intrude far upstream. For instance, in dry years with low river flow, salt water can intrude up the Mississippi River as far as New Orleans, threatening the city's water supply. During low water, an underwater sand sill is constructed across the river bottom a few kilometers downstream of the city to prevent this catastrophe. Major estuaries in the United States include Chesapeake Bay, Puget Sound, San Francisco Bay, and much of the Gulf Coast. Estuaries are one of the world's most productive natural environments and serve as the nursery areas for many commercial and sport fisheries. Typically, adult animals spawn at sea, and their eggs and larvae are carried by tide and currents to estuaries, where the young animals grow and develop.

• Intertidal Zone

Sea level is not uniform. On most coasts, sea level fluctuates between high tide and low tide twice a day. The extent of these tides also fluctuates on a monthly cycle. Once a month, there is a highest high tide, the spring tide, and two weeks later there follows the lowest low tide of the month, the neap tide. Between the levels of the spring tide and the neap tide is a gradient of conditions that are constantly varying. The intertidal zone is often subdivided into high, middle, and low subzones, each with its characteristic species. Organisms in the high intertidal zone are exposed to air virtually daily. In the low intertidal zone, organisms are exposed only a few days a month. Above the high intertidal zone is a splash zone, where terrestrial organisms must be tolerant of salt water. Below the level of low tide is the subtidal zone, which is further stratified by salinity tolerance and depth of light penetration. The consequences of tidal fluctuation are most dramatic on rocky shorelines, where the stratification of algal and invertebrate animal communities is clearly visible at low tide.

• Marine Zones

Oceans are the largest reservoir of carbon in the carbon cycle and readily dissolve carbon dioxide (CO_2) from the atmosphere. As such, they were thought to provide a buffer against global warming, with algal productivity increasing in proportion to carbon uptake from the atmosphere. However, when CO_2 dissolves in water, it forms carbonic acid, which lowers pH to the detriment of marine organisms. In addition, increasing oceanic nitrogen levels from fertilizer runoff and sewage washed by rivers into the sea and from nitrous oxide produced by burning fossil fuels is having a negative impact on marine life. Again, initial thought was that adding nitrogen to the oceans would be a good thing, fertilizing the sea and promoting algal growth. In many places, however, the resulting algal blooms deplete dissolved oxygen in the water, forming extensive dead zones, where fish cannot survive. For instance, every year a dead zone forms off the mouth of the Mississippi River, extending from Louisiana to Texas.

• Coral Reefs

Coral reefs are important coastal habitats in tropical waters, such as off the coasts of Hawaii and southern Florida. They are highly productive and, like estuaries and marshes, serve as a nursery for many marine animals. A reef itself is produced largely by calcium deposition from coral animals and coralline algae. The organisms secrete calcium carbonate, or limestone, around themselves for protection against predators, slowly building new reef upon older, dead layers. Coral animals have a narrow range of optimal temperatures, and scientists think that warming is responsible for some of the bleaching and death occurring on reefs worldwide, but particularly in the Caribbean Sea. Coral animals require symbiotic unicellular algae for calcium secretion, and animals that lose their symbionts quickly beach and die. In addition to temperature change, gradual acidification of the water due to dissolved CO_2 also may be a critical factor in coral death. As corals die, they often are replaced by carpets of algae that completely change the community composition of the reef.

Although reef walls may extend to great depths, their most productive zones are just below the surface. Fringing and barrier reefs are often visible from shore because of waves forming breakers as the water suddenly becomes shallower. As a result, the reef crest dissipates wave energy and protects the shoreline from the power of wave and storm action. A shallow lagoon,

Toxic algae, known collectively as a "red tide," threaten the coast of Dubai, United Arab Emirates, in April, 2009. (Matthias Seifert/Reuters/Landov)

with its own characteristic species, typically forms between the reef and the shore.

• **Context**

All coastal ecosystems are characterized by gradients of temperature and salinity that are directly related to water depth and distance from freshwater sources. Temperature-induced sea-level rise has an impact on seawater intrusion into freshwater and terrestrial habitats, changing the nature of the ecosystem. Higher temperatures also increase the rate of evaporation, concentrating salts and further raising salinity levels.

Marshall D. Sundberg

• **Further Reading**

Bertness, Mark, Brian Reed Silliman, and Robert Jefferies. "Salt Marshes Under Siege." *American Scientist* 92, no. 1 (2005): 54-61. Global warming is a long-term threat, but this article concentrates on some of the more immediate anthropogenic threats to this productive ecosystem.

Ricketts, Edward F., and Jack Calvin. *Between Pacific Tides.* Stanford, Calif.: Stanford University Press, 1962. Classic introduction to marine life on the shore by a friend and companion of novelist John Steinbeck. Ricketts was the model for Doc in Steinbeck's *Cannery Row.*

Silver, Cheryl Simon, with Ruth S. DeFries. *One Earth, One Future: Our Changing Global Environment.* Washington, D.C.: National Academy Press, 1990. One of the first warning calls concerning climate change. One chapter addresses coastlines and rising seas.

Wolanski, Eric, et al. "Mud, Marine Snow, and Coral Reefs." *American Scientist* 91 (2005): 44-51. Coral reefs are one of the most productive ecosystems on Earth, but they have been in noticeable decline for several decades. The authors elaborate on some of the causes, including warming oceans.

See also: Alliance of Small Island States; Barrier islands; Estuaries; Louisiana coast; New Orleans; Ocean acidification; Ocean life; Reefs; Sea-level change; Sea sediments; Sea surface temperatures.

Coastline changes

• **Categories:** Geology and geography; oceanography

• **Definition**

The term "coastline" is used to describe those places where the land and the water of the world's oceans meet. These are dynamic parts of the globe that are daily shaped by tides, winds, waves, changing sea levels, and human activities. Large river discharges may dominate them entirely, or they may be ice covered much of the time; cities may border them with their polluting activities, or they may be remote with no human presence at all. Oceanographers generally classify coastlines as being in one of two categories: either erosional or depositional.

Erosional coastlines are found where wave action is actively removing bedrock or sediment exposed along a coast, or where rivers or glaciers eroded the coast in past times when sea levels were lower than they are today. Some of the distinctive features found along erosional shorelines include rocky headlands separated by bays and coves, sea cliffs, sea caves, sea arches, sea stacks, and uplifted wave-cut terraces. Additional features such as fjords, old glacial moraines, drowned river valley systems, faults, and volcanoes may also be present.

Depositional coastlines are places where the shoreline grows outward because of the accumulation of sediment or the action of organisms such as mangroves and corals. Important features found along these coasts include beaches, sand bars, cusps, spits, sand hooks, and offshore barrier islands with long, narrow lagoons behind them. In addition, coral reefs, deltas, mangrove swamps, and salt marshes may be present. The term "coastal zone" is used in legal and legislative documents to describe the shoreline area. The landward boundary of the coastal zone is defined as the distance inland from a chosen reference point, usually the high water mark. This distance is frequently 60 meters. The seaward boundary of the coastal zone is variously defined by local, state, or federal laws.

Dauphin Island, off the coast of Alabama, has been split in two, but its halves may be growing back together. Coastal features such as these barrier islands can change relatively rapidly as a result of maritime events. (Bill Starling/The Press-Register/Landov)

• Significance for Climate Change

As sea levels rise as a result of the thermal expansion of seawater and the addition of new water from ice melting on land, a variety of coastline changes can be anticipated to take place. Areas that were previously above wave action will be subject to vigorous wave attack, and storm surges should intensify and extend farther toward the poles as a warming ocean generates more lethal storms. While cliffs made of resistant rock may remain unaffected, those coastlines built of loose sedimentary materials, such as sand or gravel, may be eroded back as much as a few meters in a single storm. Coastal areas that were never flooded before may be covered with seawater during severe storms and high tides; finally, they will be flooded by the ocean on a permanent basis. Low-lying urban areas would be particularly affected. The flooding of New Orleans during Hurricane Katrina in 2005 is an example, and already the city of Venice experiences periodic floods when tides are extremely high.

Ultimately, the people who live in high-risk areas such as these will be forced to move, and millions of people will be displaced. The Nile and Mississippi deltas are examples of high-risk places with large urban populations, as are the low-lying coastal areas in Bangladesh and the adjoining countries of Southeast Asia. As seas rise because of global warming, beaches will be forced to migrate inland or drown in place. Those in important tourist areas, such as Miami Beach and the Gulf Coast of the United States, will require expensive nourishment programs. Estuaries, which are critical for commercial fish stocks during spawning and early life stages, will gradually change from brackish to saline or in time may become part of the sea itself. The coastal wetlands behind them, which filter out pollution from the land and act as sponges for storm surges, will become saline too. These, along with the mangroves that protect the shoreline from wave attack and are also part of the nutrient system, will eventually have to migrate inland, provided the areas behind them are not already developed.

Even the offshore coral reefs that provide protection to the shoreline from wave erosion may be affected, because corals have a very low tolerance to elevated water temperatures. It can also be anticipated that salt water will gradually intrude upon coastal aquifers, making their supplies of drinking water unfit for humans, and that expensive structures such as dikes and seawalls will have to be built to protect most of the urban areas. Infrastructure, such as highways, bridges, waterways, ports, mass transit systems, airports, water supply facilities, and waste storage systems, may be threatened as well, and some of the nuclear power plants that have been built next to the ocean to make use of seawater for their cooling towers will have to be moved. Even lowly storm drains, built to carry rainwater off the land, will play a significant role in coastal flooding by providing conduits to send the seawater onshore.

Donald W. Lovejoy

• Further Reading

Bird, Eric. *Coastal Geomorphology: An Introduction.* Hoboken, N.J.: Wiley, 2008. Definitive survey of the classic shoreline types, with excellent photographs showing examples of each of these features throughout the world.

Davis, Richard A., and Duncan M. Fitzgerald. *Beaches and Coasts.* Malden, Mass.: Blackwell Science, 2004. Provides a good overview of beach and coastal features with useful information on the steps humans are taking to protect themselves from rising sea levels.

Glick, Daniel. "The Big Thaw." In *Global Climate Change*, edited by Paul McCaffrey. New York: H. W. Press, 2006. A sampling of firsthand reports illustrating the ways in which rising sea levels are affecting shorelines in many parts of the world.

See also: Bangladesh; Barrier islands; Coastal impacts of global climate change; Floods and flooding; New Orleans; Storm surges.

Collapse

- **Category:** Popular culture and society
- **Date:** Published 2005
- **Author:** Jared Diamond (1937-   )

• Background

Collapse: How Societies Choose to Fail or Succeed was published in 2005 by Jared Diamond, who previously won the Pulitzer Prize for *Guns, Germs, and Steel: The Fates of Human Societies* in 1998. Diamond's central theme is that environmental mismanagement and failure to confront emerging environmental problems have repeatedly led to the collapse of societies, even when the problems were obvious.

Collapse consists of four parts. Part 1 examines contemporary Montana, which saw the collapse of once-lucrative industries like mining and logging, and which is struggling with the need to adapt to change while grappling with ingrained hostility toward regulation and restraints on individual freedom. Part 2 is an examination of historical examples of societal collapse: Easter Island and other Polynesian examples, the Anasazi of the American Southwest, the Maya, and the Norse settlements in Iceland and Greenland. Diamond describes two contrasting paths taken by successful societies, the decentralized approach of New Guinea tribal societies and the top-down style of Japan under the Tokugawa regime (1603-1867). Part 3 is a study of several contemporary societies: Rwanda, the contrasting examples of Haiti and the Dominican Republic, China, and Australia. Finally, in Part 4, Diamond attempts to catalog the reasons some societies fail to respond to environmental threats, discusses environmental practices of several large industries, and finally attempts to apply his analysis to contemporary society. Although Diamond is critical of many Western attitudes and practices, the tone of the book is not apocalyptic. In fact it closes on a rather hopeful note by enumerating ways that modern societies have deliberately changed.

Societies may fail to deal with environmental threats for many reasons. They may not recognize the threat. The problem may be recognized, but attempts to solve it may be inappropriate. There may be no attempts to solve the problem, because certain interest groups may be adversely affected by the solution, the benefits of solving the problem may not seem to outweigh the costs, or the problem may be seen as diffuse or remote. Some environments, like Greenland or the American Southwest, may simply be unsuited for intensive human use. In contrast to errors that can be termed "rational," that is, attempts to cope with problems using what seem to be sound reasoning and methods, Diamond discusses "irrational failures," courses of action that are harmful to all. For example, many of the collapses discussed by Diamond were driven in part by maintaining the status of ruling elites, or unwillingness to abandon cultural prejudices.

• Significance for Climate Change

Several of the collapses discussed by Diamond were related directly to climate change, particularly the Anasazi, the classical Maya, and the Norse of Greenland. None of those societies was scientifically capable of recognizing the change in time or devising suitable strategies for coping with it. The relevance of *Collapse* is not in its use of past case studies of climate change, but its discussion of ways that societies have dealt with environmental degradation and either averted catastrophe or succumbed to it. Many of the patterns Diamond discusses can be seen at work in the climate change controversy.

Societies may be unaware of an emerging problem, because the onset of the problem is gradual, like the deteriorating climate of Norse Greenland. The onset of the problem may also be concealed by short-term random fluctuations, like the intermittent droughts that led to the collapse of Anasazi and Classical Maya society, which were interspersed with more normal years. In contemporary Western society, skeptics point to unusually cold episodes as evidence against climate change, while believers point to increasingly frequent evidence of warming. The "noisiness" of climate data is a major reason climate change is so controversial.

Environmental problems may be recognized but attempts to solve them may be inappropriate—for example, introduction of exotic species to control pests or total fire suppression in the American West. Similarly, some proposals for dealing with climate change may have unintended negative effects. Other attempted solutions may be costly but ineffective, impairing economic productivity and quality of life but not appreciably affecting climate change. There may be no attempts to solve the problem, because certain interest groups may be adversely affected by the solution, or because the problem is seen as diffuse or remote.

Many of the choices made by failed societies involve deep internal conflicts—for example, immediate survival needs versus long-term survival needs, or survival needs versus other deeply held values. Some of the collapses discussed by Diamond were partly caused by practices for maintaining the status of ruling elites, such as the way Easter Island society collapsed because resources were exhausted in building progressively larger statues in a prestige race among chiefs. Other fatal decisions involved unwillingness to abandon cultural prejudices. For example, the Norse settlements in Greenland died out because the Norse never adopted Inuit survival techniques, even as the climate deteriorated. Diamond notes that many of the values so-

The Metaphor of Easter Island

In Collapse, *Jared Diamond asserts that Easter Island and its tribal culture represent a useful metaphor for Earth and its inhabitants. In the excerpt below, he explains why.*

Thanks to globalization, international trade, jet planes, and the Internet, all countries on Earth today share resources and affect each other, just as did Easter's dozen clans. Polynesian Easter Island was as isolated in the Pacific Ocean as the Earth is today in space. When the Easter Islanders got into difficulties, there was nowhere to which they could flee, nor to which they could turn for help; nor shall we modern Earthlings have recourse elsewhere if our troubles increase. Those are the reasons why people see the collapse of Easter Island society as a metaphor, a worst-case scenario, for what may lie ahead of us in our own future.

cieties are most reluctant to change are often the very values that made the society successful in the past. These examples may have parallels in contemporary society, where much of the opposition to the idea of climate change is motivated by concern over impacts on lifestyle and where environmentally inefficient choices such as overly large homes and vehicles are often motivated by desire for social status.

Steven I. Dutch

See also: Arctic peoples; Carson, Rachel; Civilization and the environment; Displaced persons and refugees; Maldives; Megacities; Population growth; Preindustrial society.

Colorado River transformation

• **Category:** Water resources

Major reservoirs, constructed along the Colorado River to capture and more efficiently use annual river flow, promote evaporative water loss that concentrates the mineral salts washed into the river. Global warming will exacerbate this problem.

• Key concepts

aqueduct (irrigation canal): a human-made channel or pipe to carry water for consumption or industrial or agricultural use

desalination: the process of removing dissolved salts and minerals from water to produce freshwater for consumption or irrigation

evaporation: transformation of liquid water to water vapor, which is absorbed into the air

reservoir: a body of water that forms behind a dam

watershed: the area of land drained by a river and all of its tributary streams

• Background

The Colorado River is the major source of freshwater for a watershed of seven southwestern U.S. states and the adjacent nation of Mexico. In the early twentieth century, California was growing rapidly and the other states realized that their future development depended upon water allocation from the river. A series of compacts, laws, and court rulings legally parceled out more water than annually flows down the river, so the river no longer reaches the sea.

• Law of the River

In 1902, the Reclamation Act provided for development of irrigation projects throughout the West, including projects in California, Arizona, Utah, and Colorado in the Colorado River Basin. The Colorado River Compact (1922), the first of a series of laws and agreements specific to the Colorado River, divided the river into an upper and lower basin and allocated 9.25 cubic kilometers of water to each. The lower basin was given priority, so in dry years the upper basin allocation must be cut back to ensure that 9.25 cubic kilometers flows to the lower basin. The states in each basin negotiated a share of the water.

In a "good year," state allocations of Colorado River water are as follows: Colorado, 4.76 cubic kilometers; New Mexico, 1.04 cubic kilometers; Utah, 2.11 cubic kilometers; Wyoming, 1.28 cubic kilometers; Nevada, 0.37 cubic kilometer; Arizona, 3.45 cubic kilometers; and California, 5.43 cubic kilometers. In 1944, Mexico was granted an allocation of 1.85 cubic kilometers, and a subsequent agreement specified allowable salinity. The courts have allocated 1.11 cubic kilometers to Native Americans in the lower basin, and Native American rights in the upper basin are pending.

• Dams and Reservoirs

In the late 1800's, agricultural development of the Imperial Valley of California was watered by a canal from the Colorado River running mostly though Mexico. A major flood in 1905 caused a breach in the canal, and for two years the river emptied into the valley, creating the Salton Sea. Another major flood in 1910 stimulated planning for a flood-control dam upstream: The Hoover Dam was completed in 1935, and Lake Mead formed behind it. The Colorado River Compact between the states was made in response to plans to build the Hoover Dam, which would primarily benefit California.

Water is released from the Glen Canyon Dam in March, 2008, in an effort to restore depleted ecosystems downstream. (Tami Heilemann/UPI/Landov)

The dam provided flood control and generated electricity for Las Vegas and Los Angeles.

There are now seven major dams on the Colorado River, from Glen Canyon Dam, just north of Lees Ferry (the divide between the upper and lower basins), to the Mexican border. An additional thirty-five dams are built on tributaries within the watershed. The four largest reservoirs expose a tremendous surface area to evaporation: Lake Powell, 658 square kilometers; Lake Mead, 640 square kilometers; Lake Mojave, 114 square kilometers; and Lake Havasu, 83 square kilometers. This evaporation has driven an increase in the river's salinity. Indeed, river salinity levels have increased so much that in 1973 an agreement was reached between the United States and Mexico to mandate maximum salinity levels in the water crossing the border. The following year, Congress authorized construction of a desalination plant on the U.S. side of the border to ensure that water flowing into Mexico achieved treaty salinity specifications.

In addition to affecting water quality, dams and reservoirs change the physical characteristics of the river. Sediments normally deposited along the river banks during floods are now captured behind the dams. Absence of annual scouring by floods, with additional deposition of new sediments, has greatly altered habitat along the banks of much of the river.

• Human Uses

Shortly after completion of the Hoover Dam, plans were made to build additional dams, aqueducts, and canals. The Parker Dam, creating Lake Havasu, was completed in 1938, and within three years the Colorado River Aqueduct was supplying water to Los Angeles, 400 kilometers to the west. The following year, the Imperial Diversion Dam was completed to divert water through the All-American Canal, which replaced the old irrigation canal through Mexico. The new canal provided reliable irrigation water to the Imperial Valley and water to the city of San Diego.

Today, the Colorado River provides irrigation water for more than 809,000 hectares of agricultural land, as well as water for consumption by more than 24 million people in Denver, Phoenix, Salt Lake City, Las Vegas, Los Angeles, San Diego, and hundreds of other cities and towns. Almost from the beginning, California had been using more than its allocation of 5.43 cubic kilometers. By 1990, California had to begin reducing its consumption, as other states used greater and greater amounts of their allocations. In 1963, Denver completed the Harold D. Roberts Tunnel under the continental divide to divert water from the Colorado River watershed to the city. The Central Arizona Project, whose backbone is an aqueduct from Lake Havasu through Phoenix to Tucson, was essentially complete by 1993.

Today, California's use is still greater than its allocation at about 5.92 cubic kilometers, and further increases by any other state will require additional cutbacks by California. For more than a decade, the complete allocation of Colorado River water has been used; the river no longer reaches the sea. This situation has additional consequences in the former delta region in Mexico, where freshwater input has been cut off.

• Context

Early in the twentieth century, politicians realized the importance of achieving an equitable allocation of Colorado River water for the future development of their states and the region. Their agreements were based on available data suggesting an annual flow of 20.23 cubic kilometers per year. It would later be discovered that the average flow is considerably less, about 16.65 cubic kilometers. Future disputes over Colorado River water rights will continue to be shaped by that initial overestimation.

Marshall D. Sundberg

• Further Reading

Diamond, Jared. *Collapse: How Societies Choose to Fail or Succeed.* New York: Viking Press, 2005. Diamond's hypothesis about why five different human cultures, including the Anasazi in the American Southwest, collapsed.

Pearce, Fred. *When the Rivers Run Dry: The Defining Crisis of the Twenty-first Century.* Boston: Beacon Press, 2006. The Colorado is just one overused river that no longer reaches the sea.

Reisner, Marc. *Cadillac Desert: The American West and Its Disappearing Water.* New York: Viking Press, 1986. Overview of the history and ecology of Western water and water rights in the United States. Rated number sixty-one in the Modern Library Top 100 works of nonfiction of the twentieth century.

See also: Aral Sea desiccation; Desalination of seawater; Freshwater; Siberian river diversion proposal; Water quality; Water resources, global; Water resources, North American; Water rights.

Competitive Enterprise Institute

- **Category:** Organizations and agencies
- **Date:** Established 1984
- **Web address:** http://cei.org

• Mission

The Competitive Enterprise Institute (CEI) is a nonprofit public policy organization that promotes the principles of free enterprise and limited government. The institute contends that individuals are best served free of government intervention, allowing for unfettered progress through a free marketplace. It accepts no government funding; among its principal sponsors are Amoco, Exxon Mobil, Texaco, the Ford Motor Company, Phillip Morris Companies, and the Pfizer drug company.

CEI describes itself as a think tank and an advocacy organization. While it contends its policy positions are science based, it disputes all evidence pointing to greenhouse gases (GHGs) as the driving force behind global warming. Consistent with the CEI philosophy, the institute opposes vehicle fuel efficiency standards and calls for the elimination of the Superfund—using property rights as an argument against government regulation.

• Significance for Climate Change

CEI opposes any government action that would mitigate global warming, including mandated limits on GHGs. Consistent with this stand, it supports constitutional challenges that would impede the government's ability to intrude on corporate polluters. In an effort to garner public support, in 2006 CEI ran television ads trivializing the effects of climate change with the line: "Carbon dioxide: They call it pollution; we call it life." It insists that pollution-generated global warming is a scientific fallacy and embraces carbon dioxide as "a life-giving force" rather than a noxious waste. Consistent with its postilion, which denies the connection between GHGs and climate change, CEI participated in the opposition to the 1997 international global warming negotiations in Kyoto, insisting that global warming is a "theory not a fact."

Unlike many academic think tanks, CEI aggressively enforces its positions through the courts. It argues that climate change would create a "milder, greener, more prosperous world." A quote taken from the CEI Web site best illustrates the organization's philosophy and potential impact on climate change.

> Although global warming has been described as the greatest threat facing mankind, the policies designed to address global warming actually pose a greater threat. Luckily, predictions of the extent of future warming are based on implausible scientific and economic assumptions, and the negative impacts of predicted warming have been vastly exaggerated. In the unlikely event that global warming turns out to be a problem, the correct approach is not energy rationing, but rather long-term technological transformation and building resiliency in societies by increasing wealth. CEI has been a leader in the fight against the global warming scare.

Richard S. Spira

See also: Pseudoscience and junk science; Skeptics.

Composting

• Category: Plants and vegetation

• Definition

Composting is a natural biological process that breaks down organic materials into a stable organic substance called compost. Compost is created by combining organic wastes, such as yard trimmings, food wastes, and manures, in piles or containers. Microorganisms, primarily bacteria and fungi, decompose organic matter and form compost, a nutrient-rich material that is also called humus. Humus is dark brown or black in color and is free of most pathogens and weed seeds. Compost can either be added to soil as fertilizer or be used to support plant growth. Composting is used in landscaping, horticulture, and agriculture as an organic fertilizer to enrich soils. Compost is also used for erosion control, wetland construction, and as landfill cover. Industrial composting systems are increasingly being utilized as landfill alternatives, making them an important tool for waste management.

Composting methods range from simple backyard or onsite systems to large, commercial-scale agricultural and worm-composting centers. Composting can be either aerobic (occurring in the presence of oxygen) or anaerobic (without oxygen). Aerobic methods are generally more efficient at decomposing organic matter. Homeowners and other small-quantity generators use backyard composting systems to degrade such wastes as yard trimmings and food scraps. Larger operations utilize aerated, or turned, windrow composting, in which organic waste is piled in rows called "windrows" and aerated by periodically turning the piles. Windrow composting is suited for large quantities of waste, such as that generated by food-processing businesses including restaurants, cafeterias, and packing plants. The process can accommodate diverse wastes, including animal by-products.

In aerated, static-pile composting, organic waste is placed in one large pile rather than rows and is aerated by layering the pile with bulking agents, such as wood chips or shredded newspaper. This method is suitable for large quantities of yard waste, food scraps, and paper products, but not animal by-products or grease. It is often used by landscapers or on farms. In-vessel composting uses a container—such as a drum, silo, or similar vessel—and can be used to process large amounts of waste in a limited area. Most organic waste, including meat, manure, biosolids, and food scraps, can be decomposed using this method. Worm composting, or vermicomposting, uses worms in a container to break down organic matter. Red worms, rather than nightcrawlers or garden worms, break down food scraps, paper, and plants and generate rich compost called "castings." Vermicomposting also produces "worm tea," a nutrient-rich liquid fertilizer that can be used in gardens or for houseplants. Vermicomposting is well suited for apartments or small offices.

• Significance for Climate Change

Composting organic material has many environmental benefits, including some that mitigate some of the harmful effects associated with climate change. In general, composting leads either directly or indirectly to a reduction in the amount of greenhouse gases (GHGs) released into the atmosphere. Composting reduces the use of inorganic fertilizers and pesticides, diverts wastes from landfills, restores soil quality, and increases carbon content in soils. By regenerating nutrient-poor soils, compost increases the water-hold capacity of soils and decreases the amount of inorganic fertilizer required to grow healthy crops.

Inorganic fertilizers require a great deal of energy to produce, so reducing the demand for such fertilizers reduces energy production and consumption. Composting, moreover, slows down the depletion of existing organic matter from soils while simultaneously adding organic matter and carbon to the soil. Increasing carbon sequestration within soils contributes substantially to reduced GHG emissions. Composting also helps decrease plant diseases and pests, which reduces the amount

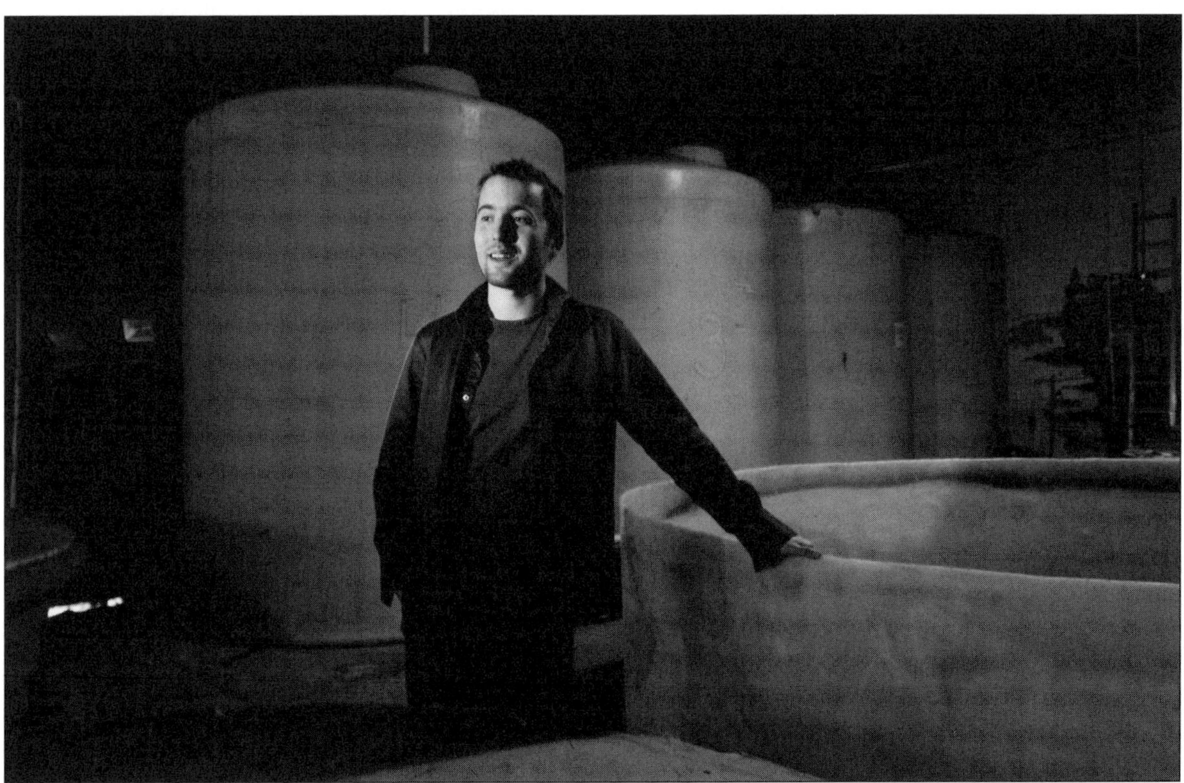

Tom Szaky's company, TerraCycle, uses earthworms to turn garbage into compost. (Jonathan Wilson/MCT/Landov)

of pesticides required. Composting also diverts organic wastes that would otherwise end up in landfills or other disposal sites. Reducing organic wastes in disposal sites reduces the amount of methane, a very potent GHG, emitted from such sites.

Although composting decreases methane emissions from landfills, some GHGs, including carbon dioxide, methane, and nitrous oxide, are produced during composting. During the intense microbial activity that occurs during composting, a significant loss of nitrogen as nitrous oxide and other nitrogen-based gases occurs. Although the worms used in composting decompose organic material very efficiently, nitrous oxide is emitted during digestion. Large, commercial-scale worm-composting plants generate GHGs in amounts comparable in global warming potential to those generated by landfills of the same scale. Nitrous oxide is believed to be many times more powerful than carbon dioxide or methane in contributing to global warming. At present, there is little research on the management of these gases during composting.

C. J. Walsh

• Further Reading

Campbell, Stu. *Let it Rot! The Gardener's Guide to Composting.* 3d ed. North Adams, Mass.: Storey, 1998. Introduction to composting for general readers. Includes descriptions of how composting works and different composting methods.

Culen, Gerald R., et al. *Organics: A Wasted Resource? An Extended Case Study for the Investigation and Evaluation of Composting and Organic Waste Management Issues.* Champaign, Ill.: Stipes, 2001. Written for students, this text provides information related to composting and waste management for a foundation in environmental science. Includes case studies and student activities.

Ebeling, Erich, ed. *Basic Composting: All the Skills and Tools You Need to Get Started.* Mechanicsburg, Pa.: Stackpole Books, 2003. A complete guide to tools and techniques of composting written by composting experts, including information on different composting methods and how to build composting bins.

Lens, Piet, Bert Hamelers, and Harry Hoitnik, eds. *Resource Recovery and Reuse in Organic Solid Waste Management.* London: IWA, 2007. Provides information on solid waste management using composting to reduce greenhouse effects and other environmental damage. Written at an advanced scientific level.

Scott, Nicky. *Composting: An Easy Household Guide.* White River Junction, Vt.: Chelsea Green, 2007. Contains a plethora of basic information about how to begin composting and was written by a chair of a Community Composting Association.

See also: Agriculture and agricultural land; Land use and reclamation; Nitrogen fertilization; Soil erosion.

Conservation and preservation

• **Category:** Environmentalism, conservation, and ecosystems

The history of conservation in the United States may be divided into three successively evolving stages: utilitarian conservation, nature preservation, and modern environmentalism. In the twenty-first century, as a result of global warming, environmentalism has adopted a more inclusive, planetary view.

• Key concepts

anthropocentric: treating human beings and human concerns as the primary source of value

biocentric: treating all living things as equal in value

natural area: a region where organisms and geological processes are undisturbed by humans, with as few controls as possible

• Background

Human abuse of nature is almost as old as recorded history. Plato lamented land degradation due to hills being denuded for lumber. Eighteenth century French and British colonial administrators understood the link between deforestation, soil erosion, and local climate change. Stephen Hales, a British plant physiologist, instigated the practice of reserving 20 percent of all green plants to preserve rainfall on the Caribbean island of Tobago. Pierre Poivre, French governor of Mauritius, appalled by forest and wildlife devastation, ordered one-fourth of the island to be preserved in woodlands. In America, conservation commenced as a pragmatic response to the excesses imbued by the nineteenth century limitless frontier mentality.

• Utilitarian Resource Conservation

George Perkins Marsh, who had witnessed the damage caused by excessive grazing and deforestation around the Mediterranean, became alarmed by the profligate waste of resources occurring on the American frontier in the mid-nineteenth century. In 1864, he published *Man and Nature*, warning of the unfortunate ecological consequences of this wanton destruction. This book had several lasting impacts, including the establishment of the National Forest Service in 1873 to protect dwindling timber supplies and endangered watersheds. In 1905, President Theodore Roosevelt, influenced by Marsh's book, moved the Forest Service from the Department of the Interior to the Department of Agriculture, and made his chief conservation adviser, Gifford Pinchot, the new head. This decision situated resource management on a straightforward, rational, and scientific basis.

Together with naturalist John Muir, first president of the Sierra Club, Roosevelt and Pinchot passed game protection laws, restructured the national park system, and reconstituted forest and wildlife refuge systems. These policies were primarily pragmatic. They believed that forests should be saved, not for aesthetic reasons or out of concern about wildlife, but to provide homes and jobs for people. Resources should be used for the greatest good for the greatest number for the longest time. Utilitarian conservation is not concerned about saving resources for future generations, but about wisely developing and using the resources for the benefit of humans now living. According to this viewpoint, there is as much waste in neglecting to develop and utilize natural resources as there is in their wanton destruction. This approach is still evident in the multiple-use policies of the Forest Service.

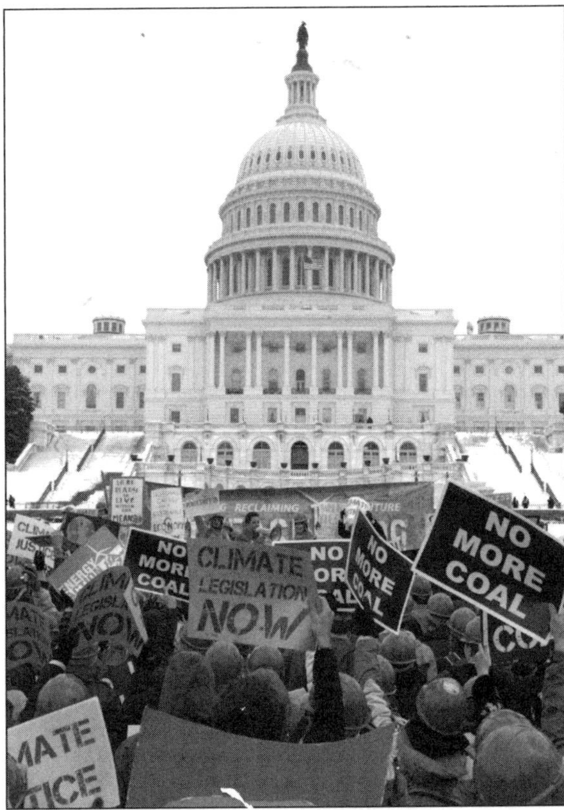

Protesters in March, 2009, demand that the power plant providing electricity to the U.S. Capitol building stop burning coal. The modern environmental movement has broadened to address the issue of climate change. (Roger L. Wollenberg/UPI/Landov)

• Nature Preservation

Muir, believing utilitarian conservation to be too anthropocentric, strenuously opposed Pinchot's influence and policies. Muir espoused the more biocentric viewpoint that all living organisms are imbued with intrinsic rights and deserve to live in nature, whether or not they are useful to humans. Every organism, as part of an ecological web, is not only entitled to continuance, but is essential to the integrity and stability of the biotic community. According to this viewpoint, humans are a miniscule component of nature; as such they have no right to value themselves above other species with whom they coexist. Humans are primarily a negative influence on nature.

In order to preserve its pristine wilderness, John

Muir fought for and achieved the establishment of Yosemite as a State Park in 1864, later incorporating additional land to become a National Park in 1890. He was also instrumental in having King's Canyon preserved until it also achieved National Park status. When the National Park Service was established in 1916, it was headed by one of Muir's disciples, guaranteeing that his ideals of attempting to preserve pristine wildernesses in their purest, unaltered state would become a guiding principle. This philosophy is often at odds with the Forest Service.

• Modern Environmentalism

Contemporary environmentalists have moved beyond the simple preservation of nature to embrace problems adversely affecting the health and well-being of all species, particularly humans. Air and water pollution began to become problematic during and particularly after World War II as a result of industrial expansion, greater use of toxic chemicals, and increased automotive traffic. One of the first books to awaken public awareness to the deleterious effects of noxious chemicals in the environment was *Silent Spring* by Rachel Carson, published in 1962. This led to an environmental movement as concerns broadened from preserving nature and using resources wisely to controlling and reducing pollution.

Two pioneers of the environmental movement were David Brower and Barry Commoner. Brower, while serving as the executive director of the Sierra Club, introduced many of the techniques now characteristic of modern environmentalism. These include litigation, intervention in regulatory hearings, and using the mass media for publicity campaigns. Commoner, a biologist, used scientific research to reveal connections among science, technology, the ecosphere, and society. Both activism and research remain defining characteristics of the modern environmental movement. By the 1970's, the movement had expanded from wilderness protection and pollution problems to include human population growth, nuclear power, fossil fuel extraction and use, and recycling. With the first Earth Day in 1970 environmentalism created public awareness and concern about health and ecological damage from pollution.

• Context

Because modern humans are interconnected in a myriad of ways, the Earth has become a global village of people sharing a common planetary environment. Attention has shifted from preserving particular landscapes or preventing pollution of a specific watershed to concern about the life-support systems of the entire world. Humans are changing planetary weather systems, increasing the extinction rate of species, and degrading ecosystems; without drastic remediation the ultimate consequences will be catastrophic. Protecting the planetary environment must become an international cause since it will take worldwide cooperation to effect the many changes necessary. Preliminary steps in this direction have been taken with the Montreal Protocol, adopted by most industrial nations in 1987, which phased out the use of ozone-depleting chlorofluorocarbons. The fledgling Kyoto Protocol is another international effort attempting to mandate carbon dioxide reduction. The agreement is still weak because the United States and Australia, two of the world's greatest emitters, refuse to sign.

Twenty-first century humans have begun to comprehend that human societies can no longer act in isolation because the Earth is an interconnected whole. Pollution and environmental problems are inextricably linked to poverty, injustice, oppression, and the exploitation of underdeveloped nations by greedy capitalists in industrialized countries. Only by working together to correct these historic wrongs and actively pursuing sustainable lifestyles can the planetary environment be conserved.

George R. Plitnik

• Further Reading

Brown, Lester R. *Plan B 3.0: Mobilizing to Save Civilization.* 3d ed. New York: W. W. Norton, 2008. Presents a worldwide plan to stabilize climate and control population. An excellent example of global environmentalism.

Commoner, Barry. *Making Peace with the Planet.* New York: Pantheon, 1990. A handbook of modern ecological conservation that treats the failures of modern environmentalism and prescribes actions that must be taken to live in harmony with the Earth.

Humphrey, Mathew. *Preservation Versus the People?* New York: Oxford University Press, 2002. Discusses humanity's interaction with nature and how one's political philosophy determines one's environmental worldview.

See also: Carson, Rachel; Civilization and the environment; Ecological impact of global climate change; Environmental economics; Environmental movement; Forestry and forest management; Marsh, George Perkins; Sierra Club; Sustainable development; Thoreau, Henry David.

Conservatism

• **Category:** Popular culture and society

Conservatism is a loosely defined political philosophy that places great emphasis on tradition, continuity, and gradualism in the development of public policy. There are many streams of conservatism, the best known of which are religious conservatism, cultural conservatism, and fiscal conservatism.

• Key concepts

Burkean conservatism: promulgated by the father of Anglo-conservatism, Edmund Burke, especially in his 1790 pamphlet *Reflections on the Revolution in France*, the classic tenets of conservatism—such as its distrust of utopian, revolutionary social reform—that laid the philosophical foundations of modern conservatism

conservationism: also known as the conservation ethic, the idea, originating in the late 1800's, that society should conserve natural resources for the purpose of ensuring continued abundance for maximum sustainable use

conservatism: a political philosophy located to the right of center on the classic political spectrum, marked by an emphasis on tradition and skepticism about radical ideas or social planning

invisible hand: an economic principle, promulgated by Scottish philosopher and author of *An Inquiry into the Nature and Causes of the Wealth of Nations* Adam Smith (considered the father of modern economics), that free markets are the best means by which to allocate goods in a society, as if guided by an "invisible hand"

• Background

The root of the term "conservatism" is "to conserve," which derives from the Latin verb *conservare*, meaning "to keep, preserve, or save." The term arose in the early 1800's, beginning in France with the writings of François-René Chateaubriand. The term then migrated to England in the writings of British statesman John Wilson Croker and was imported into America by John C. Calhoun, an early defender of states' rights.

The modern form of conservatism, however, largely derives from the works of Edmund Burke, who repudiated revolutionary government reforms in his 1790 book *Reflections on the Revolution in France*. Burke was critical of the French Revolution for its chaotic nature, though he supported the American Revolution.

Rather than being a set of strictly defined tenets, conservatism is a loose cluster of general principles which embrace nationalism and sovereignty, individual merit, free enterprise, reverence for tradition in the evolution of human institutions, strong respect for law and order, respect for family values such as the traditional nuclear family, respect for the wisdom of religious institutions, aversion to social planning, and an aversion to pure rationalism.

Conservatives are considered to be economic liberals, following the tradition of economist Adam Smith, which places great emphasis on the importance of well-defined property rights and the strict enforcement of contracts. Economic liberals (hence, conservatives) disapprove of government intervention in economic affairs beyond the provision of a highly limited set of "public goods," which the "invisible hand" of free markets would not be expected to produce but which are considered necessary for social welfare.

• Conservationism and Environmentalism

Conservatism is infused by a conservation ethic, which is distinct from the ethics of modern environmentalism. The attitude of conservatives toward the environment differs in fundamental ways from those of the modern environmental movement. Conservatives have long embraced "conservationism," which is the idea that society should conserve natural resources for the purpose of ensuring continued abundance for maximum sustainable use. Beginning with Theodore Roosevelt (1858-1919), twenty-sixth president of the United States, conservatives and many Republicans have been strong proponents of conservationism. Roosevelt's confidant and the first head of the U.S. Forest Service, Gifford Pinchot (1865-1946), was highly influential in shaping conservationism and emphasized the importance of using the greatest amount of natural resources that can be used sustainably, arguing that underutilization is just as wasteful as overutilization.

Conservationism rejects several of the ecocentric tenets of modern environmentalism, which seek to minimize natural resource use and human influences on the environment, ecosystems, biodiversity, air, water, soil, and climate. As Pinchot wrote,

> . . . the use of the natural resources now existing on this continent [is] for the benefit of the people who live here now. There may be just as much waste in neglecting the development and use of certain natural resources as there is in their destruction. . . . The development of our natural resources and the fullest use of them for the present generation is the first duty of this generation.

Pinchot further wrote:

> Without natural resources life itself is impossible. From birth to death, natural resources, transformed for human use, feed, clothe, shelter, and transport us. Upon them we depend for every material necessity, comfort, convenience, and protection in our lives. Without abundant resources prosperity is out of reach.

Pinchot's words remain at the core of the conservation ethic that infuses the philosophy of conservatism.

Major Conservative Think Tanks and Climatological Organizations

- Advancement of Sound Science Center
- American Enterprise Institute
- Cato Institute
- Competitive Enterprise Institute
- Cooler Heads Coalition
- Fraser Institute
- Friends of Science
- George C. Marshall Institute
- Heartland Institute
- Heritage Foundation
- High Park Group
- Information Council on the Environment
- International Policy Network
- Lavoisier Group
- National Center for Policy Analysis
- Nongovernmental International Panel on Climate Change
- Reason Public Policy Institute
- Science and Environmental Policy Project
- Science and Public Policy Institute
- Scientific Alliance

• Context

The conservative values mentioned above, combined with the conservation ethic, have shaped the responses of modern conservatives to the issue of climate change. In keeping with their rejection of pure rationalism, conservatives have been wary of absolutist claims regarding climate science and are inclined toward a more skeptical view, particularly of scientific pronouncements that elevate scientific rationalism above other conservative values. Averse to social planning, the expansion of government, and redistribution of wealth, conservatives have generally rejected programs to reduce greenhouse gas emissions such as the Kyoto Protocol, emissions trading, carbon taxes, and regulatory regimes.

With regard to fossil fuel consumption, the conservation ethic held by many conservatives leads them to oppose greenhouse gas control regimes on the grounds that such controls would result in the underutilization of affordable fossil fuels to the detriment of society. As explained above, this theme of maximizing sustainable utilization of natural resources differs significantly from modern environmentalism, which seeks to end the use of fossil fuels in favor of alternatives such as wind or solar power and generally to minimize human consumption of natural resources.

Finally, fiscal conservatism leads conservatives to oppose measures that they believe carry costs that will exceed their benefits, and that might compromise the economic prosperity and competitiveness of the United States.

Kenneth P. Green

• Further Reading

Pinchot, Gifford. *The Fight for Conservation.* 1910. Reprint. Whitefish, Mont.: Kesser, 2004. Pinchot makes the case for a conservation ethic in this booklet, which would later infuse the conservative philosophy as the central principle for managing natural resources and the environment.

Schneider, Gregory. *Conservatism in America Since 1930: A Reader.* New York: New York University Press, 2003. This collection of writings won great acclaim from conservative scholars and was a selection of the Conservative Book Club.

Scruton, Roger. *The Meaning of Conservatism.* Chicago: St. Augustine's Press, 2002. Scruton, an economic philosopher and leading conservative intellectual, explores the philosophical underpinnings of conservatism, such as an aversion to social planning and redistribution of wealth.

Smith, Adam. *An Inquiry into the Nature and Causes of the Wealth of Nations.* 1776. Reprint. New York: Bantam, 2003. In this successor to *The Theory of Moral Sentiments,* Smith fleshes out the economic theories that would evolve into modern economics. Smith's general thesis is that free markets are the best institution for allocating goods and services in an economy, while governmental intervention in markets leads to suboptimal outcomes and should be undertaken only reluctantly.

See also: American Association of Petroleum Geologists; American Enterprise Institute; Conspiracy theories; Cooler Heads Coalition; Falsifiability rule; Fraser Institute; George C. Marshall Institute; Heritage Foundation; High Park Group; Liberalism; Libertarianism; Media; National Center for Policy Analysis; Religious Right; Skeptics; U.S. and European politics.

Conspiracy theories

• **Category:** Popular culture and society

Each side in the climate change debate has accused the other of ulterior motives and of being part of a conspiracy to advance a hidden agenda.

• Key concepts

ad hominem argument: a logical fallacy that attempts to discredit an idea by discrediting the person espousing it

conspiracy: a secret effort by a group to achieve a (usually illegitimate) purpose

fallacy: a false belief or an argument that looks superficially convincing but is based on logical errors

hidden agenda: an ulterior motive

non sequitur: a fallacy based on an erroneous claim that a piece of evidence proves a given point

• Background

Many people believe that historic or present-day events are the work of conspiracies. Some conspiracy theories claim that specific groups are secretly working to control society. Other theories claim that major historical events were plotted by conspiracies. Finally, many people claim that unorthodox ideas, such as revolutionary inventions or medical cures, are suppressed by entrenched vested interests, although they may not identify any specific people or groups as responsible. In general, any belief that there are secret, organized efforts working behind the scenes to manipulate events can be termed a conspiracy theory.

• When Are Conspiracy Theories Valid?

There have been real conspiracies in history, such as the plot to assassinate President Abraham Lincoln, and there are real conspiracies today, such as criminal cartels and terrorist movements. In order for a conspiracy theory to be valid, a first requirement must be that the goals of the covert group be unethical or criminal. Someone who is prosecuted for selling a worthless medical remedy may claim that he is the victim of a conspiracy between the medical establishment and the government, but removing worthless remedies from the market is a legitimate goal. The efforts during World War II to keep the American landing at Normandy and the atomic bomb project secret involved large organizations acting secretly, but their objectives were legitimate. Not all secret movements are conspiracies.

Most important, for a theory to be valid, the conspiracy it describes must exist in reality. It is very hard to prove the existence of a conspiracy, unless its cloak of secrecy is removed. It may be infiltrated by law enforcement, be betrayed by a member, or be defeated and its secrets revealed. Most believers in a given conspiracy claim that the conspiracy is ongoing and therefore secret. They resort to indirect arguments in support of their theories, pointing to unusual coincidences, reasons why certain people might benefit from a conspiracy, and unanswered questions about events as evidence of their beliefs. While those lines of evidence might be consistent with a conspiracy, they are not sufficient to prove the existence of the conspiracy.

• Why Conspiracy Theories Are Faulty Reasoning

Some climate change skeptics have claimed that the idea of global climate change is part of a movement to destroy capitalism and personal freedom. Climate change activists, by contrast, claim that many skeptics are allied with front groups funded by industry or political movements. The climate change controversy is a good example of why conspiracy theories have no place in science.

In any debate, the only relevant question is whether ideas are true or false. The overriding issue in the climate change controversy is whether or not human activities are changing the climate in a dangerous fashion. The fact that someone has a personal motivation, or a hidden agenda, for supporting or opposing an idea does not make the idea true or false. This is an example of a fallacy called a non sequitur. The fact that many supporters of climate change hold liberal political beliefs and many opponents have ties to industry proves nothing about the correctness of their ideas. Merely calling something a conspiracy does not make it one.

Many people use conspiracy accusations to imply that their opponents are unethical or wrong or use conspiracy theories to explain why others reject their ideas. This is a fallacy known as an ad hominem argument. Criticism or opposition, by themselves, do not constitute a conspiracy. If an idea is generally rejected by the scientific community, that is evidence that the theory is wrong, not that scientists are conspiring against it.

• Dealing with Conspiracy Theories

There are real conspiracies in the world, so the mere fact that someone believes in a conspiracy does not make the person wrong. However, using a conspiracy argument to justify a belief commits several logical fallacies. Conspiracy allegations show that the person making them uses faulty logic. It is not logically correct to dismiss conspiracy beliefs as automatically wrong, but it is legitimate to reject them as evidence and to insist that the person using them debate ideas on their own merits rather than resorting to conspiracy claims. It is never legitimate to use conspiracy arguments to justify the acceptance or rejection of an idea.

• Context

In scientific debates, the only relevant issue is whether ideas are right or wrong. Using the idea of a conspiracy to discredit a position is a logical fallacy and, worse yet, poisons the climate of debate. For example, the title of Al Gore's celebrated documentary *An Inconvenient Truth* (2006) hints that critics of global warming are motivated not by the scientific data but by personal motives. Likewise, the name of the skeptic organization, Cooler Heads Coalition, implies that their opponents are acting irrationally and rashly. Knowing that someone has a personal motivation for a belief can certainly justify giving the person's arguments very close scrutiny, but it never constitutes scientific evidence.

Steven I. Dutch

• **Further Reading**

Hofstadter, Richard. *The Paranoid Style in American Politics, and Other Essays.* Cambridge, Mass.: Harvard University Press, 1996. Hofstadter is a prominent historian and author of a major work on anti-intellectualism in America. In this book, he explores conspiracy theories with emphasis on politically conservative conspiracy theories.

Jameson, Fredric. "Totality as Conspiracy." In *The Geopolitical Aesthetic: Cinema and Space in the World System.* Bloomington: Indiana University Press, 1995. An explication of the ideological work done by conspiracy narratives and the reason those narratives are so powerful in postmodern societies.

Newton, Michael. *The Encyclopedia of Conspiracies and Conspiracy Theories.* New York: Facts On File, 2006. Provides short articles on about five hundred persons, organizations, and historical events, arranged alphabetically. Discusses documented conspiracies (such as the Iran-Contra affair), as well as those with little or no supporting evidence.

Parish, Jane, and Martin Parker, eds. *The Age of Anxiety: Conspiracy Theory and the Human Sciences.* Oxford, England: Blackwell, 2001. Academic treatment of conspiracy theories, exploring how people use such theories to respond to anxiety and crisis.

Pipes, Daniel. *Conspiracy: How the Paranoid Style Flourishes and Where It Comes From.* New York: Free Press, 1997. Emphasizes theories revolving around secret societies and Jews and discusses conspiracy theories across the political spectrum.

Roeper, Richard. *Debunked! Conspiracy Theories, Urban Legends, and Evil Plots of the Twenty-first Century.* Chicago: Chicago Review Press, 2008. Popular discussion of about two dozen widespread conspiracy theories by a columnist for the *Chicago Sun-Times.*

See also: Gore, Al; *Inconvenient Truth, An*; Journalism and journalistic ethics; Popular culture; Skeptics.

Consultative Group on International Agricultural Research

• **Category:** Organizations and agencies
• **Date:** Established 1971
• **Web address:** http://www.cgiar.org

• **Mission**

The Consultative Group on International Agricultural Research (CGIAR) supports scientific projects at affiliated centers worldwide to address concerns regarding the detrimental impacts of extreme temperatures, altered precipitation cycles, and other erratic climatic factors on crops and livestock. The CGIAR oversees scientists' efforts to provide information, technology, seeds, plants, and resources to assist farmers, especially in developing countries. Prior to CGIAR, researchers had independently conducted plant-breeding projects to ease famines, working at the International Center for Tropical Agriculture (CIAT) in Colombia, the International Maize and Wheat Improvement Center (CIMMYT) in Mexico, the International Institute of Tropical Agriculture (IITA) in Nigeria, and the International Rice Research Institute (IRRI) in the Philippines.

Interested in centralizing their efforts, representatives of those research centers and the United Nations Food and Agriculture Organization (FAO) and Development Programme (UNDP) met at the World Bank at Washington, D.C., in May, 1971, to establish the CGIAR. By 2008, fifteen CGIAR research centers, including the initial four, were operating in Africa, Asia, Europe, the Middle East, and North and South America. Approximately one thousand researchers and seven thousand staffers pursue sustainable agricultural methods at centers for staple crops, such as wheat, rice, and maize; forestry; fisheries; and livestock. Several centers focus on agriculture in deserts or the tropics. Some centers specialize in such issues as food policy research, genetic resources, and water waste management. An alliance of governmental, public, and private groups funds the CGIAR's programs.

The CGIAR promotes scientific methods to cultivate and protect ample, nutritious food supplies and ease impoverishment in developing countries for both producers and consumers of agricultural goods. That group encourages researchers to preserve natural resources against pollution and other threats. The CGIAR hosts an annual conference, issues publications, and distributes information electronically on the Internet. CGIAR leaders adjust their research goals in response to urgent problems, such as global warming, that threaten to disrupt agriculture.

• Significance for Climate Change

The CGIAR emphasizes preparation to sustain agriculture against global warming. Since the organization's creation, CGIAR representatives have recognized the influence of weather upon agriculture, which is vulnerable to variations in temperature and precipitation. In the late twentieth century, the CGIAR stressed the dangers that changing climates posed to agriculture, particularly crops and livestock raised in developing countries. CGIAR centers initiated research to counter the detrimental impact of global warming on agriculture, highlighting the need for plants and livestock that can survive climate fluctuations. Scientists explained how increased temperatures and excess or insufficient precipitation can alter crop growth patterns and seasons by hindering photosynthesis and pollination. Increased knowledge regarding climate changes enabled people, ranging from agriculturists to government officials, to plan strategies and policies, using CGIAR resources to deal with global warming.

CGIAR scientists compile agricultural, demographic, climatic, and socioeconomic data relevant to areas being studied in order to simulate with computer models possible future conditions as climates change. Geographical-information-system (GIS) maps enhance this modeling. CGIAR researchers have used computer modeling to determine that wheat and maize crops are particularly at risk as the twenty-first century progresses. Models prepared through 2055 indicated that maize yields could decrease by 10 percent in developing countries without intervention. Also, temperature increases may cause maize cultivation to shift to highlands. The CGIAR study "Can Wheat Beat the Heat?" stated that India might lose 51 percent of its wheat fields by the mid-twentieth century. Scientists hypothesized warming climates could impede frost, enabling farmers to grow wheat near the Arctic Circle.

CGIAR research centers pursued genetic engineering in the 1970's to create more compatible and productive plants for famine-stricken countries. Changing climates resulted in revised efforts to make plants more resilient to temperature changes in air and soil and to inconsistent precipitation conditions. CGIAR scientists identified wild species of plants that exhibited resistance to heat, pests, diseases, salt, drought, or flood and evaluated samples to understand physiologically why some species are more tolerant than others. Scientists isolated genes associated with traits resistant to climate change in order to bioengineer stronger varieties of traditional crops and livestock or create hybrids for altered climates. Often, researchers determined that several genes in combination were linked to resistance to specific climate changes.

The CGIAR's successes have included the International Center for Agricultural Research in the Dry Areas (ICARDA) in Syria, which used wild *Hordeum spontaneum* to create drought-resistant barley. The IRRI produced rice resistant to high temperatures and flooding, submerging plants for several weeks and studying leaf evaporation processes to reduce temperatures. CGIAR researchers examined how to achieve successful maize pollination during drought conditions. They designed maize that grew bigger ears with more kernels than previous types.

Global warming is a catalyst for migration of some wild plant species to ecosystems where they are considered exotic. In their new habitats, these species compete for resources, crowding established species and detrimentally affecting cultivated crops. Climate change also causes insects and diseases to move to new areas, and the CGIAR has conducted research to assess how these invaders affect agriculture and how they might be controlled. When El Niño altered Peru's climate during the late 1990's, a previously unknown blight fungus attacked potato crops. The CGIAR's International Potato Center developed a hybrid that resisted blight, and it distributed those hybrid plants to farmers.

Promoting biodiversity, the CGIAR includes gene banks at its centers to preserve genetic material from agricultural resources worldwide, because computer models have indicated that many wild and domestic plant species may become extinct as a result of climate changes. Scientists estimate that one-fourth to one-half of all plant species could be extinct by 2055. CGIAR gene banks store millions of plant specimens, assuring that diverse agricultural crops from varied geographical locations will survive despite climate-provoked losses.

CGIAR scientists hypothesize 67 percent of people worldwide will experience water shortages by 2050, because changing climates will alter precipitation and melt glaciers, affecting long-term water supplies. CGIAR researchers develop water management techniques for arid locations, including techniques for storing precipitation and irrigating land. CGIAR representatives teach agriculturists techniques for dealing with inconsistent precipitation, such as applying phosphorus fertilizers to increase plants' root growth, allowing the roots to reach subsoil water. CGIAR workers also suggest agricultural adjustments to respond to depleted or eroded soil caused by climate changes. The CGIAR encourages farmers to diversify, planting alternative crops to counter climate impacts on their production and marketing possibilities.

CGIAR researchers note that some developing countries' agricultural methods, particularly the clearing of forests for fields, contribute to climate change. The CGIAR seeks ways to control global warming by urging farmers to use such agricultural techniques as limiting tillage manipulation of fields in order to retain carbon in soils. Center representatives educate agriculturists regarding forest management and biofuels, emphasizing the need to minimize emissions.

Elizabeth D. Schafer

• **Further Reading**

Ceccarelli, Salvatore, et al. "Breeding for Drought Resistance in a Changing Climate." In *Challenges and Strategies for Dryland Agriculture*, edited by Srinivas C. Rao and John Ryan. Madison, Wisc.: Crop Science Society of America, 2004. CGIAR researchers in Syria describe their work bioengineering barley for arid climates and suggest strategies to create resilient crops for diverse environments.

Fuccillo, Dominic, Linda Sears, and Paul Stapleton, eds. *Biodiversity in Trust: Conservation and Use of Plant Genetic Resources in CGIAR Centres.* New York: Cambridge University Press, 1997. Chapters feature numerous crops and forages CGIAR has identified as essential to include in its gene banks to protect agricultural variety as climate changes threaten species' survival.

Lele, Uma J. *The CGIAR at Thirty-One: An Independent Meta-evaluation of the Consultative Group on International Agricultural Research.* Washington, D.C.: World Bank, 2004. Critique by the World Bank's Operations Evaluation Department that suggests CGIAR shapes its climate change research to appease donors and should assist developing world farmers with global warming concerns not analyze its causes.

Varma, Surendra, and Mark Winslow. *Healing Wounds: How the International Centers of the CGIAR Help Rebuild Agriculture in Countries Affected by Conflicts and Natural Disasters.* Washington, D.C.: CGIAR, 2005. Profiles climate-change-related CGIAR projects. Photographs depict farmers, CGIAR representatives, and crops. Appendix provides centers' contact information.

See also: Agriculture and agricultural land; Animal husbandry practices; Composting; Food and Agriculture Organization; Land use and reclamation; Nitrogen fertilization; Pesticides and pest management; Soil erosion.

Continental climate

• **Category:** Meteorology and atmospheric sciences

• **Definition**

Defining a climate is difficult. The accuracy of weather observations and the length of time over which these data have been recorded can seriously affect the statistical understanding of a climate. Additionally, having data to produce statistical averages is no assurance that a climate's classification can accurately describe its conditions. Consider Chicago's humid, continental climate: The average temperature of the city is near 10° Celsius, but the nature of the climate in the region lends itself to extremes. Chicago's daily temperatures are usually either well above or well below the average.

The average precipitation in Chicago is about 84 centimeters per year. This is equal to the precipitation level in Seattle, which has a west coast marine climate. Precipitation in Seattle is mostly rain, however, while Chicago experiences a mix of rain and snow. Rain in Chicago falls mostly in the summer, during violent convective storms. Rain in Seattle continues through the year, with a slight increase during the winter. Seattle's maritime influence moderates the tendency toward convective storms.

Continental climates are characterized by extreme temperatures, and they are unique in that they have a global Northern Hemispheric distribution. In fact, to understand the potential impact that climate change might have on a continental climate, one must first consider its unique geographic character. Humid continental climates are found in North America, Europe, and Asia. In North America, they bound on the humid, subtropical climates of the southern and southeastern United States, and they encompass all of the northeastern quarter of the country.

The North American continental climate extends northward into Canada at a line of latitude that marks the extent of viable agricultural production. It extends westward into southern portions of the Prairie Provinces of Canada and north of the middle-latitude, dry climates that prevail just east of the Rocky Mountains. In Europe, the humid continental climate begins directly east of the marine-climate boundaries, along the west coast of the continent. It stretches into Russia, Germany, southern Sweden, and Finland, and is also found in Romania and Poland. Parts of Asia are also included in this climate. Almost all of northern China, the northern part of Japan, and North Korea fall within the humid continental climate zone.

The Köppen classification system places these climates in a group of midlatitude climates. Such climates include the west coast marine climate, the middle latitude dry climate, the humid continental-warm (long) summer climate, and the humid continental-cool (short) summer climate. The term "continental" has significance from the standpoint of the climate's paleoclimatic character. Continental climates exist presently in only the Northern Hemisphere as a result of the movement of tectonic plates, which have concentrated Earth's landmass in the northern half of the globe. The Southern Hemisphere lacks a sufficient continental landmass to form a continental climate. The Southern Hemisphere is more ocean (71 percent) than land. Most important, this region is agriculturally productive. Many soils in the region result from glaciation in the recent Pleistocene, 2 million years ago.

• Significance for Climate Change

The continental climate sustains a global breadbasket. The climate can be divided further by the growing regions of corn (maize) or wheat. An imaginary line running east through the southern half of South Dakota, and as far east as New England, defines the margin zones of the humid continental long (warm) summer climate (to the south of this line) and the humid continental short (cool) summer climate (to the north of it). A similar line divides the climate in Europe and Asia. Thus, global climatic change within this zone has the potential to affect world food supplies.

In a warming scenario affecting the continental climate zone, ancillary problems for farming may emerge. Some models suggest that warming will increase annual precipitation. The humid continental climate averages 76 centimeters of precipitation per year, with snow being the predominant form of precipitation in the winter. Precipitation in this climate zone can vary from about 51 centimeters near drier areas to 127 centimeters near the oceans. An increase in precipitation in the continental climate zone could lead to increased flooding, noxious weeds, and plant diseases. In contrast, some suggest that warming would lead to a longer growing period and a shift of the temperature toward the north.

Conversely, in a global cooling scenario, the continental climate zone might experience shorter summers, and the temperature line between warm and cool summers could move south. Drier conditions and drought would prevail. The region might expect to see more rainfall relative to snowfall, as rain might continue further into the winter months. The continental short (cool) summer climate zone is characterized by cyclonic storms in winter that can bring huge snowfalls.

During the summer, convectional storms, many of them severe with lightning, are normal in continental short (cool) summer climate zones. The temperature average during summer is 24° Celsius. During the winter, average temperatures fall below freezing, down to –12° to –11° Celsius. With the influence of cold northern air, it is not uncommon to experience temperature extremes well below –18° Celsius. With increasing temperatures, the possibility for a prolonged period of convective storms might exist. Dry-land crops, such as wheat, would need to be modified to accommodate warmer and moister conditions. More rainfall might increase erosion in already tenuous soils, especially in highly productive lõess soils. New farming methods to accommodate these changes would have to be implemented.

M. Marian Mustoe

• **Further Reading**

Aguado, Edward, and James E. Burt. *Understanding Weather and Climate.* 4th ed. Upper Saddle River, N.J.: Pearson Prentice Hall, 2007. The formation of midlatitude storms is surveyed. Continental climates and the structure of climate are discussed.

Ahrens, C. Donald. *Meteorology Today.* 9th ed. Pacific Grove, Calif.: Thomson/Brooks/Cole, 2009. Discusses global climates, climate change, and classification.

Critchfield, Howard J. *General Climatology.* Englewood Cliffs, N.J.: Prentice-Hall, 1998. Considers climate classification and discusses how climate affects agricultural productivity.

See also: Climate and the climate system; Climate change; Climate models and modeling; Climate prediction and projection; Climate zones; Climatology; Polar climate.

Contrails

• **Category:** Transportation

• **Definition**

Condensation trails, or contrails, are long, narrow cirrus clouds composed of ice crystals that form behind aircraft or rocket engines flying in the upper atmosphere. When fuels containing hydrogen, such as hydrocarbons, burn in air, the engine exhaust contains water vapor. The vapor in the hot exhaust rapidly condenses and freezes when it mixes with cold, humid atmospheric air, forming a trail of ice crystals. Contrails typically form where the air temperature is 4° to –60° Celsius and relative humidity exceeds 100 percent, with low wind turbulence. Favorable conditions typically occur above 8,530 meters. Contrails have been reported since 1915, but they became much more common as jet aircraft traffic increased. Geese flying in cold air have been seen to leave small contrails as they exhale moist air.

The condensation engendered by contrails starts to form on microscopic dust particles. Contrails become visible approximately 0.1 second after leaving an engine, as the ice particles grow large enough to scatter sufficient light for them to be seen. Although contrails start as one exhaust behind each engine, they often merge into one wingtip vortex from each wing tip. The lower pressure and temperature in the core of each tip vortex also helps accelerate condensation. Contrails move down in reaction to the aircraft's lift and the higher density of ice. During daytime ice crystals absorb sunlight. The warming air around them can convect the contrails several hundred meters up.

When the air is dry, the ice particles sublimate to vapor quickly, resulting in a short contrail. In humid air, contrails can persist for several hours or thousands of kilometers, given the speed of aircraft, and they grow into cirrus clouds as thick as 500 meters and several kilometers wide. Ice-particle size in such clouds is on the order of 200 micrometers, and their density is on the order of 1 to 50 particles per cubic centimeter of air. Many trails are over two hours old, and they continue to accumulate moisture from the air during that time. Indeed,

A plane with four jet engines leaves four contrails in its wake. (AP/Wide World Photos)

little of such a cloud comes from the original jet exhaust. A contrail cloud may contain one thousand to ten thousand times the water released by the aircraft engine itself.

• Significance for Climate Change

The cirrus cover due to contrails has been estimated to cover as much as 0.1 percent of the Earth's surface. The most famous experiment on contrails was conducted by the National Aeronautics and Space Administration (NASA) in the days following the terrorist attacks of September 11, 2001. U.S. civil air traffic was grounded for three days. During those three days, the difference between daytime high and nighttime low temperatures over the continental United States increased by roughly 1° Celsius when compared to the thirty-year average. At the same time, the trails left by six military aircraft persisted and eventually covered over 19,700 square kilometers.

The increase in cirrus cloud cover due to contrails has been studied as an anthropogenic factor in global warming. Ice crystals absorb, scatter, and reflect radiant heat. Some studies indicate that contrail cloud cover inhibits outward radiation from the Earth's surface and lower atmosphere more than it reflects incoming solar radiation, contributing to the greenhouse effect. Some argue that long-wave infrared radiation is absorbed more by ice crystals than by air or water vapor, so that cirrus clouds have a net warming effect.

Others studies suggest that night flights, which constitute only 20 to 25 percent of air traffic, may contribute to 60 percent of contrails' greenhouse effect. Published data project that a fivefold increase in air traffic would cause a net global warming effect due to contrails of 0.05° Celsius. Some argue that any detectable anthropogenic change is a cause for concern. Others point to a more severe localized effect due to the heavy traffic over industrialized nations in the temperate zone, where the air is moist and cold for a greater part of the year compared to equatorial regions.

Changing from hydrocarbon to hydrogen fuel will not reduce water vapor, but may reduce dust particles in the exhaust. A different aspect is the deposition of carbon dioxide and heat in the upper atmosphere. Aircraft emissions are believed to contribute 2 to 3 percent of all anthropogenic global warming. Contrails are highly amplified reminders of that problem.

Narayanan M. Komerath

• Further Reading

Atlas, David, Zhien Wang, and David P. Duda. "Contrails to Cirrus: Morphology, Microphysics, and Radiative Properties." *Journal of the American Meteorological Society*, January, 2006, 5-19. LIDAR, satellite, and aircraft flight track data are used to study how contrails evolve into cirrus clouds, along with their particle density, sizes, fall rates and ice content.

Hoyle, C. R., B. P. Luo, and T. Peter. "The Origin of High Ice Crystal Number Densities in Cirrus Clouds." *Journal of the Atmospheric Sciences* 62, no. 7 (July, 2005): 2568-2579. Reports measured particle size and number density in cirrus clouds. Presents hypotheses on how cirrus clouds form and evolve, including the role of wind fluctuations.

Jensen, E. J., et al. "Spreading and Growth of Contrails in a Sheared Environment." *Journal of Geophysical Research* 103, no. D24 (1998): 31,557-31,567. Compares computational predictions of how contrails evolve as a result of winds with satellite and in situ measurements. Also discusses the formation and growth of ice crystals and updrafts due to solar heating.

Penner, J. E., et al. "Aviation and the Global Atmosphere." *Intergovernmental Panel on Climate Change, Special Report.* New York: Cambridge University Press 1999. Chapter 6 includes a discussion of contrail effects. Reports global air traffic, carbon dioxide emission, and other data to estimate overall effects.

See also: Air travel; Clouds and cloud feedback; Motor vehicles; Noctilucent clouds; Polar stratospheric clouds; Transportation.

Convention on Biological Diversity

- **Category:** Laws, treaties, and protocols
- **Date:** Opened for signature June 5, 1992; entered into force on December 29, 1993

The CBD seeks to conserve Earth's biodiversity and biological resources. Climate change can affect the rate of biodiversity loss, which in turn affects the capacity of human and nonhuman biological systems to adapt to the effects of climate change.

- **Participating nations:** *1992:* Albania, Canada, Maldives, Marshall Islands, Mauritius, Monaco, Seychelles; *1993:* Antigua and Barbuda, Armenia, Australia, Bahamas, Barbados, Belarus, Belize, Burkina Faso, China, Cook Islands, Czech Republic, Democratic People's Republic of Korea, Denmark, Ecuador, European Community, Fiji, Germany, Guinea, Japan, Jordan, Mexico, Mongolia, Nauru, Nepal, New Zealand, Norway, Papua New Guinea, Peru, Philippines, Portugal, Saint Kitts and Nevis, Saint Lucia, Spain, Sweden, Tunisia, Uganda, Uruguay, Vanuatu, Zambia; *1994:* Argentina, Austria, Bangladesh, Benin, Bolivia, Brazil, Cameroon, Chad, Chile, Colombia, Comoros, Costa Rica, Cote d'Ivoire, Cuba, Democratic Republic of Congo, Djibouti, Dominica, Egypt, El Salvador, Equatorial Guinea, Estonia, Ethiopia, Finland, France, Gambia, Georgia, Ghana, Greece, Grenada, Guyana, Hungary, Iceland, India, Indonesia, Italy, Kazakhstan, Kenya, Kiribati, Kyrgyzstan, Lebanon, Luxembourg, Malawi, Malaysia, Micronesia (Federated States of), Myanmar, Netherlands, Nigeria, Pakistan, Paraguay, Republic of Korea, Romania, Russian Federation, Samoa, San Marino, Senegal, Sierra Leone, Slovakia, Sri Lanka, Swaziland, Switzerland, United Kingdom of Great Britain and Northern Ireland, Venezuela, Vietnam; *1995:* Algeria, Bhutan, Botswana, Cape Verde, Central African Republic, Guatemala, Guinea-Bissau, Honduras, Israel, Jamaica, Lao People's Democratic Republic, Latvia, Lesotho, Mali, Morocco, Mozambique, Nicaragua, Niger, Oman, Panama, Republic of Moldova, Singapore, Solomon Islands, South Africa, Sudan, Togo, Trinidad and Tobago, Ukraine, Uzbekistan; *1996:* Bahrain, Belgium, Bulgaria, Cambodia, Congo, Croatia, Cyprus, Dominican Republic, Eritrea, Haiti, Ireland, Lithuania, Madagascar, Mauritania, Niue, Poland, Qatar, Rwanda, Saint Vincent and the Grenadines, Slovenia, Suriname, Syrian Arab Republic, Turkmenistan, United Republic of Tanzania, Yemen; *1997:* Burundi, Gabon, Liechtenstein, Namibia, The former Yugoslav Republic of Macedonia, Turkey; *1998:* Angola, Tonga; *1999:* Malta, Palau, Sao Tome and Principe; *2000:* Azerbaijan, Liberia, United Arab Emirates; *2001:* Libyan Arab Jamahiriya, Saudi Arabia; *2002:* Afghanistan, Bosnia and Herzegovina, Kuwait, Serbia, Tuvalu; *2004:* Thailand; *2006:* Montenegro; *2007:* Timor-Leste; *2008:* Brunei Darussalam

- **Background**

Conservation biologists do not always agree on a common definition of biological diversity, or biodiversity. The term, however, generally refers to the breadth of variation of life on Earth at all levels of biological organization. It includes the genetic diversity within species, the species diversity within ecosystems, and the diversity of ecosystems on the planet.

Other, more specialized biodiversity protection treaties predate the Convention on Biological Diversity (CBD). These include, for example, the Ramsar Convention and the Convention on International Trade in Endangered Species (CITES). The CBD, however, is the first such broad and overarching biodiversity treaty. Pre-negotiations on the treaty began in 1987 under the auspices of the United Nations Environment Programme (UNEP). The resulting negotiated text was opened for signature in 1992 in Rio de Janeiro at the United Nations Conference on Environment and Development, where over 150 countries signed it. By 2008, the CBD had near universal membership, with 191 party nations.

• Summary of Provisions

The CBD's forty-two articles contain many provisions broadly aimed at encouraging member countries to develop national plans for protecting biodiversity and for integrating biodiversity conservation and sustainable use into sectoral or cross-sectoral plans, programs, and policies. The CBD's rules and norms overlap with those of the earlier biodiversity agreements, but they are much broader in scope. Other biodiversity protection treaties tend to focus on the protection of specific habitats (for example, the Ramsar Convention focuses on wetlands) or a specific group of species (for example, the Convention on Migratory Species focuses on species that migrate from one geographical location to another). The CBD encompasses all such conservation objectives and expands them. This breadth of focus is clearly articulated in the three stated goals of the convention: biodiversity conservation, sustainable use of biodiversity components, and equitable sharing of biodiversity benefits.

• Significance for Climate Change

Discussions about the links between biodiversity loss and climate change are not new to the CBD forum. This linkage has been a topic of discussion since 1998, when the Fourth Conference of the Parties (COP-4) to the CBD passed a decision recognizing the effect of climate change on coral bleaching. The relationship between the CBD and climate change has remained on the CBD agenda since that time, resulting in various scientific and technical papers on the issue and four key decisions relating to specific CBD work programs.

Discussion of climate change within the CBD forum also led to the creation of the Joint Liaison Group between the CBD, the United Nations Framework Convention on Climate Change (UNFCCC), and the United Nations Convention to Combat Desertification (UNCCD). The scientific and technical reports that have emerged from the CBD process have discussed various aspects of the biodiversity-climate change relationship. They have emphasized that climate change is an increasingly important driver of biodiversity loss and have identified other aspects of the relationship between biodiversity loss and climate change. For example, although some reports emphasize these issues more than others, they discuss the role that biodiversity conservation can play in climate change mitigation and adaptation, as well as the capacity of ecosystems to adapt to changing climates.

The CBD Conference of the Parties has passed

Objectives of the Convention on Biological Diversity

Article 1 of the Convention on Biological Diversity sets out the convention's objectives, which guide all of its specific provisions and define the fundamental obligations of its signatories.

The objectives of this Convention, to be pursued in accordance with its relevant provisions, are the conservation of biological diversity, the sustainable use of its components and the fair and equitable sharing of the benefits arising out of the utilization of genetic resources, including by appropriate access to genetic resources and by appropriate transfer of relevant technologies, taking into account all rights over those resources and to technologies, and by appropriate funding.

five key decisions related to the relationship between the CBD and climate change. These decisions have addressed issues including the risks climate change presents to coral reefs (Decision V/3) and forests (Decision V/4), measures to help mitigate and adapt to climate change (Decision VII/15), and attempts to promote synergy among activities related to biodiversity conservation and climate change adaptation and mitigation (Decision VIII/30). The most comprehensive decision on this topic was taken at COP-9 in May, 2008. Among other things, the conference established a new ad hoc technical expert group to develop scientific and technical advice on biodiversity as it relates to climate change.

Finally, the CBD secretariat, along with those of the UNFCCC and UNCCD, has participated in the Joint Liaison Group. Created in 2001 by parallel decisions by the CBD, UNFCCC, and UNCCD, the Joint Liaison Group provides a forum for information exchange, particularly as it relates to enhancing synergies between the three conventions.

Sikina Jinnah

• **Further Reading**

Gitay, Habiba, et al., eds. *Climate Change and Biodiversity.* Technical Paper 5. [Geneva, Switzerland]: Intergovernmental Panel on Climate Change, 2002. This technical report was developed in response to a request from the CBD's scientific body. It addresses the relationship between climate change on biodiversity with respect to impacts, mitigation measures, sustainable use, and adaptation.

Le Prestre, Philippe G., ed. *Governing Global Biodiversity: The Evolution and Implementation of the Convention on Biological Diversity.* Burlington, Vt.: Ashgate, 2002. This edited volume contains a variety of academic perspectives discussing the evolution and effectiveness of the CBD. Provides a foundation for exploring more complex issues, such as the relationship between the CBD and climate change.

Lovejoy, T., and L. Hannah, eds. *Climate Change and Biodiversity.* New Haven, Conn.: Yale University Press, 2005. Collects commentaries from a variety of leading experts that address the impacts of climate change on biological diversity.

Secretariat of the Convention on Biological Diversity. *Interlinkages Between Biological Diversity and Climate Change: Advice on the Integration of Biodiversity Considerations into the Implementation of the United Nations Framework Convention on Climate Change and Its Kyoto Protocol.* CBD Technical Series 10. Montreal: Author, 2003. This technical report explains various ways to use biodiversity conservation to enhance mitigation of and adaptation to climate change.

See also: Amazon deforestation; Biodiversity; Convention on International Trade in Endangered Species; Endangered and threatened species; Invasive exotic species.

Convention on International Trade in Endangered Species

• **Categories:** Laws, treaties, and protocols; animals; plants and vegetation
• **Date:** Opened for signature March 3, 1973; entered into force on July 1, 1975

CITES addresses unsustainable exploitation through international trade. As such, climate change is significant to CITES to the extent that such change affects populations of species that are traded internationally.

• **Participating nations:** *1975:* Brazil, Canada, Chile, Costa Rica, Cyprus, Ecuador, Madagascar, Mauritius, Morocco, Nepal, Niger, Nigeria, Peru, South Africa, Sweden, Switzerland, Tunisia, United States of America, Uruguay; *1976:* Australia, Democratic Republic of Congo, Finland, Germany, Ghana, India, Iran, Norway, Pakistan, Papua New Guinea, United Kingdom of Great Britain and Northern Ireland; *1977:* Denmark, Gambia, Guyana, Paraguay, Senegal, Seychelles; *1978:* Botswana, Egypt, France, Malaysia, Monaco, Panama, Venezuela; *1979:* Bahamas, Indonesia, Italy, Jordan, Kenya, Sri Lanka, Togo; *1980:* Central African

Republic, Guatemala, Israel, Japan, Liechtenstein, United Republic of Tanzania; *1981:* Argentina, Belize, Cameroon, China, Colombia, Guinea, Liberia, Mozambique, Philippines, Portugal, Rwanda, Suriname, Zambia, Zimbabwe; *1982:* Austria, Bangladesh, Malawi; *1983:* Congo, Saint Lucia, Sudan, Thailand; *1984:* Algeria, Belgium, Benin, Luxembourg, Netherlands, Trinidad and Tobago; *1985:* Honduras, Hungary; *1986:* Afghanistan, Somalia, Spain; *1987:* Dominican Republic, El Salvador, Singapore; *1988:* Burundi; *1989:* Chad, Ethiopia, Gabon, Malta, New Zealand, Saint Vincent and the Grenadines, Vanuatu; *1990:* Brunei Darussalam, Burkina Faso, Cuba, Guinea-Bissau, Poland, United Arab Emirates; *1991:* Bulgaria, Mexico, Namibia, Uganda; *1992:* Djibouti, Equatorial Guinea, Estonia, Russian Federation; *1993:* Barbados, Czech Republic, Greece, Republic of Korea, Slovakia; *1994:* Mali, Romania, Saint Kitts and Nevis, Vietnam; *1995:* Belarus, Comoros, Côte d'Ivoire, Dominica, Eritrea, Sierra Leone; *1996:* Georgia, Mongolia, Saudi Arabia, Turkey; *1997:* Antigua and Barbuda, Cambodia, Fiji, Jamaica, Latvia, Myanmar, Swaziland, Uzbekistan, Yemen; *1998:* Mauritania; *1999:* Azerbaijan, Grenada; *2000:* Croatia, Iceland, Kazakhstan, Slovenia, the Former Yugoslav Republic of Macedonia, Ukraine; *2001:* Moldova, Sao Tome and Principe, Qatar; *2002:* Bhutan, Ireland, Kuwait, Lithuania; *2003:* Albania, Lesotho, Libyan Arab Jamahiriya, Syrian Arab Republic; *2004:* Lao People's Democratic Republic, Palau; *2005:* Cape Verde, Samoa, San Marino; *2006:* Montenegro, Serbia; *2007:* Kyrgyzstan, Solomon Islands; *2008:* Oman

• Background

International trade in wildlife, wildlife components, and their derivatives (including pets, hunting trophies, skins, bark, food products, medicines, exotic leather goods, and timber) is estimated to be worth billions of dollars per year. The Convention on International Trade in Endangered Species of Wild Fauna and Flora (CITES) is an international treaty that regulates wildlife trade through a licensing system in order to ensure that trade in wild plants and animals does not threaten those species' survival. CITES is one of the oldest international environmental treaties in the world, and it enjoys almost universal membership, with 173 parties as of 2008.

Species Protected by the CITES

Species Type	*Species*	*Subspecies*	*Populations*
Mammals	617	36	26
Birds	1,455	17	3
Reptiles	657	9	10
Amphibians	114	0	0
Fish	86	0	0
Invertebrates	2,179	5	0
Total fauna	5,108	67	39
Plants	28,977	7	3
Total	34,085	74	42

• Summary of Provisions

CITES aims to conserve biodiversity and contribute to its sustainable use by delimiting acceptable trade practices. The treaty seeks to ensure that no species of wild animal or plant is subject to unsustainable exploitation through international trade. It thereby contributes to a significant reduction in the rate of global biodiversity loss. CITES identifies species that are or may be affected by international trade and lists them in a set of three appendixes, each affording a different level of protection. It then provides mechanisms for tracking trade in the listed species through an extensive system of trade permits and certificates. CITES's appendixes contain over thirty thousand plant and animal species.

CITES trade must generally satisfy two preconditions: scientific confirmation that trade in the species will not be detrimental to its survival and a finding that the specimens in trade were acquired legally. Within that rubric, Appendix 1 provides the most stringent level of protection. It lists those species that are both threatened with extinction and either currently or potentially affected by international trade. Appendix 1 species include the great apes, humpback whales, sea turtles, tigers, and the monkey-puzzle tree. Generally speaking, these species cannot be traded internationally for "primarily commercial" purposes, and noncommercial trade is permissible only when authorized by import-export permits.

Appendix 2 includes species such as bigleaf mahogany, queen conch, bottle-nosed dolphins, and sharks. It lists species that, although not currently threatened with extinction, may become threatened in the future unless trade in those species is strictly controlled. Commercial international trade in Appendix 2 species is allowed under an export permit, again subject to the general requirement of scientific confirmation that trade in the species will not threaten its survival and a finding that the specimen in trade was acquired legally. Finally, Appendix 3 lists species based on the input of specific member nations. Appendix 3 species are protected by law in at least one country, and that country has asked other CITES parties for assistance in controlling trade in that species.

• Significance for Climate Change

Unlike some other biodiversity conventions, such as the Convention on Biological Diversity (CBD) and the Convention on the Conservation of Migratory Species of Wild Animals (CMS; also known as the Bonn Convention), CITES does not directly address climate change issues. The CBD and CMS address drivers of biodiversity loss that are heavily influenced by a changing climate, such as habitat destruction and change. The scope of CITES, by contrast, is limited to a single driver of species loss that is not directly related to climate change: unsustainable international trade.

Thus, the relationship between CITES and climate change is limited to the potential role that CITES can play in protecting those species that are affected by climate change, but it can protect only those species that are also threatened by international trade. Thus, if there is no market in specimens or derivatives of a given species, CITES can do nothing to protect it. However, if a species is simultaneously threatened by climate change and international trade, CITES can help reduce the latter threat.

Sikina Jinnah

• Further Reading

Lovejoy, T., and L. Hannah, eds. *Climate Change and Biodiversity.* New Haven, Conn.: Yale University Press, 2005. This edited volume contains commentaries from a variety of leading experts that address the impacts of climate change on biological diversity.

Reeve, R. *Policing Trade in Endangered Species: The CITES Treaty and Compliance.* London: Earthscan, 2002. This assessment of CITES's national-level implementation and compliance provides one perspective on how CITES operates on the ground.

Wijnstekers, W. *The Evolution of CITES: A Reference to the Convention on International Trade in Endangered Species of Wild Fauna and Flora.* Geneva, Switzerland: CITES Secretariat, 2006. This guidebook, written by the CITES secretary-general, provides an introduction to the various provisions of CITES and its political process; available in electronic form via the CITES Web site.

See also: Amazon deforestation; Biodiversity; Convention on Biological Diversity; Endangered and threatened species; Invasive exotic species.

Convention on Long-Range Transboundary Air Pollution

• **Categories:** Laws, treaties, and protocols; pollution and waste
• **Date:** Signed November, 1979

LRTAP is one of the oldest international environmental treaties and one whose success has made it a model for later agreements. It addresses air pollution, including GHG emissions, and creates structures for gathering and assessing data.

• **Participating nations:** *1980:* Belarus, Hungary, Portugal, Russian Federation, Ukraine; *1981:* Bulgaria, Canada, Finland, France, Norway, Sweden, United States; *1982:* Austria, Belgium, Denmark, European Community, Germany, Ireland, Italy, Luxembourg, Netherlands, Spain, United Kingdom; *1983:* Greece, Iceland, Liechtenstein, Switzerland, Turkey; *1985:* Poland; *1991:* Cyprus, Romania, Bosnia and Herzegovina, Croatia, Slovenia; *1993:* Czech Republic, Slovakia; *1994:* Latvia, Lithuania; *1995:* Republic of Moldova; *1997:* Armenia, Malta, the Former Yugoslav Republic of Macedonia; *1999:* Georgia, Monaco; *2000:* Estonia, Kyrgyzstan; *2001:* Kazakhstan, Serbia; *2002:* Azerbaijan; *2005:* Albania; *2006:* Montenegro

• Background

In the 1960's, scientists proved for the first time that the acidification of lakes in Scandinavia was the result of industrial emissions of sulfur dioxides and nitrogen oxides throughout Europe. In 1972, the year the term "acid rain" was first used, the United Nations convened the Conference on the Human Environment, also known as the Stockholm Conference, the U.N.'s first conference on international environmental issues. This conference led to increased public awareness of air pollution and to research that confirmed the theory that air pollution could travel large distances and cause harm far away from its source.

It was clear that air pollution generated in one country could damage the environment in another and that only through international cooperation could the environment be protected. After five more years of research and diplomacy, the Economic Commission for Europe convened a meeting in Geneva, Switzerland, during which the Convention on Long-Range Transboundary Air Pollution (LRTAP) was signed. Thirty-four governments and the European Community signed the treaty between November 13 and 16, 1979, and it went into effect in 1983. The Holy See and San Marino were signatories to the convention in 1979, but they have not ratified it. Of the industrialized nations in the Northern Hemisphere, only China and Japan are not parties to the convention.

• Summary of Provisions

The convention was based on the principles that individual nations have

> the sovereign right to exploit their own resources pursuant to their own environmental policies, and the responsibility to ensure that activities within their jurisdiction or control do not cause damage to the environment of other States.

Contracting parties agreed to share research and other information and to gradually reduce air pollution. They agreed to hold talks between those nations most affected by air pollution and those generating the highest levels of it and to fully implement the work of the Cooperative Programme for the Monitoring and Evaluation of the Long-Range Transmission of Air Pollutants in Europe (generally referred to as EMEP). Finally, the convention created an executive body, which would meet annually, and a secretariat, which would convene meetings and direct the dissemination of information.

Most important, the parties agreed to take reasonable steps to reduce air pollution:

> Each Contracting Party undertakes to develop the best policies and strategies including air quality management systems and, as part of them, control measures compatible with balanced development, in particular by using the best available technology which is economically feasible.

The original convention addressed levels of sulfur dioxide only, but it recognized that further research could indicate a need to consider other pollutants.

Between 1984 and 1999, the convention was extended with eight additional agreements, or protocols, each signed by between twenty-one and forty-two parties. The first addressed financing of EMEP, but the others addressed specific pollutants and targets for their reductions: sulfur (1985 and 1994), nitrogen oxides (1988), volatile organic compounds (1991), heavy metals (1998), persistent organic pollutants (1998), and pollutants leading to acidification, eutrophication, and ground-level ozone (1999). Several of these protocols established reduction targets based on the concept of critical load.

• Significance for Climate Change

Several targets established under LRTAP have been met. The 1985 Protocol on Sulfur Emissions, for example, was entered into force in 1987, and its target of reducing emissions by 30 percent by 1993 was met or exceeded by all twenty-one parties. This success encouraged the signatories to agree to the

1994 Protocol on Further Reduction of Sulfur Emissions, which used the concept of critical load, rather than strict percentages, to establish targets. Overall, the levels of sulfur in the air decreased some 60 percent in Europe, and almost 50 percent in the United States and Canada, between 1979 and 2004. In addition, nineteen of the twenty-five parties to the 1988 Protocol Concerning the Control of Nitrogen Oxides met their targets of freezing and then reducing their emissions of nitrogen oxides and ammonia.

The success of LRTAP led the UNECE to press for four more environmental agreements: the Convention on Environmental Impact Assessment in a Transboundary Context (1991; also known as the Espoo Convention); the Convention on the Protection and Use of Transboundary Watercourses and International Lakes (1992); the Convention on the Transboundary Effects of Industrial Accidents (1992); and the Convention on Access to Information, Public Participation in Decision-Making, and Access to Justice in Environmental Matters (1998; also known as the Aarhus Convention).

Although preventing global warming was not the impetus for LRTAP, several of the pollutants regulated by the convention have direct and indirect effects on global warming. Nitrous oxide, for example, is itself a greenhouse gas (GHG). Emissions of nitrogen oxide and VOCs create ozone, another GHG. Parties to LRTAP, therefore, had already made some progress toward reducing anthropogenic emissions of GHGs before treaties such as the Kyoto Protocol (1992) were drafted.

By funding EMEP, LRTAP put into place structures for monitoring and assessing air quality, and for disseminating research findings. These structures have provided useful data for the study of global warming. As scientists study the relatively new questions about the effects of global climate change on the processes of acidification and eutrophication, they will draw on data already

Fundamental Principles of the Convention on Long-Range Transboundary Air Pollution

Articles 2 through 5 of the Convention on Long-Range Transboundary Air Pollution set out the fundamental principles of the agreement, by which all contracting parties agree to be bound.

Article 2:

The Contracting Parties, taking due account of the facts and problems involved, are determined to protect man and his environment against air pollution and shall endeavour to limit and, as far as possible, gradually reduce and prevent air pollution including long-range transboundary air pollution.

Article 3:

The Contracting Parties, within the framework of the present Convention, shall by means of exchanges of information, consultation, research and monitoring, develop without undue delay policies and strategies which shall serve as a means of combating the discharge of air pollutants, taking into account efforts already made at national and international levels.

Article 4:

The Contracting Parties shall exchange information on and review their policies, scientific activities and technical measures aimed at combating, as far as possible, the discharge of air pollutants which may have adverse effects, thereby contributing to the reduction of air pollution including long-range transboundary air pollution.

Article 5:

Consultations shall be held, upon request, at an early stage between, on the one hand, Contracting Parties which are actually affected by or exposed to a significant risk of long-range transboundary air pollution and, on the other hand, Contracting Parties within which and subject to whose jurisdiction a significant contribution to long-range transboundary air pollution originates, or could originate, in connection with activities carried on or contemplated therein.

gathered under the terms of the 1999 LRTAP protocol. Similarly, researchers will use data gathered under LRTAP to develop methodologies for determining the most cost-effective emissions reduction targets.

Cynthia A. Bily

• Further Reading

Kuokkanen, Tuomas. "The Convention on Long-Range Transboundary Air Pollution." In *Making Treaties Work: Human Rights, Environment, and Arms Control,* edited by Geir Ulfstein, Thilo Marauhn, and Andreas Zimmermann. New York: Cambridge University Press, 2007. An overview of the negotiations leading up to the passage of LRTAP, the treaty's terms and structures, and the efforts of the governing committee to encourage compliance.

Sliggers, Johan, and Willem Kakebeeke. *Clearing the Air: Twenty-Five Years of the Convention on Long-Range Transboundary Air Pollution.* New York: United Nations, 2005. A look back on the successes of the convention by scientific and political experts, many of whom participated in the original convention negotiations and drafting.

Soroos, Marvin S. *The Endangered Atmosphere: Preserving a Global Commons.* Columbia: University of South Carolina Press, 1997. Examines global public policy related to the atmosphere, with extensive discussion of the LRTAP convention and climate change.

United Nations Economic Commission for Europe. *Handbook for the 1979 Convention on Long-Range Transboundary Air Pollution and Its Protocols.* New York: United Nations, 2004. A history of the convention, and its protocols and activities, and a collection of all the major documents produced during the first twenty-five years of the convention.

See also: Air pollution and pollutants: anthropogenic; Air pollution and pollutants: natural; Air pollution history; Air quality standards and measurement; United Nations Conference on Environment and Development; United Nations Conference on the Human Environment; United Nations Framework Convention on Climate Change; Water quality.

Cooler Heads Coalition

- **Category:** Organizations and agencies
- **Date:** Established May 6, 1997
- **Web address:** http://www.globalwarming.org

• Mission

Originally a project of the National Consumer Coalition, which was itself a project of the nonprofit group Consumer Alert, the Cooler Heads Coalition strives to dispel the "myths of global warming by exposing flawed economic, scientific, and risk analysis." The coalition, consisting of about twenty-five groups, was established in May, 1997, in response to a concern that Americans were not being adequately informed about the economic impact of reductions of greenhouse gas emissions and about the pros and cons of global warming. A major focus of the group is the consumer impact of global warming policies that restrict energy use drastically and that would raise consumer costs. Cooler Heads also maintains that the science of global warming is uncertain and that global warming policies have a negative impact on consumers.

The Cooler Heads Coalition publishes a biweekly newsletter, *The Cooler Heads Newsletter,* which provides updates on scientific, economic, and political issues relating to global climate change. The coalition maintains a global warming Web site as a clearinghouse for information on global warming science and policy proposals. *The New York Times* listed this site among fifteen top environmental Web sites; it was the only market-oriented site listed.

• Significance for Climate Change

Members of the Cooler Heads Coalition hold the view that the science of global warming is uncertain, while the negative impacts of policies addressing global warming are not. One statement on the issue maintains that "global warming may or may not be a problem" and that "man may or may not be driving it." Members of the Cooler Heads Coalition publish reports, research studies, and briefs on scientific, economic, and political aspects of global warming policy and sponsor press conferences, hold rallies, conduct consumer surveys, participate in debates, make television and radio talk show appearances, and write newspaper and magazine articles. Several of its groups involve themselves in the global warming treaty negotiations at the United Nations.

The primary focus of the coalition is economic rather than environmental, as evidenced by comments by the executive director of Consumer Alert:

> Policies disguised as middle-of-the-road instead will be leading us down the road to national industrial policy and a planned economy. The losers will be American consumers, who will bear the brunt of restrictions on energy use in their everyday lives. They'll have to pay the costs, not just in higher prices, but in a drastically lower standard of living.

Victoria Price

See also: Cato Institute; Skeptics.

Coriolis effect

- **Category:** Physics and geophysics

• Definition

The Coriolis effect is an apparent acceleration of a moving object as seen in a rotating system. The acceleration is not a true change in velocity, but an illusion caused by the rotation of the system beneath the moving object.

An unconscious tendency to regard the Earth as a fixed frame of reference generates an unrecognized expectation on the part of many people that objects free of forces will move with unchanging direction and speed. That is, they will be unaccelerated. This expectation is consistent with Sir Isaac Newton's First Law of Motion, which states that a body in motion will remain in motion, in a constant direction, with constant speed, unless acted on by an outside force.

Newton's First Law of Motion, however, applies only in frames of reference that are themselves not accelerating. These so-called inertial frames must be moving in a constant direction with constant speed. This condition is not met by rotating frames, such as the surface of the Earth, where every point in the frame (though traveling at a constant speed) is constantly changing direction as it completes a circular path around the axis of rotation.

The farther an object is from the axis of rotation, the faster it will travel on its circular path. An object on the equator, for example, completes a path of 40,000 kilometers in one day, while an object at 60° north latitude travels only half that distance in the same time. Thus, the object at 60° north latitude travels at half the speed of the object on the equator.

Wind and water that leave the equator headed due north carry their equatorial speed with them. As they move north, they pass over territory that is traveling eastward more slowly than they are. As a result, the wind and water move eastward relative to the ground or the seafloor. Conversely, wind and water starting at northern latitudes and moving due south will cross ground that is moving eastward faster than they are; they will move westward relative to the ground or seafloor. This deflection from the original direction of motion is the Coriolis effect.

• Significance for Climate Change

Convection-driven currents carry warm water and air poleward from the equator and carry cool air and water from the polar regions toward the tropics. Both the tropical and the polar currents are de-

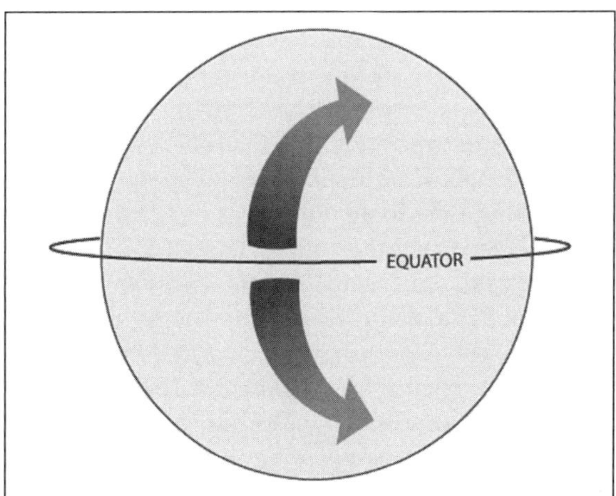

The Coriolis effect causes moving fluids to be deflected to their right in the Northern Hemisphere and to their left in the Southern Hemisphere. (NOAA)

flected to the right, relative to their direction of motion, in the Northern Hemisphere and to the left, relative to their direction of motion, in the Southern Hemisphere. In regions where the currents converge, the deflections merge into circular rotations about the point of convergence. In the Northern Hemisphere, these rotations move counterclockwise; in the Southern Hemisphere, they move clockwise.

A low-pressure weather system in the Northern Hemisphere, for example, draws in air from the surrounding terrain in all directions. The wind flowing in from the north is deflected to the west. The wind from the west is deflected to the south, the wind from the south is deflected to the east, and the wind from the east is deflected to the north. In combination, the winds form a vortex rotating counterclockwise about the center of the low-pressure area. A high-pressure system, by contrast, repels winds, creating a clockwise vortex. In the Southern Hemisphere, these directions are reversed. Similar effects occur in ocean currents.

The magnitude of the deflection caused by the Coriolis effect is proportional to the distance from the point of deflection to the rotation axis. For that reason, the Coriolis effect is most prominent at the equator. It is also proportional to the speed of the currents involved. High winds associated with hurricanes readily display the effect, generating the characteristic circular wind pattern with a calm eye at the center.

The Coriolis effect establishes the circulation pattern of major storms, trade winds, jet streams, and large-scale ocean currents. All of these convection currents transport thermal energy from the warm tropics to the temperate and polar regions, moderating the global difference in temperatures. The Gulf Stream, for example, keeps Great Britain, Ireland, and the North Atlantic coast of Europe substantially warmer than other regions of the Northern Hemisphere that are located at the same latitude. Air currents also transport large amounts of water evaporated from tropical oceans to temperate and polar regions, where the water precipitates as rain and snow.

The rate at which convection currents transport mass and heat poleward from the tropics is a function of the temperature difference between the two regions. If climate change raises average temperatures in the tropics more than it raises them at the poles, it will create more energetic and powerful currents. If climate change raises polar temperatures more than it raises equatorial temperatures, it will dampen these currents. The resulting effects on the number, type, and destructive power of storms in either case would be complex and difficult to model.

Billy R. Smith, Jr.

- **Further Reading**

Mayes, Julian, and Karel Hughes. *Understanding Weather: A Visual Approach.* New York: Oxford University Press, 2004. This introduction to the science of meteorology instructs primarily through pictures, graphs, and maps. The Coriolis effect is explained within the larger context of atmospheric movements (winds, fronts, storms, and so on).

Stommel, Henry, and Dennis Moore. *An Introduction to the Coriolis Force.* New York: Columbia University Press, 1989. Requires some understanding of basic physics and familiarity with calculus through differential equations. Includes an appendix devoted to the construction of a laboratory demonstrator of the Coriolis effect.

Walker, Gabrielle. *An Ocean of Air: Why the Wind Blows and Other Mysteries of the Atmosphere.* Orlando, Fla.: Harcourt, 2007. The Coriolis effect is included as part of a discussion of winds and storms.

See also: Atmospheric dynamics; Atmospheric structure and evolution; Earth motions; Gulf Stream; Jet stream; Ocean-atmosphere coupling; Ocean dynamics.

Cosmic rays

- **Category:** Physics and geophysics

- **Definition**

Cosmic rays are energetic particles that travel through space. Of those that strike the Earth's atmosphere, about 90 percent are protons, 9 percent are helium nuclei (also called alpha particles), and 1 percent are electrons. When high-energy protons collide with atoms in the Earth's atmosphere, they produce showers of secondary particles, such as lower-energy protons, neutrons, electrons, and various mesons. Secondary particles generally have enough energy to form positive ions by knocking electrons away from atoms. These ions can attract atoms from the air and form tiny balls that serve as nucleation sites for water vapor, known as cloud condensation nuclei. Water vapor may then condense into water droplets on the ion balls to form clouds.

• Significance for Climate Change

Thus, scientists have theorized that cosmic rays form nucleation sites that promote cloud formation, so increases or decreases in cosmic ray bombardment of the Earth would correspond with increases or decreases in Earth's cloud cover. High clouds increase Earth's albedo, reflecting sunlight back into space and cooling the planet. Lower-energy cosmic rays come from the Sun, but high-energy particles originate elsewhere in the galaxy. When the Sun's activity increases, its magnetic field pushes further outward, partly blocking galactic cosmic rays from reaching the Earth. Thus, increased solar activity should lead Earth to experience fewer high-energy cosmic rays, less cloud cover, and increased global warming.

The theory may be tested by comparing records of cosmic ray intensity to past global temperatures. The ratio of the isotopes oxygen 18 to oxygen 16 in rocks and fossils provides a proxy indicator of past temperatures, while the amount of carbon 14 and beryllium 10 found in ice cores from Greenland provides similar evidence of past cosmic-ray flux. Carbon 14 is formed when a secondary neutron collides with nitrogen 14 and exchanges a proton for a neutron. Beryllium 10 is one of the possible by-products when a high-energy proton slams into either a nitrogen nucleus or an oxygen nucleus, and knocks out several neutrons and protons. This is the only source of beryllium 10.

Supporters of the cosmic ray theory note that during the Medieval Warm Period (roughly 1100 to 1300) cosmic ray intensity seems to have decreased, based on a decrease in carbon 14 and beryllium 10 at that time. On the other hand, during the part of the Little Ice Age from about 1600 to 1800, cosmic ray intensity increased and temperatures decreased. However, the temperature rise of the past fifty years has not been accompanied by a significant increase in cosmic ray intensity. Global warming is a complex phenomenon, because it may have many different causes, and the dominant cause may be different at different times. A reasonable assessment today is that 25 percent of current warming may be due to greater solar activity, but the remaining 75 percent is most likely due to greenhouse gas concentrations, with human activities being the major contributor to those concentrations.

Charles W. Rogers

See also: Albedo feedback; Clouds and cloud feedback; Noctilucent clouds; Polar stratospheric clouds; Ultraviolet radiation.

Cryosphere

• **Category:** Cryology and glaciology

• Definition

The cryosphere (from the Greek *kryo*, meaning "too cold") refers to those parts of the Earth's surface where temperatures are sufficiently low that water is frozen solid, in the form of either snow or ice. The conditions that freeze the available water within a particular area can be seasonal or can last for years or centuries. The cryosphere includes land covered with snow in the winter; freshwater lakes and river systems that freeze over seasonally; glaciers that freely move about larger water systems and are thus prone to melting and reshaping; and permafrost, or frozen soil and rock that remains frozen year round. The places most associated with the cryosphere are the North and South Poles, but frozen surfaces are found in many high-elevation regions of both the Northern and Southern Hemispheres.

Scientists distinguish two types of formations

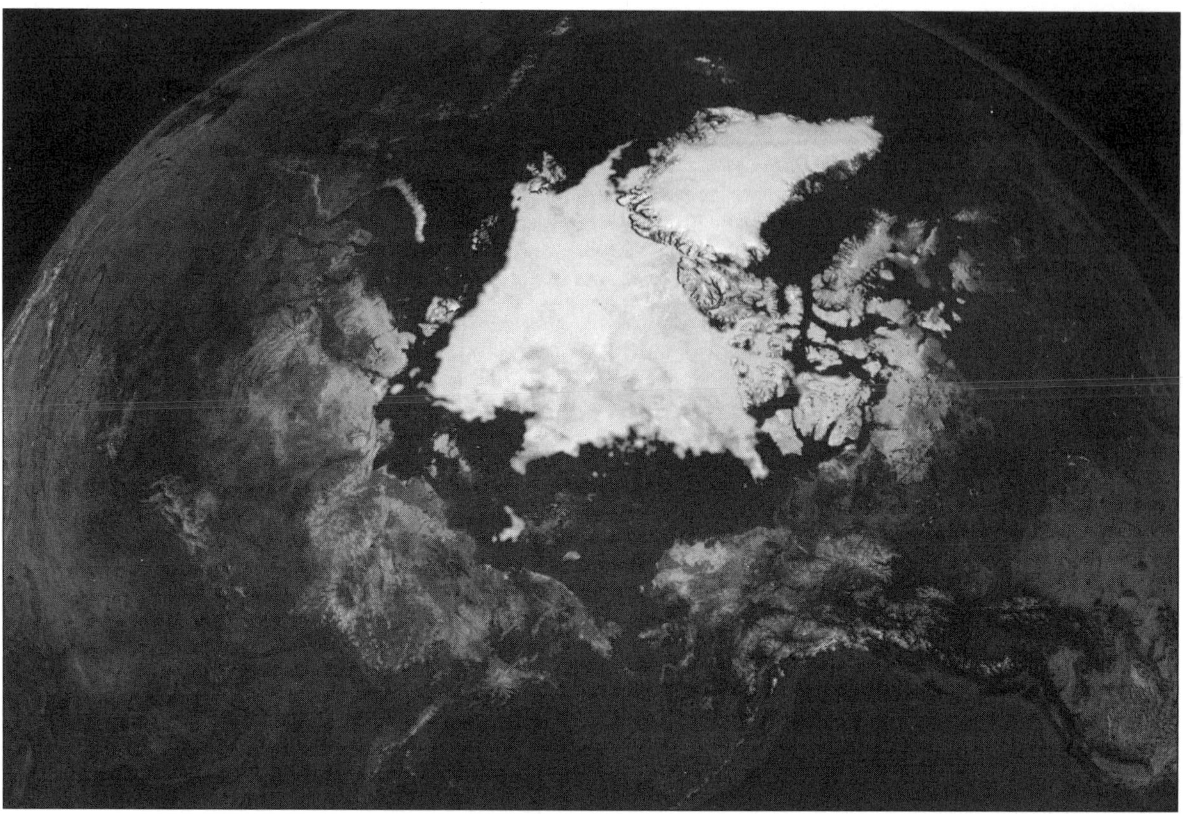

Earth's Arctic ice cover in September, 2008. Polar ice accounts for a large proportion of both the planet's cryosphere and its albedo. (Reuters/Landov)

that make up the cryosphere: land ice and sea ice. Land ice is formed slowly by compressed snow that becomes layers of ice. Land ice is thus freshwater. Perhaps the most familiar examples of land ice are glaciers, great slow-moving ice masses that store at any one time close to 70 percent of the world's available freshwater. Other examples of land ice are ice shelves left where glaciers break off and head into the open oceans as icebergs; ice shelves are found in coastal areas of Greenland, northern Canada, northern China, lower South America, southern Australia, and, of course, both poles.

Conversely, the polar oceans, both north and south, are covered with sea ice, or frozen seawater. Sea ice floats on the surface of the water and has an average thickness of 1 meter in the Antarctic and nearly 3 meters in the Arctic. Because this ice exists within a dynamic environment—that is, one subject to temperature changes, wind, and ocean currents—sea ice can be measured by its duration (generally one year or multiyear). Because navigation depends on charting these fluid conditions, climatologists measure the sea ice as it cracks and even splits into huge moving parts, particularly as it inevitably diminishes during the abbreviated summer seasons at the poles. In the most extreme reaches of both poles, sea ice survives summer melting and becomes far thicker and can measure up to 381 centimeters.

• Significance for Climate Change

Although seasonal fluctuations in mean temperatures in polar regions are to be expected and do not affect the general dynamic of the cryosphere, long-term climate shifts resulting from decades of burning fossil fuels have contributed to a significant rise in the Earth's average air temperature. This global warming, in turn, affects the thousands of square kilometers that make up the cryosphere. As that fragile environment undergoes radical changes over a relatively brief period of time, such changes, in turn, affect a variety of climate and meteorological conditions around the globe. Interest in the cryosphere has greatly increased over the last generation, as climatologists see this frozen environment as the earliest indicator of rising global temperatures. The National Snow and Ice Data Center at the University of Colorado monitors the cryosphere.

Most dramatically, the diminishing of the snow and ice cover and the shortening of the winter season at the poles means that the planet's natural insulation from the direct bombardment of solar energy is diminishing. The bright surface of the snow and ice of the cryosphere contributes to Earth's albedo and is responsible for reflecting back into space 70 percent of the Sun's energy. As that protection recedes, the Earth absorbs more solar energy, resulting in an increase in mean air temperature.

The global warming trend causes inland waterways to thaw earlier than they otherwise would, disrupting navigation lines and storm patterns and affecting the ecosystems of indigenous wildlife and plants. Groundwater levels in turn decline. Glaciers melt, and scientists must confront the possibility of significant impacts on the Earth's water system and the need for global water management. There is cause for concern: Scientists estimate that global sea levels have risen over the last two decades by 7.5 to 10 centimeters, but the loss of significant ice in the endangered Antarctic could raise ocean levels a catastrophic 9 meters in the next century, making the nearly 15 percent of the world's population who live along shorelines climate refugees.

The sea-ice shelves—which protect coastlines in both poles, Alaska, Canada, and Russia from wave erosion—are disappearing, upending peoples who have worked in that difficult environment for centuries. In turn, under the impact of rising air temperatures, the permafrost loses its integrity, a process further complicated by the drilling into the rich deposits of fossil fuels. But loss of the permafrost has a greater significance. Trapped within its thousands of frozen kilometers are centuries of decayed plant and animal detritus. As the permafrost thaws, carbon dioxide and methane, themselves greenhouse gases, are released in great volume to further influence global air temperature.

Joseph Dewey

• Further Reading

Archer, David. *The Long Thaw: How Humans Are Changing the Next 100,000 Years of Earth's Climate.* Princeton, N.J.: Princeton University Press, 2008. Accessible description of the consequences of cryosphere loss. Pitched to a nonscientific audience, resisting alarmist argument, and generally objective.

Flannery, Tim. *We Are the Weather Makers: The Story of Global Warming.* Rev. ed. London: Penguin, 2007. Investigates the widest implications of global warming. Essential for understanding how the cryosphere affects a broad range of climate phenomena across the globe.

Michaels, Patrick J. *Climate of Extremes: Global Warming Science They Don't Want You to Know.* Washington, D.C.: Cato Institute, 2009. Important conservative corrective to growing alarmist projections about crysophere damage. Moderates the predictions and indicates progress in monitoring the cryosphere.

Slaymaker, Olav, and Richard Kelly. *The Cryosphere and Global Environmental Change.* Hoboken, N.J.: Wiley-Blackwell, 2007. Groundbreaking textbook on the specific issues facing the cryosphere. Explains causes of cryosphere damage. Helpfully illustrated.

See also: Glaciations; Glaciers; Greenland ice cap; Ground ice; Ice shelves; Permafrost; Sea ice.

Damages: market and nonmarket

• **Categories:** Economics, industries, and products; environmentalism, conservation, and ecosystems; popular culture and society

• **Definition**

Climate change has the potential to inflict significant social, financial, and environmental damages. Some of these damages will be market damages, measurable as negative influences on gross domestic product (GDP), and some will be nonmarket damages, not accounted for in GDP. A clear example of a market damage that could be attributed to climate change is the loss of tourism associated with the destruction of coral reefs by sea-level rise or temperature change. An example of nonmarket damage is the loss of the ecosystem service of storm protection provided by coastal wetlands in a sea-level-rise scenario. Perhaps ironically, the value of the storm protection services of coastal wetlands increases with climate-induced increases in storm frequency and intensity.

Some impacts of climate change fall into even murkier territory with respect to assigning phenomena as net benefits or costs. Suppose increasing global temperatures cause an increased use of air conditioning around the world, which boosts GDP via increased energy revenues and increased air-conditioner sales and service. This increased use of air conditioning exacerbates climate change, however, through increased carbon dioxide (CO_2) emissions. Does this situation represent a market benefit or cost? Does it represent a nonmarket benefit or cost? Traditional economic paradigms are increasingly challenged by environmental problems, and assessing the costs of climate change from a strictly economic perspective is particularly problematic.

• **Significance for Climate Change**

Assessing the market and nonmarket damages of climate change is an active, diverse, and contested area of economic and social science research. Climate change presents many complex and problematic market-failure potentials in the categories of public goods, common property, and negative externalities. A public good is an acknowledged market failure, in that a free market will not provide appropriate supplies of it, because it is nonexcludable and nonrival in consumption. A commonly used example is lighthouses. Society benefits from lighthouses, because they prevent damage to life and property in cost-effective ways. However, the private sector will not build lighthouses, because owners of lighthouses cannot charge for their services effectively, nor can they exclude nonpayers from using those services. Governments typically provide public goods, such as street lighting and national defense, for this reason.

Institutions that provide conservation services, such as the U.S. Fish and Wildlife Agency and the Environmental Protection Agency, could be considered public goods. Climate change may necessitate the creation of new institutions that monitor and enforce policies related to various phenomena associated with global warming. Public goods such as ecosystem services are an example of a nonmarketed public good that are threatened in myriad ways by global warming. Common property is another market failure well described in Garrett Hardin's famous paper "The Tragedy of the Commons." Communally owned properties are often not used in economically optimal ways because of conflicts between public and private interests. The atmosphere is a global common property that is used as a greenhouse-gas dumping ground to varying extents by the nations of the world.

Externalities are another market failure that occurs when all of the costs or benefits of an activity are not accounted for. When a university improves its campus with new and appealing buildings and improves its reputation, raising the property values of nearby residential real estate, that represents a positive externality. When a manufacturer pollutes the environment but does not account for the resulting environmental costs by reimbursing the affected community, that represents a negative externality. Climate change will probably involve both positive and negative externalities; nonetheless, the growing consensus is that in the aggregate it will be negative.

Climate change produces negative externalities via sea-level rise, increased frequency and intensity of extreme weather events, and reduced soil moisture. As with much environmental damage, the community responsible for climate change is not necessarily the one that suffers as a result. Greenhouse gas (GHG) emissions of the developed world could cause sea-level rise that swamps small, poor, island nations. If so, serious human and environmental damage will have occurred, and it will be difficult if not impossible to put a dollar value on these damages, or to determine who should be compensated, in what way, and with whose resources. The intra- and intergenerational equity questions raised by the impacts of climate change present the global community with profound ethical questions and institutional challenges.

Monitoring and enforcement of GHG emissions will necessitate the establishment of institutions that are regarded as fair, just, accurate, and effective by the global community in order to be accepted. The establishment and maintenance of these institutions will inevitably require expenditures. These costs may be recouped in the form of improvements to the human economy, such as preventing damage to accrued and inherited wealth, rather than in the form of increased GDP. Wars, wildfires, and extreme weather events such as Hurricane Katrina all increase economic activity and consequently GDP, yet no one would recommend disasters of this nature to create market benefits. An economic paradigm that values current economic activity (GDP) over established wealth (both marketed and particularly nonmarketed) is increasingly inadequate to many of the most pressing policy questions in relation to climate change. The scale, scope, and costs of this global institutional challenge are unprecedented.

Paul C. Sutton

- **Further Reading**

Costanza, R., et al. "The Value of the World's Ecosystem Services and Natural Capital." *Nature* 387 (May 15, 1997). Economic analysis of natural resources and the damage to their value done by ecological degradation.

Daily, Gretchen, ed. *Nature's Services: Societal Dependence on Natural Ecosystems.* Washington, D.C.: Island Press, 1997. Looks at ecosystems as part of a global anthropomorphic service economy that includes nonhuman entities.

Heal, Geoffrey. *Nature and the Marketplace: Capturing the Value of Ecosystem Services.* Washington, D.C.: Island Press, 2000. Market analysis of natural resources.

See also: Carbon footprint; Conservation and preservation; Deforestation; Economics of global climate change; Employment; Environmental economics; Global economy; Sustainable development.

Dating methods

- **Categories:** Chemistry and geochemistry; science and technology

Reliable dating methods permit scientists to describe past climate change quantitatively and to establish connections between known astronomical cycles and climate cycles.

- **Key concepts**

cosmogenic isotope: an isotope—possibly radioactive—produced when a cosmic ray strikes the nucleus of an atom

decay constant: a measure of how radioactive an isotope is, determined with a Geiger counter

half-life: the time needed for half of a quantity of a radioactive isotope to decay; it is calculated from the decay constant, not measured directly

isotopes: variants of an element that are chemically identical but have different atomic mass numbers and vary in radioactivity

primordial isotope: an isotope that has been present on Earth since the planet formed 4.5 billion years ago

varve: an annual layer in a sediment, usually the result of seasonal variation in inputs

- **Background**

Because the geological and climatological history of Earth began long before recorded history, scientific dating methods are necessary to determine when many climatic events occurred. For example, such methods could be used to determine when glacial deposits formed or when a boulder was dropped on top of those deposits by a melting glacier.

• Primordial Isotopes

When the Earth formed, it inherited an inventory of radioactive elements that have been decaying ever since. The decay constant for a particular isotope can be determined by measuring the rate at which disintegrations occur in a sample of known mass. Half-lives are calculated from decay constants. Known half-lives of radioactive isotopes enable scientists to determine the age of some objects that contain those isotopes.

For example, water moving through the ground will often dissolve small amounts of uranium. A stalagmite may form from this water in a cave as the water evaporates, incorporating any uranium 238 (U^{238}) present. The uranium will decay to produce thorium 234 (Th^{234}). Thorium is insoluble in water, so it can be assumed that the stalagmite initially contained no thorium. Thorium, too, is radioactive, and may decay into U^{234}, which decays to Th^{230}, which is also radioactive. Using the decay constants and the amounts of U^{238}, U^{234}, and Th^{230} present in a specimen, scientists can calculate how long it has been since the uranium came out of solution. This technique is limited to ages less than 500,000 years.

• Cosmogenic Isotopes

Cosmic rays are subatomic particles traveling at very high velocities. When they strike the nucleus of an atom, they can eliminate nucleons, altering the identity of the atom. An atom of nitrogen 14 (N^{14}), for instance, might become carbon 14 (C^{14}), or atoms of silicon or oxygen might become beryllium 10 (Be^{10}) or aluminum 26 (Al^{26}). C^{14}, Be^{10}, and Al^{26} are all radioactive, and their decay constants are known, so they provide a means of dating organic material and the surfaces of boulders.

On Earth, C^{14} is generally created only as a result of cosmic ray bombardment in the atmosphere, so only atmospheric carbon replenishes its C^{14} level. Nonatmospheric C^{14} decays over time without replenishment. An organism will interchange carbon with the atmosphere while it is alive, maintaining a relatively constant ratio of C^{14} to carbon 12 (C^{12}), but once it dies that interchange will cease and the ratio will decrease. By assuming a historically constant ratio of C^{14} to C^{12} in the atmosphere (and thus in living organisms) and by comparing that ratio to the ratio in a sample of tissue from a deceased organism, it is possible to determine how many half-lives of C^{14} have passed since the organism died. The assumption of a constant ratio is known to be invalid, but it will produce the same errors in all samples, giving the same results for samples of the same age. If the goal of analysis is to compare different samples with one another and there is little need for actual calendar years, samples' ages are often reported in C^{14} years.

To convert results accurately to calendar years, corrections are made using calibration curves derived from other dating techniques such as may produce different calibrated ages from the same C^{14} age. The effective limit of this technique is about forty-five thousand years.

Cosmic rays also cause reactions in the outer layers of quartz-rich rocks. Be^{10} and Al^{26} accumulate in these layers at small but relatively constant rates. These isotopes are produced slowly, at a rate of about 100 atoms per gram of rock per year, requiring accelerator mass spectrometry (AMS) techniques to detect them. Cosmic rays do not penetrate solids by more than a few meters, so the exposure age of a surface can reveal when glacial ice melted away above that surface.

• Nonradiometric Methods

Dendrochronology. Dendrochronology is a method for determining the age of wood by counting and examining annual tree rings. The thickness of a given ring in a tree is determined by environmental factors obtaining during the year in which the ring was formed. Such factors as temperature and rainfall affect the rate of growth and overall health of trees. As a result, patterns of ring thickness in trees that were alive at the same time in the same area tend to resemble one another. Matching patterns of ring thickness between trees of known and unknown age can thus provide evidence that the trees were alive at the same time. The reliability of this method has been extended back to about ten thousand years.

Varves. Just as trees have annual growth cycles, so do sediments deposited in lakes in regions near glaciers. In the summer, rains bring coarse sediments into the lake. In the winter, fine clays have time to settle out. The banded sediments that result from this seasonal alternation are called varves. Just as with tree rings, patterns of thick and thin layers can be correlated in different varved sequences. Some sequences cover more than thirteen thousand years.

Lichenometry. Lichens grow at fairly constant rates in a given area. In a given area, rocks covered by larger lichens have surfaces that have been exposed longer than have the surfaces of rocks covered by smaller lichens of the same type. By calibrating measurements using tombstones and other objects of know age, the absolute exposure age of lichen-covered surfaces can be estimated.

• Context

Understanding climate change requires knowledge of Earth's climatological history, which in turn requires methods capable of dating events of climatic significance over the last few million years. As technology has improved, the precision and accuracy of these methods has increased dramatically, and the size of the samples required for accurate dating has decreased by orders of magnitude.

Scientists looking at isotope-ratio records in marine sediment cores have sometimes found that different radiometric techniques indicate different dates for the same climatic excursion. As the climatic excursions were found to be global and strongly correlated with known astronomical cycles, it became possible to determine their age with greater accuracy, validating some results over others. This correlation with astronomical cycles could also be used to calibrate radiometric dating methods, just as counting tree rings was used to calibrate C^{14} dating methods. As methods developed, it became possible to date specific geologic and climatic events, such as the encroachment or retreat of ice from a given area, the speed of uplift of a surface, or the rate of development of a valley.

Otto H. Muller

• Further Reading

Cremeens, David L., and John P. Hart, eds. *Geoarchaeology of Landscapes in the Glaciated Northeast: Proceedings of a Symposium Held at the New York Natural History Conference 6.* Albany: State University of New York: State Education Deptartment, 2003. Contains an outstanding example of dating, using the New England varve sequence, uranium-thorium calibrations, C^{14} dates, and calendar C^{14} dates; shows how all these methods can be used in concert to reinforce one another. Figures, maps, bibliography.

Ruddiman, William F. *Earth's Climate Past and Future.* 2d ed. New York: W. H. Freeman, 2008. This elementary college textbook has several sections concerning dating methods, their limitations, errors, and resolution. Illustrations, figures, tables, maps, bibliography, index.

Thurber, David L., et al. "Uranium-Series Ages of Pacific Atoll Coral." *Science* 149, no. 3679 (1965): 55-58. An early paper describing the uranium-thorium method; uses graphs to illustrate the viability of the technique. Figures, tables, bibliography.

See also: Carbon cycle; Carbon dioxide; Carbon isotopes; Climate reconstruction; Cosmic rays; Deglaciation; Earth history; 8.2ka event; Holocene climate; Medieval Warm Period; Oxygen isotopes; Paleoclimates and paleoclimate change; Pleistocene climate; Sea sediments; Tree rings; Younger Dryas.

Day After Tomorrow, The

- **Category:** Popular culture and society
- **Date:** Released 2004
- **Director:** Roland Emmerich (1955-   )

• Background

The motion picture *The Day After Tomorrow* is a fictional depiction of the advent of a modern ice age. Brandishing elaborate special effects and generic political characters who ignore the warnings of prominent scientists that a global climate change is coming upon them, the movie attempts to explain how a rapid climate change can occur. Based on the concept of the Younger Dryas period, a much-debated dramatic cooling that occurred in northern Europe approximately ten thousand years ago, *The Day After Tomorrow* attempts to engage with modern global warming concerns.

The movie begins with a slice of reality—the thawing of the glaciers. A massive chunk of the arctic ice shelf breaks away from the continent. Scientists monitoring the temperatures in the ocean note a thirteen-degree drop in water temperatures in multiple locations off Greenland. Events go awry on a worldwide scale shortly thereafter. A critical desalination of the ocean's water occurs, shifting ocean currents and wreaking havoc on the world's climate. The planet is besieged by bizarre weather changes. Snowstorms, brick-sized hail, massive wind sheers, and tornadoes begin to affect regions where such meteorological events had never occurred before. Scientists quickly develop a climate model based on these occurrences that predicts the arrival of a new ice age within six to eight weeks. Politicians ignore this prediction, thereby dooming millions.

• Significance for Climate Change

The events depicted in *The Day After Tomorrow,* while based on solid theories and conjectures, are presented in a sensationalized way. One gigantic storm covers the globe and plunges the world into a new ice age. These events take place in a matter of days, not months or years.

Most scientific experts agree that a cataclysmic climate change would not happen so quickly. Even the Younger Dryas period, with its radical climate change, is speculated to have taken seventy years to develop. Ice-core samples confirm that the changes to the climate took a significant period of time to occur. According to widely accepted beliefs, a climate change of the magnitude depicted in *The Day After Tomorrow* would be gradual by human temporal standards, evolving over the course of ten to fifty years. An increase in atmospheric carbon dioxide (CO_2) from industrialization and poor regulation is expected to double the levels of CO_2 over the course of the next century. A slow warming trend is expected, but it unlikely to lead to the type of rapid climate change depicted in the film.

Even with the realization that *The Day After Tomorrow* is more a work of fiction than of fact, one is left with the fear that the events depicted in the film are possible. A widely publicized research study on the ocean's thermohaline circulation—the flow of tropical waters to the Earth's northern polar region—has indicated that, if the water's flow should abruptly cease, it could lead to rapid and severe climate change in Great Britain and western Europe. Such an occurrence could cause a miniature ice age to befall that region in only a matter of years.

There is speculation and conjecture to both support and debunk the notion that the thermohaline circulation could stop. Climate models, such as the ones depicted in the film, cannot take into account the millions of variables required accurately to predict what will occur given specific stimuli. The Earth's climate and oceanic flows are just too complex for such accuracy.

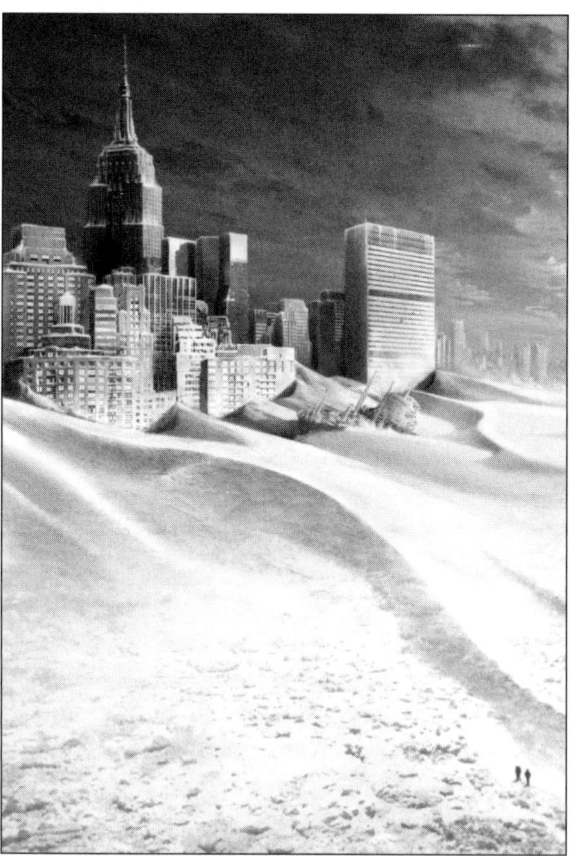

This image from The Day After Tomorrow *shows New York City after the sudden onset of a new ice age.* (AP/Wide World Photos)

Still, *The Day After Tomorrow* does represent in stylized fashion actual risks to the environment. Unlike the Younger Dryas period, during which nature alone created a climate change, human intervention into the delicate balance of atmospheric and oceanic concerns can only upset the equilibrium further. Greenhouse gases in the atmosphere have the potential to raise the Earth's temperature, causing the polar ice caps to melt and drop billions of tons of freshwater into the ocean, thereby causing the desalination depicted in the film. It is conjectured that such events are already occurring and are interfering with polar ecosystems.

As was true during the Younger Dryas period, life-forms—both animal and plant—are the first indicators that something is wrong with the ecosystem. Plant life, acclimated to specific temperatures and climates, will slowly diminish and die when climate changes occur. Animals that feed on the plant life either have to adjust to the loss of a food source or perish as well. *The Day After Tomorrow* fails to denote this fact, although it could certainly be implied from the events depicted in the film that animal life perished on the same scale as human life.

As a piece of entertainment, *The Day After Tomorrow* is an engaging and thought-provoking film. It depicts a future no one wants to come to fruition—a new ice age. This ice age occurs in a matter of days, throwing the world into chaos and killing billions. While an effective argument against the horrors of global warming and environmental abuse, *The Day After Tomorrow* depicts a rapid descent into a new ice age that is highly unlikely. It is more probable that slow change will occur and that humankind will be able to take the necessary steps to ensure its survival.

Roger Dale Trexler

• **Further Reading**

Fagan, Brian M. *The Little Ice Age: How Climate Made History, 1300-1850.* Jackson, Tenn.: Basic Books, 2001. A discussion of how climate change altered European history.

Pearce, Fred. *With Speed and Violence: Why Scientists Fear Tipping Points in Climate Change.* Boston: Beacon Press, 2008. A critical study of anthropogenic climate change written by a journalist for *New Scientist* magazine.

Weart, Spencer R. *The Discovery of Global Warming.* Rev. and expanded ed. Cambridge, Mass.: Harvard University Press, 2008. An informed history of the concept of global warming.

See also: Abrupt climate change; Little Ice Age; Media; Popular culture; Speculative fiction; Younger Dryas.

Deforestation

• **Category:** Plants and vegetation

• **Definition**

Deforestation is the long-term or permanent conversion of forested land to another use. Deforestation can be a natural phenomenon: floods, hurricanes, wildfires, landslides, and droughts, for example, can all cause widespread damage and destruction to a forest. However, deforestation is frequently anthropogenic. Such anthropogenic deforestation results from destructive logging; the transformation of forest to cropland, pasture, or developed areas; and overutilization of forest resources past the point where the ecosystem can recover. Road building, oil extraction, mining, and hydroelectric dam construction also involve deforestation.

Humankind has cleared forests since its earliest days to build shelter, obtain fuel, and make way for crops and livestock. Deforestation has inevitably accompanied human settlement, development, and commerce. Rates of forest loss are highest in developing nations, where trees are harvested in response to an international demand for wood products. They are also cut down to meet domestic needs for fuel, as wood and charcoal are still widely used for cooking and heating. Tropical forests are eliminated and the land reworked to provide more profitable, exportable commodities such as beef cattle, biofuel crops, sugar, palm oil, rubber, tea, and coffee.

In its 2005 Global Forest Resources Assessment, the Forestry Department of the United Nations

A deforested portion of the Amazon rain forest near Tailandia, Brazil. (Jack Chang/MCT/Landov)

Food and Agriculture Organization (FAO) reported a global deforestation rate of about 13 million hectares per year between 2000 and 2005. Although forest restoration efforts, afforestation (establishing forest plantations in historically unforested areas), and natural forest expansion offset part of the destruction, the net loss remained substantial: Every year during the study period, an estimated 7.3 million hectares of forest—an area roughly the size of Panama—disappeared. (This represents an improvement over the years from 1990 to 2000, when the annual net loss was 8.9 million hectares.) Forests in Africa and South America were hardest hit; North and Central America and Oceania also experienced a net loss. Europe showed a slow expansion of forested area, and China reported a net gain due to large-scale afforestation. As of 2005, the total global forested area was just under 4 billion hectares and covered about 30 percent of the planet's land area.

• Significance for Climate Change

Living plants take in carbon dioxide (CO_2) during photosynthesis and retain, or sequester, it. The carbon is returned to the atmosphere when the plant decomposes or burns. The FAO's 2005 Global Forest Resources Assessment estimates that forests worldwide store 283 billion metric tons of carbon in their biomass alone; the carbon stored in that biomass, together with that in forest deadwood, litter, and soil, is about 50 percent more than the amount of carbon in the atmosphere.

Because of the role trees and other forest plants play in the carbon cycle, deforestation is regarded as a major source of greenhouse gas (GHG) emissions. When a forest's trees are harvested or cleared, they can no longer pull carbon from the atmosphere. Furthermore, as they burn or decompose, they release their stored carbon to the atmosphere. The FAO reports that global carbon stocks retained in forest biomass dropped by 1.1 billion metric tons annually between 1990 and 2005 in response to deforestation and forest degradation.

According to *Land Use, Land-Use Change, and Forestry* (2000), a report by the Intergovernmental Panel on Climate Change (IPCC), an estimated net release of 121 billion metric tons of carbon resulted from the expansion of agriculture through conversion of forest and grasslands during the 140-year period between 1850 and 1990. Approximately 40 percent of that was emitted from middle- and high-latitude areas in the Northern Hemisphere, primarily before the middle of the twentieth century; the remaining 60 percent came from low-latitude tropical forests, mostly during the latter half of the twentieth century. The IPCC attributes more than 90 percent of net carbon emissions during the 1980's to land-use changes (chiefly deforestation) in the tropics.

It was during the 1980's that international concern about the widespread liquidation of forests mounted. At the 1992 United Nations Conference on Environment and Development in Rio de Janeiro, Brazil, the nonbinding Statement of Forest Principles was developed and made several recommendations for responsible and sustainable forestry. Since then, most of the world's countries have adopted forestry laws and policies that integrate environmental, economic, and social considerations. Some countries, including Paraguay, Costa Rica, China, Thailand, the Philippines, and much of Europe, have implemented deforestation bans or moratoria.

Deforestation is largely driven by economic considerations: The financial advantages of harvesting or clearing a forest are clear, while the benefits the forest provides in terms of carbon storage, biodiversity, water purification, and erosion control are less evident. Tax credits, subsidies, incentive programs, and carbon trading have been proposed to encourage forest conservation and preservation. In Costa Rica, forest loss has been halted through a combination of tax incentives and a program of payment for environmental services. In 2008, the United Nations launched the Reduced Emissions from Deforestation and Forest Degradation (REDD) Program, in which developed nations will pay developing nations to slow climate change by protecting and planting forests.

Karen N. Kähler

• Further Reading

Food and Agriculture Organization of the United Nations. *State of the World's Forests, 2007.* Rome: Author, 2007. Includes tables summarizing forest areas and area change; forest growing stock, biomass, and carbon; and the ratification status of international conventions and agreements. Figures, photographs, acronyms list, references.

Gay, Kathlyn. *Rainforests of the World: A Reference Handbook.* 2d ed. Santa Barbara, Calif.: ABC-CLIO, 2001. Chapter 2 is devoted to the causes and impacts of deforestation. Chapter 5 includes the text of the 1992 Statement of Forest Principles. Bibliographic information, glossary, index.

Williams, Michael. *Deforesting the Earth: From Prehistory to Global Crisis.* Chicago: University of Chicago Press, 2003. A survey of humanity's ten-thousand-year history of clearing forests. Part 3 looks at the acceleration of global deforestation during the twentieth century. Figures, tables, plates, notes, bibliography, index.

See also: Agroforestry; Amazon deforestation; Forestry and forest management; Forests; Intergovernmental Panel on Forests; Tree-planting programs.

Deglaciation

• **Category:** Cryology and glaciology

• Definition

Deglaciation is the uncovering or exposure of a land surface that was previously covered by glacial ice. It results from ice melting or subliming (transforming from solid directly into vapor). Deglaciation, therefore, accompanies the end of a glacial stage. As deglaciation occurs, several processes take effect. Among these processes are meltwater stream flow, development of meltwater lakes, addition of water to the world's oceans (raising the global sea level), exposure of the land, and rebound of the land (lifting of the land's elevation as a result of the removal of the weight of the overlying ice). In addition, faunal and floral changes accompany deglaciation in response to changes in landscape and ecology.

Glaciers have a seemingly infinite capacity to entrain and transport sedimentary material, from tiny clay particles to giant rock boulders. Glaciers move these materials within and upon the ice, but when they melt all this sediment is deposited in the area where the ice melts. For this reason, areas that have experienced deglaciation are typically covered by glacially transported sediment. Alternatively, glaciers may sweep an area clean of loose material, creating deglaciated areas of bare bedrock. Where deglaciation has formed modern shorelines, those shorelines tend either to be laden with glacial sediment or to present bare bedrock to the waves. In some places, rebound has lifted the land along the modern shore, forming sea cliffs.

Deglaciation accompanies the transition from a glacial stage to an interglacial or warm stage. There have been several such transitions over the past two million years. In addition, deglaciation—to a lesser extent—accompanies minor warming events that occur during glacial stages. Prior to the current epoch of glacial and interglacial stages, the Earth experienced several periods during which glaciers episodically covered large parts of its surface. There have been at least four such glacial periods during the past one billion years.

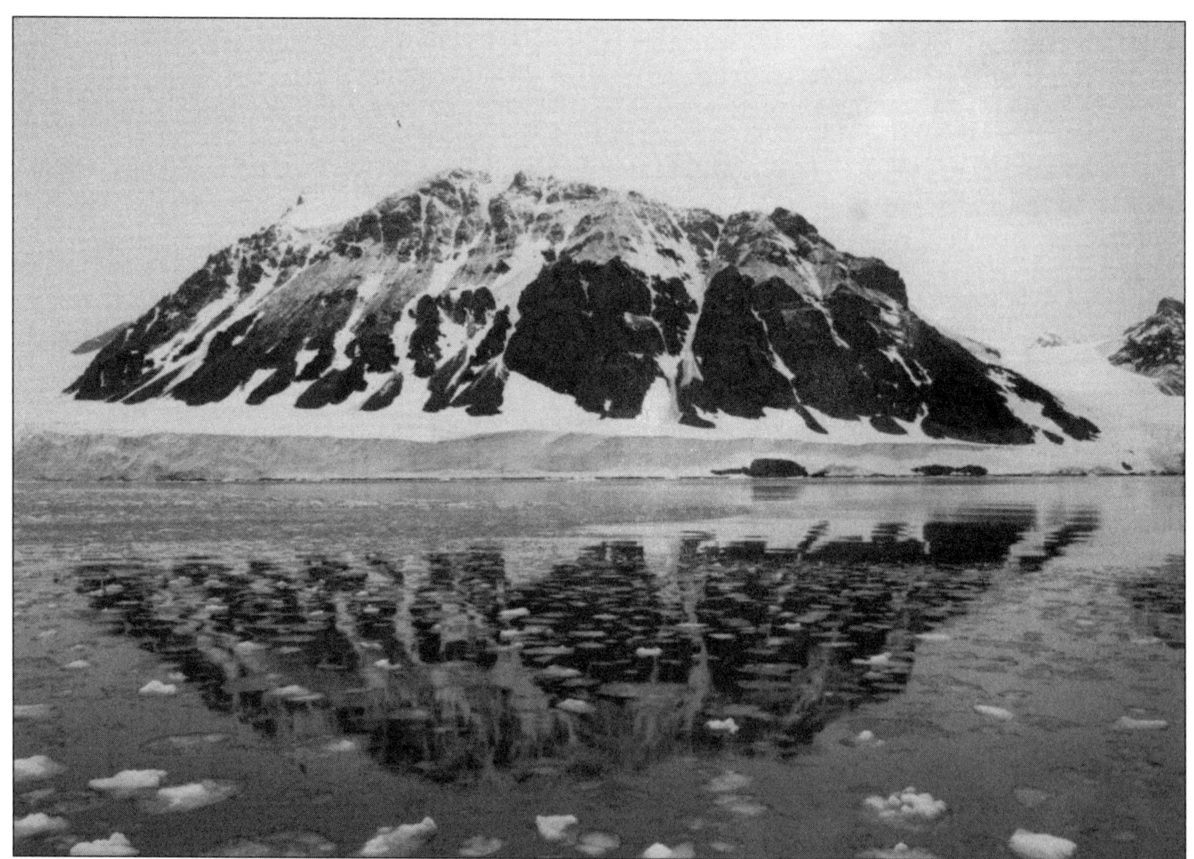

An Antarctic mountain is reflected in the surface of a bay that was once covered by a glacier. (Stuart McDill/Reuters/Landov)

• Significance for Climate Change

Deglaciation accompanies climate change and can be a cause of climate change. The geological record indicates that, when deglaciation commences, there is typically a climatic turn toward global warming. In other words, after climatic warming initiates deglaciation, the deglaciation itself can create a positive feedback loop engendering further warming. Glaciers are highly reflective of sunlight, contributing to Earth's albedo (the percentage of sunlight reflected back into space from the planet's surface). As glaciers melt, white ice and snow are replaced with darker surface elements. Earth's albedo decreases, and more solar radiation is absorbed and retained by the planet. As the planet warms, more glaciers melt, the albedo decreases even further, and the process continues. Melting glaciers also contribute water to lakes and oceans, which help retain atmospheric heat. Rising sea levels due to glacial meltwater contribute to global climate change as well.

Using the modern deglaciation as an example, loss of glacial ice cover on the land has had and continues to have profound consequences for global climate change. For example, release of water locked up in glaciers has affected the amount of water in the oceans as well as on land, in rivers and streams, and in the form of groundwater. This has affected coastal and interior ecosystems, which are dependent upon water for life. Changing patterns in the distribution of water cause both climates and ecosystems to change.

Deglaciation, as compared to glaciation, can be a relatively rapid process, once the melting triggers feedback mechanisms that increase its pace. The rapid nature of this change has led to disequilibrium conditions on land, such as unstable slopes, high gradients in rivers and streams, and unstable lakes that drain catastrophically. In the biotic realm, rapid deglaciation has led to mass death and mass extinction among plant and animal groups, as well as mass migration of animal populations.

Deglaciation leaves behind profound physical effects of ice movement, such as landscapes altered by the erosive forces of massive ice sheets, depositional landforms created by sediment released from melting ice, and lakes created by the meltwater—including waters from marooned blocks of ice that melt long after the main glacial mass is gone. The resulting altered landscapes generally have low levels of vegetation (at least initially), as well as areas of low elevation where water can accumulate. This type of landscape has a higher capacity to retain radiated heat from the Sun and therefore also contributes to atmospheric warming.

David T. King, Jr.

• Further Reading

Alley, R. B., and P. U. Clark. "Deglaciation of the Northern Hemisphere." *Annual Reviews* 28 (2000): 149-182. One of the best references for a complete understanding of the current deglaciation, its chronology, and its overall effects. Well illustrated.

Ehlers, J., and P. L. Gibbard, eds. *Quaternary Glaciations: Extent and Chronology.* 3 vols. San Diego, Calif.: Elsevier, 2004. Massive, comprehensive compendium of studies of the glaciations and deglaciations of the Quaternary period. Includes five CD-ROMS of digital maps and other supporting material.

Stanley, Steven. *Earth Systems History.* 3d ed. New York: W. H. Freeman, 2009. Lays out the history of Earth, including the factors affecting global climates and climate change, over the vastness of geological time. Selected chapters focus on climate change in terms of glaciation and deglaciation.

See also: Albedo feedback; Climate feedback; Cryosphere; Glaciations; Glaciers; Ground ice; Ice shelves; Interglacials; Sea ice; Sea-level change.

Desalination of seawater

• Categories: Water resources; science and technology

Desalination of water is an often-disputed subject where climate change is concerned. Scientists are unsure whether desalination will have a major effect on Earth's climate.

• Key concepts

distillation: use of evaporation and condensation to remove solutes from a liquid; one of the earliest forms of artificial desalination

passive vacuum technology: method that utilizes gravity and atmospheric pressure, rather than pumps, to create a vacuum, which enables evaporation to occur at lower temperatures, requiring less energy

reverse osmosis: forced passage of a liquid through a membrane to remove solutes

• Background

The removal of salt from seawater is an ages-old process that has become a multimillion-dollar industry. The demand for freshwater, especially in arid regions, has driven people to create and implement new and more effective ways to remove salt from water. Desalination occurs naturally as part of the hydrologic cycle. The Sun evaporates water from the ocean. The vapor, condensed by cooler air in the atmosphere, forms rain clouds. The rain from these clouds reaches the ground as pure liquid water. Earth's ecosystems are dependent upon this process.

All artificial desalination processes are based on the natural hydrologic cycle. For the most part, the energy requirements to desalinate seawater are heavy, making the process expensive. Still, it is estimated that 30 percent of the world's irrigated areas suffer from salinity problems that prevent crops from flourishing as they would if freshwater were available. The need for desalinated water for human and crop consumption is critical in the Middle East and other regions where freshwater is not abundant.

• Distillation

The most fundamental form of desalination is distillation, one of the earliest forms of water treatment. Ancient mariners used this process to convert seawater into drinking water on long voyages. By heating salt water and capturing the vapors, then letting them condense back into a liquid, they removed salt and other impurities. The same process is used to separate alcohol from fermented grains.

• Passive Vacuum Technology

Passive vacuum technology is used to decrease the energy requirements of desalination. By elevating a container, a partial vacuum can be created by the difference between air pressure inside and outside the container. The vacuum in turn allows water directed through the container to evaporate at a lower temperature, making it feasible to heat the water to its evaporation point with solar power. The temperature requirement for this system is less than for other methods, in part because it represents a closed system, so heat and vapors remain within it.

As pure water evaporates from salt water, the salinity of the remaining water increases, thereby decreasing its evaporation rate. Fresh salt water needs to replace the remaining brine at a rate equal to the rate of evaporation in order to maintain the overall salinity of the system. A tube-in-tube heat exchanger is used to inject new salt water while simultaneously drawing off the concentrated brine. The freshwater produced is reconstituted to its liquid state through a series of condenser coils similar to that of a moonshiner's still. The freshwater is delivered to a storage tank, while the concentrated brine can be sent to a solar tank and condensed further.

This method has the advantage of using far less energy than do typical solar evaporators, and the vacuum effect produced within the enclosed system makes a pump unnecessary. Internal pressure is developed within the closed system that is sufficient to push water through the system. It is a continuous system, yet it has its shortfalls as well. The system needs to be cleaned and restarted on a periodic basis, lest the noncondensable gases produced from the process build up and destroy the vacuum.

• Reverse Osmosis

Another method of desalinating seawater is reverse osmosis. A reverse osmosis system has four major steps: pretreatment, pressurization, membrane separation, and post-treatment stabilization. In this method, saltwater is pumped into an enclosed system, building pressure that forces it through a water-permeable membrane. The membrane prevents dissolved salts and other impurities from passing through it, thereby purifying the water. The re-

Interior of the world's largest seawater reverse osmosis desalination plant, in Ashkelon, Israel. (Yael Tzur/Israel Sun/Landov)

sulting brine is pushed through the pressurized side of the reactor and stored. The concentrated salts are then discharged in an effort to minimize the pressure build-up. Without discharging the build-up of salts and other impurities, more energy would have to be input into the system to achieve purification. The main energy expenditure of reverse osmosis is thus in the initial pressurization of the feedwater through the membrane.

The key advantage to reverse osmosis is its simplicity. The only difficulties that the process presents are in minimizing the need to clean the membrane by producing enough clean feedwater to keep the system running at top capacity and in the need to remove particulates in that feedwater by pretreating the water. The technology is viable for regions with a readily available supply of brackish groundwater or seawater.

• The Purity of Ice

Yet another way to desalinate water that is seldom used is freezing. This technique is based on the simple fact that freshwater will freeze before salt water will. By allowing water to freeze and crystallize, separating the crystals from the salty slurry, then applying heat, freshwater may be obtained from salt water. Northern states and territories are the most likely to find this method viable, because they experience more months of cold weather and harsher climates than do more southern states. The harsh climate can be harnessed to freeze water without energy input.

This method has its benefits. At freezing temperatures, scaling—the build-up of elements other than salt within the water—is minimized. Also, equipment does not suffer the corrosion it does when water vapor is involved. Corrosion is a major drawback in methods involving heating the water that, in general, does not exist when freezing is involved.

Icebergs have been considered as a potential source of a massive supply of freshwater because of their purity. Icebergs contain water that is almost as pure as distilled water. They are abundant and can be easily procured. Large towing ships could, in theory, remove an iceberg from the polar region and tow it to an area where the already desalinated water within the iceberg could be easily thawed and put to use. Because arctic icebergs are irregularly shaped, antarctic icebergs are the more suitable for transport. A suitable iceberg would not only have to be shaped correctly, but it would also need to weigh somewhere around 91 million metric tons to retain enough frozen pure water by the time it reaches its destination. The drawbacks of this method include the requisite time, erosion of the iceberg during transport, financial concerns, and the uncertainty of the ecological and climatic effects of removing icebergs from an already decaying environment.

• Context

The advantages of desalination are many and varied. In regions where freshwater is not readily available, desalination and purifying water could provide an economic and population boom. Arid regions could be irrigated, and crops could be sown to increase the food supply. Desalination will play a part in the future of mankind, and it will be up to humans to create new and improved methods of providing freshwater. Different methods are required to desalinate water in different regions of the world, but fresh, potable water is a necessity everywhere.

Roger Dale Trexler

• Further Reading

Army Corps of Engineers, U.S. *Water Desalination.* Honolulu: University Press of the Pacific, 2005. Explores and creates guidelines that are useful in both military and civilian settings for selecting a method to produce potable water from seawater and brackish water.

Lauer, William C., ed. *Desalination of Seawater and Brackish Water.* Denver: American Waterworks Association, 2006. Collection of articles taken from AWA conference proceedings, periodicals, and previously unpublished sources. Topics include seawater, brackish water, osmosis-membrane softening, disposal, and the cost involved in desalination.

Wilf, Mark, et al. *The Guidebook to Membrane Desalination Technology: Reverse Osmosis, Nanofiltration, and Hybrid Systems Process, Design, Application, and Economics.* Rehovot, Israel: Balaban, 2007. Encompassing the latest membrane desalination technology, this book provides a thorough overview of the latest systems. Explores all facets and processes in the field of membrane desalination.

See also: Ekman transport and pumping; Estuaries; Freshwater; Groundwater; Hydrologic cycle; Ocean acidification; Sea ice; Water quality; Water resources, global; Water resources, North American.

Desertification

• **Category:** Environmentalism, conservation, and ecosystems

Expansion and intensification of agriculture in dryland environments can lead to land degradation. When drought occurs in such environments, further land degradation is so severe that the ecosystem cannot recover fully after the rains return, and the region becomes a desert. Should drought frequency increase in the future, vulnerability to desertification will also increase.

• Key concepts

Inter-Tropical Convergence Zone: a meandering zone of convergence of the northeast and southeast trade winds adjacent to the equator

salinization: accumulation of soluble salts in the soil, greatly reducing fertility

soil structure: the size of soil particles and their tendency to combine in lumps or clusters

stable air: air that resists convectional mixing and uplifting and in which precipitation is unlikely to occur

summer monsoon: a summertime influx over a continent of unstable, rain-bearing air from over the ocean
unstable air: air that is readily susceptible to convectional mixing and uplifting, which often results in precipitation
winter monsoon: a wintertime, large-scale wind system that extends cool, dry, stable air from a continental interior over a large area

• Background

Desertification is a process of land degradation in arid and semiarid areas resulting from climatic variations and human activities. Degradation results from pressure from expansion of agriculture and livestock numbers, which make the land increasingly vulnerable to the impact of drought. Also, rising human populations have led people to farm on increasingly marginal land, which is even more at risk. The pressure on the land is such that, when drought occurs, the land degrades to the point that it is unable fully to recover.

• Characteristics of Desertification

Manifestations of desertification include a breakdown of soil structure, accelerated soil loss to wind and water erosion, an increase in atmospheric dust, a reduction in soil moisture-holding capacity, an increase in surface-water runoff and streamflow variability, salinization of soils and groundwater, and reductions in species diversity and plant biomass. The net result is a reduction in the overall productivity of dry-land ecosystems. This reduction leads to the impoverishment of human communities that are dependent on the land for survival.

The best examples of desertification are to be found in the Sahel region of Africa and the Rajasthan state of India, vulnerable areas on the borders of the Sahara and Thar Deserts, respectively. The underlying problem is poverty, which means few resources are available for managing the environment. Most of the people are subsistence farmers whose food supply is dependent on an adequate harvest each year. Farmers rely on summer monsoon rains. If one rainy season fails, people have very little in the way of stored food or money to see them through. The most vulnerable are the pastoralists, whose animals rapidly weaken and perish when there is nothing left to graze. Those animals that do survive will have stripped the land of vegetation so intensively that it may fail fully to recover when the rains eventually return.

• Climatic Feedback Processes

The impact of drought is in some cases linked to feedback processes between the atmosphere and a land surface that is modified and used by the very population that is at risk. Stress on the land is not the sole cause of desertification, but it weakens an ecosystem's ability to withstand drought. It is an important part of the feedback chain. In wet years, there is often an expansion and intensification of grazing and cultivation of land that is otherwise marginal. Following a relatively dry year, excessive demands may be placed on the water stored in the soil. The soil will then dry, become susceptible to wind erosion, and eventually blow away. Even if the rains were to return, what soil remains would be washed away by sheet erosion and gullying. Most important, if such changes affect a large area, positive feedback processes to which localized climate is highly sensitive are set in motion, which accentuates existing anomalies in climate.

A key factor in the role of climate is stability of air. When the atmosphere is stable, upward motion of air is suppressed. Even in humid airstreams, rainfall will not occur unless stability is overcome. Subtropical high-pressure belts at about 30° latitude on either side of the equator are associated with masses of stable air. This air accounts for most of Earth's large areas of arid and semiarid climate. Rain occurs when the air in these high-pressure belts is displaced by the advance of unstable, rain-bearing oceanic air moving with the Inter-Tropical Convergence Zone (ITCZ). This movement of rain-bearing air is known as the summer monsoon. The Sahel and Rajasthan are at the northern edge of an area that experiences drought when the summer monsoon does not reach far enough north.

Earth's atmospheric pressure systems migrate with the seasons. Should the ITCZ advance north only a few degrees of latitude less than normal, large areas of the desert borderlands will experience a reduction in rainfall. If climate changes so that the norm is redefined and the northward movement of the rain-bearing air is frequently below the previous norm, the desert would expand southward. There are various theories as to how this expansion could come to pass.

One popular theory asserts that overgrazing of the land by livestock leads to loss of vegetation, resulting in bare soil being exposed to wind erosion. Large quantities of windblown dust in the atmosphere reduce the amount of solar radiation reaching and heating the land surface. Overgrazing also increases the land surface's albedo—that is, it causes the surface to reflect more incoming energy from the Sun compared with a vegetated surface.

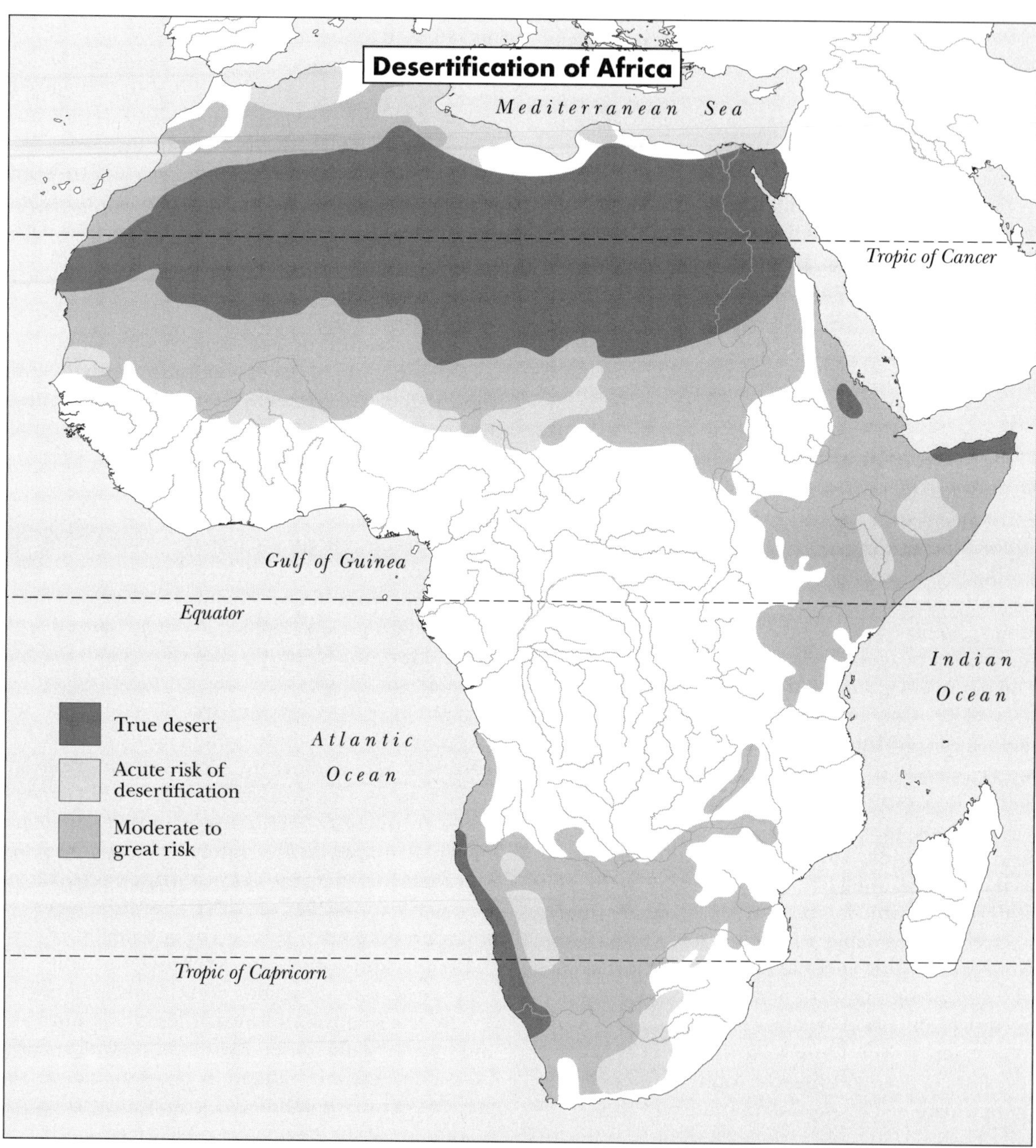

This additional reflection of solar energy results in land cooling. Thus, overgrazing has multiple consequences that increase cooling and air stability, which effectively adds to the extent of the stable air of the subtropical high-pressure belt and erodes the northern edge of the advancing rain-bearing air of the summer monsoon. The dry conditions result in a further loss of vegetation, causing increased reflectivity and subsidence of dry air even further to the south, and so on in a cycle of positive feedbacks.

• Desertification and Climate Change

The Intergovernmental Panel on Climate Change (IPCC) has commented on the possibility of increased frequency of droughts in certain areas. If the vulnerable desert borderlands are among the regions affected, then desertification may be intensified. However, connections between anthropogenic increases in greenhouse gases (GHGs) in the atmosphere and changed drought frequency and intensity are only speculative, as climate models that are used to assess these connections have not been shown to be reliable. The level of scientific uncertainty and the existence of conflicting results are such that reliable predictions of future climate are not possible at this time.

• Context

Owing to the uncertainty surrounding scientists' understanding of the global climate, neither the trends in drought occurrence nor the interannual variability of droughts can be simulated reliably in global climate models. Despite this, projections have been made about future trends in precipitation extremes linked to increases in GHGs. Vulnerability will decline if drought frequency and intensity are reduced. The salient point is that there is no clear answer to the question of what will happen to trends in drought occurrence.

C R de Freitas

• Further Reading

Delville, Philippe L. *Societies and Nature in the Sahel.* London: Taylor & Francis, 2007. Explores the links between environmental and social systems that lead to desertification across sub-Saharan Africa.

Geist, Helmut. *The Causes and Progression of Desertification.* Burlington, Vt.: Ashgate, 2005. Examines desertification on the local and international scales and assesses the role of causal processes.

Middleton, N. *Global Desertification: Do Humans Cause Deserts?* Amsterdam, the Netherlands: Elsevier, 2004. Explores and assesses various possible causes of desertification.

Williams, M. A. J., and Robert C. Balling, Jr. *Interactions of Desertification and Climate.* London: Arnold, 1996. Comprehensive account of the interactions between desertification and climate.

See also: Agriculture and agricultural land; Albedo feedback; Climate feedback; Climate prediction and projection; Deserts; Drought; Dust storms; Sahara Desert; Sahel drought; Soil erosion; United Nations Convention to Combat Desertification.

Deserts

• **Category:** Environmentalism, conservation, and ecosystems

Deserts cover some 26.2 million square kilometers, or about 20 percent, of the Earth's land surface, mainly in the subtropical to midlatitudes. The importance of water to human and natural systems in deserts makes those systems very sensitive to climate changes affecting the amount, type, timing, and effectiveness of precipitation.

• Key concepts

drought: an extended period of months or years during which a region experiences a deficit in its water supply, mainly as a result of low rainfall

dune mobility index: a measure of potential sand mobility as a function of the ratio between the annual percentage of the time the wind is above the sand transport threshold and the effective annual rainfall

subtropical anticyclonic belts: a series of high-pressure belts situated at latitudes 30° north and south of the equator

• Background

Deserts are fragile environments, easily affected by natural and human disturbance. They are being affected by a rapidly growing and increasingly urban population that is dependent on scarce surface- and groundwater. The historical and observational record indicates the great natural variability of climates in these regions, including the occurrence of periodic severe and multi-decadal droughts.

• Causes of Deserts

Desert climates are characterized by low humidity (except in cool, foggy coastal deserts such as the Namib and Atacama), a high daily range of temperatures, and precipitation that is highly variable in time and space. The most extensive deserts lie astride the tropics. Solar heating in equatorial latitudes gives rise to rising moist air, which then condenses, loses moisture as tropical rainfall, cools, and descends away from the equator. As the air descends, it warms and becomes very dry. This descending, dry air in the subtropical anticyclonic belts maintains arid conditions throughout the year. The effects of stable air masses are reinforced in some areas by mountain barriers, which block moist air masses (for example, the Himalaya and other mountain ranges prevent the penetration of the southwest monsoon to the Gobi and Takla Makan Deserts of central Asia). Deserts located on the west coasts of South America and southern Africa (the Atacama and Namib) owe their hyperarid climates to the influence of cold oceanic currents offshore. These currents reinforce the subsidence-induced stability of the atmosphere by cooling surface air and creating a strong temperature inversion.

• Effects of Climate Change

The effects and potential effects of past and future climate change on deserts are many and are influenced by the topographic and climatic diversity of desert regions. Global climate models differ in their predictions of the direction and magnitude of future change in arid regions. In some areas—such as China, southeastern Arabia, and India—increased monsoon precipitation is predicted, but its effects may be offset by higher evaporation as a result of increased temperatures. In the Sahara Desert, there is support in many climate-model predictions for increased rainfall in the southern and southeastern areas (including the Sahel), but strong drying in the northern and western areas. Some models, however, suggest a strong drying throughout the region.

The differences between varioius models' predictions for the Sahara demonstrate the complexity of forcing factors in the region, as well as the possible influence of feedbacks between land-surface conditions and the atmosphere. Such feedbacks could affect rainfall total, effectiveness, and spatial distribution. Most of the interior of southern Africa is also predicted to become drier, leading to the mobilization of sand dunes in the Kalahari Desert, as well as severe impacts on surface and groundwater resources.

In the southwestern United States, higher temperatures are predicted to increase the severity of droughts. Some models indicate that the region may already be in transition to a new, more arid state as a result of anthropogenically influenced climate change. The economies of many desert regions (including Atacama, the American Southwest, Iran, western China, and southwest Asia) depend on runoff derived from winter snow in mountain areas for domestic use and irrigated agriculture. Higher temperatures are already reducing

Deserts and Climate Change

Likely effects of climate change upon deserts include but are not limited to:

- changes in the amount, type, and seasonal distribution of precipitation
- changes in the magnitude and frequency of extreme events such as dust storms, floods, wildfires, and periods of extended drought
- changes in the mobility of sand dune areas, including reactivation of vegetation-stabilized dunes
- vegetation change, including increased vulnerability to invasive species

the amount of snowpack and changing the timing and duration of spring runoff. More precipitation is falling as rain, leading to less natural storage and an increased risk of flooding. Such changes, if continued into the future, will require costly upgrades of water management systems and possibly a reduction in available water supply.

Many desert areas experienced significant increases in temperature and reductions in rainfall during the 1990's and early twenty-first century. During that time period, droughts occurred in the Colorado River Basin, Australia, southern Africa, Iraq, and Afghanistan. Sand dunes occupy up to one-third of the area of many desert regions. Dune mobility is a function of the ratio between wind strength and effective rainfall and is measured by the dune mobility index. Increased temperatures, accompanied by decreased rainfall, are predicted to lead to remobilization of vegetated sand dunes in the Kalahari and drier areas of the Australian desert.

The effects of climate change on vegetation patterns in desert regions is difficult to separate from anthropogenic disturbance. Increased levels of atmospheric carbon dioxide (CO_2) may increase plant productivity in arid regions. Higher CO_2 levels may also favor invasive exotic species such as cheat grass, with possible effects on fire regimes in the Great Basin Desert. Models that incorporate CO_2 fertilization of vegetation indicate a reduction in desert areas in the next century, introducing an additional level of uncertainty about the future of desert ecosystems.

• Context

The great natural variability of climatic conditions, especially the distribution of rainfall in space and time, presents challenges for the prediction of the response of desert regions to future climate change. However, the experience of recent drought episodes indicates that the natural and human systems of deserts and desert margin areas are highly susceptible to soil moisture deficits. Climate change is expected to decrease water availability in all desert regions, through increased temperatures, changes in the amount of precipitation, or a combination of both. The result will be increased pressure on existing water resources for human and ecosystem use, possibly leading to higher levels of conflict over scarce resources.

Nicholas Lancaster

• Further Reading

Ezcurra, E., ed. *Global Deserts Outlook.* Nairobi, Kenya: United Nations Environment Programme, 2006. A comprehensive and well-illustrated interdisciplinary overview of desert environments, their issues, and their problems, as well as challenges and opportunities for the future.

Goudie, A. S. *Great Warm Deserts of the World: Landscapes and Evolution.* New York: Oxford University Press, 2002. Provides basic information on the physical environment of low- and midlatitude deserts, from a geomorphological perspective.

See also: Desertification; Drought; Dust storms; Sahara Desert; Sahel drought; United Nations Convention to Combat Desertification.

Dew point

• **Category:** Meteorology and atmospheric sciences

• Definition

Dew point is the critical temperature at which a parcel of air will become saturated with water vapor if it is cooled at constant pressure and constant water vapor content. Dew point is a measure of humidity, or atmospheric vapor content, since the higher the dew point temperature of an air mass, the greater the water vapor content. Compared to warm air masses, cold air masses have smaller water-holding capacities and therefore lower dew points. Dew point is also related to evaporation and condensation. The closer the dew point is to actual air temperature, the lower the rate of evaporation. When actual air temperature cools to the dew point, condensation will occur.

• **Significance for Climate Change**

A rise in the concentration of greenhouse gases (GHGs) in the atmosphere makes available more energy at the Earth's surface. This additional energy could heat the atmosphere by way of the sensible heat flux, increasing temperature, or it could evaporate water at the Earth's surface via the latent heat flux, increasing humidity. The latter course of events would lead to higher dew points. When water is freely available, the latent heat flux will always dominate, meaning that energy otherwise available to heat the atmosphere will be used in the evaporative transpiration of moisture from the surface to the atmosphere.

Most of Earth's surface is either water (approximately 70 percent is occupied by oceans, seas, and lakes) or land that is well supplied with precipitation. Thus, most of the additional energy at the surface due to an increased concentration of GHGs in the atmosphere will enhance the latent heat flux. The resulting warming of the atmosphere would be less in this case than if all the additional available energy were accounted for by an increase in the sensible heat flux alone. Dew point would then rise accordingly.

Water vapor is by far the most important GHG in the atmosphere, so a rise in global dew point would add significantly to the greenhouse effect, leading to enhanced warming and further enhanced evaporation and transpiration, and so on. This self-reinforcing cycle is known as a positive feedback effect. On the other hand, a rise in dew point could result in increased cloudiness as moisture is added to the Earth's atmosphere. Clouds, especially low clouds, act to reflect incoming energy from the Sun, energy that would otherwise be absorbed at the Earth's surface. The result is to reduce the energy available for heating the air and for evaporation and transpiration of moisture. This negative feedback, or stabilizing effect, would only partially compensate for the warming effect of higher dew points.

C R de Freitas

See also: Greenhouse effect; Greenhouse gases; Humidity; Hydrologic cycle; Hydrosphere; Ocean-atmosphere coupling; Ocean dynamics; Water vapor.

Diatoms

• **Category:** Plants and vegetation

• **Definition**

Diatoms are ubiquitous, microscopic, golden-colored algae. The most distinctive characteristic of a diatom is its rigid cell wall. The cell wall is formed in two halves, with one half fitting inside the other as a shoebox fits its lid. This rigid, glassy wall containing silica is patterned with pores, variable thicknesses, and spine-like extensions projecting from its surface.

Once a diatom dies, its cell walls either dissolve into the water or fall to the bottom of the sea or lake, where they become part of its sediments. When large numbers of diatoms are present in a body of water, their cell walls tend to accumulate in the sediments. Large deposits of diatom cell walls have been found on land in areas that were once covered by seas. This material is mined as diatomaceous earth and is used commercially as a fine abrasive material or filtering agent.

The classification system for diatoms is based on two key features, the pattern of the cell wall and the shape of the cell. Two major groups are often distinguished: Pennate diatoms are typically bilaterally symmetrical, while radially symmetrical diatoms are known as centric diatoms.

Although diatoms are found in most environments, these organisms are very important members of marine and freshwater ecosystems. As photosynthesizers, these tiny organisms harvest energy from the sun, fixing carbon dioxide (CO_2) into organic compounds that are used by the diatom and by organisms that consume the diatom. In this capacity, diatoms are at the first trophic level of the food chain, providing energy and organic compounds for the heterotrophic organisms in that ecosystem.

• **Significance for Climate Change**

Researchers have estimated that 20 to 25 percent of all carbon fixed on Earth via photosynthesis is fixed by planktonic marine organisms. In some oceans, diatoms are the most numerous members of the phytoplankton; in other areas, they are significant but not dominant. Through photosynthesis, diatoms also play a significant role in the carbon cycle, removing carbon from the atmosphere. The carbon may be made available to other organisms within the ecosystem, or it could be removed from the carbon cycle for millions of years as the dead diatoms become a part of the sediments.

Numerous environmental factors influence the growth of diatoms, including light, wind, currents, temperature, and available nutrients. The nutrients that most commonly limit growth of diatoms in aquatic environments are nitrogen, phosphorus, silicon, and iron. Since numerous factors affect the metabolism, growth, and reproduction of these organisms, the impact of climate change is likely to be complex and difficult to predict. It is unlikely that all environment factors would work in either a negative or a positive manner, so the end result is likely to be cumulative. Given the vital importance of diatoms to the local ecosystem and to the carbon cycle, it will be critical to monitor these organisms as changes occur.

Climate changes affecting diatoms could through them have a profound effect on the carbon cycle. Any factor that affects the rate of photosynthesis, the health, or the reproduction of these organisms would affect the role they play in the cycle. If there were a drastic decrease in the number of diatoms or a drop in their photosynthetic rates, less CO_2 would be removed from the air, ultimately resulting in warmer atmospheric conditions and contributing to global warming. On the other hand, massive

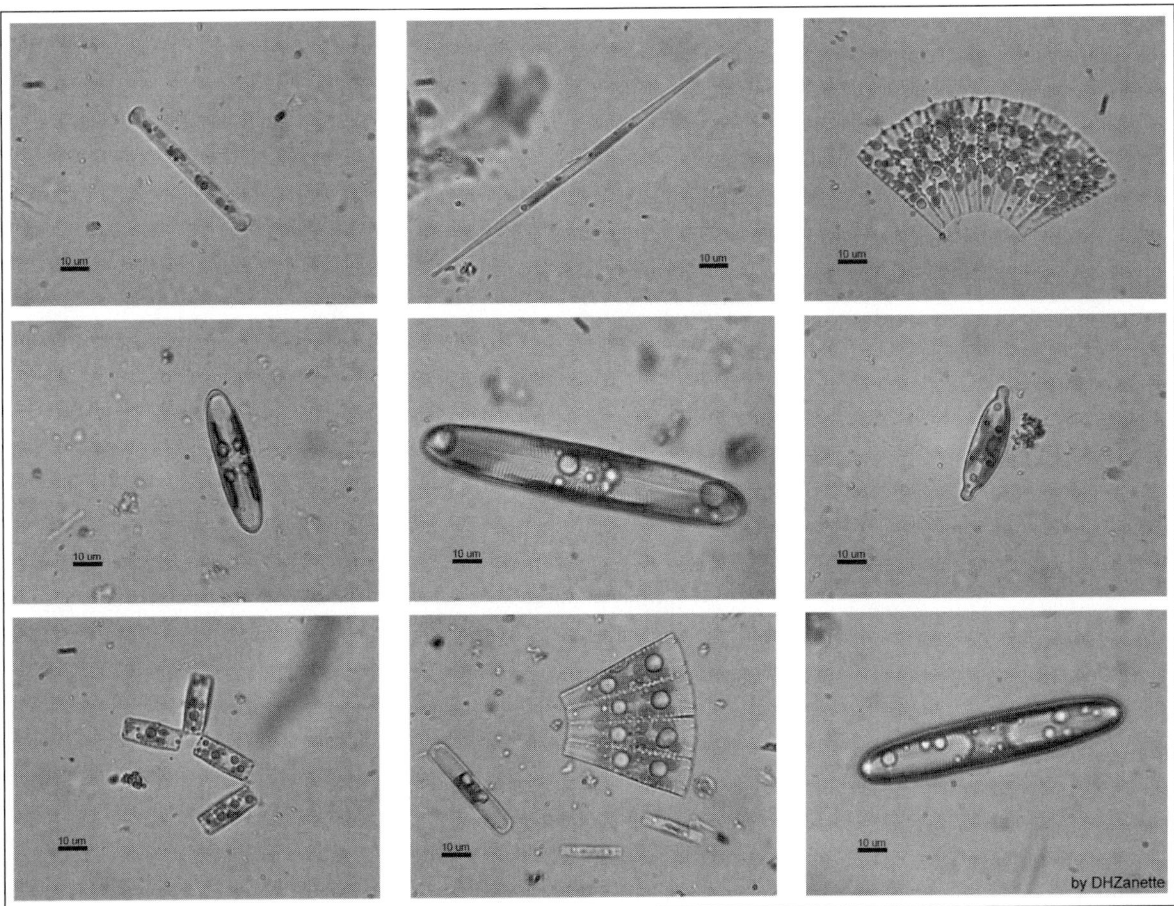

Microscopic images of diatoms. Despite their small size, these phytoplankton play a crucial role in Earth's carbon cycle. (DHZanette)

blooms of these organisms could lead to more CO_2 being removed via both photosynthesis and greater sedimentation rates, resulting eventually in a cooler atmosphere. Using historical evidence, researchers have suggested that such changes in atmospheric CO_2 levels during the glacial periods were correlated with changes in diatom abundance and carbon fixation.

Some of the best evidence linking diatoms and climate change has come from research in Antarctica. Diatoms are the dominant photosynthetic organisms in the cool southern ocean and on the ice shelves at the edges of the continent. Researchers have found that these organisms produce a compound known as dimethyl sulfide (DMS). Once airborne, DMS can serve as a nucleus for water condensation to form clouds or be converted to sulfuric acid and return to the ground as acid rain. Changes in numbers of diatoms in these areas could result in changes in cloud and moisture patterns in the Antarctic.

At least one aspect of climate change has generated concern regarding diatoms in Antarctica. These organisms are being exposed to increasing amounts of ultraviolet (UV) radiation, as the ozone layer in this part of the world continues to thin. Laboratory studies have shown that diatoms suffer damage to their photosynthetic pigments and deoxyribonucleic acid (DNA) when exposed to UV radiation.

Changes in the number or photosynthetic activity of diatoms due to climate change could have significant effects on ecosystems where diatoms contribute significantly to the first trophic level of the food chain. Areas most likely to be affected are those where diatoms tend to dominate, such as in the open ocean. The impact of climate change to shoreline marine ecosystems or to freshwaters could be less significant, as these ecosystems tend to contain a greater diversity of photosynthetic organisms that contribute to the primary productivity of the system.

Joyce M. Hardin

• Further Reading

Amsler, Charles D., ed. *Algal Chemical Ecology.* Berlin: Springer, 2008. Reviews the production and role of secondary metabolites by micro- and macroalgae. Illustrations, figures, tables, bibliography, index.

Graham, Linda E., and Lee W. Wilcox. *Algae.* Upper Saddle River, N.J.: Prentice Hall, 2000. Reviews the major groups of algae and the interactions of algae with other organisms and the environment. Illustrations, figures, tables, bibliography, index.

Miller, G. Tyler, Jr., and Scott Spoolman. *Environmental Science: Problems, Concepts, and Solutions.* 12th ed. Belmont, Calif.: Brooks Cole, 2008. Provides an overview of environmental issues, including information on the sulfur and carbon cycles. Illustrations, figures, tables, bibliography, index, maps.

Stoermer, Eugene, and John P. Smol, eds. *The Diatoms: Applications for the Environmental and Earth Sciences.* New York: Cambridge University Press, 1999. Explicates the use of diatoms as indicators of environmental change in aquatic and extreme environments. Illustrations, figures, tables, bibliography, index, maps.

See also: Antarctica: threats and responses; Carbon cycle; Carbon dioxide; Carbon dioxide fertilization; Ocean life; Photosynthesis; Plankton; Sea sediments; Sequestration.

Diseases

• **Category:** Diseases and health effects

Climatic conditions that support the successful colonization of geographical locations by human societies also tend to support the populations of pests and pathogens associated with human diseases. Abrupt climate change can destabilize trends in the distribution of diseases in populations, as well as society's ability to cope with emerging pathogens and shifting demographic patterns.

• Key concepts

air quality: normal atmospheric constituencies, such as levels of particulate matter, elements, and toxins

disability-adjusted life years (DALYs): a time-based quantitative measure of the burden of disease in a population that combines years of life lost to premature mortality and years of life lost to poor health or disability

heat wave: a long period of exceedingly hot, uncomfortable weather

pathogens: viruses, bacteria, protozoa, or other chemical or biological agents that can infect a human host to produce disease

vector-borne diseases: illnesses associated with pathogenic microorganisms whose transmission from an infected host to a new host is mediated by an insect or other agent (vector)

waterborne diseases: illnesses caused by pathogens that are transmitted through contaminated drinking water or contact with environmental waters

• Background

The most important factor in the emergence and proliferation of pathogens is the availability of susceptible hosts. Therefore, many pathogens have coevolved, not only with human biological constraints against disease, but also with socially developed constraints such as climate-controlled domiciles and disinfection. In the event of abrupt climate change, leading to excessive fluctuations in extreme weather conditions, there occurs a selection process that affects the microbial diversity of ecosystems, with some organism declining, whereas other organisms increase in population density. In addition, long-term climate change may forge new interactions among different organisms. When these biodiversity changes coincide with increasingly dynamic relocation of people in response to climatic events, epidemics can result from the resurgence of old diseases, emergence of new diseases, or exacerbation of preexisting disease conditions.

• Emergence of Climate Change as a Threat to Public Health

Seasonal trends in morbidity and mortality have long been understood by human societies. Such understandings have formed the basis of preventive health care plans in many countries. For example, preparation for the influenza (flu) season means massive vaccination campaigns during the months of September, October, and November. Similarly, respiratory conditions with no clear involvement of pathogenic agents, such as asthma and allergies, are known to follow seasonal patterns. Humans have, more or less, adapted to such seasonal inconveniences until they become so extreme to the extent that population migrations can occur. Outbreaks of contagious disease associated with scarce water supplies can destabilize communities or force local extinctions in human habitats. However, it has not been straightforward to project health impacts as part of the consequences of anthropogenic climate change.

In 1990, a World Health Organization (WHO) task group issued one of the early reports on the potential health effects of climate change. The group based its assessments on the scenario that global average temperature could increase by 3° Celsius by the year 2030; that sea level could rise by 0.10-0.32 meter; and that ultraviolet radiation, mainly UV-B, is expected to increase by a maximum of 20-25 percent in the same period. Based on these conditions, the task group anticipated both direct and indirect effects of climate change on human health.

The direct effects include those associated with thermal factors (heat disorders) and the effects of UV radiation on the incidence of skin cancer, immune response, eye function, and air quality. Indirect effects of climate change on human health are expected to include impacts on food production and nutrition, on wildlife and biodiversity, and on communicable diseases through effects on disease vectors and the incidence of infectious diseases that are not associated with specific vectors. In addition, indirect impacts of climate on health include the repercussions of human migration.

In 1997, responding to a request from the Subsidiary Body for Scientific and Technological Advice (SBSTA) of the United Nations Framework Convention on Climate Change (UNFCCC), Working Group II of the Intergovernmental Panel on Climate Change (IPCC) published a special report on the assessment of vulnerability in the regional impact of climate change. The Conference of the Parties (COP) to the UNFCCC needed information on the degree to which human conditions and the natural environment are vulnerable to the potential effects of climate change, but the regional assessment approach adopted by the IPCC revealed wide variation in the vulnerability of different populations, especially in the health sector. Different levels of vulnerability exist under similar climate and pathogen distribution patterns because of local economic, social, and political conditions, as well as the level of dependence on resources sensitive to climate variability. Therefore, instead of producing quantitative predictions of the impacts of climate change for each region, the IPCC took the approach of assessing regional sensitivities and vulnerabilities.

The adoption of "Weather, Climate, and Health" as the theme of the 1999 World Meteorological Day signified the convergence of global issue-framing strategies with the health impacts of climate change. This event emerged after more than a decade of policy formulation and scientific assessment activities by the WHO, the IPCC, and the World Meteorological Organization (WMO). Following the progress made by WHO researchers during the 1990's toward the development of quantitative methods for assessing the global burden of disease, it became possible to compare or project into the future the disease burden associated with specific risk factors such as climate change. Composite measures of disease burden such as disability-adjusted life years (DALYs) account for both mortality and morbidity, and are particularly suitable for evaluating risk factors with a broad range of disease end points. For example, in 2004, WHO estimated that global climate change accounts for approximately 5.5 million DALYs lost directly, but exacerbation of disease conditions associated with the creation of unsanitary conditions could result in a lot more DALYs lost. Not surprisingly, children younger than five years are particularly vulnerable.

To cap the evolution of health effects as a dominant frame of reference for the threats associated with climate change, in May 2008, the 193 member countries represented at the World Health Assembly adopted a resolution to protect public health from impending global climate change. This event signaled a much higher level of commitment from the health sector to strengthen the evidence for anthropogenic climate change and to better characterize the risks to public health at the regional and global levels.

• Diseases Associated with Climate Change

Communicable diseases. Vector-borne diseases, such as malaria, have dominated research on the impacts of climate change on public health. The rationale behind these studies is that increases in temperature and rainfall would support the proliferation of mosquito vectors and their ability to incubate disease-causing protozoa, leading to more infections. According to WHO, malaria infects 400 to 500 million, killing approximately 2 million people annually. Although the rate of morbidity and mortality associated with malaria in endemic zones might intensify, the real fear associated with climate change is that malaria zones will expand toward the temperate regions that have hitherto been free of the parasite. There is spotty evidence of recent incidences of malaria in Europe and North America, but it is not clear that these cases are not associ-

A swampy area in Thailand with a high rate of malaria transmission. If global warming increases the number of such areas, the global malaria-transmission rate will increase accordingly. (AP/Wide World Photos)

ated with population migration, which has led to the coining of the phrase "airport malaria." Nonetheless, a Roll Back Malaria initiative was launched in 1998 by the WHO, the United Nations Children's Fund (UNICEF), the United Nations Development Programme (UNDP), and the World Bank to provide a coordinated global approach against malaria, including scenarios associated with the influence of climate change. The IPCC predicted in 2007 that under certain climate change scenarios, the global population at risk from vector-borne malaria will increase by between 220 million and 400 million in the next century.

Other vector-borne diseases that are of concern with respect to climate change include lymphatic filariases, which are also transmitted through tropical mosquitoes, typically in urban slums. The geographical zone of these diseases may expand with increasing average global temperature, but good urban planning and hygienic conditions can limit the impact of the diseases on society. This scenario is expected to be similar for other climate-sensitive vector-borne diseases such as onchocerciasis (vector: African black fly, *Simulium damnosum*), schistosomiasis (vector: water snails such as *Biomphalaria glabrata*), African trypanosomiasis (vector: tsetse flies, *Glossina palpalis gambiensis*), leishmaniasis (vector: sandfly, *Phlebotomus* species), and dracunculiasis (vector: waterborne copepods such as *Mesocyclops leuckarti*).

Incidences of bacterial diseases that are transmitted through ticks and body lice (tick-borne relapsing fever caused by several species of spiral-shaped bacteria; tularemia, caused by *Francisella tularensis*; and louse-borne relapsing fever, caused by *Borrelia recurrentis*) are also considered to be sensitive to climate change, primarily because of the well-defined ecological conditions that support the proliferation of the vectors. Arboviral diseases also represent a major category of potentially climate-sensitive communicable diseases that can change from endemic to epidemic forms, given favorable environmental conditions. These diseases include dengue/hemorrhagic fever (caused by *Flavivirus*), Rift Valley fever (*Phlebovirus*), and Japanese encephalitis and St. Louis encephalitis (caused by viruses in the family *Flaviviridae*).

Finally, waterborne diseases that are not clearly associated with vectors have also been linked to climate change. Diarrheal diseases are at the forefront in this category that includes bacterial (for example, *Vibrio cholera*) viral (for example, Norwalk virus), and protozoan (for example, amebic dysentery) causes. Together, these diseases account for a large portion of the global burden of diseases that disable or kill children younger than five years in developing countries. The association with climate change is that in cases of drought, people tend to use contaminated sources of water, and in the absence of reliable disinfection programs, the incidence of these diseases will increase. In addition, natural disasters such as floods, hurricanes, and earthquakes can damage water supply and sewage treatment infrastructures in developed countries, leading to the contamination of potable water supplies that can increase the incidence of waterborne diseases. Hence emergency public health preparedness is a major category of planned adaptation to climate change.

Noncommunicable diseases. Heat-related diseases are the most researched category of noncommunicable diseases that have been associated with climate change. During the summer season, the frequency of extreme heat waves is predicted to increase. For example, the IPCC predicted in 2007 that the cities of Chicago and Los Angeles will experience up to 25 percent more frequent heat waves and a fourfold to eightfold increase in heat wave days by the year 2100. Based on current estimates of morbidity and mortality associated with prolonged periods of extreme heat, people with preexisting conditions such as heart problems, asthma, the elderly, the very young and the homeless will be more vulnerable. In contrast, it is also likely that under certain climate change scenarios, warmer temperatures will prevail during the winter months, leading to fewer cases of death and disability from hypothermia.

Climate change is also expected to adversely affect air quality, especially in urban areas where higher temperatures may increase the concentrations of respirable particulate matter (smaller than 2.5 micrometers) and the concentration of tropospheric ozone, which can be especially dangerous for people suffering from asthma and other chronic pulmonary diseases.

• Context

Motivating action around climate changes requires framing the issue in ways that command attention. Linking climate change to public health impacts continues to be one of the most cogent framings that have engendered research and policy questions about societal preparedness and adaptation. Ultimately, morbidity and premature mortality represent the crucial end points of most scenarios of climate change impacts. Most people are afraid of contracting diseases that were previously unknown in their communities, or for which there are no known cures. Many of the tropical diseases associated with climate change fall in these fearsome categories. However, it is also becoming increasingly clear that many of the diseases associated with climate change are preventable through well-known public health approaches, but these approaches require economic resources that may not be available to the most vulnerable populations across the world. Therefore, the roles of international organizations such as the WHO and its supporting agencies are crucial in global assessments of disease burden and future projections of climate-sensitive diseases, and in building capacity for adaptation in vulnerable societies.

Oladele A. Ogunseitan

• Further Reading

Campbell-Lendrum, Diarmid, and Rosalie Woodruff. *Climate Change: Quantifying the Health Impacts at National and Local Levels.* Environmental Burden of Disease 14. Geneva: World Health Organization, 2007. Provides comprehensive estimates of the disease burden associated with the impacts of climate change, based on WHO composite measures, including DALYs.

Intergovernmental Panel on Climate Change. *Climate Change, 2007—Impacts, Adaptation, and Vulnerability: Contribution of Working Group II to the Fourth Assessment Report of the Intergovernmental Panel on Climate Change.* Edited by Martin Parry et al. New York: Cambridge University Press, 2007. The IPCC Working Group II assesses the scientific evidence linking climate to health impacts, describes vulnerable populations, and suggests strategies for adaptation.

Ogunseitan, Oladele A. "Framing Environmental Change in Africa: Cross-Scale Institutional Constraints on Progressing from Rhetoric to Action Against Vulnerability." *Global Environmental Change* 13 (2003): 101-111. Assesses the state of knowledge about health support systems and identifies impediments to the development of policies that can protect the most vulnerable populations from climate change.

See also: Asthma; Health impacts of global warming; Skin cancer; World Health Organization.

Displaced persons and refugees

• Category: Ethics, human rights, and social justice

• Definition

Humans have always migrated in response to climatic changes, and a rapid global climate change in the twenty-first century would be no exception. In some parts of the world, climate change may make the environment inhospitable or unsuitable for human habitation, leading to exodus of the affected populations.

People who migrate for climatic reasons are sometimes referred to as "climate refugees." However, refugees, as defined by the 1951 U.N. Refugee Convention, are those who flee their home country under justified fear of persecution due to their religion, ethnicity, nationality, or social or political affiliation. Migrants fleeing the effects of climate change do not usually fit this definition. Rather, they fall into the broader category of displaced persons, those who migrate internally or internationally to escape intolerable conditions such as civil strife, economic collapse, or land degradation. Such movement is termed "forced migration."

Climate change can cause forced migration in several different ways. Perhaps the most obvious is sea-level rise, which can obliterate homes, communities, and entire countries. Climate-related natural disasters such as hurricanes and floods can also

Afghani refugees, who had been driven from their country by drought and warfare, wait to return home. If drought becomes more common, it will displace more people globally. (AP/Wide World Photos)

necessitate permanent migration. "Slow disasters," such as drought, desertification, and glacier loss, may gradually render an area uninhabitable. Finally, violent conflict may arise from climate-induced resource scarcity.

• Significance for Climate Change

International law confers refugee status only upon those who cross international borders to flee violent persecution. Climate "refugees" who are internally displaced, and those who are escaping nonviolent conditions, currently have no formal recourse to international aid. However, many countries and organizations have begun to acknowledge the need for a concerted response to climate-induced migration.

An impending problem with no legal precedent is the disappearance of an entire country due to sea-level rise, leaving the citizens of that country stateless. With projected rates of sea-level rise, several island states will lose their entire territory within the twenty-first century. One such country is Tuvalu, from which citizens have already begun evacuating. Although other countries have no legal obligation to accept Tuvaluan migrants, New Zealand has formally invited them since 2001 (at the rate of 75 per year). Australia has declined to enter into a similar agreement. Tuvalu Overview, a Japanese NGO, has sought to call attention to Tuvalu's plight—and that of other small island states—by documenting the life of each of Tuvalu's ten thousand citizens in photos and stories.

Other climate-related causes of human migration—such as floods, droughts, storms, and land degradation—are even more ambiguous in their legal implications, since these events cannot be definitively attributed to climate change. The Office of the United Nations High Commissioner for Refugees (UNHCR) has expressed concern about the impending problem of populations displaced by climate change, but its mandate does not extend to most of those populations. New legal and humanitarian arrangements may prove necessary. Since attribution of blame will often be difficult or impossible, the best solution may be voluntary aid from nations that have the resources to assist.

The number of people likely to migrate because of climate change is highly uncertain. Predictions of such migration over the next century vary from a few tens of millions of people (mainly due to the direct effect of sea-level rise) to over a billion people (due to drought, crop failure, storms, conflict, and other indirect effects). More accurate estimates are needed to help institutions prepare for the likely consequences.

Despite the great uncertainties surrounding climate-related forced migration, there is a general consensus that developing countries will be the most negatively affected. This is due to their greater climatic vulnerability, their lower adaptive capacity, and the fact that they already host most of the world's displaced persons and refugees. However, vigorous adaptation efforts may be able to prevent the most serious consequences of climate change and reduce the need for migration. There will still be instances of unavoidable migration, such as disappearance of a country below sea level, in which case strengthened institutional and legal frameworks will be needed to help resettle the affected population as promptly and equitably as possible.

Amber C. Kerr

- **Further Reading**

Black, Richard. "Forced Migration and Environmental Change: The Evidence." Chapter 2 in *Refugees, Environment, and Development.* New York: Addison Wesley Longman, 1998. Discusses the evidence for and controversies surrounding environmental causes of forced migration, including natural disasters, desertification, and sea-level rise.

Reuveny, Rafael. "Climate Change-Induced Migration and Violent Conflict." *Political Geography* 26 (2007): 656-673. Reviews thirty-eight case studies of climate-related migration, concluding that climate and other environmental factors can catalyze violent conflict, especially in the developing world.

Salehyan, Idean. "From Climate Change to Conflict? No Consensus Yet." *Journal of Peace Research* 45, no. 3 (May, 2008): 315-326. Urges pursuit of adaptation efforts to forestall potential conflicts and rejects the notion that climate change will inevitably cause political violence.

United Nations High Commission for Refugees. "Climate Change, Natural Disasters and Human Displacement: A UNHCR Perspective." Geneva, Switzerland: Author, 2008. Outlines UNHCR's role in tackling the problem of climate-induced displacement, asserting that "refugee" is often the wrong term for these displaced persons.

Unruh, Jon D., Maarten S. Krol, and Nurit Kliot, eds. *Environmental Change and Its Implications for Population Migration.* Dordrecht, the Netherlands: Kluwer Academic, 2005. Based on 2001 workshop proceedings, this book includes overview chapters and topic-specific chapters on technical, legal, and economic aspects of environmentally induced migration.

See also: Coastline changes; Islands; Sea-level change; Tuvalu.

Dolphins and porpoises

- **Category:** Animals

- **Definition**

Dolphins and porpoises are aquatic, carnivorous mammals found worldwide from polar to tropical waters. Most species are marine, with a few having freshwater populations, but only the Indian River dolphin (*Platanista gangetica*), La Plata River dolphin (*Pontoporia blainvillei*), and the Yangtze River dolphin (*Lipotes vexillifer*) are strictly relegated to freshwater habitats.

The term dolphin can be used to refer to the Odontoceti (toothed whales), but here it will be used to refer to the dolphin family Delphinidae (approximately thirty-six species), the four families of river dolphins (one species apiece), and the porpoise family Phocoenidae (six species). Taxonomic uncertainty makes the exact number of species an open question.

Dolphins and porpoises face a variety of threats, including targeted hunting, entanglement in fishing gear, pollution, disease, habitat degradation, water diversion, acoustic disturbance, and competition with fisheries. The 2008 IUCN Red Lists identify twenty-three of forty-eight species and subspecies of dolphins and porpoises as threatened, and extensive surveys suggest that the Yangtze River dolphin is almost extinct.

• Significance for Climate Change

Global warming is expected to induce biogeographic range shifts in dolphins and porpoises, as water temperature is often a good predictor of the presence of these species. For example, off the coast of Scotland an increase in the number of warm-water species since the 1980's was coincident with an increase in water temperatures over the same period, while the abundance of cold-water species diminished. Around the British Isles in general, the white-beaked dolphin (*Lagenorhynchus albirostris*) prefers colder waters, and the short-beaked common dolphin (*Delphinus delphis*) prefers warmer waters. If these waters warm further, the short-beaked dolphin could expand its range at the expense of the white-beaked dolphin.

On global scales, the diversity of whales (including dolphins and porpoises) in deep water is maximum at sea surface temperatures of 21° Celsius, which occur commonly at midlatitudes. When applied to warming scenarios for the next few decades from the Intergovernmental Panel on Climate Change, these results suggest that biogeographic ranges would shift so that the diversity of whales in the tropics would decline while the diversity in higher latitudes would increase.

Climate change could induce range shifts in dolphins and porpoises by affecting species lower in the food web and thereby making preferred food items less abundant. This could also change competitive relationships between species. For instance, the 1982-1983 El Niño-Southern Oscillation event caused a reduction in squid off of California and coincided with the displacement of short-finned pilot whales (*Globicephala macrorhynchus*) by Risso's dolphins (*Grampus griseus*), which are able to dive more deeply to obtain scarce prey.

Climate change might also fragment dolphin and porpoise populations, which has implications for genetic diversity and generally reduces population persistence. For instance, water diversions for shipping and irrigation have already fragmented the populations of the three freshwater river dolphins. If precipitation patterns change, so too could the demand for water withdrawals and impoundments, meaning that the ranges of these species could become further fragmented. Climate change could also fragment marine populations. For example, warming could enhance the degree to which the warm-water current flowing through the Bay of Biscay divides northern and southern

Three dolphins await a postmortem examination after becoming stranded and dying in a river creek in Cornwall, England. (Barry Batchelor/PA Photos/Landov)

populations of the harbor porpoise (*Phocoena phocoena*).

Ocean acidification, a result of carbon dioxide (CO_2) dissolving into marine waters, could affect the food supply of dolphins and porpoises. Dolphins and porpoises are often the top predators in food webs based on organisms that have calcium carbonate ($CaCO_3$) body structures, which are susceptible to dissolution in acidified waters. For example, the pantropical spotted dolphin (*Stenella attenuata*) consumes pteropods, which are known to be sensitive to acidification. Moreover, many species of fish that are prey to dolphins and porpoises are dependent on coral reef habitats, which are susceptible to warming-induced bleaching and decalcification from acidification.

Another consequence of ocean acidification is that acidified water absorbs sound less effectively, making the environment "noisier." Dolphins and porpoises depend upon echolocation to locate prey, discern their environment, and locate one another. Many species already suffer from acoustic disturbances caused by boat traffic, seismic exploration, and military exercises which can propagate tens of kilometers underwater, and such disturbances are associated with behavioral changes, strandings, internal injury, and death. Deep water and waters at high latitudes are expected to be especially affected since the acidification reaction is temperature-dependent.

Global warming could exacerbate the threats of disease and pollution. Warmer temperatures and shifting currents are expected to extend the range of disease-causing agents. For example, warmer temperatures are associated with outbreaks of the bacterium *Vibrio* known to infect dolphins. Climate change might also influence the frequency and severity of harmful algal blooms.

Adam B. Smith

• **Further Reading**

Burns, William C. G., and Alexander Gillespie, eds. *The Future of Cetaceans in a Changing World.* Ardsley, N.Y.: Transnational, 2003. Collection of twelve essays on current and future threats to whales and on international organizations and conventions designed to manage and ameliorate these threats.

Culik, Boris M. *Review of Small Cetaceans: Distribution, Behavior, Migration, and Threats.* Bonn, Germany: United Nations Environment Programme/Convention on Migratory Species Secretariat, 2004. Review, illustrations, and range maps of almost all toothed whales, including dolphins and porpoises. Each species is described in an essay detailing its conservation status, distribution, and behavior.

Reeves, Randall R., et al., comps. *Dolphins, Whales, and Porpoises: 2002-2010 Conservation Action Plan for the World's Cetaceans.* Gland, Switzerland: International Union for Conservation of Nature, 2003. A detailed account of conservation actions and research required to address threats to many species of small whales.

Soulé, Michael E., Elliot A. Norse, and Larry B. Crowder. *Marine Conservation Biology: The Science of Maintaining the Sea's Biodiversity.* Washington, D.C.: Island Press, 2005. Twenty-five essays on conservation of marine species, including issues germane to small cetaceans, such as pollution, algal blooms, fisheries, and synergy of threats.

See also: El Niño-Southern Oscillation; Maritime climate; Ocean acidification; Ocean life; Whales.

Downscaling

• **Category:** Science and technology

• **Definition**

Downscaling is a technique for applying information derived from larger-scale models or data analyses to smaller-scale models. In the context of global warming research, downscaling is used to link large-scale atmospheric variables with local climates. Global climate models, also known as general circulation models (GCMs), are data-rich, three-dimensional computer models of the climate mapped on a grid. GCMs employ variable atmospheric data to make estimates of future climate change on a global, hemispheric, or continental scale. Their resolution is generally about 150 to 300 kilometers by 150 to 300 kilometers. However, smaller-scale, local models require data at scales of 10 to 100 kilometers. To translate the larger-scale information to a regional scale requires downscaling into higher spatial and temporal resolution grids. With this finer resolution, the effect of subgrid topographical features such as clouds, mountain ranges, and wetlands can be incorporated. There are two main methods of downscaling: dynamical downscaling and statistical, or empirical, downscaling.

• Significance for Climate Change

Downscaling is an important tool for the study of anthropogenic climate change caused by increased carbon dioxide (CO_2) emissions. GCMs allow for predictions of large-scale climate change based on assumed patterns of greenhouse gas (GHG) emissions and other impacts. However, this scale can lack accuracy for regional and sub-regional models where differences in climate occur at a scale below GCM resolutions. Downscaling is used to make more accurate predictions at this local-to-regional scale.

Downscaling has been used in atmospheric forecasting for several decades. In dynamical downscaling, limited area models (LAMs) and regional climate models (RCMs) are nested into GCMs using GCM data as boundary conditions. In statistical downscaling, statistical relationships are calculated between factors simulated by the GCM at the large-scale level and variable data measured at the local level, such as rainfall occurrences and surface air temperatures. Dynamical techniques produce more data but are computationally expensive and depend on the accuracy of the GCM grid-point data used to calculate boundary conditions.

Dynamical downscaling has been used, for example, to calculate the number of extreme temperature days per year, as well as changes in their distribution. Such data can then be used to calculate the increased drought risk in dry climate areas, such as prevail in Australia and New Zealand. Statistical downscaling has been used, for example, to chart projected temperature ranges in various regions through the remainder of the twenty-first century, as well as mean monthly temperatures and precipitation. A combination of dynamical and statistical methods will probably obtain the best results in the future, allowing local planners to better adapt strategies and policies for coping with climate changes in their regions.

Howard Bromberg

See also: Bayesian method; Climate models and modeling; Climate prediction and projection; General circulation models; Parameterization.

Drought

• **Categories:** Meteorology and atmospheric sciences; environmentalism, conservation, and ecosystems

Many parts of the world are expected to suffer more frequent, more widespread, and more severe droughts as a result of climate change. Drought will likely be one of the most significant impacts of climate change on ecosystems and human society, but adaptation can lessen its effects.

• Key concepts

agricultural drought: lack of sufficient soil moisture for crop growth, leading to partial or total loss of yield

El Niño-Southern Oscillation (ENSO): a periodic fluctuation in Pacific sea surface temperatures, causing an alternating pattern of above- and below-average precipitation on either side of the Pacific

hydrologic drought: significantly below average water levels in lakes, rivers, reservoirs, and aquifers

megadrought: a drought lasting for years, decades, or centuries, often leading to permanent changes in ecosystems and human societies

meteorological drought: prolonged deficiency of precipitation in an area, as compared to the historical average

socioeconomic drought: a drought severe enough to cause disruption to human societies and economies

• Background

A drought is a period when the water supply of a given area is insufficient for the needs of humans or of ecosystems, usually resulting from below-normal precipitation. If defined simply as a precipitation deficit, such an event is called a meteorological drought. If a drought depletes soil moisture and harms crops, it is an agricultural drought. Eventually, lack of precipitation will diminish water levels in lakes, rivers, and aquifers; this is a hydrological drought. If a drought affects human well-being, it can be called a socioeconomic drought. These categories overlap, and there are many other ways to define droughts based on their causes or effects.

• Causes and Effects of Drought

Precipitation patterns result from large-scale atmospheric and oceanic processes, which in turn are affected by solar forcing, atmospheric composition, land-surface characteristics, and other factors. Scientific understanding of these processes is imperfect but improving; for example, in the 1980's, it was discovered that a cyclical variation in Pacific sea surface temperature called the El Niño-Southern Oscillation (ENSO) has a profound effect on rainfall around the world. Paleoclimate records show that short- and long-term rainfall fluctuations are common and predate human history, though the causes are often unclear.

Another important cause of drought is above-average temperatures, which increase evaporation and hasten loss of soil moisture. Warm temperatures can also cause precipitation to fall as rain instead of snow, causing it to run off rather than being stored and gradually released. Human land use, such as deforestation or poor soil manage-

A Time Line of Historic Droughts

1270-1350	SOUTHWEST: A prolonged drought destroys Anasazi Indian culture.
1585-1587	NEW ENGLAND: A severe drought destroys the Roanoke colonies of English settlers in Virginia.
1887-1896	GREAT PLAINS: Droughts drive out many early settlers.
1899	INDIA: The lack of monsoons results in many deaths.
1910-1915	SAHEL, AFRICA: First in a series of recurring droughts.
1933-1936	GREAT PLAINS: Extensive droughts in the southern Great Plains destroy many farms and create the Dust Bowl during the worst U.S. drought in more than 300 years.
1968-1974	SAHEL, AFRICA: Intense period of drought; 22 million affected in four countries, 200,000-500,000 estimated dead, millions of livestock lost.
1977-1978	WESTERN UNITED STATES: Severe drought compromises agriculture.
1981-1986	AFRICA: Drought in 22 countries, including Angola, Botswana, Burkina Faso, Chad, Ethiopia, Kenya, Mali, Mauritania, Mozambique, Namibia, Niger, Somalia, South Africa, Sudan, Zambia, and Zimbabwe, results in 120 million people in 22 countries affected, several million forced to migrate, significant loss of life and of livestock.
1986-1988	MIDWEST: Many farmers are driven out of business by a drought.
1986-1992	SOUTHERN CALIFORNIA: Drought brings increased water prices, loss of water for agricultural production, water rationing.
1998	MIDWEST: Drought destroys crops in the southern part of the Midwest.
1999	LARGE PART OF UNITED STATES: Major drought strikes the Southeast, the Atlantic coast, and New England; billions of dollars in damage to crops.
Beg. 2006	WESTERN UNITED STATES: Drought-induced wildfires cause significant property and environmental damage in both urban and rural areas.

ment, can contribute to drought. Droughts may have multiple causes; for example, the 1930's Dust Bowl event in the U.S. Great Plains was a result of both meteorological drought and erosion-prone farming techniques.

Drought has both direct and indirect effects. The direct effects most relevant to human well-being include reduced water supply (both in quantity and quality); crop failure, especially of rainfed crops; loss of livestock; soil degradation, such as through dust storms and desertification; death of forests, often resulting in wildfires; and electricity shortages (if hydropower is used). These effects, in turn, can have profound consequences for human society, including mortality (due to starvation, thirst, fire, or dust); large-scale migration, often to urban centers; economic depression and entrenchment of poverty; and, under very adverse conditions, violent conflict. Ecosystems can usually survive and may even benefit from periods of drought; however, an especially severe drought may cause biodiversity loss and long-term disruption of ecosystem processes.

This bridge in northern Spain, shown in the midst of a severe drought in October, 2005, is normally submerged in the Pisuerga River. (AP/Wide World Photos)

• Effects of Climate Change on Drought Patterns

There is general agreement that climate change is likely to worsen droughts worldwide, but the location, timing, and magnitude of these effects are highly uncertain. Averaged globally, climate change will increase precipitation by speeding ocean evaporation. However, rising temperatures will affect oceanic and atmospheric circulation patterns, changing the global distribution of rainfall. Some regions will become wetter; others, drier. Even in regions where total precipitation does not decrease, drought risk may increase as a result of faster land-surface evaporation, loss of water stored in snow and ice, and greater variability of precipitation events (causing not only droughts but also floods).

Climate models generally agree that several regions of the world are likely to suffer from decreased precipitation and increased drought risk under future climate. These regions include the Mediterranean (southern Europe and northern Africa); southern Africa, especially its southwest corner; the southwestern United States; parts of Central America; and southern Australia. Semiarid and arid regions will probably be hardest hit by drought, with subsistence farmers in developing countries being especially vulnerable. In addition, climate change may cause water shortages wherever populations depend upon glaciers or snowmelt for their water supply, such as the Andes in South America, the Himalayas in central Asia, and the Sierra in California.

• Drought Monitoring, Response, and Adaptation

Close monitoring of atmospheric conditions can reveal an impending drought before its agricultural or socioeconomic symptoms become severe, allowing protective measures to be taken. This is the basis of drought early-warning systems, for which there are several regional and global networks. Once a drought has begun, relief efforts may be necessary. Such efforts may include allocation of water to communities and farms, cash payments to farmers, and relocation assistance. However, the availability of relief can ultimately increase vulnerability to drought by encouraging inappropriate settlement and farming practices.

Although drought cannot be prevented, its effects can be minimized through adaptation. Examples of drought adaptation include selecting drought-tolerant crops; improving agricultural soil management; reducing the size of livestock herds; controlling fuel loads in forests; building dams and reservoirs; and encouraging water conservation. A society can insulate itself against drought (or any hazard) by reducing poverty, diversifying its economy, and building strong social institutions.

• Context

Humans have always needed to cope with drought; it has been a cause of hardship, conflict, and migration throughout recorded history. Megadroughts may have led to the disappearance of some civilizations, such as the Maya. More than 10 million people are thought to have been killed by drought during the twentieth century. Climate change will likely not only increase the incidence of drought but add other stresses to human society as well.

Drought is a natural phenomenon, and a given drought cannot be attributed definitively to climate change. Furthermore, there is much uncertainty about when and where climate change will increase drought incidence. This does not, however, imply that adaptation efforts should be delayed. Many drought-adaptation actions will have benefits in the present as well as the future.

Amber C. Kerr

• Further Reading

Boken, Vijendra K., Arthur P. Cracknell, and Ronald L. Heathcote, eds. *Monitoring and Predicting Agricultural Drought: A Global Study.* New York: Oxford University Press, 2005. Technical volume discussing drought monitoring and its use in different world regions. Includes a chapter on climate change. Figures, tables, maps, index.

Botterill, Linda C., and Melanie Fisher, eds. *Beyond Drought: People, Policy, and Perspectives.* Collingwood, Vic.: Commonwealth Scientific and Industrial Research Organisation, 2003. Evaluates the social, cultural, political, and economic aspects of drought and drought relief. Focuses on Australia. Figures, tables, maps, index.

Cooley, Heather. "Floods and Droughts." Chapter 4 in *The World's Water, 2006-2007: The Biennial Report on Freshwater Resources,* edited by Peter Gleick. Washington, D.C.: Island Press, 2006. Discusses causes and effects of drought; provides statistics on historical droughts; distinguishes between drought management, mitigation, and response. Figures, tables.

Intergovernmental Panel on Climate Change. *Climate Change, 2007—The Physical Science Basis: Contribution of Working Group I to the Fourth Assessment Report of the Intergovernmental Panel on Climate Change.* Edited by Susan Solomon et al. New York: Cambridge University Press, 2007. Forecasts weather and climate, including precipitation and drought, for each region of the world; provides an integrated overview of the most likely severe effects worldwide. Figures, maps, tables.

Kallis, Giorgos. "Drought." *Annual Review of Environment and Resources* 33 (2008): 85-118. Concise review of all aspects of drought. Provides definitions and metrics; discusses natural science and social science perspectives; emphasizes climate change. Tables, glossary.

Trenberth, K. E., J. T. Overpeck, and S. Solomon. "Exploring Drought and Its Implications for the Future." *Eos* 85, no. 3 (January 20, 2004): 27. This brief article reviews evidence that severe droughts have increased in the past century, compares contemporary climates to the paleoclimate record, and discusses effects of climate change on drought severity.

See also: Desertification; Floods and flooding; Freshwater; Glaciers; Hydrosphere; Rainfall patterns; Sahel drought; Water resources, global; Water resources, North American.

Dust storms

• **Category:** Meteorology and atmospheric sciences

• **Definition**

Dust storms are meteorological events in which visibility is reduced to 1 kilometer or less as a result of blowing dust (defined as material less than 63 microns in size—the size of silt and clay). In many areas, wind conditions conducive to dust storms are associated with the passage of frontal systems (especially cold fronts) and downdrafts from convectional storms. Elsewhere, strengthened pressure gradients give rise to sustained high wind speeds capable of raising dust. Examples include the Shamal and Hundred Days winds in the Arabian Gulf and Seistan regions of Iran and Afghanistan and the Santa Ana winds of Southern California.

The dust emission process involves both the horizontal transport of coarse, sand-sized material and the vertical flux of fine, silt- or clay-sized particles. Coarse particles abrade fine materials (as, for example, on the surfaces of dry lake beds) and eject the fine particles into the air, where atmospheric turbulence transports them in suspension, often for many tens or hundreds of kilometers. Plumes of dust from the Sahara Desert regularly reach southern Europe, and they occasionally cross the Atlantic Ocean to reach the Caribbean Sea and South America. Dust plumes from China reach Japan on a regular basis and occasionally cross the Pacific Ocean to North America.

• **Significance for Climate Change**

Dust emitted and transported from deserts represents a major linkage between deserts and other environments. Dust deposition has important effects on ocean and terrestrial productivity by contribut-

A massive cloud of dust blows over Griffith, Australia, in November, 2002. (AP/Wide World Photos)

ing nutrients to ecosystems (especially iron, which is limiting to marine productivity). Air quality is affected by particulate loading. Atmospheric radiative properties are changed through the scattering and absorption of solar radiation by mineral aerosols. Such processes may result in either warming or cooling effects, depending on the size and composition of the dust particles. Dust is trapped by vegetation and adds to soils on desert margins and elsewhere. Saharan dust is a major component of soils throughout the Mediterranean Basin and even as far afield as the Caribbean. Human health is affected by high dust concentrations, which can lead to respiratory disease. Changes in the dust loading of the atmosphere have also been linked to rainfall changes in areas adjacent to the Sahara and to the intensification of drought conditions.

Most dust source areas are located in arid climate zones and are associated with topographic lows, flat surfaces, finely textured soils, and sparse or no vegetation cover. Such areas have been identified by orbital sensors, particularly the Total Ozone Mapping Spectrometer Aerosol Index (TOMS AI), as well as ground observations. The majority of mineral dust is derived from natural surfaces; the contribution of human activities (such as agriculture) to the dust loading of the atmosphere is uncertain; global estimates vary from anywhere between 50 percent to less than 10 percent. Major dust source areas include the Bodelé Depression in Tchad, the Aral Sea area, southeast Iran, the Taklamakan Desert of China, Inner Mongolia, and the loess plateau of China. The Sahara region is believed to be the largest single dust source, accounting for as much as 690 million metric tons of dust per year. Satellite data show that three of the world's most important dust source areas lie in this region.

The interannual frequency and magnitude of dust storms is strongly linked to climatic variability: In many areas, dust emissions are inversely correlated to rainfall, although in a complex nonlinear manner, in part because a supply of fine sediment is required for dust emission. For example, periods of increased rainfall cause runoff that may flood playas and contribute sediment that is mobilized by wind in subsequent dry periods. In this way, dust emissions appear to correlate with El Niño-Southern Oscillation (ENSO) cycles in the southwestern United States. Elsewhere, there are strong anti-correlations between the flux of dust from the Sahara and rainfall.

On a longer time scale, important millennial and centennial variations in global and regional dust loadings are recorded by dust deposits in marine sediments and ice cores. For example, dust flux from the Sahara and many other desert regions (such as Australia) was high during the Last Glacial Maximum, as a result of increased aridity and wind strength at this time. Dust flux also increased sharply following the desiccation of the Sahara some five thousand years ago.

Future changes in dust emissions are therefore likely to be influenced by climate change, most notably as the effect of vegetation-cover change, with human activities affecting conditions locally. The direction of change in dust emissions, as predicted by different models, is, however, uncertain. In one model, Saharan dust emissions are predicted to increase by 11 percent as a result of higher wind speeds. Other simulations predict a decrease of 4 percent as a result of increased monsoonal rainfall and vegetation cover.

Nicholas Lancaster

• Further Reading

Christopher, Sundar A., Donna V. Vulcan, and Ronald M. Welch. *Radiative Effects of Aerosols Generated from Biomass Burning, Dust Storms, and Forest Fires.* Springfield, Va.: National Technical Information Service, 1996. The contribution of dust storms to radiative forcing is compared to those of natural and anthropogenic conflagrations.

Engelstaedter, S., I. Tegen, and R. Washington. "North African Dust Emissions and Transport." *Earth Science Reviews* 79 (2006): 73-100. Review of the processes of dust emissions and transport for the Saharan region.

Goudie, A. S., and N. J. Middleton. *Desert Dust in the Global System.* New York: Springer, 2006. Comprehensive overview of dust processes and their relation to environmental conditions.

See also: Aerosols; Desertification; Deserts; Sahara Desert; Sahel drought; Soil erosion; Sulfate aerosols.

Earth history

• **Categories:** Geology and geography; astronomy

The key to understanding present-day climate change lies in interpreting the geologic past. Present in the rocks of the Earth's crust is a continuous record of change. Scientists use the evidence found in those rocks to help understand current Earth processes.

• Key concepts

accretion: the final process of planetary formation, in which smaller objects are pulled into the nascent planet

continents: large sections of an earthlike planet's surface with high elevations and lower-density rocks

core: the central part of a planetary body, usually composed of high-density metal alloys

crust: the outer portion of an earthlike planet, consisting of lower-density silicate minerals and rocks

magnetic field: the flow patterns of magnetism that surround a celestial body as a result of an electrodynamo effect

mantle: the middle section of an earthlike planet, consisting of higher-density silicate minerals and rocks

ocean basins: large sections of an earthlike planet's surface with low elevations and higher-density rocks

planetary nebula: the gas and dust that is required to form a star and its family of planets

plate tectonics: the mechanisms responsible for creating movement in large sections of a planet's surface

radiometric dating: an analytical technique using radioactive isotopes to determine the age of rocks and minerals

• Background

The Earth is one of eight major planets and a host of minor planets that circle a fairly average, middle-aged, main-sequence star. It formed from the accretion of residual material from the gravitational collapse of the solar nebula that produced the Sun. Based on their specific distances from the Sun, the planets formed into groups of similar chemical compositions, sizes, and densities. The four smaller inner planets are of higher density and are composed of a mixture of rock and metal. In contrast, the next four planets are massive gas giants with lower densities and larger sizes. The final grouping consists of small, low-density, icy worlds. Unlike the other inner planets, the Earth has remained a geologically active planet and has undergone continuous change since its formation. It is a planet dominated by the presence of liquid water. Throughout its long history, the Earth has evolved from a lifeless world into one that is populated by an uncountable number of species.

• Early History of the Earth

Based on radiometric dating analyses of moon rocks and meteorites, the Earth is believed to be about 4.5 billion years old. Many scientists believe that the Earth originated as a cold, undifferentiated body that internally heated up from the energy released through giant impacts, radioactive isotope decay, and the mass of the Earth itself. Once the appropriate melting temperatures were reached, heavy metallic elements sank to the planet's center of gravity, as lighter elements were displaced upward toward the surface. This process may have taken place in as little as 50 million years. It is fairly certain that the present core-mantle-crust structure was in place by 4 billion years ago.

Earth's core consists of both solid and liquid metal, presumably of nickel-iron composition. The outer, liquid metallic core revolves around the inner, solid metallic core and, in the process, generates an electric current. This electric current is responsible for the Earth's magnetic field. Sandwiched between the core and the crust is the mantle, a region of high-density iron- and magnesium-rich silicate rock material. Depending upon specific temperatures and pressures, this material can behave either as a solid or as a liquid.

Scientists closely link the history of the Earth to that of the Moon. The Earth and Moon are geologically similar in many ways, but there are significant differences in their respective elemental abundances, and lunar specimens exhibit an apparent lack of certain volatiles. One origin theory suggests

In a graphic illustration of Earth's geologic history, the Creteceous-Tertiary boundary is clearly visible in this rock formation near Drumheller, Alberta. The transition from light to dark rock marks the point at which dinosaurs became extinct, probably as a result of abrupt climate change. (G. Larson)

that the Moon is the product of a huge collision between a Mars-sized object and the primordial Earth. The resulting debris from this impact later accreted and formed the Moon. This theory mainly draws its support from computer impact simulations and a great mnany assumptions, but it lacks the physical evidence necessary to support it.

Finding a common theory of origin for the Earth and Moon has proven to be quite difficult. In fact, when comparing size, density, and internal structure, the Moon seems to have more in common with Mars than it does with the Earth. What is certain is that the Moon has a definite influence on the Earth's environment. The Moon's daily tidal effects on the world's oceans are obvious, but the Moon also has a gravitational influence on the Earth's axial rotation. Without the gravitational pull of the Moon, the Earth's axial tilt would fall outside of its normal range of between 21° and 25° and literally fall over. Without the Moon "holding the Earth in place," the equator of today could easily become the polar regions of tomorrow.

• Evolution of the Atmosphere and Oceans

The primitive Earth did not have the same atmosphere that it does today. Many scientists believe that the Earth's original atmosphere may have been a very dense, hot mixture of ammonia and methane. This is consistent with the conditions present during Earth's proto-planetary stage that probably dissipated during the Sun's T-Tauri phase. The first "permanent" atmosphere formed as a product of volcanic degassing. Through this process, carbon dioxide (CO_2) gradually became the dominant gas, along with a significant amount of water vapor. As the Earth cooled, water vapor condensed and fell as rain. This rain later filled the lowlands and created the first primitive oceans. Chemical reactions occurring in seawater slowly began to extract CO_2 from the atmosphere and form large amounts of carbonate rocks. Somehow, during the first 500 million years of Earth history, life crossed over the threshold between being a collection of complex organic molecules and living organisms. With life-forms such as blue-green algae at work, the CO_2-rich atmosphere slowly transformed into a nitrogen and oxygen-rich atmosphere that could support animal life. This transition is clearly marked in the Earth's geological record when previously dark, iron-bearing sediments turned red from oxidation. It is shortly after this geological benchmark that the first marine animals appear in the fossil record.

• Supercontinents and Continental Drift

The formation of supercontinents and continental drift is essentially tied to the internal mechanisms of the Earth's upper mantle. There, convection cells provide the energy necessary to split apart the crust into both large and small sections that can move relative to one another. The direct evidence for the effects of this convective energy is the large number of volcanoes and earthquakes that occur along plate boundaries. No one is certain how long plate tectonics has been a part of Earth history, but it is certainly responsible for the continent-ocean basin relationship, which forms the present crust.

In the geologic past, supercontinents have existed only to be broken apart and distributed across the face of the Earth. This movement is continuous, and the formation of supercontinents seems to be inevitable, as is their eventual breakup. The formation of the world's great mountain chains is the direct result of colliding continents transforming marine sediment into hard rock. New crustal rock is created by volcanic activity at mid-oceanic ridges that pushes plates apart from one another. Such plate movement can also carry older, crustal rock to its destruction in an oceanic trench or be welded together into a new continental mass.

• The History of Life

There is strong evidence to support the hypothesis that life has existed on Earth for more than 4 billion years. Initially, single-celled life-forms dominated the planet, and they gradually evolved into more complex forms. Scientists theorize that life could have originated on Earth in two possible ways: either as an indigenous form, created from the organic compounds and conditions present in the primordial Earth, or from organic compounds or even living bacteria that were transported to Earth by comets or meteorites. This later "panspermia" theory suggests that organic compounds and living bacteria may have come to Earth from space and served as the seeds for life.

Regardless of its origin, life has flourished on Earth for billions of years and has adapted to an ever-changing variety of environmental conditions. For over 2.5 billion years, cyanobacteria were the dominant life-forms and were responsible for the gradual buildup of an oxygen-rich atmosphere. In the Cambrian geological period, there was an explosion in the diversity of marine animal life. Various species seem to have come and gone as if they were the products of some biological experiment to see which could survive best. The survivors continued to evolve into more complex organisms that gradually found their way to land.

It seems that the end of one species's dominance and the beginning of another's is usually marked by a dramatic change in global climate. The best evidence to support such a theory occurs at the Cretaceous-Tertiary boundary, approximately 65 to 70 million years ago. An iridium-rich layer of sediment, attributed to the impact of an asteroid, marks the boundary between the two geological periods. Dinosaur fossils are found below this layer but are notably missing from the layers above. Scientists interpret this as evidence for global climate change that led to the mass extinction of a large number of species. An event such as this is not limited to the Cretaceous-Tertiary periods but may also be responsible for several other mass extinctions.

• Ice Ages and Interglacial Periods

There are many scientific theories that suggest the Earth is usually a cold planet covered by vast amounts of ice. Periodically, warm periods emerge that last for several thousand years and eventually phase back into long ice ages. It is during these warm periods that land animals flourish and perhaps even speed up their evolutionary processes. There are no certainties as to when ice ages begin or end. Global climate change is the underlying reason, but what initiates this change? Scientists have suggested such possibilities as changes in the Earth's axial tilt, increased volcanic activity blocking incoming sunlight, shifts in the world's ocean currents, impact debris from collisions with comets or asteroids, and the effects of human pollution. No one is certain which of these possibilities holds the answer, and the truth is probably found in some combination of factors. What is certain is that ice ages and their interglacial warm periods are features inherent to planet Earth, and they will continue to occur with or without human influence.

• Context

Astronomers have presented evidence to support the existence of hundreds of other planets orbiting distant and even exotic stars. These planets range from massive gas giants to a few with nearly earthlike masses. All evidence tends to indicate, however, that Earth is a very unusual planet. It has evolved from a hot gaseous world to one that is dominated by liquid water. Its atmosphere changed from one that was poisonous to animal life to one that is oxygen-rich and supports uncountable species of life, both on land and in the oceans. In the geological past, the processes of plate tectonics may have been beneficial to the evolution of life, while at other times they may have caused mass extinctions. Perhaps the one constant throughout Earth's history has been that change is inevitable. Present-day humans are the result of the ever-changing conditions of the Earth's surface. Adaptation to climatic changes means survival for some species, while others that cannot change face extinction. Unique to this period in Earth history is the fact that a particular species, human beings, possesses the ability to have either a positive or a negative effect on the environment. One path may lead to a better world, while the other may lead to extinction.

Paul P. Sipiera

• Further Reading

Canup, R. M., and K. Righter, eds. *Origin of the Earth and Moon.* Tucson: University of Arizona Press, 2000. A compilation of twenty-nine scientific papers dealing with terrestrial planetary formation with emphasis on the Earth-Moon system. Excellent reference source for undergraduate and graduate students.

Hartmann, William K. *Moons and Planets.* 5th ed. New York: Brooks/Cole-Thomson Learning, 2005. A comprehensive examination of the solar system from the planetary geologist's perspective. Suitable for advanced high school and undergraduate students.

McSween, Harry Y., Jr. *Meteorites and Their Parent Planets.* 2d ed. New York: Cambridge University Press, 1999. A concise yet comprehensive work on the importance of meteorite studies to understanding planetary formation. Chapter 9 is particularly relevant to Earth history.

Russell, Michael. "First Life" *American Scientist* 94, no. 1 (2006): 32-39. A detailed and well-illustrated description of the emergence of life on the early Earth. Brings together all of Earth's dynamic processes as vital contributors to the origin of life. Suitable for a general readership and science students alike.

Schubert, Gerald, Donald L. Turcotte, and Peter Olsen. *Mantle Convection in the Earth and Planets.* New York: Cambridge University Press, 2001. An essential reference for the serious student of Earth history. Clearly explains how tectonic mechanisms in the Earth's upper mantle affect all Earth processes, including the evolution of life.

See also: Atmospheric structure and evolution; Climate reconstruction; Dating methods; Deglaciation; Earth motions; Earth structure and development; Earthquakes; Global energy balance; Paleoclimates and paleoclimate change; Planetary atmospheres and evolution; Plate tectonics; Sun.

Earth motions

- **Category:** Astronomy

- **Definition**

Earth is not fixed in space. Rather, Earth orbits the Sun in a nearly circular orbit once every 365.26 days, a period of time called the sidereal year. As Earth orbits the Sun, it rotates once every 23 hours 56 minutes about a rotational axis that is tilted at approximately 23.5° with respect to its orbital motion. This axial tilt is called obliquity, and Earth's daily rotation is called diurnal motion.

Earth's orbit is nearly circular, but it is not a perfect circle. The measure of how far an orbit deviates from being circular is called its eccentricity. Earth's orbital eccentricity is about 0.017, meaning that its orbital distance from the Sun deviates by 1.7 percent from its average distance of 149,597,871 kilometers. Interactions between the Sun, Earth, and other planets can cause the Earth's eccentricity to change somewhat. The entire orbit shifts a little each time around, causing the point in the orbit where the Earth is closest to the Sun to shift a little each year. This motion is called precession.

Though the Earth's orbit is slightly elliptical, it is the planet's obliquity that is responsible for the seasons. That obliquity, however, is not constant. The tilt, or inclination, varies somewhat over time, increasing and decreasing. This change in inclination is called nutation. There are many causes for nutation, with the largest being tidal interactions between the Earth, Sun, and Moon. These interactions create an 18.6-year nutation cycle, shifting Earth's obliquity by up to 9 arcseconds. Other causes of nutation also exist, with periods ranging from days to thousands of years. Furthermore, Earth's rotational axis, too, gradually shifts the direction of its obliquity. This shift, also called precession, takes about twenty-six thousand years to go through a complete circle. The combination of the precession of the orbit and the precession of the poles causes the seasons to repeat every 365.24 days, a period of time known as the tropical year, instead of repeating in a sidereal year.

- **Significance for Climate Change**

Earth's climate is governed by the interaction between the Earth and the Sun. Sunlight warms the Earth, and Earth radiates heat into space. Anything that affects this process can have an effect on Earth's climate. Thus, some long-term climate change can be attributed to the motion of the Earth as a planet.

Changes in Earth's obliquity can affect seasons by changing the angle of the sunlight reaching

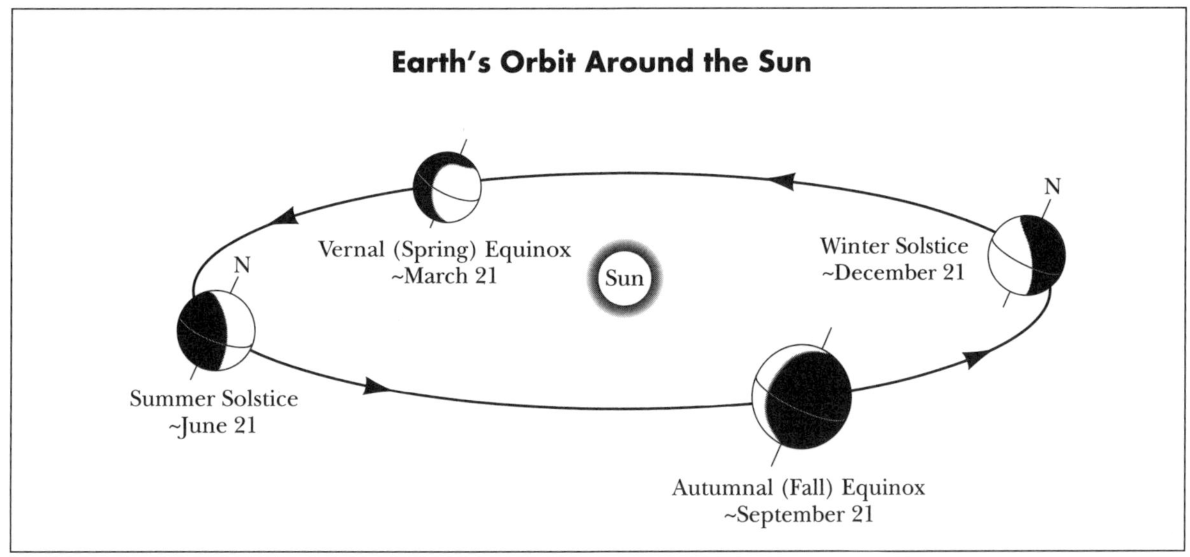

Earth's surface. However, these changes in obliquity change only the distribution of solar energy reaching Earth, not its total amount. As the obliquity decreases, the difference between summer and winter decreases. Conversely, as obliquity increases, the difference between summer and winter becomes more pronounced. This represents a change in climate, but it does not directly result in global warming or cooling.

At present, about two weeks before Earth reaches its aphelion, its greatest distance from the Sun, Earth's Northern Hemisphere is tilted most toward the Sun, creating summer in the Northern Hemisphere and winter in the Southern Hemisphere. Likewise, Earth's Southern Hemisphere is pointed most toward the Sun, creating summer in the Southern Hemisphere and winter in the Northern Hemisphere, about two weeks before Earth reaches its perihelion, its closest distance to the Sun. Because of this, the Southern Hemisphere receives more intense solar radiation during its summer and less intense solar radiation during its winter than does the Northern Hemisphere.

Normally, this difference in radiation intensity would be expected to produce more extreme seasons in the Southern Hemisphere than in the Northern Hemisphere. However, the Northern Hemisphere has a larger landmass than does the Southern Hemisphere. The large bodies of water in the Southern Hemisphere act, in part, as a heat sink to moderate the effects of the differences in solar radiation. Over time, though, the difference between the sidereal year and the tropical year causes seasons to shift the point along Earth's orbit at which they occur. This process is called the precession of the equinoxes. Eventually, the seasons will reverse, with the Northern Hemisphere experiencing the more intense differences in summer and winter solar radiation. Without the buffering effect of the large bodies of water of the Southern Hemisphere, this shift will result in more extreme seasons. These changes, though, do not result in a change in the total solar radiation incident upon Earth, only in how that radiation is distributed.

Eccentricity changes, however, can change the total solar energy incident upon Earth, and such changes can have a dramatic impact on Earth's climate. Furthermore, global warming or cooling can result in a decrease or increase in ice on the planet. Ice and snow reflect sunlight rather than absorbing it, amplifying the effects of orbital changes. Warming results in less ice, causing more solar energy to be absorbed by the planet's surface and warming Earth more. Cooling results in more ice and less solar energy absorbed, cooling Earth more. All of these effects, though, are expected to occur on fairly long timescales compared with human experience.

Raymond D. Benge, Jr.

• Further Reading

Drake, Frances. *Global Warming: The Science of Climate Change.* London: Arnold, 2000. Excellent introduction to the principles of climatology and climate change. Contains a good description of the role of the planetary motions of the Earth and how they affect climate. Each chapter is well referenced.

Leroux, Marcel. *Global Warming: Myth or Reality? The Erring Ways of Climatology.* Chichester, West Sussex, England: Praxis, 2005. Good description of both natural and anthropogenic climate change, with many references. Some discussion is given to the role of changes in Earth's motions in producing climate change.

Mathez, Edmond A., and James D. Webster. *The Earth Machine: The Science of a Dynamic Planet.* New York: Columbia University Press, 2004. Introduction to Earth as a planet, with a good description of the motions of the Earth. Three chapters are dedicated to climate and climate change.

See also: Albedo feedback; Climate and the climate system; Climate change; Climate feedback; Earth history; Earth structure and development; Global energy balance; Planetary atmospheres and evolution; Seasonal changes; Sun.

Earth structure and development

• **Categories:** Astronomy; geology and geography

Climate depends strongly on the geometry of the Earth's surface: Where the continents are, and where the mountain ranges are within them, controls winds, currents, and continental glaciers. The Earth's surface geometry is a result of its structure and development.

• Key concepts

asthenosphere: the ductile layer of the Earth, beneath the lithosphere, defined by mechanical behavior; although solid, it deforms horizontally over tens of millions of years

crust: the uppermost layer of the Earth as defined by seismic properties inferred to represent composition

inner core: the innermost layer of the Earth, extending from 5,150 kilometers to 6,370 kilometers below ground and composed mostly of solid iron

lithosphere: the uppermost layer of the Earth, about 150 kilometers thick, as defined by mechanical behavior; it does not deform horizontally

mantle: the layer beneath the crust, known to be solid and defined by seismic properties inferred to represent composition

outer core: a liquid region extending from the base of the mantle at a depth of 2,900 kilometers down to the inner core, at a depth of 5,150 kilometers; mostly molten iron

• Background

The Earth is a dynamic planet. Heat from its formation and radioactive decay moves to the surface and radiates to outer space. In the process, pieces of the surface are moved around, and mountain ranges are produced. The result is a mosaic of plates, some with continents and mountain ranges on them, that shift around over geologic time periods. The pattern of oceans, continents, and mountains controls ice-sheet formation, ocean currents, and atmospheric circulation patterns that, in turn, control the climate.

• Layers Based on Composition

By studying the waves generated by earthquakes, seismologists have been able to determine the elastic properties and density of the Earth as a function of depth. This technique has revealed four major layers: the crust (5 to 70 kilometers thick), the mantle (2,900 kilometers thick), the outer core (2,300 kilometers thick), and the inner core (1,200 kilometers thick). The outer core is liquid; all the other layers are solid. These layers are thought to represent differences in composition.

The continental crust, typically 30 to 70 kilometers thick, is made of granitic rocks, while the somewhat denser oceanic crust, typically 0 to 7 kilometers thick, is made of basaltic rocks. The mantle, a silicate rock called peridotite, is denser than the crust. The core, both inner and outer, is mostly iron, and is about twice as dense as the mantle. The inner core is solid, because it is subjected to higher pressure than the outer core.

• Layers Based on Behavior

When Alfred Wegener struggled to convince his skeptics that the continents drifted around, he assumed that the compositional layering described above would also be important mechanically and that the continents moved like giant ships through the mantle and oceanic crust. At the temperatures present in the crust and upper mantle, however, horizontal deformations like this are not possible.

As the evidence for drifting continents accumulated in the 1960's, a different approach developed. Although the peridotite is denser than granite or basalt, the mechanical behavior of these materials is similar. This behavior depends on temperature, which is known to increase with depth. The cooler, outer part of the Earth (about 150 kilometers thick), which does not deform horizontally, was defined as the lithosphere, and the hotter, deeper part, which does deform horizontally, as the asthenosphere. The theory of continental drift became the theory of plate tectonics, where the continents were seen to be passively riding on much thicker, "rigid" plates of the lithosphere. Plates deform vertically beneath glacial loads, but not horizontally when pushed by convection, because they are only about 150 kilometers thick but thousands of kilometers wide. There are about a dozen major lithospheric plates, and their interactions produce most of the mountains on Earth.

• Mantle Convection

In the Earth, heat moves from the hot interior to the surface. Heat from the core enters the lower mantle, expanding some of it, making it buoyant. The mantle is solid, but, because of its immense size, it behaves like silicon putty, deforming to permit the rise of this buoyant, heated rock. As the rock approaches the crust, tens of millions of years later, it spreads out, cools, and sinks. This process is called convection.

The convection cells on Earth rise at mid-ocean ridges, where lithospheric plates grow, and descend at subduction zones. The presence of plates influences the convection and limits the movements it can produce at the surface.

• Plate Tectonics and Climate

As convection moves the continents around, their latitude will change, making it more or less likely that continental glaciers will form on them. Currently, glaciers can only form at high latitudes, above 45°. If the rate of convection changes, the atmospheric carbon dioxide (CO_2) concentration is likely to change. Volcanism transfers CO_2 from within the Earth to the atmosphere; faster convection will result in more volcanism and thus higher levels of atmospheric CO_2.

Ocean circulation patterns depend on the shapes of the ocean basins. Before the closing of the Isthmus of Panama, warm, salty water from the Atlantic Ocean was flushed into the Pacific Ocean. After the isthmus's closure, this water remained in the Atlantic Basin, changing circulation patterns.

Collision of the Indian subcontinent with Asia caused the uplift of the Himalayan plateau, which led to the development of the monsoon winds. These are likely to have resulted in an unusually high rate of weathering, removing CO_2 from the atmosphere. The Rocky Mountains have been shown to deflect prevailing westerlies as they move across North America, bringing great quantities of heat to western Europe.

• Context

Scientists' confidence in being able to predict future climate change depends on their understanding of past climate change. Inputs that vary too slowly to affect the historical record, such as the locations of continents and mountain ranges, need to be evaluated from a geological perspective. These formations affect global climate, and their changes in the future will affect the evolution of Earth's climate on a geological timescale. One way to gauge the accuracy of existing models of geological history is to compare climate signals from times

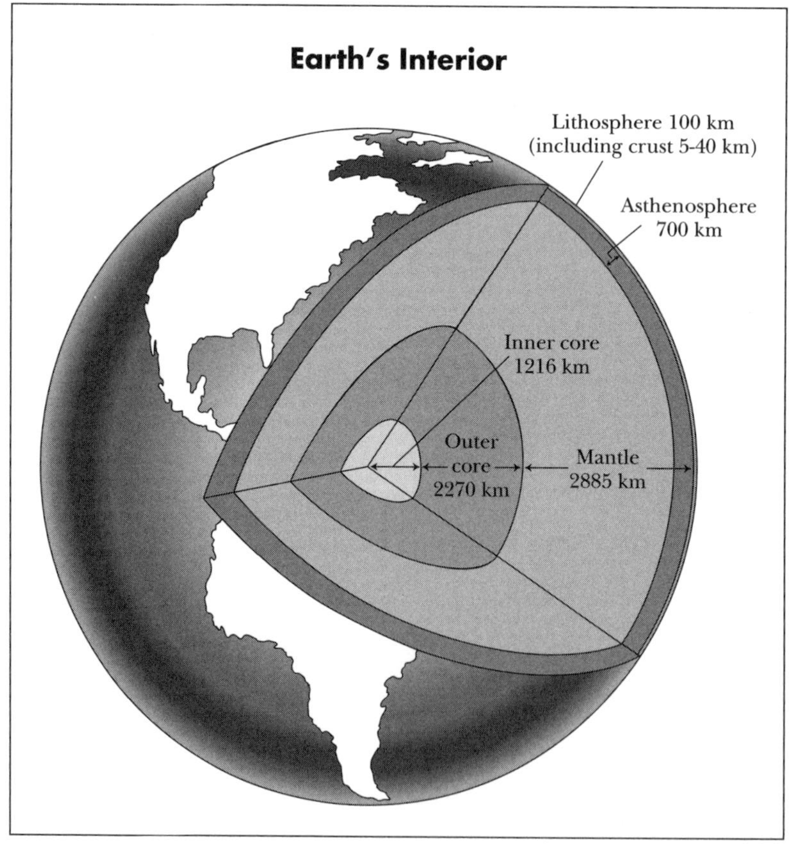

when the elevation and location of a geological formation are known to have differed from their current values. Knowledge of the structure and development of the Earth provides a path to obtain this information.

Otto H. Muller

• Further Reading

Fowler, C. M. R. *The Solid Earth: An Introduction to Global Geophysics.* 2d ed. New York: Cambridge University Press, 2005. Excellent discussion of current plate motions, plate rheology, and so on, but somewhat technical in approach. Illustrations, figures, tables, maps, bibliography, index.

Grotzinger, John. *Understanding Earth.* New York: W. H. Freeman, 2007. Introductory college-level textbook. Good overview of the plate tectonic paradigm. Illustrations, figures, tables, maps, bibliography, index.

Ruddiman, William F. *Earth's Climate Past and Future.* 2d ed. New York: W. H. Freeman, 2008. Chapter 6 of this elementary college textbook describes how the locations of continents and mountain ranges may have been responsible for ice ages. Illustrations, figures, tables, maps, bibliography, index.

Seager, R. "The Source of Europe's Mild Climate: The Notion That the Gulf Stream Is Responsible for Keeping Europe Anonymously Warm Turns Out to Be a Myth." *American Scientist* 94 (2002): 340-41. Good description of how the mountains in the American West control the climate of Europe by deflecting prevailing westerlies. Illustrations, maps.

Van der Pluijm, Ben. *Earth Structure: An Introduction to Structural Geology and Tectonics.* New York: W. W. Norton, 2004. The second half of this book concerns tectonics and describes in detail the formation of many mountain ranges on Earth. Illustrations, figures, tables, maps, bibliography, index.

See also: Continental climate; Earth history; Earth motions; Earthquakes; Plate tectonics; Volcanoes.

Earth Summits

• **Category:** Conferences and meetings

The Earth Summits held by the United Nations have presented forums for the creation and maintenance of important international treaties and agreements relating to climate change. They bring with them both the strengths and the weaknesses of the United Nations as a body for organizing international cooperation among nations with very different interests and agendas.

• Key concepts

greenhouse gases (GHGs): gases that trap heat in the atmosphere, preventing it from escaping into space and increasing the temperature of the planet

Kyoto Protocol: a treaty that established specific and legally binding limits on GHG emissions for the global community, ratified by 182 nations

sustainable development: the growth of population, industry, and agriculture in a fashion that does not deplete global resources

U.N. Framework Convention on Climate Change (UNFCCC): the agreement underlying the Kyoto Protocol and all other subsequent U.N. agreements relating to climate change

• Background

During the twenty years following the 1972 United Nations Conference on the Human Environment, in Stockholm, Sweden, the global environment continued to deteriorate, as the ozone layer and natural resources were depleted, while global warming and pollution increased. Importantly, little had been done to integrate environmental issues with economics and development. In 1983, Gro Harlem Brundtland of Norway, head of the World Commission on Environment and Development, put forth a report that defined sustainable development as "the growth of population, industry, and agriculture occurring in a manner that allows the current generation to fulfill its own needs without jeopardizing those of future generations." A sustainable way of life was said to depend on "equitable economic growth, conservation of natural resources, and the environment," in addition to social development.

As a result of the Brundtland Report, the United Nations requested that a conference be convened that focused equally on the environment and development. The First Earth Summit (formally, the United Nations Conference on Environment and Development, or UNCED) was held in 1992. It served as the foundation for U.N. activities to coordinate sustainable development and resist global climate change. Two subsequent summits, Earth Summit +5 and Earth Summit +10, functioned as reviews of the progress made since UNCED after five and ten years, respectively.

• The First Earth Summit

Held in Rio de Janeiro, Brazil, from June 3 to 14, 1992, UNCED set out to provide a basis for global collaboration between the developed and the developing nations so as to bolster socioeconomic development and halt the deterioration of the planet's environment. Some 108 participating countries adopted five significant agreements with the aims of improving the environment and redefining the traditional concept of development to one that included sustainability.

Agenda 21 was the sole product of UNCED that covered all aspects of sustainable development, from goals and responsibilities to financing. It included proposals for social and economic action, such as fighting poverty and balancing production with consumption, while integrating environmental and developmental concerns. The Rio Declaration on Environment and Development was a statement of principles in support of Agenda 21, designed to guide nations in their efforts to protect the environment. It proclaimed that humans are entitled to life in harmony with nature via sustainable development and that reducing global disparities and eradicating poverty are essential for such development. The Rio Declaration also emphasized the importance of strengthening the role of women, youth, farmers, and indigenous peoples in contemporary society.

The Statement of Forest Principles described the principles that underlie the sustainable management of forests. The Convention on Biodiversity focused on sustaining and conserving the plethora of species inhabiting the planet. Finally, the United Nations Framework Convention on Climate Change (UNFCCC), a nonbinding treaty on the environment, was ratified on June 12, 1992, by 154 countries with the intent of stabilizing their greenhouse gas (GHG) emissions at 1990 levels by 2000. The principle of "common but differentiated responsibilities" for developed and developing countries was agreed upon, with the developed countries bearing the greater burden of accountability.

• Earth Summit +5

From June 23 to 27, 1997, delegates from 165 nations, including over 53 heads of state, attended a special session of the U.N. General Assembly at the U.N. headquarters in New York City to "review and appraise" the implementation of Agenda 21 and to assess whether there had been global progress toward sustainable development. The participants of Earth Summit +5 determined that the global environment had deteriorated since UNCED, concluding that while some progress had been achieved in forestalling climate change and the loss of forests and freshwater, global poverty continued its downward spiral, and there were few commitments from the global community to reduce GHG emissions and help fund sustainable development. The north-south divide between developed and developing nations was seen as largely to blame for these problems.

• Earth Summit +10

Earth Summit +10 was held in Johannesburg, South Africa, from August 26 to September 4, 2002, ten years after UNCED, in order to explore the progess made in promoting sustainable development. Known as the World Summit on Sustainable Development (WSSD), the meeting gathered leaders from the business sector, nongovernmental organizations (NGOs), and other concerned groups with the goal of enhancing the quality of life for all humankind and conserving Earth's natural resources. During Earth Summit +10, a number of parallel activities were simultaneously held by independent organizations and groups that believed that agreements reached by WSSD were full of good intentions concerning access to potable water, biodiversity, and fishing resources, but were devoid of objectives for promoting renewable ener-

Fellow world leaders and diplomats applaud as Brazilian president Fernando Collor de Mello signs the United Nations Framework Convention on Climate Change at the first Earth Summit in June, 1992. (AP/Wide World Photos)

gies, and lacking specific commitments for funding. These groups echoed similar concerns expressed following Earth Summit +5. Some believe that the most positive outcome of Earth Summit +10 was the commitment garnered from several governments to ratify the Kyoto Protocol.

• Significance for Climate Change

The First Earth Summit, UNCED, set the tone for the two Earth Summits that followed by emphasizing sustainability as vital for environmental progress in the fight to contain climate change and forestall global warming. In addition, UNCED produced the most important treaty for achieving this goal, the UNFCCC, with a goal of stabilizing GHG emissions at a level that might prevent anthropogenic disruption of the Earth's climate. Originally a legally nonbinding document, because it set no mandatory limits on GHG emissions, the treaty included provisions for updates called "protocols," the best known being the Kyoto Protocol, which established specific, enforceable limits on GHG emissions. Parties to the UNFCCC were asked to adopt legally binding targets for developed countries.

The UNFCCC created a "national GHG inventory" to keep track of GHG emissions and their removal from the atmosphere. It was ratified by the United States in October, 1992, when it was opened for signing. While the UNFCCC was ratified by 166 countries, few developed countries met the goal of reducing GHG emissions to 1990 levels by 2000. However, since the UNFCC entered into force in 1994, the parties to the treaty have been meeting yearly in the Conferences of the Parties (COP) in order to evaluate their progress and to establish and implement legal obligations for developed nations to decrease their GHG emissions under the terms of the Kyoto Protocol.

Cynthia F. Racer

• Further Reading

Bradbrook, A. J., et al., eds. *The Law of Energy for Sustainable Development.* New York: Cambridge University Press, 2005. Based upon lectures given at a conference on sustainable energy; focuses on the legal implications of employing energy resources required for sustainable development.

Gagain, J. R., Jr. "The United Nations and Civil Society: A New Step in the Right Direction." *UN Chronicle* 40 (June, 2003). Argues that UNCED demonstrated the need for countries to interact with the United Nations framework, particularly on borderless issues such as the environment and sustainable development.

Strong, Maurice. *Where on Earth Are We Going?* London: TEXERE, 2001. Strong, a major organizer of the Rio Earth Summit, describes his pivotal role in various environmental movements and speaks about what has to be done to save the planet in an historical context.

See also: Annex B of the Kyoto Protocol; Kyoto lands; Kyoto mechanisms; Kyoto Protocol; Sustainable development; United Nations Conference on the Human Environment; United Nations Division for Sustainable Development; United Nations Framework Convention on Climate Change.

Earthquakes

• **Category:** Geology and geography

• Definition

An earthquake is a shaking or trembling of the ground caused by energy coming from volcanic eruptions or from rocks breaking during the geologic process known as faulting. This process takes place as a result of differential movements of segments of the Earth's crust known as plates. The plate tectonics theory has identified seven major crustal plates, as well as several minor plates, which are thought to be colliding with one another, separating from one another, or sliding past one another like books on a tabletop. The sudden movements of these plates, as they separate, collide, or slide past one another, are believed to cause most of the planet's earthquakes.

Volcanoes can cause earthquakes too, but these earthquakes are generally minor and occur only during periods of heightened volcanic activity. Some additional causes of earthquakes include landslides that drop large quantities of earth suddenly on underlying surfaces, and human activities, such as blasting during surface or underground mining operations. The shock waves from underground nuclear tests can generate vibrations similar to earthquakes as well.

• Significance for Climate Change

Concerns have been raised that climate change may lead to earthquakes in Alaska, or elsewhere, and these concerns may not be entirely unwarranted. Many glaciers and ice sheets are located in regions of past or continuing geologic activity, where crustal movements are still taking place. Removal of ice cover during periods of global warming might be sufficient to activate ancient faults in these areas and set off earthquakes of considerable magnitude. The rate at which the ice melts may not need to be rapid to cause such activity. Rocks can build up strains over a period of years without breaking at all, until the tension is suddenly released in a devastating shock. Scientists describe this process as elastic rebound. While the rocks on either side of a fault slowly slide past each other, the fault itself may remain "locked," with no movement taking place. Meanwhile, the rock masses on both sides of the fault begin to stretch, in a fashion similar to the stretching of an elastic band. When the strain in the rocks becomes too great, the fault will break, and the rocks on either side will snap back to their original dimensions.

There is clear evidence that the removal of ice from the Earth's surface following the ice ages is causing significant upward movement of the crust in several places. In parts of Scotland, Scandinavia, and Canada, the amount of rise has been dramatic. Interior Scandinavia has risen as much as 250 meters since the ice disappeared, and elevated beaches around Hudson Bay in northern Canada indicate the land there has risen more than 330 meters during the same period. Scandinavia is still rebounding at a rate of one meter per century, and some coastal cities have been uplifted so rapidly that docks constructed several hundred years ago are now inland, while harbors have grown so shallow that they can no longer accommodate the ships that once docked there.

Scientists refer to this uplifting process as isostatic rebound. When the weight of a continental ice sheet is added to the Earth's crust, the crust responds by subsiding. When the ice is removed, the crust begins to rise again, just as a cargo ship that sank lower in the water when it was loaded floats higher after its cargo is removed. The weight of the three-kilometer-thick ice sheets that once covered North America and Europe apparently caused down-warping of the crust by hundreds of meters. Now that the ice has been removed, the crust is rebounding to its original height.

Whether faults that were once active can be reactivated sufficiently by this rebounding process to cause earthquakes is difficult to determine, but such rebounding may well explain a series of mysterious earthquakes that plagued upper New York State during the late 1900's, as well as earthquakes that rattled Boston at the same time. Both areas were once buried beneath the North American ice sheet, and Boston's jolts have continued into the twenty-first century. By contrast, the continent of Antarctica still has its ice cover, and Antarctica's continental shelves, which would normally be found 120 meters below sea level, are now found 330 meters down. Presumably, the shelves were formed in water depths of 120 meters when Antarctica was ice free, and they will rise again to normal

A woman mourns her relatives at the site of a devastating 2009 earthquake in Sichuan, China. (Aly Song/Reuters/Landov)

depths if the weight of the ice now covering the continent is someday removed.

Donald W. Lovejoy

• **Further Reading**

Anderson, David E., Andrew S. Gaudie, and Adrian G. Parker. *Global Environments Through the Quaternary: Exploring Environmental Change.* New York: Oxford University Press, 2007. Analysis of environmental changes during the last two million years of Earth history, with particular attention to the post-glacial uplift of Scotland, Scandinavia, and North America.

Hough, Susan Elizabeth. *Earthshaking Science: What We Know and Don't Know About Earthquakes.* Princeton, N.J.: Princeton University Press, 2002. Excellent basic description of how the various kinds of plate movements can cause earthquakes and a helpful explanation of the elastic rebound theory.

Zeilinga de Boer, Jelle, and Donald Theodore Sanders. *Earthquakes in Human History: The Far-Reaching Effects of Seismic Disruptions.* Princeton, N.J.: Princeton University Press, 2005. Summarizes ways in which earthquakes affect human lives; includes an important section on "induced earthquakes," which are caused when weight is added to the Earth's crust.

See also: Coastline changes; Deglaciation; Earth structure and development; Ice shelves; Lithosphere; Mass balance; Plate tectonics; Volcanoes.

Easter Island

• **Category:** Nations and peoples

• **Definition**

Easter Island is a small island, occupying 163 square kilometers in the Eastern Pacific Ocean. It is considered to be the most remote place on Earth—it is the spot furthest away from any other populated landfall. The island is perhaps most famous for the stone megaliths found there, huge heads with impassive faces called "moai" carved for unknown reasons by the islanders. The island was named Easter Island by a Dutch explorer who landed there on Easter Sunday, 1722. It is called Rapa Nui in the language of those who inhabited the island (who are also called the Rapa Nui) and Isla de Pascua in Spanish. The island is currently Chilean territory.

Some historians believe that Polynesian islanders were the first humans to populate Easter Island in about 700 C.E., while others argue that the island was inhabited as early as 300 C.E. It has had a turbulent history involving European visitors who brought smallpox, the plague, slave raids, and war to the island, decimating the population and contributing to the collapse of its culture and economy. The island's history is complex and subject to ongoing archaeological interpretation.

• **Significance for Climate Change**

Easter Island is often given as an example of a societal collapse that was due, in part, to mismanagement of natural resources. The population of the island has risen and fallen in conjunction with the fluctuations in its ecosystem. Seafaring people from Polynesia arrived on Easter Island perhaps as early as the fourth century, coming to an island that is believed to have been covered in forest. The land was very fertile, and the ease with which food grew led to a rapid increase in the island's population. The Easter Island palm, part of the island's subtropical forest, was tall enough to provide timber suitable for homes and canoes and was also available for fuel.

Present-day Easter Island is deforested; many of its unique tree species are now extinct. Competing theories abound as to how the deforestation occurred—some blame the islanders for using trees to erect their statues or for firewood; others speculate that the Little Ice Age, which occurred between 1650 and 1850, may have affected the island's forests. (The term "Little Ice Age" is used differently by different writers. Many use it to refer to the climate cooling from about 1300 to 1850, while others use it for the latter half of that interval, when cooling was greatest, beginning around 1550 or 1600.) Perhaps because of deforestation, the island's soil

Some of the famous megaliths of Easter Island. (©Galina Barskaya/Dreamstime.com)

has eroded over the years. This erosion may have also been sped along by agricultural development, including sheep farming.

As the island became deforested and the soil became less fertile, the islanders' entire community structure disintegrated. Resources were consumed steadily with no conservation plan. People began to live in caves rather than in homes, because wood was no longer available, and it is thought that the islanders were also unable to build canoes, although they may have built canoes out of reeds rather than trees.

As deforestation and soil erosion led to extinction of the trees and reeds, the Rapa Nui people were unable to build new canoes, so they began to eat birds and mollusks rather than fish and sea mammals. Eventually, the birds on the island, both native and migratory, neared extinction. With no birds to pollinate the trees and other forest plants, the deforestation of the island increased. With food and fuel diminishing rapidly, the islanders' quality of life suffered dramatically.

By the time of European contact, Easter Island was nearly completely deforested, and the islanders were malnourished by their limited diet. The ships that arrived from Europe brought smallpox, which led to a severe decline in the island's population. European contact also led to slave raids on the islanders. Interisland wars over food and other resources may have contributed to the island's population decline as well. These factors, in addition to the mismanagement of the island's ecosystem, eventually led the island's population to shrink dramatically. In 1877, Easter Island was home to only

111 humans. In 2002, the population had rebounded to 3,700.

European ships also contained rats, which helped destroy many of the native seabird and landbird species on the island. They also brought chickens, which helped destroy many of the insect and plant species. Rats also contributed to the extinction of the Easter Island palm by eating its seeds.

Marianne M. Madsen

- **Further Reading**

Arnold, C. *Easter Island: Giant Stone Statues Tell of a Rich and Tragic Past.* New York: Clarion Books, 2004. Includes information about decimation of natural resources on the island. Includes color photos, maps, a complete glossary, a chronology, a bibliography, and a pronunciation guide.

Bahn, P. *The Enigmas of Easter Island.* New York: Oxford University Press, 2003. Discusses Easter Island as an example of "eco-stupidity"; includes analysis of destruction of the ecosystem.

Fischer, S. R. *Island at the End of the World: The Turbulent History of Easter Island.* London: Reaktion Books, 2006. Discusses social and cultural history of the island.

Loret, J., and J. Tanacredi, eds. *Easter Island: Scientific Exploration into the World's Environmental Problems in Microcosm.* New York: Springer, 2003. Discusses the effects of human population growth and climate change on the island on its ecology.

Routledge, Mrs. Scoresby. *The Mystery of Easter Island.* Reprint. New York: Cosimo Classics, 2007. An account by an archeologist studying the island in 1913. Includes native folklore, culture, and traditions.

Van Tilburg, J. *Easter Island: Archaeology, Ecology, and Culture.* London: British Museum Press, 1994. Offers an account by the current leading authority on Easter Island. Discusses the cultural, ecological and archaeological history of the island and its people.

See also: Alliance of Small Island States; Coastal impacts of global climate change; Islands; Maldives; Mean sea level; Sea-level change; Sea surface temperatures.

Eastern Europe

- **Category:** Nations and peoples
- **Key facts**

Population: 120,154,616 (2008-2009 estimates)

Countries: Albania, Bulgaria, Czech Republic, Hungary, Poland, Romania, Slovakia, and the former Yugoslavia (now consisting of Bosnia-Herzegovina, Croatia, Kosovo, Macedonia, Montenegro, Serbia, and Slovenia)

Area: 1,166,322 square kilometers

Gross domestic product (GDP): $1.919 trillion (purchasing power parity, 2008-2009 estimates; several countries in the former Yugoslavia have large informal economies up to 50 percent as large as their official economies)

Greenhouse gas (GHG) emissions in millions of metric tons of carbon dioxide equivalent (CO_2e): 955.6 in 2006 (excluding former Yugoslavian nations)

Kyoto Protocol status: Accepted or ratified by Czech Republic (2001), Hungary (2001), Romania (2001), Bulgaria (2002), Poland (2002), Slovakia (2002), Slovenia (2002), Albania (2005), Bosnia-Herzegovina (2007), Croatia (2007), Macedonia (2005) Montenegro (2007), and Serbia (2007); not signed by Kosovo as of 2009

- **Historical and Political Context**

Prior to the twentieth century, most of Eastern Europe, with the exception of Poland, was part of the Austro-Hungarian or Ottoman (Turkish) Empire. Bulgaria, Romania, and Serbia gained their independence from the Ottoman Empire during the nineteenth century. Following the defeat of Austria-Hungary during World War I (1914-1918), the Austro-Hungarian Empire was broken into several countries by the victors. Some of the resulting nations, such as Austria itself and Hungary, are defined by fairly natural and stable cultural and geographic boundaries. The Slavic portions of the empire were grouped into two collective countries, Czechoslovakia (reflecting its Czech and Slovak ethnicity) and Yugoslavia ("land of the southern Slavs," from the Slavic root *yug,* meaning "south"). Neither of these two entities remained intact.

A further legacy of imperial rule in Eastern Europe is that there are many cultural minorities living in enclaves surrounded by other ethnic groups. When these regions were ruled from a distant imperial capital, an ethnic patchwork made little difference. However, when the empires dissolved, many ethnic groups found themselves in countries ruled by hostile groups, or outside their ethnic homelands. In Czechoslovakia, ethnic differences led to peaceful division, but in the former Yugoslavia, ethnic tensions erupted in civil war. The tendency of countries in the Balkan region (the portion of Eastern Europe between the Adriatic and Black Seas) to fragment has been noted for a long time and has given rise to the term "Balkanization."

All the Eastern European countries were either allied with or occupied by Nazi Germany during World War II. The Soviet counterattack led to Soviet military occupation of most of Eastern Europe. Yugoslavia and Albania expelled German forces on their own and were not occupied, but installed their own communist regimes. By 1948, communist governments had been established in all Eastern European countries. Communism was an attempt to implement the political and economic theories of Karl Marx. Under communism, the state owned all means of industrial production, though in practice varying degrees of legal and extralegal private enterprise existed as well. As an economic system, communism proved to be inefficient, since there were few incentives for efficiency. Moreover, there were no adequate means of accurately determining market needs or setting appropriate prices. Ultimately, however, the most destructive feature of communism was its stifling of dissent, which served to conceal corruption and poor environmental practices. More important, the suppression of criticism prevented communist governments from knowing how urgently reform was needed until it was too late.

Since most of the Eastern European governments were highly subservient to the Soviet Union, they were commonly called "satellites." In 1955, all the countries of Eastern Europe except Yugoslavia joined the Soviet Union in a military alliance commonly called the Warsaw Pact. Albania later left the alliance as well. An attempt by Hungary to overthrow its communist government was crushed by Soviet invasion in 1956, and Czechoslovakia was occupied by the Soviet Union in 1968, when reforms by the Czech government were seen as threatening. Nevertheless, countries such as Poland, Czechoslovakia, and Hungary tried to liberalize their economies within the limits set by Soviet tolerance. Hungary, in particular, instituted economic policies that accommodated a great deal of capitalism. The Cold War—the name given to the period of tension between the United States and the Soviet Union—ended with the dissolution of the Soviet Union in 1991. Over the next few years, communist governments were dissolved in all the countries of Eastern Europe.

• Impact of Eastern European Policies on Climate Change

Because Eastern Europe was economically and militarily dominated for decades by the Soviet Union, its policies during the Cold War were largely dictated by the Soviet Union. The economic emphasis in the years after World War II ended (in 1945) was on reconstruction, followed by an emphasis on heavy industry. With repression of dissent and little opportunity for public participation in government, environmental issues received little attention; greenhouse gas (GHG) emissions began to be seen as an important issue only about the time of the breakup of the Soviet Union, in 1991. Since 1991, the countries of Eastern Europe have tended to adopt Western European practices and policies.

In 1997, Poland, the Czech Republic, Slovakia,

Ranking Eastern European Corruption

Nation	*Corruption Perceptions Index*
Czech Republic	45
Hungary	47
Slovakia	52
Poland	58
Croatia	62
Romania	70
Bulgaria	72
Macedonia	72
Albania	85
Montenegro	85
Serbia	85
Bosnia-Herzegovina	92

Source: Transparency International.

Hungary, Slovenia, Romania, and Bulgaria applied for admission to the European Union. All except Romania and Bulgaria were admitted in 2004. Romania and Bulgaria were admitted in 2007. Macedonia and Croatia have applied for admission. All Eastern European countries, except Kosovo, have ratified or accepted the Kyoto Protocol.

• Eastern European Nations as GHG Emitters

The Kyoto Protocol calls for signatories to reduce their GHG emissions relative to 1990 emissions—that is, relative to levels before the dissolution of the Soviet Union. The economic turmoil following the breakup of the Soviet Union caused all countries of the former Soviet Union and the countries of Eastern Europe to suffer serious economic declines. The former Yugoslavia, though not a Soviet satellite, nevertheless suffered its own breakup in the 1990's and also experienced serious economic decline. As a result, despite having a great deal of inefficient and antiquated infrastructure, the nations of Eastern Europe are well below their Kyoto emissions targets. Their total emissions, roughly the equivalent of a billion tons of carbon dioxide, are approximately a third lower than their targets, which are about 1.3 billion tons. The economic cloud over Eastern Europe has its silver lining: Eastern European countries have large emissions allowances that can be traded, as well as a substantial margin for growth.

• Summary and Foresight

Eastern Europe is characterized by several themes. First is the legacy of domination by the Soviet Union: inefficiency, corruption, and in some areas organized crime. Inexperience with market economies has led to conditions such as severe inflation in Bulgaria in 1997 and the collapse of the Albanian economy through pyramid investment schemes that same year. Transparency International, which publishes a Corruption Perceptions Index based on a variety of reports by business, government, and human rights groups, gave Slovenia the highest (least corrupt) ranking, at number 26 of 180 countries surveyed in 2008—comparable to some Western European countries. Several other Eastern European countries had much lower ratings. Nevertheless, all Eastern European countries had substantially less corruption than Russia, which ranked 147th.

Unresolved ethnic tensions continue to be important in Eastern Europe. Ethnic Albanians are split between Albania itself and the neighboring region of Kosovo in former Yugoslavia. Kosovo, in turn, contains Serbian enclaves that object to separation from Serbia. There is a significant Hungarian population in northern Serbia and southern Slovakia. The independent country of Moldova is a former slice of Romania annexed by the Soviet Union. Greece objects to the name Macedonia out of concern that it might serve to justify claims on territory in Greece that is also called Macedonia. As a result, the nation of Macedonia is officially called by the cumbersome title Former Yugoslav Republic of Macedonia.

Some Eastern European countries, especially those that had democratic and capitalist traditions before Soviet domination, have successfully resumed their previous courses. Considering the repressive governments of the Cold War era, the countries of Eastern Europe have made remarkable progress toward democratization. The Czech Republic, Poland, Slovakia, Slovenia, and Hungary are all rated 1 for political and civil liberty by Freedom House, the highest rating. Bulgaria, Croatia, and Romania are rated 2, that is, highly free but with some restrictions. Albania and the countries of former Yugoslavia are mostly rated 3 in each area, that is, partly free. The lower ratings in these countries stem from weak democratic traditions and the aftereffects of strife during the breakup of Yugoslavia. In contrast, Russia in 2009 rated 6 in political freedom and 5 in civil liberties.

All the Eastern European countries, except disputed Kosovo, have signed and ratified the Kyoto Protocol. They are less developed than the countries of Western Europe, and as they seek to grow economically they will face the interlocking problems of meeting increased consumer demand while simultaneously attempting to replace inefficient infrastructure and reform environmental safeguards. However, like Russia, the comparatively low level of industry in Eastern Europe, plus the drop in productivity during the transition from communist to capitalist economies, gives these countries a significant buffer zone that will enable them to grow but still remain within their Kyoto targets. Eastern Europe is about one-third below its

Kyoto emissions target. Also, replacement of aging and obsolete infrastructure will enable the nations of Eastern Europe to install more environmentally friendly infrastructures rather than having to grapple with modifications of existing ones. Finally, membership in the European Union gives the member countries of Eastern Europe access to funding for development.

Steven I. Dutch

• Further Reading

Bideleux, Robert, and Ian Jeffries. *A History of Eastern Europe: Crisis and Change.* New York: Routledge, 2007. Traces the history of Eastern Europe from Greco-Roman times to the present. Roughly a quarter of the book deals with the Cold War and its aftermath.

Douma, W. Th., L. Massai, and M. Montini, eds. *The Kyoto Protocol and Beyond: Legal and Policy Challenges of Climate Change.* West Nyack, N.Y.: Cambridge University Press, 2007. Includes chapters on the Russian Federation and on non-Annex I Eastern European nations.

United Nations Industrial Development Organization. *Regional Forum on Industrial Cooperation and Partnership in Central and Eastern Europe and the Newly Independent States.* [Vienna]: Author, [2006]. The second of the three sessions documented here addressed environmental management, climate change, and industrial energy efficiency.

See also: Europe and the European Union; Kyoto Protocol; Poland; Ukraine.

Ecocentrism vs. technocentrism

• **Category:** Environmentalism, conservation, and ecosystems

• Definition

"Ecocentrism" is a label for views that place at the center of moral concern the relationship of human beings to the Earth. "Technocentrism" is a label for views that explicitly or implicitly locate the center of moral concern in human technical and technological capabilities. Technocentrists see the solution for woes brought on by human ingenuity as more ingenuity, while ecocentrists in contrast believe that humans must find and respect their proper place in the world, rather than seeking to use technology to transcend it. Technocentric ambitions seem to ecocentrists to be expressions of arrogance or hubris, and ecocentric pieties seem to technocentrists to acquiesce in mysticism and passivity in the face of nature's hard challenges.

• Significance for Climate Change

In the light of the prospect of global climate change, ecocentrists tend to seek solutions that involve changing destructive practices and undoing the lifestyle choices that have brought humans to the present crisis. Technocentrists, on the other hand, tend to look for a technical fix. Julian Simon is an example of a technocentric enthusiast. For him, human beings are the ultimate resource, so the more of them there are, the better the prospects for solving the world's problems, including environmental challenges. For those who agree with Simon, popular claims about resource scarcity—and, accordingly, calls for scaling back enterprise—sell short the fecundity of human powers of imagination and invention. To technocentrists, calls to sacrifice ambitions of development are premature, because they ignore the likelihood that seemingly overwhelming problems will yield to unforeseeable technical innovations.

The value of civilization, if not the skills that make society itself possible, has been in question at least since the reaction to the Enlightenment of the Romantic movement (growing out of the work of the Swiss-French philosopher Jean-Jacques Rousseau and the English poet William Blake, among others). The German philosopher Martin Heidegger traces the woes of the modern worldview, which he refers to as "the enframing," back to Plato and other ancient Greek thinkers. In more recent times, such thinkers as Henry David Thoreau, Aldo Leopold, Wendell Berry, Arne Naess, and various other exponents of deep ecology have called for a radical reconsideration of the place of human experience, and even of animal sentience itself, in the vast, interconnected web of nature.

Though their call for a better relation to nature predates concern with global warming, the dire consequences of climate change constitute for ecocentrists an object lesson in human arrogance and exploitation. On this view, salvation from human folly is to be found only in a pared-down existence: a life simpler, closer to the soil, and more local in scale and scope. What is wrong can only be fixed by relinquishing human presumption and acquiescing in limits that are natural, if not divine. This view accordingly holds that "small is beautiful," and urges people to reduce their carbon footprint.

For their technocentric opponents, such attitudes not only underestimate human creativity in meeting challenges and solving problems but also sacrifice too readily the prospects of socially disadvantaged individuals and groups: Economic opportunity and social justice, they believe, will require continuation on the path of development and discovery. Moreover, it may seem presumptuous to believe humans could possibly encompass within their time-bound minds all of the relevant aspects and possibilities. For technocentrists, the imperative is to continue to re-create the world in humanity's own, reasonable image, while for the ecocentrists, it is to rethink fundamentally humanity's place in the world.

Edward Johnson

• Further Reading

Ehrenfeld, David. *The Arrogance of Humanism.* New York: Oxford University Press, 1978. A sweeping critique of human-centered (including technocentric) attitudes.

Kunstler, James Howard. *The Long Emergency: Surviving the Converging Catastrophes of the Twenty-first Century.* New York: Atlantic Monthly Press, 2005. Bleak and sobering forecast of the collapse of the world's "cheap-oil fiesta" and of the ensuing postindustrial constraints and disasters.

Naess, Arne. *The Ecology of Wisdom: Writings by Arne Naess.* Edited by Alan R. Drengson. New York: Counterpoint, 2008. Collection of essays by the Norwegian philosopher often regarded as the founder of the deep ecology movement.

O'Riordan, Timothy. *Environmentalism.* London: Pion Books, 1981. The original application of the specific terms "ecocentric" and "technocentric" to the contrasting views described.

Ponting, Clive. *A New Green History of the World: The Environment and the Collapse of Great Civilizations.* Rev. and expanded ed. New York: Penguin, 2007. Provides a stunning long-term view of human history that sees the original wrong turning as the development of agriculture.

Quinn, Daniel. *Beyond Civilization: Humanity's Next Great Adventure.* New York: Crown, 2000. The famous ecocentric novelist proposes that the human future lies in a rejection of civilization and the embrace of a "new tribal revolution."

Rodman, John. "The Liberation of Nature." *Inquiry* 20, no. 1 (1977): 83-145. An early philosophical examination of what an alternative to sentience-based ethics might look like.

Simon, Julian. *The Ultimate Resource 2.* Princeton, N.J.: Princeton University Press, 1998. Upbeat, technocentric assessment of environmental problems, including a skeptical approach to global warming, in light of the author's conviction that having more people means better prospects for discovering ingenious solutions to human problems.

Zerzan, John. *Twilight of the Machines.* Port Townsend, Wash.: Feral House, 2008. A radical ecocentric comparison of productionism and the primitive future as fates for the human species, which argues for voluntary abandonment of the industrial mode of existence.

Zimmerman, Michael. *Contesting Earth's Future: Radical Ecology and Postmodernity.* Berkeley: University of California Press, 1994. A Heideggerian philosopher's sympathetic critique of deep ecology and other ecocentric approaches to environmental issues.

See also: Anthropogenic climate change; Civilization and the environment; Conservation and preservation; Economics of global climate change; Environmental economics; Environmental movement; Human behavior change; Technological change.

Ecological impact of global climate change

• **Category:** Environmentalism, conservation, and ecosystems

Anthropogenic climate changes, especially global warming, have affected terrestrial, freshwater, and marine ecosystems in the past few centuries. If the Earth's average global temperature continues to increase, these ecosystems will change even more radically in the future.

• Key concepts

anthropogenic: caused by humans

biodiversity: the entire variety of living organisms in a given location, from the local to the global

biosphere: the ecological zones on Earth where life is found, whether on land, in the water, or in the air

ecological niche: the relational position of a species in an ecosystem with respect to all other species, resources, and physical and chemical factors affecting life and reproduction within that ecosystem

ecosystem: a community of different species interacting with each other and with the nonliving environment

extinction: the total disappearance of a species from the Earth

greenhouse gases (GHGs): lower atmospheric gases, such as water vapor and carbon dioxide, that trap heat radiated from the Earth's surface

• Background

Long before the appearance of *Homo sapiens*, periodic momentous climate changes had devastating effects on Earth's ecosystems. For example, 245 million years ago, a colossal effusion of greenhouse gases (GHGs) from massive volcanoes raised average global temperatures by 5° Celsius, leading to the demise of over 90 percent of living things in the great Permian-Triassic extinction. Humans, in the few million years they have lived on Earth, have not had as cataclysmic an effect on the biosphere as have natural climatic catastrophes. However, over time and particularly since the development of advanced industrialized societies, they have been having an increasingly potent effect on the Earth's climate, which in turn has led to the decline and even the extinction of many species.

• Effects of Global Climate Change on Plants

The response of the Earth's land and water vegetation to anthropogenic climate change is extremely complex and varied, extending from photosynthesis in relatively simple microscopic plants to the global distributions of highly variegated plant species. Scientists researching climate-induced environmental changes and the effects of those changes have concentrated on a careful selection of flora in the distant and recent past. They have then extrapolated from those focused studies of the past to draw conclusions about present and future time periods, mainly through computer models. Paleobotanists have discovered how some large ecosystems dominated by plants responded to global climate changes in prehistoric ages, but this understanding has not been easily applicable to modern conditions. More secure understanding has resulted from studies of how vegetation responded to the climate changes during the ice ages: Migration proved a means for many species to survive glacial and interglacial climates.

While, in recent centuries, the greatest threat to plant species has been from their unsustainable harvesting by humans and the elimination of their habitats, anthropogenic global warming has also had an influence on certain plant distributions. For example, with a temperature rise of nearly 2° Celsius in the boreal zone of North America, conifer forests have moved closer to polar regions. Certain computer models relating global warming and plant survival have predicted future changes in the species composition and locations of forests, while still other forests may desiccate, burn, and disappear completely. In tropical rain forests, certain orchids are rare, because they have highly unusual ecological niches. If their environments are climatically transformed, these orchids often prove to be mediocre migrants. Some plant species, unable to tolerate warmer fresh and salt waters, will become extinct, while climate-caused habitat loss will lead to other plant extinctions. Flooding of coastal estuaries and wetlands may further the destruction of plants unable to adapt to the changed conditions.

• Effects of Global Climate Change on Animals

The abundance, distribution, and ecological niches of animal species will continue to be affected by changing climatic factors. As the climate warmed at the end of the ice ages, some species expanded their ranges while other species migrated to new regions. This postglacial warming resulted in the extinction of many species of large mammals, including the wooly mammoth. Many scientists believe that climate played a role in these extinctions. Analogously, many scientists believe that in the past few centuries, when human activities have contributed to global warming, elevated global temperatures have affected certain animal species.

The declining numbers of amphibians is an important example of this phenomenon. From 1975 to 2000, certain Costa Rican forests warmed significantly, causing shifts in the habitats of birds, reptiles, and amphibians. Some frogs, unable to adapt to these changes, became extinct, while others became endangered or critically threatened. Not all scientists are convinced that climate change is to blame for amphibian declines, however. Some cite pollution and fungus as possible alternate causes. Another often-cited example of a declining species is the polar bear. From 1975 to 2005, Arctic sea ice shrank by about 20 percent, reducing the bears' access to seals, their main prey. In the two decades from 1985 to 2005, polar bear populations fell by over 20 percent.

Because of the many variables contributing to the decline of various animal species, it has been difficult to determine with precision the role that global warming has played in these declines. Nevertheless, certain scientists have used computer models to issue warnings that accelerated global warming will inevitably lead to the extinction of many animal species. In 2004, the United Nations published a report based on the expertise of many scientists that estimated a million species of plants and animals could become extinct by 2050, unless global warming can be stopped. Environmentalists think that the ecosystems most likely to experience the greatest species loss are polar seas, arctic tundra, and coastal wetlands.

• How Species Cope with Climate Change

Nature has provided species of plants and animals with a powerful mechanism of coping with environmental changes, including climate changes. Charles Darwin, who discovered this mechanism and showed how it could account for the origin and evolution of species, called it natural selection. Throughout the long history of life on Earth, plants and animals have had to adapt to changing environmental conditions, often gradually, sometimes dramatically. After the appearance of *Homo sapiens* and human technology, plants and animals were forced to accommodate themselves to various anthropogenic changes, chief of which were habitat destruction and global climate change.

After the period of the ice ages, many species of plants and animals had to adjust to warmer climates, and paleobotanists, paleozoologists, and paleoclimatologists have studied how various species responded. In general,

A polar bear stands on summer pack ice, hunting for seals. As such pack ice becomes sparse, the bears have difficulty finding their prey. (B & C Alexander/Photoshot/Landov)

because of shifting climate zones, species often migrated to live in ecological niches similar to those to which they had become accustomed. Temperate-zone species moved north, and polar species migrated more deeply into arctic regions. By studying numerous species, scientists have been able to quantify the extent of migration as zonal temperatures rose. Specifically, pollen records from northern Europe and eastern North America have revealed migration rates between 0.02 and 2 kilometers per year, with an average of about 0.4 kilometer per year, though scientists note that migration rates differed widely from species to species. These variations are evidence that some species were unable to keep up with climate change, and this probably played a role in their extinction rates.

Although migration routes for species other than trees are not so well known, the evidence that has been gathered and studied indicates that such organisms as snails, mayflies, and beetles responded to climate changes much faster than did trees. During the past few centuries of anthropogenic climate warming, examples of both range expansions and range contractions have occurred. Ornithologists have observed range extensions for some bird species, and zoologists have found rates of nearly 50 kilometers per year for certain mammals. Those species that are able to expand their midlatitude ranges may be able to cope with the future rise of global temperatures.

On the other hand, some plants and animals have experienced problems in adapting to warming temperatures, especially when new conditions have exceeded their physiological limits or destroyed their ecological niches. For example, multiyear droughts in the southern United States have caused the contraction and destruction of various salt marshes along the southeast and Gulf coasts, precipitating the loss of many species (though some species, such as certain snails, increased their numbers when their predators declined). Anthozoan polyps of coral reefs, which are very sensitive to increases in ocean temperatures, have experienced widespread declines. For example, in 1998, a 1° rise in the Indian Ocean's temperature led to the destruction of over 80 percent of the coral reefs. Nearly all the reefs surrounding the Maldives and Seychelles disappeared.

• Possible Future Scenarios

Scientists who use computer models to predict the possible ecological effects of global warming generally issue caveats about the uncertainty of their predictions, since these models necessarily involve gross oversimplifications. Nonetheless, this practice has not prevented some researchers from warning of massive extinctions produced by rising sea levels and rapidly transformed ecological zones. There are also more hopeful scientists, who, basing their models on the abilities of ancient plants and animals that adapted to disastrous climate changes, believe that global warming may bring about the flourishing of many species, particularly plants that will find high carbon dioxide levels and warm temperatures favorable to their development. Between these extremes are scientists who, like medical doctors employing the principle of *primum non nocere* ("first, do no harm"), urge their fellow humans to be very careful as they modify the Earth's temperature, since living things will most likely continue to interact in complex and unexpected ways.

In constructing their scenarios for the future, scientists have used the growing body of evidence about how plants and animals have responded to climate change in the distant as well as recent past. Several scenarios predict that global warming will inevitably lead to large elevations of sea level and the massive melting of ice in the polar regions. This would lead in turn to both habitat destruction and habitat creation, during which some species adapt and increase in numbers, while others, because they fail to adapt, become extinct. If global warming continues to accelerate, some scientists predict, extinction rates will increase, leading to a loss of plants and animals second only to the gargantuan extinctions during the Cretaceous-Tertiary cataclysm of 65 million years ago. Some pessimists contend that this massive extinction cannot be halted, only slowed. Optimists, however, point out that extraordinary changes also create extraordinary opportunities for adaptive innovations.

• Context

Much of the analysis of the effect of global warming on biodiversity has taken place in the context of a human population whose numbers have dramatically increased in the past few centuries and may continue to increase in the future. If certain plant species (such as trees used for lumber and paper) expand their ranges, then humanity may benefit, but if habitats for edible fish are destroyed by warmer ocean temperatures, then the effects on humanity will be negative. Physicians at the Center for Health and the Global Environment at Harvard Medical School have shown how human health profoundly depends on biodiversity. For example, many important drugs derive directly or indirectly from plant species. Furthermore, if global warming increases the numbers of insects that carry such viral diseases as dengue, encephalitis, and yellow fever, this would seriously endanger human health. Ecosystems have served humanity well throughout its history, but many concerned scientists now believe that numerous plant and animal species in these ecosystems are threatened by human activities that these scientists feel sure have been definitely linked to global warming. However, what humans have done, humans can undo.

Robert J. Paradowski

• Further Reading

Chivian, Eric, and Aaron Bernstein, eds. *Sustaining Life: How Human Health Depends on Biodiversity.* New York: Oxford University Press, 2008. This lavishly illustrated, oversize book is aimed at both general readers and scholars. One of its themes is the threat posed by global warming to plants and animals that are beneficial to humans. Appendixes, references, index.

Cowie, Jonathan. *Climate Change: Biological and Human Aspects.* New York: Cambridge University Press, 2007. This introductory text has sections on the past, present, and future effects of climate change on plants and animals. Illustrations, figures, tables, references, appendixes, index.

Flannery, Tim. *We Are the Weather Makers: The Story of Global Warming.* Rev. ed. London: Penguin, 2007. A scientist and conservationist surveys the history of climate change for the general reader. Beside dealing with its effects on plants and animals, Flannery also offers specific suggestions for politicians and citizens to reduce global warming. Illustrations, chapter notes, index.

Woodward, F. I., ed. *Global Climate Change: The Ecological Consequences.* San Diego, Calif.: Academic Press, 1992. Collects articles by climatologists, atmospheric scientists, and ecologists that discuss the interconnections between climatic and biological systems. Though written by academics for academics, the articles are sufficiently general to be understandable by readers with modest scientific backgrounds. Maps, charts, graphs, references, index.

See also: Amphibians; Convention on International Trade in Endangered Species; Endangered and threatened species; Invasive exotic species; Polar bears.

Economics of global climate change

• **Category:** Economics, industries, and products

Policy makers rely on natural scientists to explain the causes and likely physical consequences of climate change, but they also need economists to provide estimates of the costs of mitigation strategies and to offer guidance on balancing costs in the present against potential damages in the future.

• Key concepts

cap and trade: a system in which the government limits total carbon dioxide emissions but allows individual emitters to buy and sell permits giving legal permission to portions of the total cap

carbon tax: a tax on activities that lead to carbon dioxide emissions, designed to "internalize the externalities" of such emissions

discount rate: the percentage discount to be applied per unit of time in order to account for the higher weight given to costs and benefits that occur sooner

inefficiency: preventable waste in the use or distribution of resources

leakage: the shortfall in targeted greenhouse gas emissions reduction caused by industries relocating to jurisdictions with weaker regulation

negative externality: a situation in which market prices do not reflect the full social costs of some behavior

• Background

Many scientists argue that human activities release greenhouse gases (GHGs) that cause climate changes that will impose harms on future generations. Economists classify such activities as negative externalities that may require corrective policies. Measures that reduce carbon dioxide (CO_2) emissions will cause economic hardship in the present, but their benefit will be lower damages from climate change in the future. Economists already have tools to analyze trade-offs between present costs and future benefits.

• Benefits of Reducing GHG Emissions

The primary benefit of reducing GHG emissions is the expected reduction in damages from climate change in the future. In order to combine various types of damage—such as increased mortality from heat stress, reduction in crop yields, and loss of coastal property—economists must reduce the impacts to a common denominator, namely money.

The monetary damages from future climate change can be quite broad. For example, if climate change leads to more intense hurricanes, the total damages could include not just the money spent rebuilding damaged property, but also the monetary value of the lives lost and even an estimate of the psychological harm suffered by people living with a greater fear of hurricanes. In general, economists prefer cost estimates tied to objective measures, to limit the bias of an individual analyst.

Many economists would not include the creation of new, green industries and jobs to be a net benefit from policies that reduce GHG emissions. While a tax on CO_2 may create jobs in the manufacture of electric cars and solar panels, it would also destroy jobs in the manufacture of SUVs and coal-fired power plants.

• Costs of Reducing GHG Emissions

The costs of climate change policies are the reduced outputs of various goods and services. For example, if the government levies a tax on every ton of CO_2 emitted by a factory, this will increase the cost of production, leading to smaller output and higher prices for the goods produced in the factory. If the factory produces components used in other processes, then the tax will ultimately lead to higher prices for those goods as well.

When a carbon tax or a cap-and-trade program is put in place, the economy will adjust to the new incentives. Producers will strive to maximize their profits, workers will flock to the jobs with the highest salaries, and consumers will shop for the cheapest products. However, any new policy that causes GHG emissions to be lower than they otherwise would have been will necessarily impose economic hardship in the near term. In the absence of such climate policies, the economy would have produced goods and services that emitted a greater amount of GHGs.

Unregulated business does not purposely seek to cause climate damage; it rather seeks to maximize profits by providing consumers with the products they desire at the lowest prices. If government policies force businesses to change their behavior, this results in a lower quantity or quality of output. For example, if it truly were more efficient (disre-

Economics of Climate Change Mitigation

Below is a table of the reduction levels in GHG emissions that have been projected by the Intergovernmental Panel on Climate Change (IPCC) to be achievable in a cost-effective manner: These are the levels of reduction that the IPCC believes can be achieved at a cost of no more than $50 per metric ton of CO_2 equivalent.

Source Sector	*Reduction*
Buildings	5.50
Industry	3.50
Energy supply	3.25
Agriculture	2.75
Forestry	2.00
Transportation	1.75
Waste	0.75

Source: Intergovernmental Panel on Climate Change.

*Amounts are expressed in billions of metric tons of CO_2 equivalent, rounded to the nearest 0.25 billion metric tons.

garding climate change issues) for African nations to build an infrastructure reliant on solar and wind power, then industry would not need special incentives to cause this outcome. The present costs of policies to mitigate climate change may well be justified, in light of the future benefits (reduced climate change damages). Even so, policies leading to lower GHG emissions are costly.

- **Proper Discount Rate to Compare Future Benefits with Present Costs**

In order to determine the relative merits of various climate change policies, including the policy of doing nothing about climate change, economists can first reduce the varied and often qualitative harms of future climate change into an aggregate figure of total monetary damage at each future date. They can perform a similar calculation for the total expected monetary costs of compliance with a given policy for every time period. At each interval of time (for example, every year or every decade) from the onset of the new policy into the future, an economist would then have an estimate of the net costs or benefits from the proposed policy.

Typically, climate change policies impose net costs in the beginning, but at some point in the future yield net benefits, because technology allows for easier compliance with low emission targets, and also because the benefits from reduced climate change grow larger as time passes. The more aggressive the policy—in other words, the harsher the penalties it places on GHG emissions—the higher the net costs in the earlier years, but also the higher the net benefits in later years. By definition, a baseline laissez-faire policy imposes no compliance costs, but eventually yields the largest damages from climate change because of unrestricted GHG emissions. The problem of choosing the most efficient or optimal policy can thus be reduced to choosing the most preferred stream of net costs and net benefits accruing at respective time intervals.

A central concept in economics is the discount rate, which allows one to translate costs and benefits accruing at a future date into their present value. For transactions in the marketplace, the relevant market rate of interest is often used. For example, the owner of a building might consider the installation of insulation that will cost $10,000 up front, but that will reduce heating expenses by $1,200 per year for the next ten years, after which time the insulation will have to be replaced. If there were no discount on future dollars, the insulation would clearly be profitable; it would yield a lifetime savings of $12,000, in contrast to its initial expense of $10,000. However, if the owner could alternatively invest his money in low-risk bonds earning 3.5 percent or more per year, then the insulation would be a poor investment, based purely on monetary considerations. In general, the higher the discount rate, the more present-oriented investors will be; projects that require large upfront expenses, or that do not yield benefits until distant dates in the future, will be penalized, because the early expenses will weigh more heavily than the later benefits.

Once the appropriate discount rate is selected, the various climate change policies can be ranked according to their present value. The lower the discount rate used, the more aggressive will the optimal policy be. This is because aggressive GHG reductions impose higher upfront costs in exchange for greater future benefits, and so their present value increases as a lower discount rate is applied. In contrast, if a high discount rate is adopted, then the optimal policy typically will involve only modest GHG reductions in early years, gently increasing over time as the atmosphere becomes more concentrated with GHGs and climate change damages become less distant in the eyes of policy makers in those time periods.

There are two major controversies among economists regarding this framework for analyzing climate change policies. The first involves adequately dealing with the problem of risk or uncertainty. Unlike stylized scenarios involving cost-saving insulation, the case of climate change involves many unknowns. Many economists argue that the method described above of selecting the best policy may be too risky, and instead they would recommend erring on the side of caution by implementing a more aggressive policy to reduce GHG emissions.

A second controversy lies in the choice of discount rate to be used in the context of climate change trade-offs. Many economists believe market interest rates are inappropriate for matters of public policy, especially ones involving future generations. After adjusting for inflation and specific risks, the "real" rate of return on capital investment in Western economies is estimated at roughly 4 percent per year. This discount rate implies that a climate change policy that imposes compliance costs of $21,000 to people living in the year 2020, but that reduces environmental damages for the people in the year 2120 by $1 million, will be dismissed as too expensive by the earlier generation. Outcomes such as this strike many as selfish.

On the other hand, many economists defend the use of market discount rates. Some discount must be applied, because future generations may be far wealthier than people in the present; it strikes some analysts as unethical to impose compliance costs on developing nations today where many do not have clean water or electricity, in order that (the equivalent of) multimillionaires in the year 2120 might live in a slightly cooler world. There is also the remote possibility that an asteroid collision, nuclear war, or other catastrophe unrelated to climate change could reduce future populations, in which case the total benefits to future humans from reduced GHG emissions would be much lower. In light of these considerations and others, a future dollar in benefits must be weighed less heavily than a present dollar of compliance costs. Many economists favor the choice of the market's discount rate, despite its possible flaws, because it is objectively measurable.

• Context

One of the earliest yet comprehensive works on the economics of climate change was William Nordhaus' 1979 book, *The Efficient Use of Energy Resources*. Since its publication, a growing number of economists have devoted their attention to the many issues involved, ranging from estimates of compliance costs and future damages to fairly technical issues of modeling uncertainty.

Economists can generally be divided into three broad groups in terms of their attitude to climate change. The first group believes that GHG emissions constitute a negative externality, and they recommend a modest carbon tax or other penalty on emissions, which grows over time as climate change damages become more severe. A second group of economists believes the situation is far more serious, and they therefore urge more drastic penalties on emissions. The difference could be due to their reliance on more severe forecasts of future damages or their use of a lower discount rate than the first group. Finally, a third group of economists does not believe the government should take any particular actions regarding climate change, either because they dispute the science underlying the policies or because they believe that politicians cannot be trusted to implement an efficient solution. For example, these skeptics might argue that without worldwide controls on GHG emissions, industries will simply move to unregulated regions, a process known as leakage.

Robert P. Murphy

• Further Reading

Bradley, Robert. *Climate Alarmism Reconsidered*. London: Institute of Economic Affairs, 2003. Bradley holds a Ph.D. in political economy but shows a command of the scientific literature on climate change. He is not a "denier" but argues that a wise understanding of government behavior and unintended consequences should lead to a "look before you leap" attitude regarding massive new programs to combat climate change. Bibliography.

Mendelsohn, Robert, and James Neumann, eds. *The Impact of Climate Change on the United States Economy*. New York: Cambridge University Press, 1999. This collection represents a more optimistic forecast of the impacts of climate change, suggesting that there will be net benefits from a modest warming of 1° or 2° Celsius. The editors argue that other, more pessimistic forecasts make very limiting assumptions on humans' adaptability to climate change.

Nordhaus, William. *The Efficient Use of Energy Resources*. New Haven, Conn.: Yale University Press, 1979. Nordhaus is a pioneer in the economics of climate change, and this is an early yet comprehensive economic analysis of climate change externalities.

_______. *A Question of Balance: Weighing the Options on Global Warming Policies*. New Haven, Conn.: Yale University Press, 2008. Summarizes the latest results of his DICE model of the global climate and economy. Nordhaus favors a carbon tax at an initially modest level that increases over time. Graphs, charts, bibliography.

Stern, Nicholas. *The Economics of Climate Change: The Stern Review.* New York: Cambridge University Press, 2007. Former chief economist for the World Bank, Stern was commissioned by the British government in 2005 to assess the economics of climate change. His aggressive recommendations for rapid reductions in GHG emissions gained worldwide attention and criticism because they deviated from most previous economic analyses. Charts, bibliography.

See also: Air pollution and pollutants: anthropogenic; Basel Convention; Baseline emissions; Carbon dioxide; Carbon footprint; Carbon taxes; Certified emissions reduction; Emissions standards; Employment; Environmental economics; Greenhouse gases; Industrial emission controls; Offsetting; Polluter pays principle.

Ecosystems

• **Category:** Environmentalism, conservation, and ecosystems

Ecosystems influence climate in at least three ways: altering energy balance, regulating water-vapor dynamics via evapotranspiration, and changing GHG cycling in the atmosphere. The ecosystem processes that influence climate change are also influenced by climate, forming ecosystem-climate feedback systems on local, regional, and global scales.

• Key concepts

albedo: the percentage of the solar radiation of all wavelengths reflected by a body or surface

biome: a geographically defined area of similar plant community structure shaped by climatic conditions

canopy: the upper layers of vegetation or uppermost levels of a forest, where energy, water, and greenhouse gases are actively exchanged between ecosystems and the atmosphere

evapotranspiration: processes through which water on surfaces or in plants is lost to the atmosphere

greenhouse gases (GHGs): atmospheric gases that trap heat, preventing it from escaping into space

latent heat flux: the flux of thermal energy from land surface to the atmosphere that is associated with evaporation and transpiration of water from ecosystems

photosynthesis: a metabolic pathway that absorbs inorganic carbon dioxide from the atmosphere and converts it to organic carbon compounds using sunlight as an energy source

respiration: metabolic reactions and processes to convert organic compounds to energy that release CO_2 as a by-product

sensible heat flux: the flux of thermal energy that is associated with a rise in temperature

stoma: a pore in the leaf and stem epidermis that is used for gas exchange

• Background

An ecosystem is a functional system, encompassing all organisms (plants, animals, and microorganisms) and all elements of the nonliving physical environment that interact together in a given area. Organisms extract chemical elements (including water, carbon dioxide, and nutrients) as substrates from the physical environment, using these substrates for their own survival, growth, and reproduction. Physical processes and chemical reactions in the environment are catalyzed by organisms so as to influence energy balance and to form biogeochemcial cycles of carbon, water, and other elements within the system. Ecosystems can be bounded on various scales, from a microcosm to the entire planet.

• Climate and Geographical Distributions of Ecosystems

Various types of ecosystems exist on Earth, including ocean ecosystems, land ecosystems, and freshwater ecosystems on a broad scale. Within land ecosystems, vegetation displays different patterns, forming different ecosystems at regional scales, such as forests, deserts, grasslands, and croplands. Except for artificial ecosystems, patterns of natural ecosystems are primarily shaped by climate conditions (such as temperature and precipitation).

Aldabra, in the Indian Ocean, is the world's largest raised coral atoll and constitutes a unique and fragile ecosystem. (Reuters/Landov)

Along a precipitation gradient from wet to dry regions, ecosystem types change from forests, woodlands, and grasslands to deserts. Along a temperature gradient from the equator to the polar region, ecosystems vary from tropical forests, subtropical forests, temperate deciduous forests, temperature mixed forests, and boreal forests to tundra. In polar climate zones with average temperatures below 10° Celsius in all twelve months of the year, ecosystems include tundra and ice cap in Antarctica and in inner Greenland. Thus, climate and other physical environmental characteristics determine the distribution of ecosystems on the globe.

• Ecosystem Responses to Climate Change

Ecosystems are very sensitive to changes in temperature, atmospheric carbon dioxide (CO_2), and precipitation. Rising atmospheric CO_2 primarily stimulates carbon influx, leading to increases in carbon sequestration and thus potentially mitigating climate change. Rising atmospheric CO_2 concentration has relatively minor impacts on canopy energy balance and water exchange at the surface. Climate warming influences ecosystem feedback related to climate change in several ways, such as exchange of greenhouse gases (GHGs), surface energy balance, and water cycling. It is generally assumed that warming affects carbon release more than carbon uptake, leading to net carbon loss from land ecosystems to the atmosphere. Temperature also affects phenology and length of growing seasons, nutrient availability, and species composition. All these processes influence carbon balances, potentially leading to the net carbon uptake from the atmosphere and negative ecosystem feedback to climate warming.

Increasing temperature also stimulates evapotranspiration, resulting in cooler land surfaces in wet regions and thus negative feedback to climate change. The ecohydrological feedback to climate warming via altering land surface energy balance is weak in dry regions. Altered precipitation regimes (that is, alterations in amount, seasonality, frequency, and intensity) under climate change modify ecosystem carbon cycling, energy balance, and water exchange with the atmosphere. Increased precipitation, for example, usually stimulates plant productivity and ecosystem carbon uptake from the atmosphere. Decreased precipitation generally causes land surfaces to be warmer and generates a higher albedo than does ambient precipitation. Impacts of altered precipitation seasonality, frequency, and intensity are complex and region-specific. In addition, precipitation regimes have long-term impacts on soil development, nutrient availability and vegetation distributions, which can be different from short-term impacts of precipitation on ecosystem processes. Moreover, climate change involves a suite of changes in temperature, precipitation, and GHGs. Those global change factors can interactively influence ecosystem processes and their feedback to climate change.

• Ecosystem Regulation of Climate Change Via Energy Balance

Land surface energy balance influences the climate system by causing fluctuations in temperature, winds, ocean currents, and precipitation. The surface energy balance, in turn, is determined by fractions of absorbed, emitted, and reflected incoming solar radiation. One of the key parameters to determine the energy balance at the land surface is albedo, which regulates differences between the amount of absorbed shortwave radiation (input) and the outgoing longwave radiation (output). Different vegetation covers have different albedo values. When land use and land cover changes occur due to either climate change or anthropogenic activities, land surface energy balance is altered. Overgrazing, for example, may increase albedo. As a consequence, evapotranspiration decreases with associated decline in energy and moisture transfer to the atmosphere. In general, vegetation absorbs more solar energy, transpires more water, drives more air circulation, and results in more local precipitation in a region with low than high albedo. Thus, ecosystems influence energy balance in the atmosphere and feed back to climate change.

• Ecohydrological Regulation of Climate Change

Water vapor exchange at the land surface significantly affects climate dynamics at local, regional, and global scales. Ecosystems receive water input via precipitation and lose water via evapotranspiration. Plant vegetation is the primary regulator of evapotranspiration. Thus, types of ecosystems significantly affect energy and water transfers from ecosystems to the atmosphere.

Because water transpired through leaves comes from the roots, rooting systems play a critical role in ecohydrological regulation of climate. Woody encroachment to grasslands, for example, can accelerate the ecosystem hydrologic cycle and then influence climate dynamics because trees usually have deep taproots to take up water from deep soil layers. Conifer forests can transpire water from the soil to the atmosphere in early spring and late fall and have longer seasons of transpiration than deciduous forests. Conversion of grasslands to winter wheat croplands accelerates evapotranspiration in winter and early spring when wheat actively grows and grasses are dormant. However, evapotranspiration is lower in fallow fields after wheat harvest than in grasslands in summer and fall. In addition, rooting systems are highly adaptive to climate change. When climate warming increases soil temperature and water stress, plants grow more roots to take up water. The adaptive rooting systems can significantly regulate climate change.

• Carbon-Climate Feedback

Ecosystems can regulate climate change via changes in uptake and releases of GHGs. The GHGs involved in ecosystem feedbacks to climate change include CO_2, methane (CH_4), nitrous oxide (N_2O), and ozone (O_3). Their uptakes and releases are modified by changes in temperature, precipitation, atmospheric CO_2 concentration, land use and land cover changes, and nitrogen deposition. For example, ecosystems absorb CO_2 from the atmosphere by photosynthesis and release it back to the atmosphere via respiration. Photosynthetically fixed carbon from the air is converted to organic carbon compounds. Some of the carbon compounds are used to grow plant tissues while others are used for plant respiration. Plant tissues die, adding litter to soil. Litter is partly decomposed by microorganisms to release CO_2 back to the atmosphere and partly incorporated to soil organic matter. The latter can store carbon in soil for hundreds and thousands of years.

Many factors and processes can alter the carbon cycles and then influence carbon-climate feedback. For example, deforestation usually results in net release of carbon from ecosystems to the atmosphere, enhancing climate change. Rising atmospheric CO_2 usually stimulates plant growth and ecosystem carbon sequestration, mitigating climate change. Climate warming can stimulate both photosynthesis and respiration. Most models assume that respiration is more sensitive than photosynthesis to climate warming and predict a positive feedback between terrestrial carbon cycles and climate warming. Field experiments, however, suggest much richer mechanisms driving ecosystem responses to climate warming, including extended growing seasons, enhanced nutrient availability, shifted species composition, and altered ecosystem-water dynamics. The diverse mechanisms likely define more possibilities of carbon-climate feedbacks than projected by the current models.

• Context

Ecosystems are basic units of the biosphere. The latter is the global ecological system integrating all living organisms and their interaction with the lithosphere, hydrosphere, and atmosphere. Biosphere-atmosphere interactions occur via exchanges of energy, water, and GHGs in ecosystems. Specifically, ecosystems interact with the atmosphere via emission and absorption of GHGs so as to influence energy balance in the atmosphere; variations in albedo to influence the amount of heat transferred from ecosystems to the atmosphere; and changes in evapotranspiration to cool the land surface, to influence water vapor dynamics, and to drive atmospheric mixing. In addition, ecosystems can influence climate dynamics by changes in production of aerosols and surface roughness and coupling with the atmosphere. Thus, understanding ecosystem processes that regulate energy balances, water cycling, and carbon and nitrogen dynamics is critical to Earth-system science.

Yiqi Luo

• Further Reading

Chapin, F. Stuart, III, et al. "Changing Feedbacks in the Climate-Biosphere System." *Frontiers in Ecology and the Environment* 6 (2008): 313-320. Provides an overview of interrelationships between ecosystems and the climate system in terms of energy balance, water cycling, and greenhouse gas release and uptake in ecosystems.

Chapin, F. Stuart, III, Harold A. Mooney, and Pamela Matson. *Principles of Terrestrial Ecosystem Ecology.* New York: Springer, 2002. Written by three prominent ecologists, this text provides a good introduction for beginners in ecosystem ecology. Comprises four major sections: context, mechanisms, patterns, and integration.

Field, C. B., D. B. Lobell, and H. A. Peters. "Feedbacks of Terrestrial Ecosystems to Climate Change." *Annual Review of Environmental Resources* 32 (2007): 1-29. Reviews major ecosystem processes that potentially result in either negative or positive feedbacks to climate change. Discusses regional differences between those ecosystem-climate feedback processes.

Luo, Y. Q. "Terrestrial Carbon-Cycle Feedback to Climate Warming." *Annual Review of Ecology Evolution and Systematics* 38 (2007): 683-712. Provides an overview of modeling results and experimental evidence regarding carbon cycle-climate change feedback. Summarizes major regulatory mechanisms underlying the carbon-climate feedbacks, including extended growing seasons, enhanced nutrient availability, shifted species composition, and altered ecosystem-water dynamics in response to climate warming.

See also: Coastal impacts of global climate change; Deserts; Estuaries; Evapotranspiration; Islands; Tundra; Wetlands.

Ecotaxation

• **Category:** Laws, treaties, and protocols

• Definition

Ecotaxation is the use of tax policy to discourage economically inefficient behavior that harms the environment. According to most economists, an unregulated market economy would fail to provide socially ideal or efficient outcomes because of negative externalities. A negative externality is the harm caused to unrelated third parties when two people engage in a market transaction, such as a customer buying a car made in a factory that releases pollution. To deal with negative externalities, most economists endorse a tax on the behavior causing the harm, referred to as a Pigovian tax after economist Arthur Pigou (1877-1959). If correctly calibrated, a Pigovian tax causes the producer to "internalize the externality." A Pigovian tax is thought to promote economic efficiency, because the tax forces individuals to take into account the full consequences of their behavior. After imposing the correct Pigovian tax, the government can allow citizens to make their own decisions without further regulations. Further measures to achieve environmental goals are unnecessary, because market prices, corrected by the Pigovian tax, guide individuals' behavior and lead to a socially desirable outcome.

• Significance for Climate Change

If industrial processes and land-use changes are causing climate change that imposes harm on people in the present or future, then these activities suffer from large negative externalities. For example, when motorists drive vehicles with poor gas mileage or when an electric utility builds a coal-fired power plant, they are not adequately accounting for the extra climate change damages that others will suffer because of their behavior. But if the government raised taxes on gasoline or imposed a tax on utilities based on the amount of carbon dioxide (CO_2) released into the atmosphere, then motorists and utilities would alter their behavior in ways that ultimately reduced anthropogenic climate change. People would still act in their own interest, but the new incentives would ensure that their selfish interests did not jeopardize the welfare of others or of future generations.

Because taxation discourages an activity, most economists favor a policy of "taxing bads, not goods." For example, if the government raised taxes on activities that emit CO_2, and used the new revenues to reduce personal income taxes, then citizens would have the incentive to produce more output but in an environmentally friendlier way. Most economists believe that a correctly calibrated Pigovian tax on environmentally harmful activities, accompanied by dollar-for-dollar tax cuts on economically productive activities, would promote economic efficiency. Many proponents of ecotaxation recommend such offsetting tax cuts on other activities in order to overcome political opposition to tax hikes.

Robert P. Murphy

See also: Air quality standards and measurement; Carbon taxes; Clean Air Acts, U.S.; Emissions standards; Environmental law.

Ecoterrorism

• **Category:** Environmentalism, conservation, and ecosystems

Ecoterrorists commit crimes, primarily arson and vandalism, against entities that they think are harming the environment. Such acts have generally represented reactions to local problems. Because climate change is a global problem, however, there is concern that ecoterrorism could also become a global problem.

• Key concepts

Earth Liberation Front: a loosely organized movement often claiming or associated with ecoterrorism acts in the United States

ecotage: vandalism and direct action taken against corporate polluters

environmental terrorism: acts of violence intended to harm the environment or deprive people of environmental benefits or resources *tree spiking:* a form of vandalism that involves hammering a metal or ceramic spike into a tree to discourage logging

• Background

Ecoterrorism refers collectively to a variety of criminal acts undertaken in the name of protecting nature while specifically not intending to harm humans. In response to a series of high-profile acts of arson connected to the Earth Liberation Front (ELF), in 2002 the U.S. Federal Bureau of Investigation (FBI) created a new definition of ecoterrorism: "the use or threatened use of violence of a criminal nature against innocent victims or property by an environmentally oriented, subnational group for environmental-political reasons, or aimed at an audience beyond the target, often of symbolic nature."

• Evolution of Ecoterrorism

The contemporary construct of ecoterrorism evolved from three terms: ecotage, environmental terrorism, and terrorism. Environmental terrorism has come to mean acts of terrorism against the environment, whereas ecoterrorism refers to terrorist acts undertaken in the perceived interests of the environment. Ecotage entered the American national lexicon in 1972, with the publication of the book *Ecotage!* (1972). The book, published by the national environmental group Environmental Action, was based on a 1971 national contest in which suggestions of ecotage were elicited to make "corporate polluters shape up," garnering the attention of the national media. Ecotage dissipated and fell out of favor, in part because a series of landmark federal antipollution laws were enacted between 1972 and 1975. Prior to 1971, few federal environmental laws existed.

In 1975, Edward Abbey's book *The Monkey Wrench Gang* was published, describing the exploits of the fictional character George Washington Hayduke III, who returns from Vietnam to the desert to find his beloved wilderness threatened by development. Hayduke and a cast of characters join to wage war on construction equipment, dam construction, and road builders. The publication of this book fueled a growth in ecotage, then referred to as "monkeywrenching," especially in the southwestern United States through the vandalizing of billboards and construction equipment. In the early 1980's, tree spiking, a form of vandalism that involves hammering a metal or ceramic spike into a tree to discourage logging, became popular in response to logging old-growth trees in national forests. Although originally the realm of individuals, ecotage began to be seen as a legitimate organizational philosophy through such groups as Earth First!, Greenpeace, and the Sea Shepard Conservation Society.

Beginning in 1990, a number of high-profile events changed the construct of ecotage and monkeywrenching into a more radical, violent, and socially unacceptable act. This included the arrest of Dave Foreman, cofounder of Earth First! In 1991, during the Gulf War, Saddam Hussein intentionally ordered two large oil spills in the Gulf and the detonation of more than twelve hundred oil wells, resulting in numerous fires. In response, President George H. W. Bush branded Saddam Hussein's actions as ecoterrorism. In the mid-1990's, the infamous Unabomber, Ted Kaczynski, received national attention as a labeled ecoterrorist resulting from his mailing of bombs to various corporate offices for anti-environmental actions. His bombing campaign resulted in three deaths and multiple injuries.

Time Line of Ecoterror Incidents

Year	*Incident*
1998	Arson of a U.S. Department of Agriculture animal damage control building near Olympia, Washington, resulting in $2 million in damages
1998	Arson of a Vail, Colorado, ski resort, resulting in $12 million in damages
1999	Arson at a genetic-engineering research office at Michigan State University
2003	Arson at a HUMMER dealership in West Covina, California, destroying 125 SUVs and resulting in an estimated $1 million in damages
2005	Arson of five townhouses under construction in Hagerstown, Maryland
2008	Arson at the Street of Dreams housing development in Woodinville, Washington, resulting in $12 million in damages

Types of Ecoterrorism

Category of Ecoterrorism	*Percent of Incidents, 1993-2004*
Vandalism	77
Arson	12.6
Assault and bodily harm	2
Bombings	1.1
Other	7.3

• Earth Liberation Front

Also during this period, the ELF was founded in Brighton, England, and began engaging in violent and destructive acts increasingly referred to as ecoterrorism. These acts included the 1998 burning of the Vail Resort in Colorado, which garnered the attention of the FBI and the Bureau of Alcohol, Tobacco, and Firearms, who focused on eliminating the ELF. In February, 2002, the FBI declared that the Earth Liberation Front was one of the country's greatest domestic terrorism threats. According to the FBI, between 1996 and 2002, the ELF and the Animal Liberation Front (a violent, direct action animal rights group) committed more than 600 criminal acts in the United States, resulting in damages in excess of $43 million.

• Context

The ELF is the primary entity associated with ecoterrorism; however, it is not an organization, but a movement. It has been very difficult to eliminate because of a lack of organization and hierarchy. A 2008 arson in Woodville, Washington, has raised concern that a resurgence in ELF-associated violent activity is a potential. Although ecoterrorist targets related specifically to global climate change have been limited, primarily SUVs and SUV dealerships, as the concern over the effects of global climate change increases, there is a potential for acts of ecoterrorism to increase.

Travis Wagner

• Further Reading

Abby, Edward. *The Monkeywrench Gang.* New York: Avon, 1975. This novel about a cast of characters performing acts of ecotage to reclaim the wilderness is the most influential book on the subject of ecoterrorism.

Long, Douglas. *Ecoterrorism.* New York: Facts On File, 2004. Well-written reference resource and useful research guide that provides useful background information, definitions, and concepts.

Taylor, Bron R., ed. *Ecological Resistance Movements: The Global Emergence of Radical and Popular Environmentalism.* Albany: State University of New York Press, 1995. Compilation of writings related to the examination and analysis of contemporary movements of ecological resistance.

Wagner, Travis. "Reframing Ecotage as Ecoterrorism: News and the Discourse of Fear." *Environmental Communication: A Journal of Nature and Culture* 2, no. 1 (2008): 25-39. Discusses the evolution of the concept of ecoterrorism as viewed in the mass media.

See also: Amazon deforestation; Anthropogenic climate change; Biodiversity; Civilization and the environment; Human behavior change.

Education about global climate change

• Category: Popular culture and society

Evidence for climate change is found in the geologic record well before the emergence of human beings. However, the notion of anthropogenic influences upon climate has led government policy makers to educate the public about human impacts upon and potential control of Earth's climate.

• Key concepts

anthropogenic climate change: changes in climate due to human activities

Environmental Protection Agency (EPA): the U.S. agency charged with enforcement and regulation of environmental policies

Intergovernmental Panel on Climate Change (IPCC): an international organization of government leaders and scientists

International Journal of Climatology: a professional

journal published by the Royal Meteorological Society

International Union for Quaternary Research (INQUA): an international scientific organization devoted to research on climate change during the Quaternary period, the last 2.6 million years of the Earth's history

National Research Council (NRC): a private, nonprofit organization, founded by congressional charter in 1916, that functions under the auspices of the National Academy of Sciences and provides policy information

National Resources Defense Council (NRDC): a nonprofit environmental action group established in 1970

nonprofit organizations (NPOs): organizations that use their profit for a particular cause or goal

scientific journal article: a report written for a scientific journal and reviewed by a panel of peers

United Nations Environment Programme (UNEP): the official U.N. educational and informational outlet for environmental concerns

World Meteorological Organization (WMO): the official voice on climate and weather issues within the United Nations

• Background

Educating the public regarding the notion of global warming somewhat evolved with the establishment of the Intergovernmental Panel on Climate Change (IPCC) in 1988. Since then the dissemination of the idea of global warming has occurred through the news media, via Internet sites, and through the publishing of formal and informal literature, along with other forms of media. Formal academic sources such as the International Union for Quaternary Research (INQUA) and the *International Journal of Climatology* are examples of scientific information sources. In contrast, books and films such as Al Gore's *An Inconvenient Truth* (2006) are popular vehicles used to educate the public on the issue. Getting the word out about climate change and in particular global warming has transcended a variety of media and organizations.

• Bringing the Issue to the Forefront

In 1962, before the notion of "global warming" was an environmental concern, Senator Gaylord Nelson, during the Kennedy administration, conceived a way to bring public attention to environmental issues. The idea ultimately came to fruition in 1970 as the first official Earth Day. Earth Day began as a grassroots movement bringing environmental concerns such as water and air pollution as well as the impact of population on the environment to an agenda at the national level. Since then Earth Day has evolved as a vehicle for educating the public about a wide range of environmental concerns, including the notion of anthropogenic influences on climate change. Earth Day, celebrated in the United States on April 22, and its U.N. counterpart, celebrated worldwide in March on the date of the spring equinox, have been among the best examples of successful environmental campaigns to promote awareness of global warming issues. Earth Day has served as a rallying point, especially for students, from the elementary to the university levels, concerned about environmental issues.

• Governmental Sources

The responsibilities of the U.S. Environmental Protection Agency (EPA) include addressing environmental issues, setting environment policy, and providing public information on the environment through hearings, press releases, and Web resources. Additionally, the EPA is an example of a U.S. government agency concerned with informing citizens about environmental policy and the impacts these laws may have on the environment. The EPA publishes a wide range of information on climate change, greenhouse gas emissions, climate and health issues, and climate economics via the Internet. It also provides curricular materials on climate change for educators and students and disseminates advance notices to the public regarding new regulations.

• Intergovernmental Panels

The agency best known for its dissemination of climate change information is no doubt the Intergovernmental Panel on Climate Change (IPCC). The IPCC emerged from the World Meteorological Organization and United Nations Environment Programme and is composed of scientists and government officials. It has been suggested that the concept of the IPCC was born from concerns first expressed by Swedish meteorologist Bert Bolin at an environmental meeting held in Stockholm in 1972. At that meeting, Bolin presented a hypothesis suggesting that an apparent rise in CO_2 levels from 1850 to 1970 had contributed to global warming. Ultimately in 1988 the IPCC, with Bolin as its first president, was established. Today, as an environmental organization concerned with informing the public about global warming issues, the IPCC acts as an intergovernmental clearinghouse assessing and disseminating scientific and other forms of research regarding natural and anthropogenic climate change.

- **Scientific Sources**

An example of an organization producing research and publishing professional literature on issues surrounding climate change is the International Union for Quaternary Research (INQUA). This organization, established in 1928, is a nonprofit council of scientists from around the world, publishing and supporting research in areas such as sea-level change and the evaluation of paleoclimatic environments. The research published in INQUA's scientific journal and newsletter is distributed to a wide range of readers.

Based on the peer-evaluated scrutiny they endure, professional journal articles are the voice of scientific research and disseminate information to both the scientific and nonscientific communities. The *International Journal of Climatology*, published under the direction of the Royal Meteorological Society of the United Kingdom, provides articles on the research on climate in general, including themes dealing with climate change. Journals such as this provide the reader with a source of objective climate research. Additionally, it is through journals such as the *International Journal of Climatology* that further evaluation of the research can be made through review of the methods used and replication of the scientific investigations reported on within these articles.

- **Sources Serving Public Policy**

The intent of the United States' National Research Council (NRC) is to assist in educating the public and to help facilitate the government in establishing public policy around a variety of issues pertaining to science. The NRC has established a set of studies entitled "America's Climate Choices" to help assist the public in anticipating problems arising as a result of climate change. Additionally, the NRC sponsors the Summit on America's Climate Choices, an open meeting to help establish a dialogue on climate change. The NRC has been successful in disseminating its information via the Internet, with podcasts and Web sites.

- **Nonprofit Organizations**

The National Research Defense Council (NRDC) is one example of an environmental action group. The NRDC was established in 1970 by law students and attorneys. Its membership exceeds one million. In addition to drawing attention to global warming issues, it provides its members with information pertaining to the establishment of environmental policy and news. Addressing such issues as energy efficiency, health, and pollution, the NRDC's Web site educates the public on global warming and suggests steps for preventing or ameliorating it.

- **Alternative Sources**

The Pennsylvania Council of Churches might sound like an unlikely source for climate education, but its Interfaith Climate Campaign is an example of one of many grassroots attempts at the community level to educate its members and the general public about the impacts of climate change. The campaign instructs members on climate topics through a Web site and community workshops. It also encourages outreach to public officials and governmental agencies and disseminates its program through videotapes and in-church bulletin papers.

- **Global Warming Cinema**

In 2006, the award-winning film *An Inconvenient Truth*, presented by former U.S. vice president Al Gore and directed by Davis Guggenheim, was released as a documentary on global warming. Although the film has undergone much scrutiny and criticism from school boards, scientists, and global warming skeptics, and in the classrooms of the United Kingdom, it is shown with a disclaimer and continues to be made available as an influential educational source on the impacts of global warming. Regardless of the scientific debate over some of the data presented, the film and book have served to increase public awareness of climate change and mobilize public opinion.

• Literature

An amazing array of literature has emerged from both the positive and negative perspectives on global warming and climate change. During the 1970's, popular nonfiction such as John Gribbin's *What's Wrong with Our Weather? The Climatic Threat of the Twenty-first Century* (1979) described the potential for the return of ice-age conditions—global cooling.

Publications for the general audience since the beginning of the twenty-first century, however, have emphasized the climatic warming trend, reflecting concerns over the growth of air pollution (CO_2 and other greenhouse gas emissions), world population, and anecdotal experience of increasing temperatures, borne out by meteorological records, albeit short-lived. Notable contemporary titles include *Greenhouse: The Two-Hundred-Year Story of Global Warming*, by Gale E. Christianson (1999) and Al Gore's *Earth in the Balance: Ecology and the Human Spirit* (2006).

The counter viewpoint also persists in the popular literature. Books such as Christopher Horner's *Red Hot Lies: How Global Warming Alarmists Use Threats, Fraud, and Deception to Keep You Misinformed* (2008) and Thomas Moore's *Climate of Fear: Why We Shouldn't Worry About Global Warming* (1998) are examples of critiques and criticisms of popular views of global warming.

• Educating with a Theme

Additionally, the theme of global warming has entered science-fiction literature and film. Films such as *The Day After Tomorrow* (released 2004), directed by Roland Emmerich, and books such as Jay Kaplan's *A Chilling Warmth: A Tale of Global Warming* (2002) continue to keep the theme of climate change active in the minds of the public.

M. Marian Mustoe

Education specialist C. J. Rea of Kenain Fjords National Park instructs a bipartisan group of thirty U.S. mayors about Alaska's experience of climate change. (AP/Wide World Photos)

• Further Reading

Alley, Richard B. *The Two-Mile Time Machine: Ice Cores, Abrupt Climate Change, and Our Future.* Princeton, N.J.: Princeton University Press, 2000. Alley is a climate research specialist. Details research he completed in Greenland and discusses the process of ice-core drilling.

Christianson, Gale E. *Greenhouse: The Two-Hundred-Year Story of Global Warming.* New York: Walker, 1999. Christianson is a history professor of the College of Arts and Sciences at Indiana State University. Reviews the foundations of the thought behind global warming by describing and characterizing the historical scientific personalities tied to the sciences surrounding the phenomena of climate.

Douglass, D. H., et al. "A Comparison of Tropical Temperature Trends with Model Predictions." *International Journal of Climatology* 28, no. 13 (2007): 1693-1701. Compares atmospheric temperatures derived from models with those observed in the upper atmosphere. Exemplifies the process of validating and checking scientific data and procedures within a study or experiment. Finds a statistical difference between model predictions and observed temperatures in the troposphere above the tropics.

Gribbin, J. *What's Wrong with Our Weather.* New York: Charles Scribner's and Sons, 1978. Describes some of the climate concerns, both warming and cooling, coming to the forefront of the public's attention during the 1970's. Includes reviews of climate patterns and data during these years and discusses some of the research up to that time.

Kench, P., and P. Cowell "Erosion of Low-Lying Reef Islands." *Tiempo* 46 (December, 2002). Geographers Kench and Cowell discuss the dynamics of sea-level rise as it pertains to small islands in the Pacific. They introduce an evaluation instrument that considers the sedimentation and erosion of islands and consider some of the management issues surrounding sea-level rise in small island nations.

Lovelock, J. *The Rough Guide to Climate Change.* 2d ed. London: Rough Guides, 2008. Guide to major themes in global warming. Considers the mechanism behind the process and details its human implications.

Mörner, Nils-Axel. *The Greatest Lie Ever Told.* Stockholm, Sweden: Author, 2007. Mörner, professor emeritus of palegeophysics and geodynamics from Stockholm University, is internationally known as a sea-level expert. Describes research in sea-level measurements and provides a detailed description of the variables that control sea level. Concludes that sea level is not rising as suggested by global warming models.

See also: Basel Convention; Bennett, Hugh Hammond; Brundtland Commission and Report; Canaries in the coal mine; Carson, Rachel; Catastrophist-cornucopian debate; Civilization and the environment; Climate Project; Conspiracy theories; Environmental movement; Falsifiability rule; Friends of Science Society; Friends of the Earth; Gore, Al; *Inconvenient Truth, An*; Human behavior change; Journalism and journalistic ethics; Peer review; Popular culture; Scientific proof; Skeptics; United Nations Climate Change Conference.

8.2ka event

• **Category:** Climatic events and epochs

• **Definition**

The 8.2 kiloyear (8.2ka) event was a sudden decrease in mean global temperatures that started approximately eighty-two hundred years ago. The 8.2ka event lasted between two hundred and four hundred years. This cooling event was not as cold as the previous strong cooling event in Earth's history, the Younger Dryas (spanning the period between 12,800 and 11,500 years ago). The 8.2ka event was, however, colder than the strong cooling event that followed it, the Little Ice Age (approximately 400 to 150 years ago). (The term "Little Ice Age" is used differently by different writers. Many use it to refer to the climate cooling from about 1300 to 1850, while others use it for the latter half of that interval, when cooling was greatest, beginning around 1550 or 1600.)

Periodic cooling events such as the 8.2ka event have left a geologic record of sediments in the northern Atlantic Ocean. These sediments are gravelly sand layers, which were laid down by numerous icebergs as they melted and the icebergs released trapped sediments, which then fell to the seafloor. The 8.2ka event was first discovered by European scientists. When ice cores were drilled in Greenland and studied for their climate record, the occurrence of the 8.2ka event was confirmed.

• **Significance for Climate Change**

The 8.2ka event appears to be one occurrence of global cooling within a repeating pattern of cooling events that have occurred periodically in the Northern Hemisphere. These cooling events started after the end of the last glacial cycle on Earth, about thirteen thousand years ago. The genesis of these events is thought to be related to changes in solar output, changes in circulation patterns in the atmosphere, or other atmospheric and oceanic factors. There have been at least eight such cooling events since the end of the last glacial cycle; however, there is evidence that such cooling events also occurred during the interglacial cycle that preceded the most recent glacial cycle. Cooling events like the 8.2ka event are spaced at intervals of five hundred to two thousand years.

The significance for global climate change of these events is that even though the Earth is in an interglacial or post-glacial warming cycle, there is a history of periodic cooling events superimposed on this current warming trend. It remains to be seen if the current trend of global warming will continue and thus prevent or delay the recurrence of such a cooling event as the 8.2ka or if the future return of such a global cooling event will temporarily reverse the trend of global warming effects.

David T. King, Jr.

See also: Alleroed oscillation; Climate reconstruction; Dating methods; Deglaciation; Holocene climate; Little Ice Age; Medieval Warm Period; Paleoclimates and paleoclimate change; Pleistocene climate; Younger Dryas.

Ekman transport and pumping

• **Category:** Oceanography

• **Definition**

When the wind blows over a body of water, it exerts a force that causes the water on and near the surface to move. Each layer of moving water pulls or drags the layer immediately underneath it. This continues into the depth of the body of water, until the drag reaches the bottom or becomes vanishingly small, whichever occurs first. This process is known as Ekman transport, and it is related to the so-called Ekman spiral. Although this spiral had been observed earlier by Fridtjof Nansen (1861-1930), the Norwegian explorer, diplomat, scientist, and 1922 Nobel Peace laureate, both the transport and the spiral were named after Vagn Walfrid Ekman (1874-1954), the Swedish oceanographer. It was the latter who conducted the first scientific study of this phenomenon and published his results. The research project itself was identified and assigned to Ekman by his mentor and teacher, Vilhelm Bjerknes (1862-1951), the Norwegian physicist and meteorologist.

The Ekman spiral is a twisting structure of liquid or gas currents that arises near a horizontal boundary. The net effect is that, as the flow moves away from the horizontal boundary, its direction rotates, thereby creating the physical structure of a spiral. The Ekman spiral is related to the Coriolis effect (named for Gaspard-Gustave de Coriolis, 1792-1843), a phenomenon that is due to the rotation of the Earth and used to explain why it is that objects moving on the surface of the Earth, or in its atmosphere, do so at an angle to the forces that one applies directly to them. Theory and experiment show that, in the Northern Hemisphere, objects move to the right of applied forces, while in the Southern Hemisphere, they move to their left. The Coriolis effect helps explain part of the Ekman spiral.

When a wind that blows on the surface of the sea varies in the horizontal direction, it induces horizontal variability in the Ekman transports, which creates vertical velocities at the top of the Ekman layer. The creation of vertical velocities is necessary, because the mass of ocean water that flows through a fixed region of space must be conserved (displaced water is replaced). This effect forces water to move up the Ekman layer, against the downward pull of gravity. That action is known as Ekman pumping. In other words, the existence of horizontal divergence in the Ekman transports creates vertical velocities in the upper boundary layer of the ocean.

• **Significance for Climate Change**

When the wind blows on the ocean's surface, the direction of surface currents that are so created do not line up with the direction of the wind. Instead, they move at an angle to it: In the Northern Hemisphere, they move to the right of the wind, and in the Southern Hemisphere, they move to its left. As the effect of the wind moves deeper and deeper into the water, the angle between the direction of the wind and that of the water current of each succeeding layer increases in size.

Consequently, if the currents at the different levels of depth could the viewed from above the ocean surface in the time sequence of their occurrence, then, in the Northern Hemisphere, one would see a water current twisting itself progressively to the right with deeper and deeper penetration, while in the Southern Hemisphere, a similar current would twist itself progressively to the left with deeper and deeper penetration of the ocean. In each hemisphere, the average angle across all depths between the direction of the wind and that of the water current is 90°. Given that the surface of the Earth is covered mostly by oceans, these large-scale movements of ocean water have an influence on the climate.

Patterns of large-scale climate variability in each hemisphere of the Earth are studied using annular modes: a Northern annular mode (NAM) and a Southern annular mode (SAM). They are used to explain the variance in atmospheric flow with time that is not associated with the changes in seasons. The El Niño-Southern Oscillation (ENSO) is a third example of a large-scale pattern of climate variability that is historically tied to the interactions between the ocean and the atmosphere in the tropical Pacific. All three patterns are affected by what happens in the oceans, as well as in the atmosphere, and they particularly reflect the sea surface temperature fields of their respective hemispheres, which are affected by the surface fluxes of latent heat, sensible heat, and heat due to Ekman pumping.

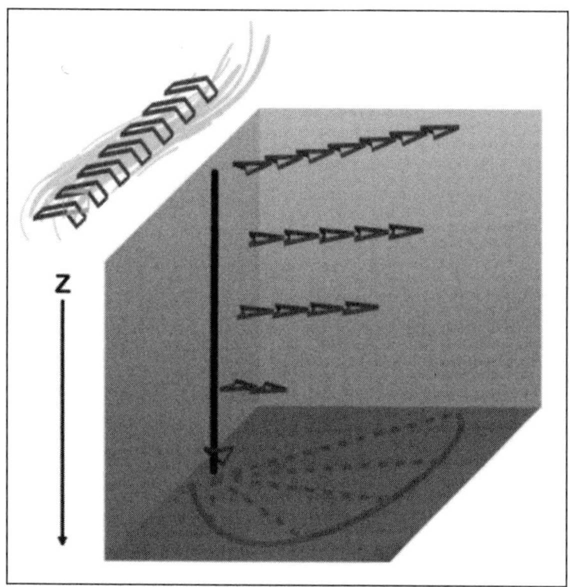

Ekman pumping results from a combination of ocean surface winds, drag, and the Coriolis effect, which cause seawater to move in a spiral pattern.

Ekman transport and pumping are very important in the study of general circulation in the world's oceans, because they create upwelling and downwelling of ocean water. Upwelling occurs when water from below the surface of the ocean is forced to come to the top. Downwelling is the reverse of upwelling. Spatial variability in wind current leads to upwelling near the shore. In the open ocean, it leads to both upwelling and downwelling, which redistribute the mass of water in the ocean. Therefore, upwelling and downwelling due to Ekman transport and pumping are leading mechanisms in the variability of the heat contents in the upper oceans.

Furthermore, simulations indicate that Ekman pumping and oceanic wind-driven circulations respond to increases in atmospheric carbon dioxide (CO_2), one of the greenhouse gases. These simulations have shown that modest increases in CO_2 in the atmosphere have several effects on wind-driven circulations in the oceans in the Southern Hemisphere. They change and intensify the distribution of wind stresses; increase the rate of water circulation, Ekman pumping, and deep-water upwelling in the southern oceans; and expand the subtropical gyres toward the poles in both hemispheres.

Josué Njock Libii

- **Further Reading**

Ahrens, C. Donald. *Meteorology Today*. 9th ed. Pacific Grove, Calif.: Thomson/Brooks/Cole, 2009. A general introduction to the meteorological sciences. Chapters 10, 16, and 17 are particularly pertinent to oceanographic studies.

Price, James F., Robert A. Weller, and Rebecca R. Schudlich. "Wind-Driven Ocean Currents and Ekman Transport." *Science* 238, no. 4833 (1987): 1534-1538. Technical discussion of the topic that is accessible to the general reader.

Sverdrup, Keith A., Alison Duxbury, and Alyn C. Duxbury. *Fundamentals of Oceanography*. New York: McGraw-Hill, 2006. A general introduction to oceanography that covers Ekman transport.

See also: Atlantic heat conveyor; Atlantic multidecadal oscillation; Coriolis effect; El Niño-Southern Oscillation; Gulf Stream; La Niña; Meridional overturning circulation; Ocean-atmosphere coupling; Ocean dynamics; Sea surface temperatures.

El Niño-Southern Oscillation

• **Categories:** Oceanography; meteorology and atmospheric sciences

El Niño and the Southern Oscillation are linked atmosphere-ocean phenomena that occur in the tropical Pacific, but their influence on climate can be seen globally. They represent a cyclical recurrence of warm ocean currents that cause large-scale changes in Earth's weather.

• Key concepts

Hadley cell: an atmospheric circulation system of air rising near the equator, flowing poleward, descending in the subtropics, and then flowing back toward the equator

La Niña: the cooling half of the cycle of which El Niño is the warming half

ocean-atmosphere coupling: the interaction between the sea surface and the lower atmosphere that drives many patterns and changes in Earth's weather systems

paleoclimates: climates of the distant past

Walker circulation: an atmospheric circulation pattern in the Pacific and elsewhere in which hot, moist air rises, travels eastward, cools and dries, descends, and returns westward

• Background

El Niño is a sporadic warming of sea surface water in the central and eastern equatorial Pacific Ocean, adjacent to the Peruvian coast. This warming is part of a cycle, and the cycle's opposite, cooling phase is called La Niña. The Southern Oscillation (SO) is a "seesaw" of air pressure and air circulation between the eastern Pacific and the Indonesian region. The terms El Niño and La Niña are used to denote the extremes of the oscillation. The hyphenate El Niño-Southern Oscillation (ENSO) describes the range of atmospheric and oceanic processes and their accompanying changes. Although ENSO is based in the tropics, it influences weather throughout the Northern Hemisphere and possibly the globe as a whole.

• Characteristics of ENSO

Definitions of ENSO vary, but a common aspect of El Niño is the irregular warming of sea surface water off the coasts of Ecuador, northern Peru, and occasionally Chile. This warming is linked to irregular changes in air pressure at sea level across the Pacific Ocean. During El Niño conditions, the westward-flowing trade winds slacken. During La Niña conditions, by contrast, the westward-flowing trade winds are stronger than normal. A common measure of the SO is the Southern Oscillation Index (SOI), which is usually based on changes in sea-level air pressure at locations on opposite sides of the tropical Pacific. The most common basis for the SOI is the mean sea-level air pressure difference between Tahiti and Darwin, expressed as the long-term difference of their monthly pressures.

Normally, there is a low-pressure zone of warm air in the western Pacific and a high-pressure zone of cool air in the eastern Pacific. This pressure differential drives a loop of warm air from over the western Pacific that rises just east of Indonesia, travels eastward, and descends over the eastern Pacific. The loop then flows in a westerly direction at the surface back toward the west. The strength of this circulation, known as the Walker circulation, is heavily influenced by the seesaw-like sea-level pressure differences between the eastern and western Pacific. The pattern is named after Gilbert Walker, whose work led to the discovery of the Southern Oscillation.

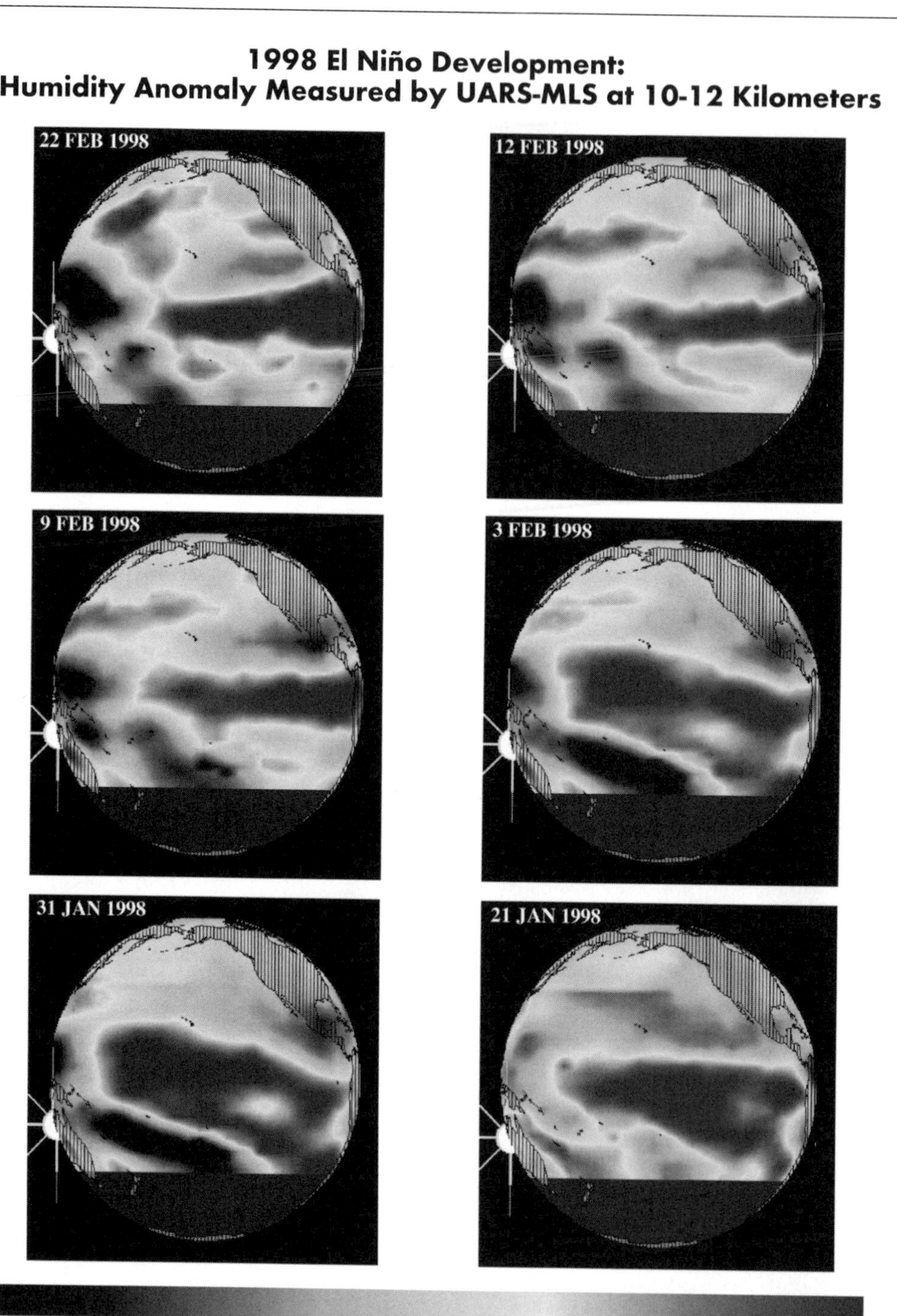

This series of images shows the development of water vapor over the Pacific Ocean during the El Niño event of January and February, 1998. Warmer water temperatures result in greater-than-normal water evaporation, warmer, moister air, and finally altered global weather patterns. (NASA/JPL/Caltech)

The formation and breakdown of the Walker circulation cell is reflected in the pressure difference across the Pacific: When pressure is low in Tahiti, it is high in Darwin, and vice versa. This periodic yearly-to-decadal seesaw of atmospheric and oceanic circulation is the SO. When the SOI is negative, sea surface temperatures (SSTs) are warmer than usual in the eastern equatorial Pacific, off the coats of Ecuador and northern Peru (and occasionally Chile). A negative SOI is associated with El Niño conditions. When the SOI is positive, SSTs are cooler than usual in the eastern equatorial Pacific. A positive SOI is associated with La Niña conditions. During La Niña events, the east-west movement of air in the Walker circulation is enhanced with well-defined and vigorous rising and sinking branches.

• Links to Global Climate Change

During El Niño events, there is an increase in Hadley cell circulation, the circulation cell of air that rises over the equator and descends in the subtropical latitudes on both sides of the equator. A more vigorous overturning of the Hadley cell circulation leads to an increase in heat transfer from tropical to higher latitudes in both hemispheres. Often, the climatic effects of increased Hadley circulation can be seen globally as above-average temperatures and extreme precipitation. Various other climatic consequences of ENSO events have been reported: Warmer ocean waters have bleached coral, and vigorous atmospheric circulation has driven ocean currents northward, warming the Arctic Ocean and decreasing the amount of sea ice there.

• Assumptions About ENSO and Anthropogenic Global Warming

The Intergovernmental Panel on Climate Change (IPCC) has commented on the possibility of connections between ENSO and anthropogenic increases in atmospheric greenhouse gases. Based on the output of complex but unvalidated global climate models, the IPCC indicates that as tempera-

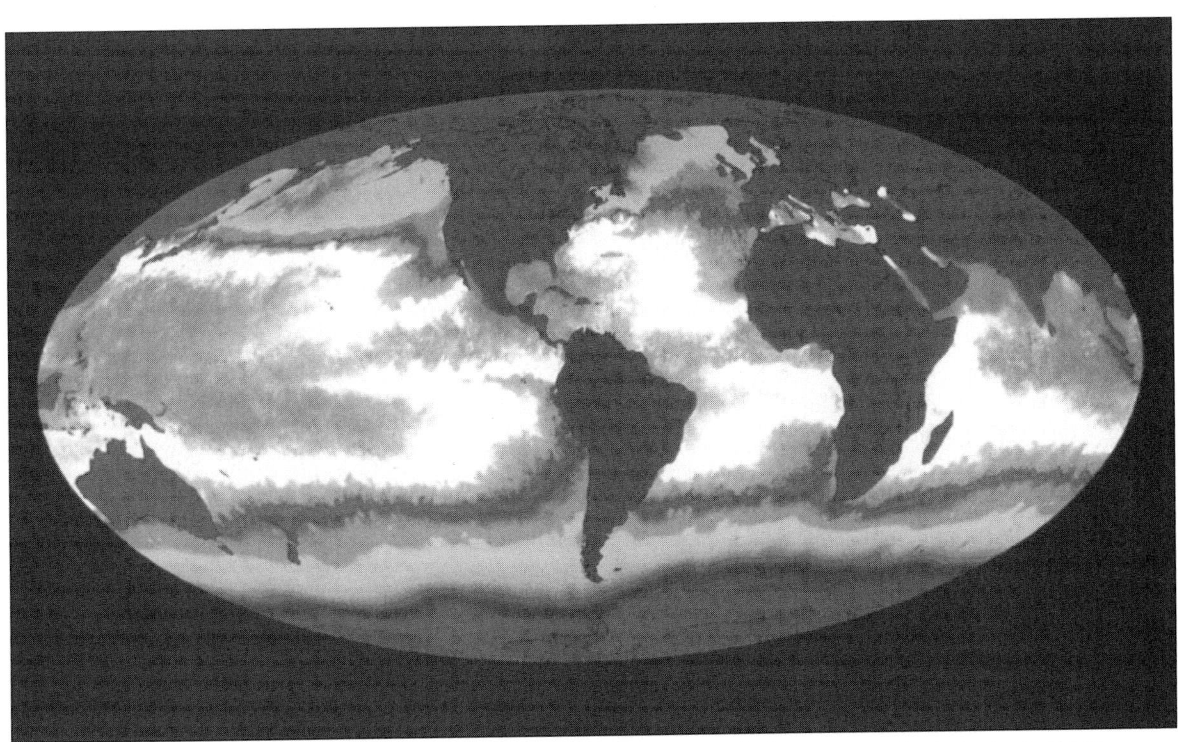

This satellite image depicts the average variation in sea surface temperatures (lighter is warmer), which alter in El Niño years. (NASA)

tures increase, the average Pacific climate could more consistently emulate El Niño conditions. However, the IPCC accepts that some climate models point to a more La Niña-like response to global warming, because Hadley cell circulation may decrease with increasing global temperature. Paleoclimatic studies support the view that global warming is likely to foster weaker and less frequent El Niño events. However, the level of scientific uncertainty and the existence of conflicting results are such that reliable prediction of future climate is not possible at this time.

• Context

Owing to the complexity and uncertainty surrounding global climate and climate change, neither the mean annual values of ENSO nor the interannual variability of ENSO can be reliably simulated in global climate models. Despite this, projections have been made about future trends in precipitation extremes linked to ENSO. ENSO has a noticeable influence on mean global temperature, and shifts in temperature are consistent with shifts in the SOI, but the relationship between temperature and ENSO effects has not been consistently strong. On one hand, strong El Niño events create significant spikes in mean global air temperature. On the other hand, there is evidence that long-term warming depresses El Niño activity. For this reason, the mutual effects of climate change and ENSO upon each other remain difficult to predict.

C R de Freitas

• Further Reading

Couper-Johnston, R. *El Niño: The Weather Phenomenon That Changed the World.* London: Hodder and Stoughton, 2000. The author is a wildlife and science documentary maker who has examined impacts of ENSO around the world. Presents a historical perspective on the topic in a popular form. Focuses mainly on the historical impacts of ENSO on human society and civilization. Illustrations, maps.

D'Aleo, J. S., and P. G. Grube. *El Niño and La Niña.* Westport, Conn.: ORYX Press, 2002. Two climate scientists describe what is known about ENSO phenomena and underlying processes in twelve short chapters. Themes dealt with include a history of ENSO, how scientists study and track ENSO phenomena, and how ENSO information is used. Glossary, large bibliography, charts, maps, and diagrams and color plates.

Glantz, M. H. *Currents of Change: Impacts of El Niño and La Niña on Climate and Society.* 2d ed. New York: Cambridge University Press, 2001. The author, a climate scientist, describes in an easily understood fashion what ENSO is and what its negative and positive consequences on weather and climate are around the world. Written for the lay audience; covers the increases in information and knowledge gained from the major El Niño event of 1997-1998. Illustrations, figures, tables, maps, bibliography, index.

See also: Average weather; Drought; Extreme weather events; La Niña; Monsoons; Tropical storms; Weather vs. climate.

Elton, Charles Sutherland

English ecologist

Born: March 29, 1900; Withington, Manchester, England
Died: May 1, 1991; Oxford, England

Elton was the first person to describe ecosystem invasions by animals and plants. As the climate changes, all indications are that such invasions will be more frequent and have greater effects.

• Life

Charles Sutherland Elton was the youngest of three sons born to Oliver Elton, a professor of English literature at Liverpool University, and Letitia Maynard Elton (nee MacColl), a children's writer. Charles attended Liverpool College and graduated from Oxford University with a degree in zoology in 1922. In 1921, he had the opportunity to join an Arctic expedition to Spitsbergen to study Arctic vertebrates. He returned to the Arctic in 1923, 1924, and 1930 to continue the study. From 1926 to 1931,

Elton was a consultant with the Hudson's Bay Company, studying species that affected the fur trade. In 1923, he was appointed departmental demonstrator at Oxford.

Elton married Rose Montagne in 1928. After five years, they separated and divorced amicably. His second marriage, a few years later, to Edith Joy Scovell was happy. He found great pleasure in his family life. In 1932, Elton established the Bureau of Animal Population at Oxford and became the editor of the new publication *Journal of Animal Ecology*. He would continue as editor for nineteen years. He was appointed reader in animal ecology at Oxford University and senior research fellow at Corpus Christi College in 1936. During World War II, the Bureau of Animal Population was given the task of controlling the loss of foodstuff to rats, mice, and rabbits. For twenty years after the war, Elton studied the animals on the Wytham estate of Oxford University. He retired in 1965, but he remained active, managing a study of tropical America in his retirement.

• Climate Work

Elton did not publish a large number of research papers, but his books left a dynamic legacy. In *Animal Ecology* (1927; rev. ed., 1946), his clear, concise writing defined ecology for future scientists. He was one of the first people to apply scientific methods to the study of animals in the wild, their interrelationships, and their relationship with the environment. He discussed ecological niches and their relationship to larger ecosystems, particularly the pyramid of numbers—the notion that each level up the food chain has a smaller population than the one below it.

Voles, Mice, and Lemmings: Problems in Population Dynamics (1942) discussed the subject of population fluctuation. Elton felt that understanding the rodent population was a major factor in controlling human plague. In his studies of the population fluctuations of animals, he came to the conclusions that the population of wild animals is never constant, that the change in number is also neither constant nor periodic, and that the ever-changing population of one species has both direct and indirect effects on the populations of other species. Elton was very concerned with conserving biodiversity. He felt that preserving biodiversity established the correct relationship of humanity with nature, provided greater opportunities for people to experience nature, and increased ecological stability.

The Ecology of Invasions by Animals and Plants (1958) brought together many different fields—such as biogeography, human history, epidemiology, population ecology, and conservation biology—to show how important the concept of invasions can be. Elton predicted that in the future there would be a smaller total number of species because some of the native species would be eliminated by invading species. Increasing human travel, he believed, would facilitate the transportation of invasive species. Invasive species in general are likely to experience explosive population growth in a new ecosystem, because they will have fewer natural predators and parasites.

The Pattern of Animal Communities (1966) was published about the time Elton retired, and the work's greatest impact on the field of ecology may still lie in the future. Some of Elton's ideas have been ahead of their time. For example, Elton posited that, when a species is under stress, it will often search for a new environment. Thus, the selection of environment by animal is more likely than the selection of animal by environment. This proposition has significant implications for climate change, which will cause major changes in the animal and plant populations of specific ecosystems. These changes may cause some species to migrate in search of new habitats, turning them into invasive species.

C. Alton Hassell

• Further Reading

Crowcroft, Peter. *Elton's Ecologists: A History of the Bureau of Animal Population.* Chicago: University of Chicago Press, 1991. Informal history of the group who worked at the Bureau of Animal Population with Elton. Illustrations, index.

Elton, Charles S. *The Ecology of Invasions by Animals and Plants.* Reprint. Chicago: University of Chicago Press, 2000. Elton's classic book on invasions by nonnative species. Illustrations, maps, bibliography.

_______. *Voles, Mice, and Lemmings: Problems in Population Dynamics.* Oxford, England: Clarendon

Press, 1942. Another of Elton's classic books that define ecology. Illustration, maps, bibliography.

Elton, Charles S., Mathew A. Leibold, and J. Timothy Wootton. *Animal Ecology*. Reprint. Chicago: University of Chicago Press, 2001. Leibold and Wootton have added introductions to each chapter of this reprint, elucidating the scientific and historical context of the chapter. Illustrations, bibliography.

See also: Budongo Forest Project; Carbon dioxide fertilization; Forestry and forest management; Invasive exotic species.

Emiliani, Cesare

Italian Ameican paleoceanographer

Born: December 8, 1922; Bologna, Italy
Died: July 20, 1995; Palm Beach Gardens, Florida

Emiliani developed the oxygen isotopic abundance method to study the history of the oceans and their climate changes.

• Life

Cesare Emiliani was born to Luigi Emiliani and Maria Manfredidi Emiliani. Cesare studied geology at the University of Bologna, and in 1945 he received a doctorate in micropaleontology from the same university. From 1946 to 1948, he worked for the Societa Idrocarburi Nazionale in Florence as a micropaleontologist. In 1948, he moved to the University of Chicago, where he had been awarded the Rollin D. Salisbury Fellowship. He received a second doctorate in 1950 from the University of Chicago, this time in geology (isotopic paleoclimatology). In Chicago, he met and, on June 28, 1951, married his wife, Rosita. They had two children: Sandra and Mario. Emiliani's family was important to him. He was a wonderful husband, father, and grandfather.

Emiliani worked in Harold Urey's geochemisty laboratory at the Enrico Fermi Institute for Nuclear Studies at the University of Chicago as a research associate from 1950 to 1956. In 1957, he moved to the University of Miami's Institute of Marine Science, which would later become the Rosenstiel School of Marine and Atmospheric Sciences. There, he developed programs in marine geology and geophysics. In 1967, he became chair of the Department of Geological Sciences, which he had organized. He remained the chair until his retirement in 1993. He was at home in 1995 when he was felled by a sudden heart attack.

• Climate Work

In Urey's lab, Emiliani learned how the two stable isotopes of oxygen were related to temperature change. He used isotopic analysis to study the carbonate remains of foraminifera, microscopic protists that live close to the ocean's surface, falling to the seafloor only after they die to become part of the sedimentary layer. Emiliani found foraminifera remains both on the ocean floor and in cores drilled into the floor. From the data he acquired, he determined that the ocean had been much warmer in the early Tertiary period, about 65 million years ago. The discovery that the ocean was not always the same temperature opened up the new field of paleoceanography to study changes in what had previously been thought to be a constant environment.

The isotopic abundance method of comparing ratios of oxygen isotopes is based on the fact that, when in thermodynamic equilibrium, calcite and seawater differ in their ratios of oxygen 18 (O^{18}) to oxygen 16 (O^{16}). Calcite forms a part of the body of foraminifera that is preserved in fossilized remains of the creatures. The ratio of O^{18} to O^{16} in these creatures and their remains increases with cooler water temperatures and with increasing volumes of ice, so it is an indicator of temperature changes.

Looking at the oxygen abundance ratio of fossilized calcite, Emiliani found a sawtooth pattern in deep-sea cores indicating that there had been periodic ice ages. His work also showed that the Pleistocene period had experienced several ice ages, at a time when most scientists believed there had been only four. Emiliani found evidence of seven ice ages in Caribbean deep-ocean bed cores, and in Pacific cores, he found evidence of fifteen ice ages.

His work introduced the notion that ice ages were cyclic phenomena.

Emiliani concluded that cycles of glaciers were caused by several different factors. The albedo of, or the amount of light reflected by, surface ice affected how much heat the ice and its surroundings absorbed and thus the amount of ice that melted. (Ice surrounded by heat-absorbing dark water will melt faster than ice surrounded by more heat-reflecting ice.) Glacier cycles also seemed to be affected by changes in the stress placed on the ocean floor by the weight of ice on land, by orogenic (mountain-forming) uplift, and by changes in the amount of solar radiation reaching Earth. Emiliani's research transformed the study of the oceans and their history.

Emiliani also discovered that different species of plankton live at different levels of the ocean. The temperature of the ocean drops quickly with increasing depth, so species living at different depths have different ratios of O^{18} to O^{16}. Emiliani's assertion that different species live at different depths was initially ridiculed but has since been proven to be true.

Emiliani's move to the University of Miami was motivated by his belief that deep-sea drilling cores provided the best evidence of ocean history. He was disappointed when one core-drilling project, the Mohole project, folded, so he sent a proposal to the U.S. National Science Foundation to fund the Long Cores project (LOCO). The cores drilled by this project produced enough data to motivate the creation of the Joint Oceanographic Institutions for Deep Earth Sampling (JOIDES). Its three subsequent projects operated continuously from 1966 until 2003. The data from deep-sea cores not only provided evidence of ocean history but also proved the theory of seafloor spreading and plate tectonics.

C. Alton Hassell

• Further Reading

Elderfield, Harry. *The Oceans and Marine Geochemistry.* Amsterdam, the Netherlands: Esevier Pergamon, 2006. Dedicated to Emiliani, among others. Shows how impotant his work was and is in understanding the oceans and ocean geology. Illustrations, bibliography, index.

Emiliani, Cesare. *Planet Earth: Cosomology, Geology, and the Evolution of Life and Environment.* New York: Cambridge University Press, 1992. Emiliani's synthesis of different sciences to describe how the world was formed.

_______. *The Scientific Companion: Exploring the Physical World with Facts, Figures, and Formulas.* New York: Wiley, 1988. Emiliani uses his vast knowledge of different topics to correlate what is known of today's world.

Hill, Maurice N., et al. *The Sea: Ideas and Observations on Progress in the Study of the Seas.* Reprint. New York: Wiley, 2005. Reissue of an earlier series of books cowritten by Emiliani and demonstrating his influence upon his coauthors. Illustrations, graphs.

See also: Climate reconstruction; Dating methods; Deglaciation; Earth history; Paleoclimates and paleoclimate change.

Emission scenario

• **Categories:** Meteorology and atmospheric sciences; pollution and waste

• Definition

An emission scenario is a means of assessing the future impact of global greenhouse gas (GHG) emissions through different story lines that use a variety of assumptions. The Intergovernmental Panel on Climate Change (IPCC), established by the World Meteorological Organization (WMO) and the United Nations Environment Programme (UNEP), uses various emission scenarios to periodically review the science, impacts, and socioeconomics of climate change. The IPCC developed the projected scenarios using a variety of sources including literature searches, alternative modeling approaches, and feedback from informed groups and individuals. These models include emissions from all types of GHGs and their driving forces.

• Significance for Climate Change

The Special Report on Emissions Scenarios (SRES) allows a forward-looking assessment of the causes and impact of emissions as viewed from demographic, technological, and economic developments. Armed with information provided by the emission scenarios, the IPCC is able to advise the Conference of the Parties to the United Nations Framework Convention on Climate Change (UNFCCC). The SRES team included fifty members from eighteen countries representing a range of scientific disciplines. In order to provide the UNFCCC with useful information covering a variety of possibilities, the IPCC used forty different scenarios; each scenario makes different assumptions for future GHG pollution. Included in the report are assumptions about possible technological and future economic development for each scenario, as well as options to reduce the predicted environmental hazards.

The reports come in the form of story lines, with each story line representing a different future possibility. For example, one of the story lines allows for a future of rapid economic growth, advanced technologies and global population peaking in mid-century and then declining. Another story line presents continuously expanding global population with slower economic growth, while another assumes the introduction of cleaner, more efficient technologies that are environmentally friendly and assumes growth of service and information technologies. A fourth story line presents variations on the other three possible outcomes.

As with any attempt to divine the future, emission scenarios have their critics. The critics argue that the scenarios don't include climate-specific initiatives and that they miss the mark on their assessment of economic growth and degree of future GHG emissions. For example, overestimating economic growth might point to higher GHG emissions than reality would suggest. Also, the scenarios do not assume implementation of the UNFCCC or the emissions targets of the Kyoto Protocol. However, emission scenarios are a useful tool used throughout the world, and many models exist independent of the IPPC model.

Richard S. Spira

See also: Climate models and modeling; Climate prediction and projection; WGII LESS scenarios.

Emissions standards

• Category: Pollution and waste

Emissions standards seek to protect the environment and the climate by setting limits on the amount of pollutants and GHGs that may be emitted by motor vehicles, factories, and other sources of anthropogenic air pollution.

• Key concepts

catalytic converters: devices that reduce vehicular carbon monoxide exhaust by converting it into carbon dioxide

emissions trading: a practice in which the right to pollute is turned into an exchangeable commodity, motivating polluters to reduce their emissions so they can profit by selling the right to others

Kyoto Protocol: a binding international agreement that includes a detailed plan to reduce greenhouse gas emissions

maximum achievable control technology (MACT) standards: standards designed to limit air pollution from stationary sources (usually heavy industries)

smog: the noticeable brown haze created by vehicular and industrial emissions and other fossil emissions, especially prevalent in large cities

volatile organic compounds: carbon-based substances that can easily enter the atmosphere through vaporization

• Background

The evolution of emissions standards began largely in response to the heavy vehicular traffic and industrialization of major urban areas. At the beginning of the twentieth century, the word "smog" was used to describe the grim atmospheric haze created by smoke and sulfur dioxide from burning coal. This haze was especially noticeable in large cities such as

London, which had already suffered from the effects of burning coal for centuries. As automobiles became popular, pollutants from coal were joined by photochemical smog, created by emissions from gasoline engines and other sources, releasing volatile organic compounds that were acted upon by sunlight.

Over time, with increased scientific research, the global scope and long-range impact of air pollution became more apparent, and local efforts to prevent further damage were joined by national and international policies and agreements. In the United States, the Environmental Protection Agency (EPA) sets National Ambient Air Quality Standards, which regulate carbon monoxide, lead, nitrogen dioxide, particulate matter, ozone, and sulfur dioxide. These key pollutants are measured by both volume and weight, and measurements are made based on the amount of fuel input. Regulations based on fuel input have been criticized as being less relevant to overall energy efficiency than regulations based on output.

One of the most significant international agreements regulating emissions is the Kyoto Protocol of 1997, which implemented the suggestions of the 1992 United Nations Framework Convention on Climate Change. An important feature of the protocol is that industrialized nations agreed to cover more of the initial costs of reducing greenhouse gases (GHGs) than did developing nations. Also, the Kyoto Protocol promoted emissions trading and other viable economic incentives that make international cooperation in combating global warming more practical.

• Stationary Emissions Standards

Factories and power plants have been subject to varying emissions standards, with a focus on regulating major sources. In the United States, the Clean Air Act Extension of 1970 was amended in 1990 to define major sources as individual or grouped stationary sources with the capacity to release 9 metric tons or more per year of a single pollutant, or 23 metric tons or more per year of combined pollutants. Coal, which is still used all over the world, is a significant source of dangerous mercury pollution, and many nations, including China, are still heavily dependent on coal for power. In 2005, the EPA issued the Clean Air Mercury Rule in an attempt to reduce mercury emissions from coal-burning power plants in the United States.

• Road-Based Vehicular Emissions Standards

One of the most significant sources of global air pollution is the exhaust from cars and trucks. While the image of factory smokestacks is often used as a symbol for air pollution, the cumulative effect of millions of vehicles, even with catalytic converters, is potentially catastrophic. In 2004, carbon dioxide emitted by personal vehicles in the United States alone reached 314 million metric tons. Within the European Union, emissions standards differentiate between diesel and gasoline fuel, as well as between vehicles of different weights and sizes. Testing also takes the temperature of an engine into account. Europe's standards have been adopted in some Asian countries, including India and China.

• Nonroad Vehicular Emissions Standards

Locomotives, boats, aircraft, and farm and lawn equipment are also sources of air pollution, and attempts have been made to control their toxic emissions. Along with particulate matter and other byproducts of combustion, diesel engines used in locomotives and boats emit sulfur, and the piston engines in aircraft emit lead. In 2007, tougher emissions standards for marine diesel engines that had been set by the EPA began to take effect, with the goal of reducing the level of sulfur in fuel by 99 percent, and stricter standards were set for newly constructed engines. Smaller boats often use gasoline engines similar to those found in cars and trucks, and although not as prevalent, they release the same pollutants as road-based vehicles. Engine emission standards for aircraft have been set by the United Nations International Civil Aviation Organization, and there is a trend toward greater regulation of this mode of transportation.

• Context

The creation and enforcement of emissions standards is one of several long-term strategies for reducing GHGs and combatting global climate change. Other strategies reject the use of fossil fuels altogether and seek to find and promote alternative sources of energy. Many legal and political complexities have arisen in determining which corporate entities or governments are responsible for preventing environmental damage to the air in another community, since the the air is not restricted by political borders of any kind.

Alice Myers

• Further Reading

Ellerman, A. Denny. *Markets for Clean Air: The U.S. Acid Rain Program.* New York: Cambridge University Press, 2000. Details the first three years of the program, including its political and economic background and its impact on the problem. Illustrated. Bibliography and index.

Hansjürgens, Bernd, ed. *Emissions Trading for Climate Policy: U.S. and European Perspectives.* New York: Cambridge University Press, 2005. Includes economic analysis and theory, as well as consideration of options for future policies. Bibliography, index.

Reed, Alan. *Precious Air: The Kyoto Protocol and Profit in the Global Warming Game.* Overland Park, Kans.: Leathers, 2006. Comprehensive study describing the United Nations' efforts to deal with global warming, the Kyoto Protocol's effects on world economics, and the development of the emissions trading industry. Illustrations, bibliography.

Sperling, Daniel, and James S. Cannon, eds. *Driving Climate Change: Cutting Carbon from Transportation.* Boston: Academic Press, 2007. An outgrowth of the 2005 Asilomar Conference on Transportation and Energy in California, this book discusses strategies for reducing GHG emissions. Illustrated. Bibliography, index.

Tietenberg, T. H. *Emissions Trading: Principles and Practice.* Washington, D.C.: Resources for the Future, 2006. Classic study of the use of transferable permits or economic incentives to control pollution. Illustrated. Bibliography, index.

See also: Carbon dioxide; Emission scenario; Greenhouse gases; Industrial emission controls; Motor vehicles; U.S. legislation.

Employment

• **Category:** Economics, industries, and products

As Earth's climate changes, some workers may leave adversely affected regions and migrate to other areas. Other workers will remain in place but may be forced to seek different employment as industries become less viable or are damaged by extreme weather events. In some cases, new economic opportunities will arise from mitigation and adaptation efforts.

• Key concepts

environmental refugees: people forced to move because of climate change

extreme weather events: major, potentially damaging weather occurrences, such as hurricanes, snowstorms, or floods

Intergovernmental Panel on Climate Change (IPCC): a group of scientists charged by the United Nations to develop various scenarios examining the impact of climate change

migration: the movement of people from place to place

scenario: outline of the impact of climate change under certain conditions

• Background

At present, human communities are more vulnerable to extreme weather events than to gradual change. Nonetheless, over time communities will be affected by global warming. As is usually the case, poorer communities will be more vulnerable than wealthier ones. Agriculture will be more affected than industry by gradual climate change, because industry is more adaptable. Extreme weather events, a likely product of global warming, will often negatively affect industries. Workers will often be forced to seek employment in new jobs when their old employers lose their viability. In some regions, workers may be forced to migrate from severely affected regions to those less affected, even when this means crossing national borders.

• Extreme Weather Events and Employment

Global warming is expected to produce several impacts. Some, such as increasingly warmer weather or decreased precipitation, would be gradual. Most scientists agree, however, that global warming is also likely to produce more extreme weather, particularly in the tropics. The scenarios produced by the working group of the Intergovernmental Panel on Climate Change (IPCC) indicate that warmer weather will likely produce more and larger hurricanes and cyclones in the tropics, as well as more intense rainstorms or droughts in some areas. The occurrence of specific extreme events is difficult to predict on a long-term basis, so it will be difficult to plan for such events in any one specific area. Nonetheless, it is possible to indicate some regions that will be likely to experience more extreme weather events, such as Category 5 hurricanes (those with wind speeds over 241 kilometers per hour).

In the United States, for example, the IPCC scenarios indicate that Florida and the Gulf Coast will experience more hurricanes and, more important, hurricanes of greater magnitude. Worldwide, several cities, such as Miami in the United States and Mumbai in India, are situated along coasts vulnerable to hurricanes and cyclones; consequently, industries in these cities are vulnerable to the flooding and wind damage produced by large storms. It may be possible to adapt to stronger storms, especially in the industrial nations, but they will often produce short-term disruption that leads some businesses to close temporarily or eventually to move elsewhere. Even in industrialized countries, increased insurance costs may lead some businesses to move from coastal areas to inland locations. Businesses in less industrialized countries may simply close or move elsewhere rather than trying to rebuild when faced with major devastation. When manufacturing plants close because of extreme weather, skilled workers are forced to mi-

In March, 2007, European Union commissioner for employment and social affairs Vladimir Spidla, left, observes a discussion between European Commission president Jose Manuel Barroso and German chancellor Angela Merkel, who is pushing her fellow E.U. members to improve the European economy by embracing clean energy initiatives. (AP/Wide World Photos)

grate elsewhere, although they may be able to find comparable jobs.

The increase in number and intensity of hurricanes will have a negative impact on low-wage employers in hurricane-prone areas in both industrial and less industrialized nations, as these businesses often do not have the capital to rebuild. Also, some people who are forced to relocate by severe weather events never return to their old homes. Hurricane Katrina, which hit New Orleans in 2005, left many people without jobs or homes. Some of the people who left New Orleans returned to the city, but many moved elsewhere, never to return.

Climate change and extreme weather events are likely to harm the tourist industry in several areas. The presence of more powerful storms in the tropics (a likely event in most climate scenarios) may deter some tourists from visiting beachfront resorts. Resorts often will be faced with added costs of operation, such as insurance costs, that will force some to close. In mountain areas, higher temperatures that come from gradual warming will affect snowpack, making it difficult to sustain ski resorts. Already, some ski resorts in the North Carolina mountains are being forced to close by the lack of enough snow to sustain their business. Many resort workers are seasonal and will be able to find work elsewhere, but many are permanent residents who will have to relocate in order to find work.

• Gradual Climate Change and Employment

Even when regions do not suffer from extreme weather, warming can lead to a gradual rise in sea level, flooding some coastal areas. In other cases, increased precipitation or snow melt will cause rivers to flood, affecting nearby communities. When combined with extreme events, this pattern can lead to substantial migration. Already, this is beginning to happen in Bangladesh. Farmers are losing productive land to water incursions and are crossing the border into the neighboring Indian province of Assam. These new immigrants often cause a reduction in wage levels, producing hostility between natives and the new arrivals. More extreme climate change scenarios indicate that some Pacific Islanders, such as those on Nauru, may be forced to evacuate their home islands before the end of the century when increases in sea level and more powerful storms make the low-lying islands uninhabitable. These migrants will have an impact on labor supplies wherever they settle. Inundations by seawater are already having an impact on the parishes south of New Orleans. Fishermen and fur trappers are being forced to move elsewhere in search of work as their livelihoods are being destroyed.

In some cases, it is the lack of water rather than too much water that leads to environmental migration. As desert regions in northern Africa become drier, herding people are finding it difficult to maintain their herds with decreased water supplies. In some areas, drought conditions are preventing farmers from growing crops, and some scenarios indicate increasing drought for parts of Africa, as well as the American Southwest. Some cities, such as Las Vegas, may have to face the possibility of water shortages under some climate change scenarios. In the case of Las Vegas, this situation would have a potentially negative impact on casino employment. As some lakes, such as Lake Chad in Africa, dry up, people who earned their livelihood from fishing in those lakes will be forced to move. In many cases, these people become desperate, as they move to cities or across borders looking for employment. In some cases, such as Darfur, this situation is already contributing to economic and political distress, as climate-based migration is being coupled to other, long-standing points of dispute. Although not as severe, the migration from Bangladesh into Assam is beginning to lead to problems, as some natives fear the potential changes occurring because of the new arrivals.

As countries try to find means to adapt to global warming and mitigate its impact, some industrial sectors will be affected in both negative and positive fashions. Various critics of the impact of carbon dioxide (CO_2) on the atmosphere call for the curtailment of the use of oil and coal (both sources of CO_2) for energy and a concomitant decrease in production. If oil and coal production decreases, there will be fewer jobs in these industries, as well as those that supply equipment or expertise such as oil well drilling. On the other hand, energy demand is likely to remain high, so new industries will grow to provide energy or old industries such as the nuclear power industry will begin to grow once again.

• Climate Change Fosters Employment Growth

Global warming does appear to be creating employment opportunities in some industries and regions. As annual temperatures rise in some temperate areas, such as northern Europe and Canada, new crops will become viable as growing seasons become longer. For the most part, this will mean simply an exchange of one sort of job for another for agricultural workers. As it becomes easier to navigate in the Arctic, companies are beginning to eye various opportunities to exploit natural resources there. The Arctic seabed is believed to hold several minerals, as well as having the potential for oil field development. Exploiting both of these resources will provide opportunities for skilled workers, and sailors will be needed to navigate the Northwest Passage from the Atlantic Ocean to the Pacific Ocean.

• Context

As countries strive to adapt to global warming, new industries will develop, creating jobs, at least in the industrial countries. Solar panel and wind turbine fabrication and construction are two new industries that are beginning to take shape in several countries as people turn to renewable energy sources. In some countries, infrastructure projects, such as the construction of higher seawalls in the Netherlands, are already under way. Unfortunately, poorer countries will share less in this sort of job creation because of lack of capital. People in less industrialized nations will simply be forced off the land into the cities, where few jobs exist. Even so, there will be some potential for low-wage jobs in cleaning up storm damage in all regions.

These sorts of impacts of global warming appear to be quite straightforward, yet as the IPCC panel of scientists indicates, the reality is more complex. People may also move from one area to another in search of higher wages or better working conditions. Climate change plays a role in their decision, but so do other factors, making it difficult to assign an impact to climate change alone.

Predicting the impact of climate change on global employment patterns will continue to be of interest but will be exceedingly difficult to do. There are no neat trade-offs from one industry to another that can be assigned to climate change. Because people move for a variety of reasons, it is hard to assign values to the impact of climate change on population movement.

It is apparent that people in developing countries will bear a large share of the employment (and other) costs of climate change. The economies of industrial nations are more adaptable, but they too will face some employment dislocations by the end of the twenty-first century. As always, the more extreme climate change scenarios (those predicting average global warming of 5° to 7° Celsius) will lead to larger impacts on the world economies, including employment patterns.

John M. Theilmann

• Further Reading

Agnew, M. D., and D. Viner. "Potential Impact of Climate Change on International Tourism." *Tourism and Hospitality Research* 3 (2001): 37-60. Examines the impact of climate change on several tourist areas worldwide.

Faris, Stephan. *Forecast.* New York: Henry Holt, 2009. Several chapters deal with the economic impacts of climate change, especially population migration.

Intergovernmental Panel on Climate Change. *Climate Change, 2007—Impacts, Adaptation, and Vulnerability: Contribution of Working Group II to the Fourth Assessment Report of the Intergovernmental Panel on Climate Change.* Edited by Martin Parry et al. New York: Cambridge University Press, 2007. Comprehensive analysis of the impact of climate change worldwide by a group of leading climate scientists.

Jorgenson, Dale W., et al. *U.S. Market Consequences of Global Climate Change.* Washington, D.C.: Pew Center on Global Climate Change, 2004. Focuses on the effects of climate change on the U.S. economy.

Stern, Nicholas. *The Economics of Climate Change: The Stern Review.* New York: Cambridge University Press, 2007. Excellent analysis of the economic impact of climate change, sponsored by the British government. Some of Stern's assumptions are questioned by some other economists.

Unruh, J., M. Krol, and N. Kliot. *Environmental Change and Its Implications for Population Migra-*

tion. New York: Springer, 2004. Useful treatment of the impact of climate change on population movements.

See also: Human migration; Poverty; Renewable energy; Solar energy; Wind power.

Endangered and threatened species

- **Categories:** Animals; plants and vegetation

Climate change has overtaken other threats such as deforestation and pollution as a primary danger to the survival of plant and animal species.

• Key concepts

biodiversity: the variety of species in a given location
biome: a major biological community adapted to a particular climate or area
ecosystem: all living organisms within an area, as well as the nonliving area's environment, understood as a coherent functioning unit
endangered species: a species that is in danger of extinction throughout all or a significant portion of its range
habitat: the place where a biological population normally lives
threatened species: a species likely to become endangered within the foreseeable future

• Background

Although the endangerment and extinction of species has taken place continuously since life on Earth began, human activities have intensified the process. In the past century, an increasing human population has built more cities, towns, and roadways; sacrificed forests for agricultural land; increased pollution; and produced other trends that disrupt stable ecosystems. In addition, symptoms of the Earth's gradual warming have been evident since the mid-1970's. An number of plant and animal species unprecedented on the human timescale faces conditions that threaten their survival.

• Identifying the Dangers

The U.S. Endangered Species Act was enacted in 1973 in an effort to protect wildlife and counter those processes that endanger the survival of various animal and plant species. The World Conservation Union issues its Red List of species deemed to be at risk. Its 2008 report classified 16,928 species as threatened, out of a total of 44,838 surveyed. Half of all the mammal species surveyed were in decline. The report listed one in eight bird species, one in three amphibians, and 70 percent of all plant species as likely to perish in the foreseeable future.

Because so many anthropogenic hazards threaten the planet's life-forms, proof of the prime role of climate change has been difficult to establish. As time goes on, though, the effects of warming temperatures and related weather phenomena become more clear. Startling events in several different biomes bear this out.

• Extinction Observed in a Costa Rican Rain Forest

Climate change leading to extinction probably has taken place many times, but out of the view of biologists or of any human observer. Zoologist Tim Flannery recounts a series of events in the Monteverde Cloud Forest Preserve in Costa Rica which, in retrospect, demonstrate a species suddenly dying out as a result of climate change.

The preserve is a mountain area whose forests are typically cloaked in mist. It is the home of an intriguing variety of rain-forest species. Among these was the golden toad, whose bright orange-gold males were so spectacular they became a symbol for happiness in tribal legends. These toads were seen gathered in their forest pools to mate in April, 1987. Then, the pools rapidly dried up, along with the golden toad eggs just deposited there. It was an El Niño year. The dry conditions were the culmination of a decade when, each year, the cloud cover drifted higher up the mountainside. As a result of this drift, the toads lost the mist in which they thrived.

The higher cloud levels were later traced to a rise in sea surface temperatures of the central Pacific over the preceding decade. In May, 1987, more toad eggs were laid, perhaps an instinctual effort to replace the dried-up ones. Few tadpoles

An exhibition of sixteen hundred panda sculptures in Nantes, France, draws attention to the plight of endangered species. (Bertrand Bechard/Maxppp/Landov)

hatched or survived, however. During the next two years (1988 and 1989), the researcher found only one lone toad where many had lived before. Since that time, no toads have been seen. Because the golden toads live only in this region and have been meticulously studied, their disappearance became the first documented case of extinction from climate change.

Golden toads were not the only victims of this quiet catastrophe. The Cloud Forest Preserve, a relatively accessible rain-forest area, was home to some fifty species of frogs and toads. At the end of the disastrously dry 1987 season, thirty of the fifty species had vanished. Among these was the Monteverde harlequin frog. Its demise was also connected to shifting climatic conditions, via the outbreak of a fungus that thrived only as the surrounding weather became drier and warmer. In fact, the continual decline of amphibians has been a worldwide occurrence. It seems likely that these animals' sensitivity to changes in temperature and humidity has been a significant factor in their decline. Their loss is an early warning of the ravages of a warming world.

Other animal life in the preserve also suffered. Several species of lizards living there have disappeared or become rarer. The keel-billed toucan, a lowland bird, has moved onto the mountainside, where it threatens the eggs of the quetzal, a spectacular bird famous in Mayan folklore. Altogether, this one small area, crowded with species adapted to its unique climatic conditions, has been an object lesson in what happens when these conditions abruptly change.

• The Warming of Polar Regions

Among Earth's most rapidly changing habitats are those in the Arctic. For some time, each Arctic winter has been milder than the winter before. Sea ice in Hudson and Baffin Bays, for example, typically broke up in the early twenty-first century some three weeks earlier than it did in the 1970's. This trend disrupts the polar bears' annual trek onto ice to find their main food, the ringed seals that live there. There are now malnourished adult polar bears, and fewer cubs are being born.

Large stretches of the sea lack ice chunks big enough to support a bear. There have been reports of polar bears marooned on ice 640 kilometers away from any land or food source. The higher winter temperatures bring rains, which collapse the bears' birthing dens, killing the mothers and cubs inside. In short, each season of the polar bears' life cycle is threatened by a warmer Arctic, and this region has been warming almost twice as rapidly as most of the world.

The ice is home to some four species of seals that are equally threatened. The Gulf of Saint Lawrence has had several years when, because of the scarcity of ice, no harp seal pups were born. In fact, the whole Arctic biosphere is at risk. Walruses, caribou, and other animals all are threatened as winters average four to five degrees warmer than in generations past. Caribou herds, for instance, have drastically decreased in size. The main factor seems to be newly occurring autumn rains, which freeze their lichen supply so it is hard to browse. The rains also create swollen rivers which are fatal to many caribou cubs.

• Endangered Coral Reef Communities

Coral reefs flourish in shallow tropical seas. Known for their intricate, many-colored forms and their role in island formation, the reefs also serve as a sort of nursery for fish and other marine organisms. A typical reef in the Indian Ocean may contain over five hundred coral species and provide food and shelter for more than two thousand different fish species. Coral itself is a phylum of invertebrate animals. The coral formations are an exoskeleton built out of calcium carbonate which supports a tiny animal living inside, called a polyp. Corals live in symbiosis with algae, the zooxanthellae. The algae give the reefs their spectacular color. They provide the polyp with food produced from photosynthesis.

Coral reefs require a delicate balance of temperature, water chemistry, and sunlight to stay healthy. When the surrounding sea's temperature rises above a certain level, the algae-polyp partnership breaks up. Extended warm spells make the algae disappear, and the coral polyps starve. The reef becomes bleached and dead. In the two El Niño years of 1998 and 2002, this happened on Australia's Great Barrier Reef, damaging a total of 60 percent of the reef's area in whole or in part. Many coral reefs on the outskirts of civilization have been damaged by surface runoff of silt, sewage, and trash and by harvesting of the coral. The added destruction from warming tropical waters means the loss not only of the reef structures themselves, but also of the habitat and breeding grounds of much marine life. Fish, crustaceans, anemones, and other creatures, some not even known to science, lose their home. Coral reefs are a complex ecosystem in delicate balance, susceptible to ruin from small changes in the planet's weather patterns.

• Habitat Loss in Slow Motion

The foregoing scenarios are notable for their exotic settings and fauna, and for the rapid pace with which global warming took a toll on the species involved. However, no part of the Earth is unaffected. Even nonscientists living in temperate regions have noticed such symptoms as the scarcity of butterflies, which in the earlier twentieth century were abundant.

One of the markers of ecosystem threat is the earlier occurrence of spring. University of Texas biologist Camille Parmesan examined nearly two thousand studies showing this happening on six continents and in all of the world's oceans. In the majority of cases, life cycles were disrupted. Insects and animals in a given ecosystem regulate their dormancy, reproduction, and growth in tandem with the growth and blossoming of plants. Temperature and day length appear to be the two signals which key their life cycle events. When the two become de-synchronized, the web of interactions involving food availability, pollination, seed dispersal, and other daily events does not stay intact.

When this happens, animals tend to migrate when they can. Shifts of habitat northward have been documented for multiple species of birds and butterflies since 1960, in both Europe and America. Red foxes have expanded their range northward, driving the arctic fox further toward the Arctic Ocean. This can be a successful survival strategy, but it also disrupts existing ecosystems and may displace native species that filled the same ecological niche. Eventually a species may run out of spaces to migrate to, or exceed its own ability to adapt.

Migrating to higher elevations is a variant of this strategy, and it is subject to the same risks. The Edith's Checkerspot butterfly in California has shifted its range upward some 100 meters in the Sierra Nevada. A closely related species, the Quino Checkerspot, has become endangered because it is unable to cross the Los Angeles metropolis and establish itself in cooler, wetter environment.

Past experience in reversing the plight of endangered species is not always relevant to the global warming situation. The Endangered Species Act emphasizes using law and human efforts to counter damage threats from hunting, disruption of habitats by urbanization or agriculture, and pollution. It has had notable successes in the recovery or reestablishment of species such as the gray wolf, the peregrine falcon, and the humpback whale. But there are still eighteen hundred species on its list, and thousands of other threatened species needing protection. Moreover, the act was not designed to counter threats to an entire biome, much less those of planetwide scope.

An officer of the Kenyan Wildlife Services displays ivory seized from travelers seeking to smuggle it out of the country. Trade in the materials of endangered species increases the threat of their extinction. (Reuters/Landov)

• Context

There have been five great extinctions in Earth's geologic past in which most existing species vanished. Their causes appear to be varied, but rapid climatic change is implicated in most. The most recent, culminating with the dinosaurs' extinction and the slow ascendancy of the age of mammals, happened 55 million years ago. The immediate cause was probably Earth's collision with an asteroid, but it was the aftereffects that affected the climate so drastically as to overturn all existing ecosystems. By colliding with limestone-rich rock, the asteroid created an explosion that put enough carbon dioxide into the air to warm the planet by an estimated 4° to 10° Celsius. No creatures that weighed over 35 kilograms survived; there were also major changes in the vegetation.

A case can be made that ever since human activity began to alter the planet, Earth's other fauna and flora have been threatened. In the past decades, however technologies have been developed that can document and measure the planet's warming trend, and these technologies may prove sufficient to convince humanity to alter its behavior. Still, unless humans find ways systematically to counter the trend toward anthropogenic climate change and environmental degradation, the survival of threatened species can be accomplished only on a piecemeal basis.

Emily Alward

• Further Reading

Flannery, Tim. *We Are the Weather Makers: The Story of Global Warming.* Rev. ed. London: Penguin, 2007. Wide-ranging survey of the ways human life impinges upon the rest of nature. The author, director of the South Australian Museum, ranges back through time and across the globe, and urges action to slow global warming.

McGavin, George C. *Endangered: Wildlife on the Brink of Extinction.* Buffalo, N.Y.: Firefly Books, 2006. Lavishly illustrated coffee-table type book showing the diversity of animal life on Earth. Climate change is discussed as one among several threats to diverse species. Excellent charts and explanations.

Spicer, John. *Biodiversity: A Beginner's Guide.* Oxford, England: Oneworld, 1996. Topical guide to the interaction of life-forms. Contains hard-to-find nuggets of information on many topics, from extinction cascades to the place of the biosphere in religious and secular value systems.

Ward, Peter D. *Under a Green Sky.* New York: HarperCollins, 2007. Highly personalized account of the search for extinctions of the past, and consideration of what they mean for the present climate change.

See also: Amazon deforestation; Budongo Forest Project; Convention on International Trade in Endangered Species; Extinctions and mass extinctions; Fishing industry, fisheries, and fish farming; Invasive exotic species; Whales.

Energy efficiency

• **Category:** Energy

• Definition

The efficiency of an energy-producing system is defined as the benefit derived from operating the system divided by the cost necessary to operate it. It can theoretically vary between 0 and 100 percent. The benefit is the net amount of work produced by the system during its operating time, while the cost is determined by how much energy must be provided in order to generate this work. According to the laws of thermodynamics, only part of the input energy will be converted into work, so the energy efficiency will always be less than 100 percent.

• Significance for Climate Change

Market strategies, economic policies, and political interests have generally assigned a lower priority to the implementation of energy-efficiency measures to reduce global warming than they have to increased energy production. If energy efficiency were properly addressed, promoted, and increased, world demands for energy production could be decreased and climate change stabilized. By increasing the energy efficiency of homes, vehicles, and industries, fossil fuel usage and global warming could be reduced, and the energy security of a nation could be increased by decreasing the demand for imported petroleum. About 75 percent of the electricity consumed in the United States could be saved by implementing energy-efficiency measures that cost less than the electricity itself.

As an example, energy-efficient strategies that have continually been implemented in California since the mid-1970's are projected to reduce greenhouse gas (GHG) emissions to at least 1990 levels by 2020 while keeping per capita energy consumption approximately the same. The top priority in California's energy policy is to reduce fossil fuel usage by promoting and providing more energy-efficient homes, vehicles, and businesses, which will reduce GHG emissions that generate climate change. High-performance windows, properly installed insulation, and energy-efficient cooling, heating, and lighting systems can reduce energy use in a standard home by at least 15 percent. Similar policies could be adopted worldwide.

In 1992, the Energy Star program was created in the United States by the Environmental Protection Agency in an effort to reduce energy consumption and GHG emissions. Initially established to identify and promote energy-efficient products, the program was expanded to include energy-efficiency labels for residential heating and cooling systems, new homes, and commercial and industrial buildings. Energy Star has also promoted the use of more efficient power supplies, home appliances, electronic systems, light-emitting-diode (LED) traffic lights, efficient fluorescent lights, and energy management equipment for office systems in order to reduce energy consumption, global warming effects, and air pollution.

Alvin K. Benson

See also: Biofuels; Carbon taxes; Clean energy; Energy from waste; Energy resources; Ethanol; Fossil fuels; Fuels, alternative; Renewable energy; U.S. energy policy.

Energy from waste

• **Categories:** Energy; environmentalism, conservation, and ecosystems

Energy-from-waste technologies are designed to reduce or eliminate waste that otherwise contributes to global warming, while simultaneously generating energy and reducing the need to consume nonrenewable resources.

• Key concepts

incinerator: a facility used to burn waste until it is reduced to ash

landfill: a site where municipal or industrial wastes are buried in the ground

methane: a greenhouse gas with the molecular formula CH_4

microalgae: photosynthetic microorganisms similar to plants

municipal solid waste (MSW): the trash, refuse, and garbage thrown away from homes and commercial establishments

• Background

Energy from waste, or waste-to-energy, refers to any waste treatment that generates energy from a waste source. Most waste-to-energy technologies make electricity directly through combustion, or they produce fuels, such as methane, hydrogen, biodiesel, or ethanol. These technologies reduce the amount of waste on the planet and also reduce the need to produce energy using technologies that create more waste.

• Refuse-to-Energy Conversion

As human populations and industrialization have increased, the amount of generated municipal solid waste (MSW) has grown steadily. In the early twenty-first century, the United States produced around 1.8 kilograms of such waste per person per day. Because it has high energy content, refuse material can be burned to release energy. The amount of energy generated by burning waste is about half that produced by burning the same mass of coal.

Several cities around the world use incinerators to convert refuse to electricity. Incinerators are huge furnaces, capable of handling 15 metric tons of waste per hour, in which temperatures are high enough to allow waste to be burned completely. The value of the electricity generated by these incinerators offsets the costs of MSW handling and burning. However, incineration has serious environmental consequences, including the production of greenhouse gases (GHGs). Burning MSW produces nitrogen oxide as well as carbon dioxide (CO_2), the primary GHG. There is a worldwide awareness that CO_2 emissions contribute to global warming. Some of the CO_2 produced in incinerators is biomass-derived and is considered to be part of the Earth's natural carbon cycle, however.

Incineration is particularly popular in countries where land is a scarce resource. Sweden and Denmark have been leaders in using the energy generated from incineration for more than a century. Although incinerators reduce the volume of the original waste by 95-96 percent, the ash produced after incineration must still be disposed of in landfills. Ash often contains high concentrations of hazardous heavy metals, such as lead or cadmium. Ash may also include precious metals, such as aluminum, gold, copper, and iron. These metals are recycled before ash deposition into the landfills. Alternatively, ash can be used for road work or building construction, provided that it does not contain hazardous substances.

• Methane Production

Buried in landfills, wastes do not have access to oxygen. The resulting anaerobic waste decomposition produces biogas, which is 30 percent methane. Methane is a very powerful GHG, but it is also a very good fuel. In order to avoid releasing methane into the atmosphere, a number of cities install "gas wells" in landfills to capture the methane they generate and use it as fuel. There are several such landfill gas facilities in the United States that generate electricity.

Biogas can also be generated from wastewater and from animal waste. Domestic wastewater consists of substances such as ground garbage, laundry water, and excrement. All these components are biological molecules that microbes can eat. Thus, one of the most common treatments of wastewater and manure employs microbial anaerobic digestion. Such digestion is very similar to the process of landfill waste decomposition, and it yields biogas.

In this process, wastewater or manure is fed into digesters (bioreactors), where microorganisms metabolize it into biogas. Biogas can be used to fuel engines connected to electrical generators to produce electricity. The nutrient-rich sludge remaining after digestion can be used as fertilizer. In many countries, millions of small farmers maintain simple digesters at home to generate energy. The only side effect of this technology is that burning methane in combusting engines produces CO_2. Because methane has a greater global warming potential than does CO_2, however, the process potentially results in a net decrease in GHG emissions.

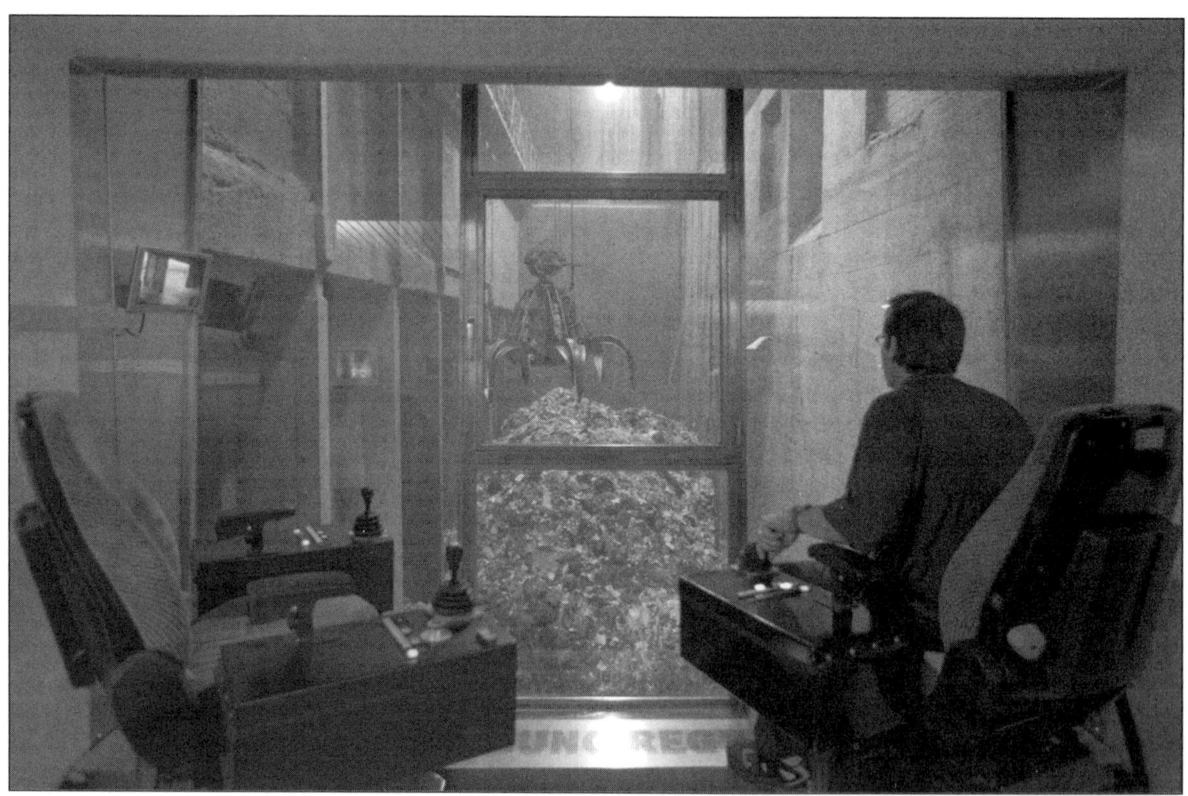

A worker feeds waste into an incinerator in Aargau, Switzerland. (Alessandro Della Bella/Keystone/Landov)

• Agricultural Waste to Energy

Where energy is generated as a by-product of waste disposal, agricultural waste may have considerable merit. A great number of cellulosic wastes result from the cultivation of crops such as corn. This waste can be turned into ethanol. Use of ethanol as fuel has been vigorously promoted. Ethanol is mainly produced by fermentation of sugars derived from food crops with the help of baker's yeast. However, making ethanol from leftover materials such as corn stover is highly desirable, because such materials consist largely of sugars but they have no direct food use. Other cellulosic waste material such as sawdust, wood chips, cane waste (bagasse), and wastepaper can be converted into ethanol as well. In contrast to food-to-ethanol conversion, converting farm waste to ethanol involves little or no contribution to the greenhouse effect.

Plant waste material may also be gasified to produce syngas, a mixture of carbon monoxide (CO) and hydrogen (H_2). Syngas is considered an alternative fuel, because it generates electricity with coproduction of water and CO_2 when burned. Syngas can also be converted by certain microbes into other alternative fuels, such as ethanol and H_2.

• Harnessing Photosynthesis

Exhaust streams from power plants and other manufacturing units contain high levels (up to 20 percent) of CO_2. Typical coal-fired power plants account for up to 13 percent of anthropogenic CO_2 emissions. Researchers are exploring the application of photosynthetic microalgae to remove CO_2 from the emissions of power stations and other industrial plants. Algae utilize CO_2 and, at the same time, produce oil and H_2 as part of their growth process. Therefore, they can be used to generate environmentally friendly biofuels such as biodiesel and H_2.

The biological potential of photosynthetic microalgae for CO_2 removal and biofuel generation is determined by their cultivation techniques. Current industrial production of microalgae is achieved in open "raceway" ponds of some thousand square meters in size. These systems suffer from severe limitations, such as lack of temperature control, low attainable cellular concentrations, and difficulty in preventing contamination. The need to overcome these limits led to design and development of photobioreactors.

Photobioreactors are closed systems that are made of an array of tubes or tanks, in which microalgae are cultivated and monitored. The main challenge in photobioreactor design is to create a simple, inexpensive, high-cell-density, energy-efficient reactor that is scalable to meet the needs of industrial production. Several U.S. companies (GreenFuel Technologies, GreenShift, Solix, and Valcent Products) have created pilot-scale photobioreactors for CO_2 mitigation and biofuel production by microalgae.

• Context

Contemporary society generates significant amounts of waste that affect the climate on Earth as the result of GHG release, as well as polluting the environment generally. One of the solutions to this situation is to turn this waste into energy. Energy-from-waste technologies are valuable for energy generation and may represent efficient means for removing waste while minimizing environmental and climatological side effects. However, energy-from-waste technologies should be used under strict environmental regulations to avoid generating additional pollutants or otherwise exacerbating the problems they seek to solve. For instance, responsible application of energy-from-waste technology would require recycling the CO_2 released from incinerators.

Another solution to waste management is recycling. Making products such as paper from recycled materials requires 65 percent less energy and generates 75 percent less CO_2 and methane emissions than does similar production using virgin raw materials. Therefore, waste recycling, although it does not generate energy, saves considerable amounts of energy and reduces GHG emissions.

Sergei Arlenovich Markov

• Further Reading

Haag, Amanda L. "Algae Bloom Again." *Nature* 447 (May 31, 2007): 520-521. An introductory review on algal biotechnology for CO_2 removal from the atmosphere and biofuel generation.

Nebel, Bernard J., and Richard T. Wright. *Environmental Science: Towards a Sustainable Future.* Englewood Cliffs, N.J.: Prentice Hall, 2008. Introductory environmental education textbook. Several chapters describe waste management and energy-from-waste issues.

Williams, Paul T. *Waste Treatment and Disposal.* Chichester, West Sussex, England: John Wiley and Sons, 1998. Comprehensive book on waste-treatment technologies.

See also: Carbon dioxide; Clean energy; Energy efficiency; Energy resources; Fuels, alternative; Greenhouse gases; Methane; Renewable energy.

Energy Policy Act of 1992

- **Category:** Laws, treaties, and protocols
- **Date:** Signed into law October 24, 1992

The Energy Policy Act of 1992 covers numerous resource conservation initiatives, but Title 16 in particular puts into place mechanisms for monitoring and addressing climate change and global warming.

• Background

The 1973 oil crisis in the United States created a sense of urgency for action to reduce America's dependence on foreign oil while simultaneously reducing carbon dioxide (CO_2) emissions. The 1975 Energy Policy and Conservation Act (EPCA) established corporate average fuel economy (CAFE) standards. In 1978, the Public Utility Regulatory Policies Act (PURPA) promoted development of alternative energy sources. The National Appliance Energy Conservation Act (NAECA) of 1987 envisioned a 21.6 million metric ton reduction in carbon emissions by the year 2010. Title VI of the 1990 Clean Air Act Amendments responded to the domestic ozone depletion issue, which is related to reducing greenhouse gas (GHG) emissions. Then, in 1992, the United States ratified the United Nations Framework Convention on Climate Change (UNFCCC), which challenged industrialized countries to become leaders in reducing GHGs. The Energy Policy Act (EPACT) of 1992 represents the major initial response to that initiative.

• Summary of Provisions

The Energy Policy Act of 1992, an extensive and varied act with twenty-seven titles, addresses various aspects of overriding issues such as improving energy efficiency, implementing measures to reduce GHG generation, and incentives to increase renewable energy and clean coal technology. Title XVI of the act, however, contains specific stipulations that address global climate change and global warming and provides partial fulfillment of obligations which the United States incurred upon ratification of the UNFCCC. The act contains four main mandates.

The act directed the secretary of energy to submit a report to the Congress by October 24, 1994, that was to assess a wide variety of implications for implementing policies that would enable the United States to comply with its obligations under the UNFCCC. These include potential implications relating to the economy, energy, society, the environment, and competition. The report was also to assess the feasibility and implications for jobs and of stabilizing GHG generation in the United States by the year 2005; of stabilizing GHG generation by the year 2005; of reducing the generation of GHGs; and of successfully reducing 1988 levels of CO_2 by 20 percent by the year 2005.

Section 1602 of Title XVI of the act required that each National Energy Policy Plan that the president submitted to Congress include a least-cost energy strategy prepared by the secretary of energy. The strategy had to consider economic, energy-related, social, environmental, and competitive costs and benefits. The strategy design had to be one that could achieve specific things at least cost to the country: federal energy production; utilization and energy conservation priorities; stabilization and eventual reduction in GHG generation; an increase in the efficiency of the total energy use in the United States by 30 percent over 1988 levels by the year 2010; an increase of 75 percent over 1988 levels in the percentage of energy derived from renewable resources by the year 2005; and a reduction in U.S. oil consumption from about 40 percent (1990 level) of total energy use to 35 percent by the year 2005. The strategy also had to identify federal priorities to include policies that implemented standards for more efficient use of fossil fuels, increased the energy efficiency of existing technologies, and encouraged new ones that included such things as clean coal technologies and others that lower GHG levels.

The secretary of energy was directed to establish a director of climate protection within the Department of Energy who would have three major responsibilities: to serve as the secretary's representative for global change policy discussions, including activities of the Committee on Earth and Environmental Sciences and the Policy Coordinating Committee Working Group on Climate Change; to monitor domestic and international policies as to their effects on GHG generation; and to have the authority to participate in planning activities of relevant programs of the Department of Energy.

Section 1609 of the 1992 Energy Policy Act directed the secretary of the treasury, in consultation with the secretary of state, to set up the Global Climate Change Response Fund to be used for American contributions toward global efforts to effect global climate change. Management and uses of the fund were clearly stipulated: No fund deposits would be made before the United States had ratified the UNFCCC; the money was to be used by the president as authorized and appropriated under the Foreign Assistance Act of 1961, exclusively for matters related to the UNFCCC; and $50 million was to be appropriated for deposit for fiscal year 1994, with amounts as deemed necessary for fiscal years 1995 and 1996.

Title XVI: Global Climate Change

Title XVI of the Energy Policy Act of 1992 addresses efforts to mitigate climate change. Section 1601, reproduced below, calls for the production of a report on the ability of the United States to respond to climate change and to limit greenhouse gas emissions.

Not later than 2 years after the date of the enactment of this Act, the Secretary shall submit a report to the Congress that includes an assessment of—

(1) the feasibility and economic, energy, social, environmental, and competitive implications, including implications for jobs, of stabilizing the generation of greenhouse gases in the United States by the year 2005;

(2) the recommendations made in chapter 9 of the 1991 National Academy of Sciences report entitled 'Policy Implications of Greenhouse Warming', including an analysis of the benefits and costs of each recommendation;

(3) the extent to which the United States is responding, compared with other countries, to the recommendations made in chapter 9 of the 1991 National Academy of Sciences report;

(4) the feasibility of reducing the generation of greenhouse gases;

(5) the feasibility and economic, energy, social, environmental, and competitive implications, including implications for jobs, of achieving a 20 percent reduction from 1988 levels in the generation of carbon dioxide by the year 2005 as recommended by the 1988 Toronto Scientific World Conference on the Changing Atmosphere;

(6) the potential economic, energy, social, environmental, and competitive implications, including implications for jobs, of implementing the policies necessary to enable the United States to comply with any obligations under the United Nations Framework Convention on Climate Change or subsequent international agreements.

• Significance for Climate Change

The Energy Policy Act of 1992 began a debate which resulted first in developing a process for deregulating the electric industry. The act mandated open access for transmitting energy to wholesale customers, but not to retail customers. It addressed disposal standards for nuclear waste so as to protect individuals from abnormal amounts of exposure, specifically those in the Yucca Mountain repository site area. It initiated increased research on the production of alternative-fuel vehicles (electric, ethanol, natural gas, and propane) as well as on hybrid electric and hydrogen fuel cell vehicles. The act further resulted in new efficiency standards for faucet fixtures, commercial air conditioning and heating equipment, electric motors, and lamps. Modifications were made in commercial and residential building energy codes. A program for federal support for renewable energy technologies was established, and standards were revised for long-term power purchases, in addition to a host of other provisions.

While some of the energy efficiency measures indirectly affected climate change and global warming—for example, reduction of GHG emissions—as a whole, performance reports on the Title XVI provisions (those concerning climate control and global warming) were disappointing; implementation deadlines, for one thing, were not met. Also, a number of measures were to be voluntary, thus weakening the impact of the law. The act did, however, provide a framework for pursuing its mandates and result in subsequent amendments that could enhance the odds of future realization.

Victoria Price

• Further Reading

Broecker, Wallace S., and Robert Kunzig. *Fixing Climate: What Past Climate Changes Reveal About the Current Threat—and How to Counter It.* New York: Hill and Wang, 2008. Explains why climate change may be getting out of control, but offers possibilities for reversing the trend.

Gore, Al. *An Inconvenient Truth: The Planetary Emergency of Global Warming and What We Can Do About It.* Emmaus, Pa.: Rodale, 2006. The climate crisis is seen as an opportunity for a generational mission and moral purpose. Addresses the numerous areas in which Americans, in particular, have contributed to a global warming crisis.

Leroux, Marcel. *Global Warming, Myth or Reality? The Erring Ways of Climatology.* New York: Springer, 2005. Outlines the history of global warming. Discusses climate models and their limitations, and postulates alternative causes of climate change.

See also: Air pollution and pollutants: anthropogenic; Clean Air Acts, U.S.; Clean energy; U.S. energy policy; U.S. legislation.